Heavy Highway Construction

Level One

Trainee Guide
Second Edition

nccer

P **Pearson**

Boston   Columbus   Indianapolis   New York   San Francisco   Amsterdam
Cape Town Dubai London Madrid Milan Munich Paris Montreal Toronto Delhi
Mexico City Sao Paulo Sydney Hong Kong Seoul Singapore Taipei Tokyo

NCCER

President: Don Whyte
Vice President: Steve Greene
Chief Operations Officer: Katrina Kersch
Heavy Highway Construction Project Managers: Luke Tia, Patty Bird
Senior Development Manager: Mark Thomas

Senior Production Manager: Tim Davis
Quality Assurance Coordinator: Karyn Payne
Desktop Publishing Coordinator: James McKay
Permissions Specialists: Kelly Sadler, Adrienne Payne
Production Specialist: Kelly Sadler
Editor: Graham Hack

Writing and development services provided by Topaz Publications, Liverpool, NY

Lead Writer/Project Manager: Thomas Burke
Desktop Publisher: Joanne Hart
Art Director: Alison Richmond

Permissions Editors: Andrea LaBarge
Writer: Thomas Burke

Pearson

Director of Alliance/Partnership Management: Andrew Taylor
Editorial Assistant: Collin Lamothe
Program Manager: Alexandrina B. Wolf
Director of Marketing: Leigh Ann Simms
Senior Marketing Manager: Brian Hoehl

Composition: NCCER
Printer/Binder: LSC Communications
Cover Printer: LSC Communications
Text Fonts: Palatino and Univers

Credits and acknowledgments for content borrowed from other sources and reproduced, with permission, in this textbook appear at the end of each module.

1 17

Pearson

ISBN-13:   978-0-13-448247-7
ISBN-10:   0-13-448247-6

Preface

To the Trainee

Highways and bridges are a crucial part of our nation's infrastructure, allowing for the efficient transportation of people and goods to locations both near and far. The construction of highways and bridges, known as *heavy highway construction*, is an exciting and complex field, requiring the work of a wide variety of craft professions. Jobs available within the industry include grade checker, form carpenter, concrete finisher, heavy equipment operator, crew leader, project manager, and many more.

Craft professionals in heavy highway construction must learn both the fundamentals of the industry as well as specific skills and knowledge for their craft. Many are trained through craft apprenticeship programs, which provide both classroom instruction and on-the-job learning. Workers in the industry are well paid and, according to the Bureau of Labor Statistics, constitute a workforce of over 200,000 workers. Regardless of where your career may take you, NCCER's *Heavy Highway Construction* curriculum will provide you a solid foundation on your road to success.

New with *Heavy Highway Construction*

NCCER is pleased to release the second edition of *Heavy Highway Construction* in full color, with new photographs and figures. Level One and most of Level Two are in NCCER's improved instructional systems design. Now each module's sections align with its learning objectives in both order and content.

The curriculum has also been thoroughly redesigned, updated, and expanded. Level One provides an overview of the fundamentals of heavy highway construction. Level Two is a survey of more advanced topics relevant to different crafts within field. In addition, trainees who complete the four new crane and rigger modules in Level One will earn NCCER's *Basic Rigger* training credential.

Stackable credentials can be earned for various crafts found in the heavy highway construction industry. Trainees earn a given credential by completing the prescribed modules from both *Heavy Highway Construction* and other NCCER craft curricula. Stackable credentials that are available for Heavy Highway Construction include the following:

- Heavy Highway Equipment
  - Grade Checker
  - Compacting and Grading Equipment
  - Excavation and Loading Equipment
  - Leveling and Spreading Equipment

- Heavy Highway
  - Asphalt and Concrete
  - Ironworker

  - Carpenter
  - Electrician
  - Mobile Crane Operator
  - Surveyor

For more details on requirements for stackable credentials requirements and other information about this curriculum, visit the "Heavy Highway Construction" craft details page in the Program Resources section on NCCER's website.

We wish you success as you progress through this training program. If you have any comments on how NCCER might improve upon this textbook, please complete the User Update form located at the back of each module and send it to us. We will always consider and respond to input from our customers.

We invite you to visit the NCCER website at **www.nccer.org** for information on the latest product releases and training, as well as online versions of the *Cornerstone* magazine and Pearson's NCCER product catalog.

Your feedback is welcome. You may email your comments to **curriculum@nccer.org** or send general comments and inquiries to **info@nccer.org**.

NCCER Standardized Curricula

NCCER is a not-for-profit 501(c)(3) education foundation established in 1996 by the world's largest and most progressive construction companies and national construction associations. It was founded to address the severe workforce shortage facing the industry and to develop a standardized training process and curricula. Today, NCCER is supported by hundreds of leading construction and maintenance companies, manufacturers, and national associations. The NCCER Standardized Curricula was developed by NCCER in partnership with Pearson, the world's largest educational publisher.

Some features of the NCCER Standardized Curricula are as follows:

- An industry-proven record of success
- Curricula developed by the industry, for the industry
- National standardization providing portability of learned job skills and educational credits
- Compliance with the Office of Apprenticeship requirements for related classroom training (*CFR 29:29*)
- Well-illustrated, up-to-date, and practical information

NCCER also maintains the NCCER Registry, which provides transcripts, certificates, and wallet cards to individuals who have successfully completed a level of training within a craft in NCCER's Curricula. *Training programs must be delivered by an NCCER Accredited Training Sponsor in order to receive these credentials.*

Special Features

In an effort to provide a comprehensive and user-friendly training resource, this curriculum showcases several informative features. Whether you are a visual or hands-on learner, these features are intended to enhance your knowledge of the construction industry as you progress in your training. Some of the features you may find in the curriculum are explained below.

Introduction

This introductory page, found at the beginning of each module, lists the module Objectives, Performance Tasks, and Trade Terms. The Objectives list the knowledge you will acquire after successfully completing the module. The Performance Tasks give you an opportunity to apply your knowledge to real-world tasks. The Trade Terms are industry-specific vocabulary that you will learn as you study this module.

Figures and Tables

Photographs, drawings, diagrams, and tables are used throughout each module to illustrate important concepts and provide clarity for complex instructions. Text references to figures and tables are emphasized with *italic* type.

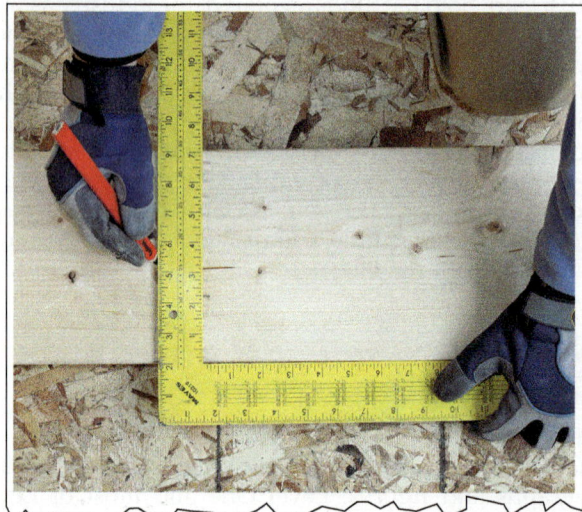

Notes, Cautions, and Warnings

Safety features are set off from the main text in highlighted boxes and categorized according to the potential danger involved. Notes simply provide additional information. Cautions flag a hazardous issue that could cause damage to materials or equipment. Warnings stress a potentially dangerous situation that could result in injury or death to workers.

Trade Features

Trade features present technical tips and professional practices based on real-life scenarios similar to those you might encounter on the job site.

Bowline Trivia

Some people use this saying to help them remember how to tie a bowline: "The rabbit comes out of his hole, around a tree, and back into the hole."

NOTE
Nameplates must be posted on each material handling device. The nameplate must indicate t...

CAUTION
It is essential to note the revision designation on a construction drawing and to use only the latest

WARNING!
Saw teeth are very sharp. Use gloves and do not handle the saw teeth with bare hands. When cutting with a saw, ensure that your fingers remain clear of the teeth at all times.

Case History

Case History features emphasize the importance of safety by citing examples of the costly (and often devastating) consequences of ignoring best practices or OSHA regulations.

> **Case History**
>
> **Requesting an Outage**
>
> An electrical contractor requested an outage when asked to install two bolt-in, 240V breakers in panels in a data processing room. It was denied due to the 24/7 worldwide information processing hosted by the facility. The contractor agreed to proceed only if the client would sign a letter agreeing not to hold them responsible if an event occurred that damaged computers or resulted in loss of data. No member of upper management would accept liability for this possibility, and the outage was scheduled.
>
> **The Bottom Line:** If you can communicate the liability associated with an electrical event, you can influence management's decision to work energized.

Going Green

Going Green features present steps being taken within the construction industry to protect the environment and save energy, emphasizing choices that can be made on the job to preserve the health of the planet.

> **GOING GREEN**
>
> **Reducing Your Carbon Footprint**
>
> Many companies are taking part in the paperless movement. They reduce their environmental impact by reducing the amount of paper they use. Using email helps to reduce the amount of paper used,

Did You Know

Did You Know features introduce historical tidbits or interesting and sometimes surprising facts about the trade.

> **Did You Know?**
>
> **Safety First**
>
> Safety training is required for all activities. Never operate tools, machinery, or equipment without prior training. Always refer to the manufacturer's instructions.

Step-by-Step Instructions

Step-by-step instructions are used throughout to guide you through technical procedures and tasks from start to finish. These steps show you how to perform a task safely and efficiently.

> Perform the following steps to erect this system area scaffold:
>
> *Step 1* Gather and inspect all scaffold equipment for the scaffold arrangement.
>
> *Step 2* Place appropriate mudsills in their approximate locations.
>
> *Step 3* Attach the screw jacks to the mudsills.

Trade Terms

Each module presents a list of Trade Terms that are discussed within the text and defined in the Glossary at the end of the module. These terms are presented in the text with bold, blue type upon their first occurrence. To make searches for key information easier, a comprehensive Glossary of Trade Terms from all modules is located at the back of this book.

> During a rigging operation, the load being lifted or moved must be connected to the apparatus, such as a crane, that will provide the power for movement. The connector—the link between the load and the apparatus—is often a sling made of synthetic, chain, or wire rope materials. This section focuses on three types of slings:

Section Review

Each section of the module wraps up with a list of Additional Resources for further study and Section Review questions designed to test your knowledge of the Objectives for that section.

> **Additional Resources**
>
> *Materials Handling Handbook*, The American Society of Mechanical Engineers (ASME) and The International Material Management Society (IMMS), Raymond A. Kulwiec, Editor-in-Chief. 1985. New York, NY: Wiley-Interscience.
>
> *Manufacturing Facilities Design & Material Handling*, Matthew P. Stevens, Fred E. Meyers. 2013. West Lafayette, IN: Purdue University Press.
>
> **1.0.0 Section Review**
>
> 1. For material handling tasks, it is just as important to be mentally fit as it is to be _____.
>   a. physically fit
>   b. physically aggressive
>   c. closely supervised
>   d. over 200 pounds
>
> 2. Which of the following is a type of knot that is often used to join the ends of two ropes in non-critical, low-strain applications?
>   a. Bowline
>   b. Clove hitch
>   c. Half hitch
>   d. Square knot

Review Questions

The end-of-module Review Questions can be used to measure and reinforce your knowledge of the module's content.

> **Review Questions**
>
> 1. Identification tags for slings must include the _____.
>   a. type of protective pads to use
>   b. type of damage sustained during use
>   c. color of the tattle-tail
>   d. manufacturer's name or trademark
>
> 2. The type of wire rope core that is susceptible to heat damage at relatively low temperatures is the _____.
>   a. fiber core
>   b. strand core
>   c. independent wire rope core
>   d. metallic link supporting core
>
> 3. Synthetic slings must be inspected _____.
>   a. once every month
>   b. visually at the start of each work week
>   c. before every use
>   d. once wear or damage becomes apparent
>
> 4. An alloy steel chain sling must be removed from service if there is evidence that _____.
>   a. the sling has been used in different hitch configurations
>   b. replacement links have been used to repair the chain
>   c. the sling has been used for more than one year
>   d. strands in the supporting core have weakened
>
> 5. A piece of rigging hardware used to couple the end of a wire rope to eye fittings, hooks, or other connections is a(n) _____.
>   a. eyebolt
>   b. hitch
>   c. shackle
>   d. U-bolt
>
> 6. A lifting clamp is most likely to be used to move loads such as _____.
>   a. steel plates
>   b. piping bundles
>   c. concrete blocks
>   d. plastic tubing
>
> 7. Chain hoists are able to lift heavy loads by utilizing a _____.
>   a. rope and pulley system
>   b. rigger's strength
>   c. stationary counterweight
>   d. gear system
>
> 8. Before attempting to lift a load with a chain hoist, make sure that the _____.
>   a. hoist is secured to a come-along
>   b. load is properly balanced
>   c. tag lines are properly anchored
>   d. tackle is connected to its power source
>
> 9. A hitch configuration that allows slings to be connected to the same load without using a spreader beam is a _____.
>   a. double-wrap hitch
>   b. choker hitch
>   c. bridle hitch
>   d. basket hitch
>
> 10. To make the emergency stop signal that is used by riggers, extend both arms _____.
>   a. horizontally with palms down and quickly move both arms back and forth
>   b. directly in front and then move both arms up and down repeatedly
>   c. vertically above the head and wave both arms back and forth
>   d. horizontally with clenched fists and move both arms up and down
>
> 00106-15 Introduction to Basic Rigging   Module Six 31

NCCER Standardized Curricula

NCCER's training programs comprise more than 80 construction, maintenance, pipeline, and utility areas and include skills assessments, safety training, and management education.

Boilermaking
Cabinetmaking
Carpentry
Concrete Finishing
Construction Craft Laborer
Construction Technology
Core Curriculum: Introductory
 Craft Skills
Drywall
Electrical
Electronic Systems Technician
Heating, Ventilating, and Air
 Conditioning
Heavy Equipment Operations
Heavy Highway Construction
Hydroblasting
Industrial Coating and Lining
 Application Specialist
Industrial Maintenance Electrical
 and Instrumentation Technician
Industrial Maintenance Mechanic
Instrumentation
Ironworking
Masonry
Mechanical Insulating
Millwright
Mobile Crane Operations
Painting
Painting, Industrial
Pipefitting
Pipelayer
Plumbing
Reinforcing Ironwork
Rigging
Scaffolding
Sheet Metal
Signal Person
Site Layout
Sprinkler Fitting
Tower Crane Operator
Welding

Maritime

Maritime Industry Fundamentals
Maritime Pipefitting
Maritime Structural Fitter

Green/Sustainable Construction

Building Auditor
Fundamentals of Weatherization
Introduction to Weatherization
Sustainable Construction
 Supervisor
Weatherization Crew Chief
Weatherization Technician
Your Role in the Green
 Environment

Energy

Alternative Energy
Introduction to the Power Industry
Introduction to Solar Photovoltaics
Introduction to Wind Energy
Power Industry Fundamentals
Power Generation Maintenance
 Electrician
Power Generation I&C
 Maintenance Technician
Power Generation Maintenance
 Mechanic
Power Line Worker
Power Line Worker: Distribution
Power Line Worker: Substation
Power Line Worker: Transmission
Solar Photovoltaic Systems Installer
Wind Turbine Maintenance
 Technician

Pipeline

Control Center Operations, Liquid
Corrosion Control
Electrical and Instrumentation
Field Operations, Liquid
Field Operations, Gas
Maintenance
Mechanical

Safety

Field Safety
Safety Orientation
Safety Technology

Supplemental Titles

Applied Construction Math
Tools for Success

Management

Construction Workforce
 Development Professional
Fundamentals of Crew Leadership
Mentoring for Craft Professionals
Project Management
Project Supervision

Spanish Titles

Acabado de concreto: nivel uno
 (*Concrete Finishing Level One*)
Aislamiento: nivel uno
 (*Insulating Level One*)
Albañilería: nivel uno
 (*Masonry Level One*)
Andamios (*Scaffolding*)
Carpintería: Formas para
 carpintería, nivel tres
 (*Carpentry: Carpentry Forms, Level
 Three*)
Currículo básico: habilidades
 introductorias del oficio
 (*Core Curriculum: Introductory Craft
 Skills*)
Electricidad: nivel uno
 (*Electrical Level One*)
Herrería: nivel uno
 (*Ironworking Level One*)
Herrería de refuerzo: nivel uno
 (*Reinforcing Ironwork Level One*)
Instalación de rociadores: nivel uno
 (*Sprinkler Fitting Level One*)
Instalación de tuberías: nivel uno
 (*Pipefitting Level One*)
Instrumentación: nivel uno, nivel
 dos, nivel tres, nivel cuatro
 (*Instrumentation Levels One through
 Four*)
Orientación de seguridad
 (*Safety Orientation*)
Paneles de yeso: nivel uno
 (*Drywall Level One*)
Seguridad de campo
 (*Field Safety*)

Acknowledgments

This curriculum was revised as a result of the farsightedness and leadership of the following sponsors:

Balfour Beatty Infrastructure, Inc.
Blythe Development Co.
Bridgerland Applied Technology College
R.E. Burns & Sons Co., Inc.
Carolina Bridge Company, Inc.
Charah

Construction Education Foundation, Inc.
Crowder Construction Company
Hubbard Construction Company
John Deere
Sundt Construction

This curriculum would not exist were it not for the dedication and unselfish energy of those volunteers who served on the Authoring Team. A sincere thanks is extended to the following:

Gerald Andrews
Kevin Burns
Russell Copley
Jon Goodney
Thomas Horrell
Mark Jones
Randy McSherry

Dan Nickel
Sean Ray
Roger Richards
Antonio Vazquez
Darry Welker
Rick Wilson

The following subject matter experts generously gave their time and recommendations to help develop the Stackable Credentials:

Justin Johnson
John Lupacchino
Scott Mitchell

Richard Nickel
Nelson Plumb
Mike Powers

NCCER Partners

American Council for Construction Education
American Fire Sprinkler Association
Associated Builders and Contractors, Inc.
Associated General Contractors of America
Association for Career and Technical Education
Association for Skilled and Technical Sciences
Construction Industry Institute
Construction Users Roundtable
Design Build Institute of America
GSSC – Gulf States Shipbuilders Consortium
ISN
Manufacturing Institute
Mason Contractors Association of America
Merit Contractors Association of Canada
NACE International
National Association of Women in Construction
National Insulation Association
National Technical Honor Society
National Utility Contractors Association
NAWIC Education Foundation
North American Crane Bureau
North American Technician Excellence
Pearson

Prov
SkillsUSA®
Steel Erectors Association of America
U.S. Army Corps of Engineers
University of Florida, M. E. Rinker Sr., School of
    Construction Management
Women Construction Owners & Executives,
    USA

NCCER Business Partners

ISNETWORLD ISN DATA PROVIDER

JUDGMENT INDEX
MEASURING, BUILDING AND STRENGTHENING GOOD JUDGMENT

NACB
NORTH AMERICAN CRANE BUREAU, INC.
PROFESSIONALISM, QUALITY & INTEGRITY SINCE 1986

Pearson

Prov

Contents

Module Ten

Crane Safety and Emergency Procedures*

Emphasizes safety as the highest priority. Crane operators and other members of a lift team must embrace their responsibility as the manager of a powerful machine that can both accomplish great things and destroy property and lives. Thousands of successful crane operations occur each day without incident; all of the lifts in the future can end the same way. The goal of this module is to present a wide variety of safety information related to crane operation and prepare lift team members for their role in a safe workplace. (Module ID 21106-18; 25 Hours)

Module Eleven

Basic Principles of Cranes*

Focuses on the types of mobile cranes and their common components. A fundamental understanding of crane types and how they move from place to place is important to riggers and signal persons as well as to crane operators. Cranes also have different types of lifting booms, each of which has its own strengths and weaknesses. This module presents specific features of various cranes and booms, and introduces the basic principles of lifting and leverage. (Module ID 21102-18; 12.5 Hours)

Module Twelve

Crane Communications*

Provides a formal introduction to methods, modes, and specifics of communications required in crane operations. The principles that crane signal persons must use to guide safe and efficient crane operations are addressed. (Module ID 53101-18; 10 Hours)

Glossary

Index

HEAVY HIGHWAY CONSTRUCTION LEVEL ONE

Module Twelve
Crane Communications *
(53101-17)

Module Eleven
Basic Principles of Cranes *
(21102-17)

Module Ten
Crane Safety and Emergency Procedures *
(21106-17)

Module Nine
Rigging Practices *
(38102-17)

Module Eight
Interpreting Civil Drawings
(22209-03)

Module Seven
Excavation Math
(22207-03)

Module Six
Site Work
(22210-13)

Module Five
Soils
(22308-13)

Module Four
Work Zone Safety
(75104-13)

Module Three
Highway and Bridge Safety
(36110-17)

Module Two
Heavy Highway Construction Equipment
(36111-17)

Module One
Orientation to the Trade
(36101-17)

**Core Curriculum:
Introductory Craft Skills**

This course map shows all of the modules in *Heavy Highway Construction Level One*. The suggested training order begins at the bottom and proceeds up. Skill levels increase as you advance on the course map. The local Training Program Sponsor may adjust the training order.

*Completion of these four modules results in a Basic Rigger credential

36101-17

Orientation to the Trade

OVERVIEW

Roads and bridges are a crucial part of the nation's economy, connecting population centers and allowing people and goods to be moved efficiently over long distances. Modern roads and bridges are complex, and this complexity calls for several types of specialized craft professionals, each fulfilling separate but necessary tasks in the construction of roads and bridges.

Module One

Trainees with successful module completions may be eligible for credentialing through the NCCER Registry. To learn more, go to **www.nccer.org** or contact us at 1.888.622.3720. Our website has information on the latest product releases and training, as well as online versions of our *Cornerstone* magazine and Pearson's product catalog.

Your feedback is welcome. You may email your comments to **curriculum@nccer.org**, send general comments and inquiries to **info@nccer.org**, or fill in the User Update form at the back of this module.

This information is general in nature and intended for training purposes only. Actual performance of activities described in this manual requires compliance with all applicable operating, service, maintenance, and safety procedures under the direction of qualified personnel. References in this manual to patented or proprietary devices do not constitute a recommendation of their use.

Objectives

When you have completed this module, you will be able to do the following:

1. Describe the trade.
   a. Describe the types of work performed in the trade.
   b. Describe the opportunities available in the trade.
2. Describe the basic processes used in heavy construction.
   a. Describe the general steps used to plan and prepare for a heavy construction project.
   b. Identify the processes in highway construction.
   c. Identify the processes in bridge construction.
   d. Identify steps taken when a heavy construction project is complete.
3. Describe apprenticeship training.
   a. Explain how apprenticeship programs are structured.
   b. Describe the basic principles of apprenticeship training.
   c. Explain licensing requirements for apprenticeship training.

Performance Tasks

This is a knowledge-based module; there are no performance tasks.

Trade Terms

Apprenticeship
Rebar
Right-of-way

Substructure
Superstructure

Industry Recognized Credentials

If you are training through an NCCER-accredited sponsor, you may be eligible for credentials from NCCER's Registry. The ID number for this module is 36101-17. Note that this module may have been used in other NCCER curricula and may apply to other level completions. Contact NCCER's Registry at 888.622.3720 or go to **www.nccer.org** for more information.

Contents

Figures

SECTION ONE

1.0.0 INTRODUCTION TO THE TRADE

Objective

Describe the trade.
  a. Describe the types of work performed in the trade.
  b. Describe the opportunities available in the trade.

Trade Terms

Apprenticeship: A system of providing training and mentoring for entry-level trade workers.

Rebar: A contraction of reinforcing bars. Used to reinforce concrete structures.

Right-of-way: A type of easement which designates land as reserved for transportation or other restricted use.

Substructure: The part of a bridge below the top of the bearings. The substructure of a bridge supports the superstructure.

Superstructure: The part of a bridge above the top of the bearings. The superstructure of a bridge supports the traffic load.

Bridges and highways are critical parts of our nation's infrastructure. In addition to providing a means of crossing waterways, bridges are critical elements of our highway system. Bridge and highway construction have many things in common, but there are also distinct skill sets associated with each of them. Common elements include the fact that the bridges and highways have surfaces paved with concrete or asphalt. Also, while a bridge is a structure that typically spans a body of water, its approaches (*Figure 1*) are essentially roads that have to be prepared in the same way as highways. What some people may think of as one process actually involves three processes: road construction, bridge construction, and paving. While the tasks associated with the three processes are closely related, each requires its own set of knowledge and skills in order to become proficient. *Figure 2*, *Figure 3*, and *Figure 4* show examples of the kinds of work performed in these disciplines.

Road construction involves everything from clearing the right-of-way to preparing the road bed, which is the structure that will support the road itself. When construction of a new road begins, the site will be uneven and must be trimmed. Rough grading is the process of taking an existing surface and removing its irregularities, then shaping it into the desired surface at a specific elevation. This requires cutting, pushing, and spreading the material. *Figure 2* (*A*) shows a bulldozer being used to rough-grade a road surface. A compactor is visible in the background of *Figure 2* (*A*). Compacting is done to make the soil more dense, which makes it more able to support heavy loads. Grading and compacting are important and unique aspects of road construction.

Bridge construction involves the creation of elevated road platforms, usually over water. Bridges are typically supported by pilings that are driven into the soil. *Figure 3* shows the pilings for a new bridge that is being built to replace the old one next to it. Note in the background that concrete pile caps have been poured to serve as supports for the bridge superstructure. In the foreground, workers are building forms for pile caps on the next set of pilings.

Paving involves the placement and finishing of the finished surface of the road or bridge. This surface can be asphalt or concrete. *Figure 4* shows a paving crew at work. In the background, the concrete paving has been poured and is being finished, while in the foreground new concrete is being poured over the rebar mats that are used to reinforce the concrete.

Did You Know?

Between new road construction and the widening of existing roads, highway construction in the United States averages more than 30,000 lane-miles of road per year. Historically, more than 60 percent of the new roads in the United States have been created for new residential development.

1.1.0 Craft Professionals in the Industry

Highway and bridge construction requires a variety of skilled craft professionals, so there are a wide variety of roles available to those entering the field. Some of these roles include the following:

- *Rigger* – Where materials must be moved by hoisting, such as by a crane, the rigger is responsible for making sure that the load is secured, stable, and safe to move. Rigging is an important part of bridge construction. *Figure 5* shows a crane lifting a bridge girder that was prepared for hoisting by riggers.

Figure 1 Bridge under construction.

(A) ROUGH GRADING WITH A DOZER

(B) FINISH GRADING

Figure 2 Grading and compacting a road surface.

Figure 3 Preparing the substructure for a new bridge.

Figure 4 Bridge paving in process.

- *Signal person* – A signal person guides a crane operator when loads are being lifted. The signal person uses specialized hand signals or electronic communications to give direction to the crane operator. A rigger may also serve as a signal person.
- *Crane operator* – A fully trained and certified crane operator is responsible for all facets of the crane's use, such as hoisting loads up to a bridge substructure or superstructure (*Figure 5*) and driving or drilling the piles that support bridges and other structures. Trainees and uncertified operators will perform other jobs, such as helping set up the crane and making sure that areas under hoisted loads are free of workers. *Figure 6* (*A*) shows a crane being used to drive piles. Most pile driving work is

Figure 5 A crane hoisting a load prepared by riggers.

performed with cranes. Excavators are also used to drive piles in some cases, usually in areas with low overhead space (*Figure 6* [*B*]).
- *Pile driver mechanic* – Pile driving equipment requires specialists who know how to set up, use, and maintain the equipment.
- *Concrete finisher* – Concrete finishers are responsible for pouring and finishing concrete, as well as repairing damaged concrete. They apply the finish surface to roads, bridges, and parking lots, which is generally comprised of materials such as concrete or asphalt. Concrete finishers ensure that the concrete surface is level. They work with a variety of tools and equipment to finish the concrete. *Figure 7* shows a concrete finisher working with a bull float to level freshly poured concrete. *Figure 8* shows a finish operator floating a concrete surface with a rotary finishing machine. Floating is a process used to remove imperfections in the surface of the concrete and embed the aggregates (stone, sand, etc. that are part of the concrete mix).

(A) CRANE DRIVING PIPE PILES

Figure 7 Worker using a bullfloat to finish a concrete slab.

(B) EXCAVATOR WITH A PILE DRIVER ATTACHMENT

Figure 6 Pile driving.

- *Grader operator* – Grader operators are responsible for bringing a site to its finished grade. Highway rights-of-way are graded in at least two stages: rough grade and finish grade. Rough grade operators use heavy equipment such as scrapers or bulldozers to bring a right-of-way to rough grade. *Figure 9 (A)* shows a bulldozer being used to bring a right-of-way to rough grade. Finish grading creates a smooth surface at the required elevation in preparation for paving. The purpose of finish grading is to finish the surface to the final elevation or slope required by the specifications. Motor graders are used for finish grading because the operator can precisely control the blade height and

position in order to achieve the required elevation with a high degree of accuracy. *Figure 9 (B)* shows a motor grader bringing a road to finish grade in preparation for paving.

- *Grade checker* – Grade checkers examine elevations and grades using manual or laser instruments. During the grading process, a grade checker gives directions regarding the depth of cut or other trimming requirements. They do this by observing elevations and grades, and then giving hand signals to the grader operator.

- *Form carpenter* – Once the right-of-way has been finish graded, a form carpenter will build or assemble forms to support the pouring of concrete piers and cast-in-place bridge elements. They may also create forms for road beds. *Figure 10* shows a form built in preparation for pouring concrete to form a bridge girder.

> **NOTE**
>
> One-time-use forms built on site from lumber, sheet material, and hardware used to be commonly used for the shaping of concrete structures. Rising labor and materials costs have led to the increasing use of pre-fabricated, re-usable forms on many jobs.

- *Carpenter helper* – An entry-level position that entails some of the most basic work at a site. Helpers are often found cleaning sites, preparing materials for more advanced workers, or performing simple construction tasks.

Figure 8 Finishing a concrete surface with a rotary finisher.

- *Heavy equipment operator* – Just as a pile driver operator is in charge of the pile driver, heavy equipment operators are in charge of the site's heavy equipment, such as bulldozers, graders, and backhoes. Some major categories of heavy equipment are grading equipment, excavation equipment, and skid steers / dozers. An operator must be trained in, and sometimes certified in, each piece of equipment on an individual basis. *Figure 11* shows a backhoe operator using a backhoe to dig a trench.
- *Mechanical maintenance specialist* – Cranes and heavy equipment used in highway and bridge construction require periodic inspection and maintenance in order to remain in good working condition. In addition, the equipment needs to be repaired quickly when a failure occurs. Mechanical maintenance specialists keep the equipment in operating condition.
- *Lubrication technician* – Heavy equipment needs frequent lubrication in order to operate efficiently. The lubrication technician is responsible for keeping the equipment well lubricated.

- *Welder* – Welders are specialists in joining materials together (*Figure 12*). Welders work primarily with steel, and sometimes with aluminum. On a construction site, welders may repair damaged equipment such as dozer blades and may join steel components used in bridge structures.

Did You Know?

The nation's infrastructure consists of highways, bridges, and tunnels. There are over 47,000 miles of highway in the US Interstate highway system alone. In addition, there are about 160,000 miles of US-designated highways (US 1, US 66, etc), along with hundreds of thousands of miles of state highways.

(A) BRINGING A SITE TO ROUGH GRADE

(B) FINISH GRADING

Figure 9 Bringing a road to grade.

- *Ironworker* – Ironworkers erect or repair the steel frame and other metal parts of bridges and overpasses, and direct crane operators in hoisting steel members into the correct position.
- *Reinforcing ironworker (rodbuster)* – Before concrete is poured, steel mesh or steel reinforcing rods known as rebar are placed in the structure to reinforce the concrete. The workers who install this material are known as *reinforcing ironworkers* or *rodbusters*. In highway and bridge work, the reinforcing material usually consists of rebar mats made by tying lengths of rebar together. An example of a rebar mat can be seen in the foreground of *Figure 13*.
- *Surveyor* – Surveyors inspect the areas through which highways are routed to detect characteristics such as changes in elevation and the locations of natural and man-made objects along the route. These surveys are called *route surveys*, and often involve staking out the proposed route and calculating earthwork quantities.
- *Electrician* – There are a variety of electrical devices installed along highways and bridges, including traffic signals, highway illumination, and roadway video/camera systems. Electricians must plan out and install the metal electrical conduit, wiring systems, and overhead and underground distribution systems along highways.

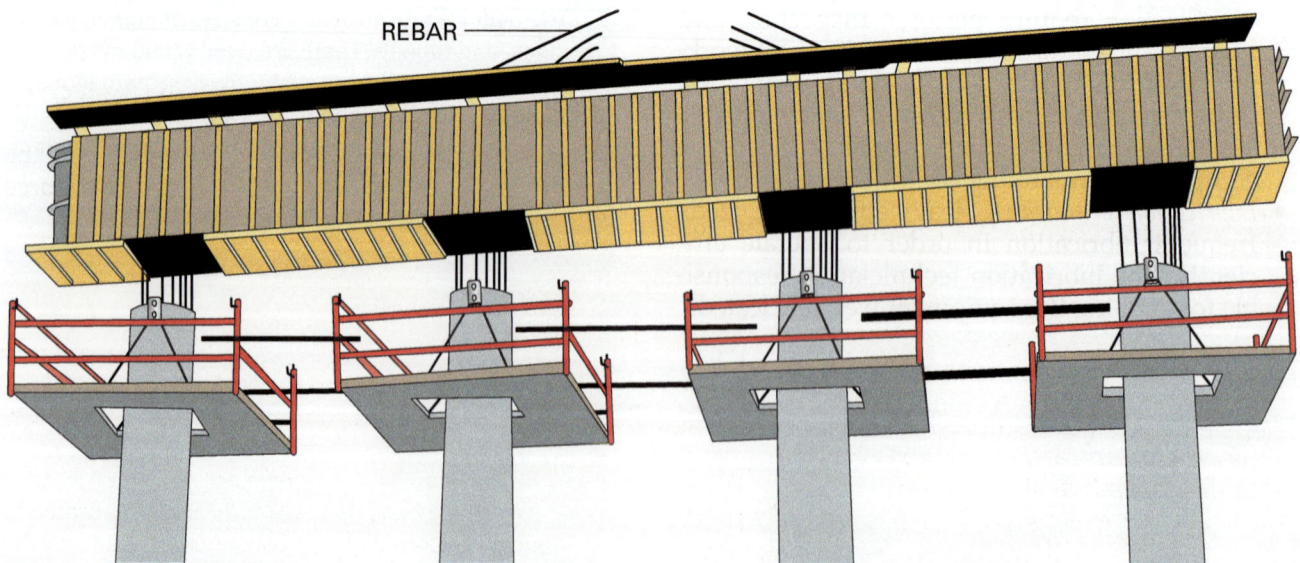
Figure 10 Prefabricated beam form.

Figure 11 Backhoe being used to dig a trench.

1.2.0 Opportunities for Advancement

Once you have a start in the construction industry, effort and application can help you move up. Some higher positions still involve hands-on craftworking, while others allow you to take what you've learned in the trade and apply that knowledge in other ways. Some of the opportunities available to craft professional in highway and bridge construction include the following:

- *Crew leader* – Supervises a crew composed of several trade workers, typically up to about five workers.
- *Foreman* – Supervises all members of a given trade on a project.
- *Maintenance supervisor* – Ensures the availability of both heavy equipment and trained equipment operators and is responsible for maintenance of heavy equipment.
- *Superintendent* – Oversees the entire project on a construction site.
- *Project manager* – Has overall responsibility for one or more projects. The project manager interfaces with owners and company management.
- *Scheduler/planner* – These two roles are very similar, and on smaller projects they may be filled by the same employee. Both positions involve forecasting the labor hours and materials required for the various stages of a project.
- *Instructor* – Schools that provide apprenticeship programs need skilled instructors and mentors to deliver their training. Instructors are generally seasoned craft professionals who have achieved journey-level status by demonstrating their mastery of the knowledge and skills associated with the craft.
- *Safety and environmental specialist* – Verifies compliance with federal (and sometimes state and local) safety guidelines and laws for

Figure 12 Welder at work.

Figure 13 Rebar application.

worker safety and environmental protection. Companies should have policies and procedures that cover all aspects of safety on the job. The job of the safety specialist is to enforce those policies and procedures.

GOING GREEN

Highway and bridge construction occurs around bodies of water and land that may be environmentally sensitive. Everyone working on these projects needs to be aware of environmental concerns and the methods used to protect the environment during construction.

1.2.1 Professionalism

Professionalism is a broad term that describes the desired overall behavior and attitude expected in the workplace. Professionalism is too often absent from the construction site and the various trades. Most people would argue that professionalism must start at the top in order to be successful. It is true that management support of professionalism is important to its success in the workplace, but it is more important that individuals recognize their own responsibility for professionalism.

Professionalism includes honesty, productivity, safety, civility, cooperation, teamwork, clear and concise communication, being on time and prepared for work, and regard for one's co-workers. It can be demonstrated in a variety of ways every minute you are in the workplace. Most important is that you do not tolerate the unprofessional behavior of co-workers. This is not to say that you shun the unprofessional worker; instead, you work to demonstrate the benefits of professional behavior.

Professionalism is both a benefit to the employer and the employee. It is a personal responsibility. Our industry is what each individual chooses to make of it; choose professionalism and the industry image will follow.

Did You Know?

Some companies encourage their employees to learn more than one job. This is referred to as *cross-training.* The employer benefits because the employee is more versatile. The employee benefits by becoming more valuable to the employer.

Additional Resources

The following websites offer resources for products and training:

Build Your Future (initiative of NCCER), **www.byf.org**

Occupational Information Network (O*NET), sponsored by the US Department of Labor / Employment and Training Administration. "My Next Move." **www.mynextmove.org**

1.0.0 Section Review

1. A rigger is responsible for preparing _____.
   a. heavy equipment for use
   b. loads to be hoisted
   c. heavy equipment for travel
   d. material for shipment

2. A person who supervises all members of a trade on a project is a _____.
   a. superintendent
   b. project manager
   c. foreman
   d. pit boss

SECTION TWO

2.0.0 HEAVY CONSTRUCTION PROCESSES

Objective

Describe the basic processes used in heavy construction.

  a. Describe the general steps used to plan and prepare for a heavy construction project.
  b. Identify the processes in highway construction.
  c. Identify the processes in bridge construction.
  d. Identify steps taken when a heavy construction project is complete.

Highway and bridge construction are complex jobs that can be broken down into several smaller processes. They share many of the same processes, but also have significant differences from one another. *Figure 14* shows the steps involved in both highway and bridge construction, including those for paving.

2.1.0 Planning and Preparation

The following steps outline the general procedures used to plan and prepare for a highway or bridge construction project:

Step 1   *Safety planning* – Once the design for a project is complete, the construction firm can plan for necessary safety equipment and provide the required training for workers. Existing safety policies and procedures deal with the specific safety issues associated with the project.

Step 2   *Preconstruction meeting* – The preconstruction meeting is a planning session in which the various contractors, along with utilities representatives and owners, discuss how the project can be safely and efficiently completed.

Step 3   *Surveying the right-of-way* – Surveying (*Figure 15*) verifies right-of-way alignment and proximity to utilities. Surveying work continues throughout construction projects, as survey workers verify that the roadways and bridges are being built in compliance with the drawings and specifications that govern the job.

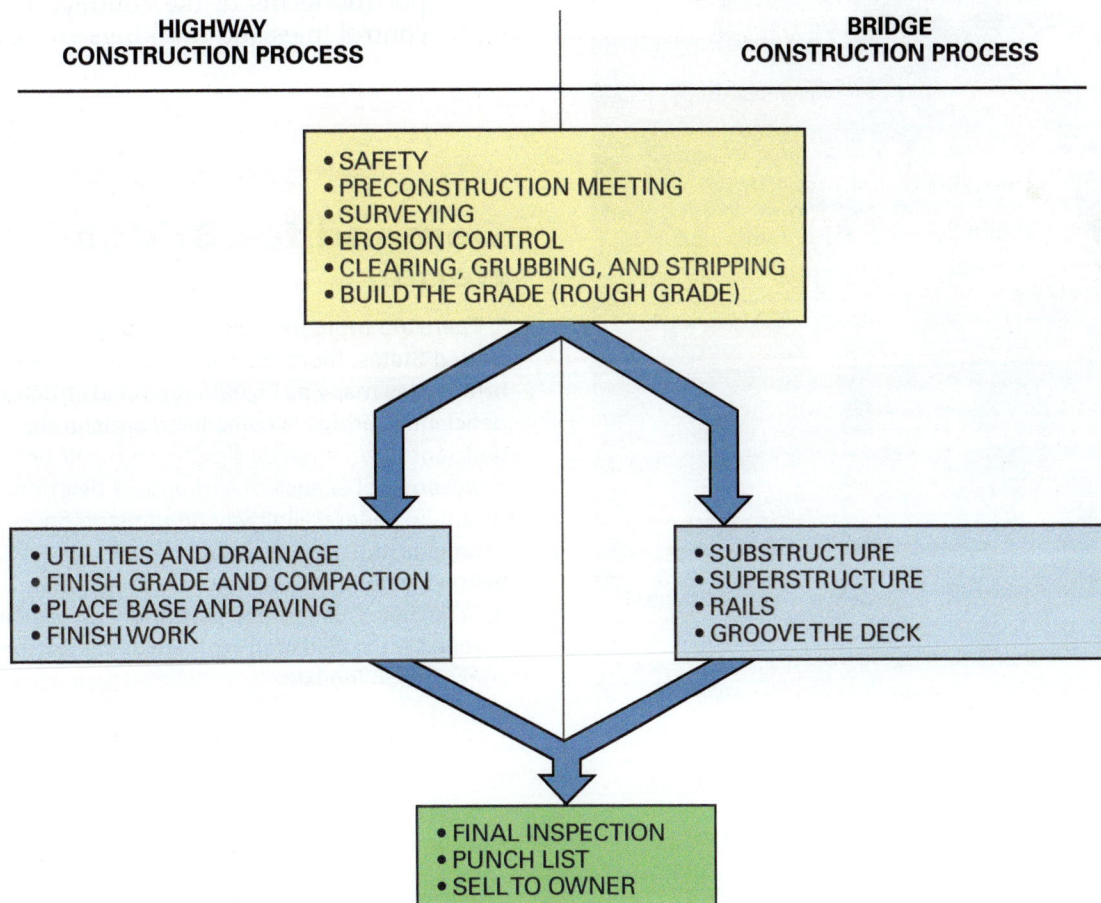

HIGHWAY CONSTRUCTION PROCESS | **BRIDGE CONSTRUCTION PROCESS**

- SAFETY
- PRECONSTRUCTION MEETING
- SURVEYING
- EROSION CONTROL
- CLEARING, GRUBBING, AND STRIPPING
- BUILD THE GRADE (ROUGH GRADE)

- UTILITIES AND DRAINAGE
- FINISH GRADE AND COMPACTION
- PLACE BASE AND PAVING
- FINISH WORK

- SUBSTRUCTURE
- SUPERSTRUCTURE
- RAILS
- GROOVE THE DECK

- FINAL INSPECTION
- PUNCH LIST
- SELL TO OWNER

Figure 14 Steps in highway and bridge construction.

Step 4 *Erosion control planning* – Construction firms are legally responsible for soil erosion at construction sites, and for consequences of failing to control erosion, which can be severe. Planning to control soil erosion and groundwater is therefore an early step in the construction process. *Figure 16* shows erosion controls in the form of a silt fence in place at a highway improvement site.

Step 5 *Clearing right-of-way* – Once erosion control procedures are in place, the site must be cleared of any existing structures as well as large rocks, trees, and shrubs. This work is called *clearing* and *grubbing*. Dozers are commonly used for this work.

Step 6 *Rough-grading the right-of-way* – Once the right-of-way is cleared, the site can be graded. At this stage the site will be rough-graded using a scraper or dozer.

Figure 15 Surveying during construction.

2.2.0 Highway Construction Steps

As illustrated in *Figure 14*, there are some steps in the heavy construction process that apply only to highway construction, and others that apply only to bridge construction. Specific steps to highway construction (following the planning and preparation stage) are as follows:

Step 1 *Utility and site drainage planning* – Controls are implemented to prevent damage to utilities and interruption of service. Other controls are put in place to control rainwater runoff. New utilities may be added, depending on site requirements.

Step 2 *Finish grading and compaction* – In this step, the site is graded to the final level and the soil is compacted to help support the road base. *Figure 17* shows a compactor in use.

Step 3 *Placing road base and paving* – The road base is poured and cured and pavers add the concrete or asphalt finish surface. *Figure 18* shows a paving crew at work placing a concrete road surface with a slipform paving machine.

Step 4 *Finish work* – The road surface is finished per the terms of the contract and traffic control lines and markings are added.

Demand for Bridge Repair

There are more than 600,000 bridges in the United States. More than 12 percent of these bridges (as many as 75,000) are rated structurally deficient. A bridge is considered structurally deficient when it needs significant repair or replacement because of component deterioration. Based on Federal Highway Administration estimates, the country would need to spend $20.5 billion a year until 2028 in order to eliminate these deficiencies. This represents a huge increase in the demand for skilled craft workers in this sector of the construction industry.

Figure 16 Erosion control features.

Figure 18 Paving crew at work.

Figure 17 Compacting the road base.

2.3.0 Bridge Construction Steps

The heavy construction steps that apply only to bridge construction (following the planning and preparation stage) are as follows:

Step 1 *Construct substructure* – The substructure supports the working elements of the bridge and must be built first. It contains the foundations, pilings, and piers that support the superstructure.

Step 2 *Construct superstructure* – After the substructure has been built, the superstructure can be added to it. The superstructure contains the roadway and railings. *Figure 19* shows the components of a substructure and superstructure for a bridge.

Step 3 *Place the roadway and groove the deck* – To help vehicles maintain good traction, the deck (road surface) of a bridge is grooved.

Step 4 *Install railings and drainage* – Protection to keep vehicles from driving off the side of the bridge is added.

Aging Transportation Infrastructure

Federal funding for maintenance of the United States transportation infrastructure, including roads, bridges, and railroads, has been below requested amounts for several years, which means that work that needs to be done is not getting done. Some experts have attributed the collapse of Bridge 9340, which carried Interstate 35 West across the Mississippi River near Minneapolis, to a lack of inspection and maintenance.

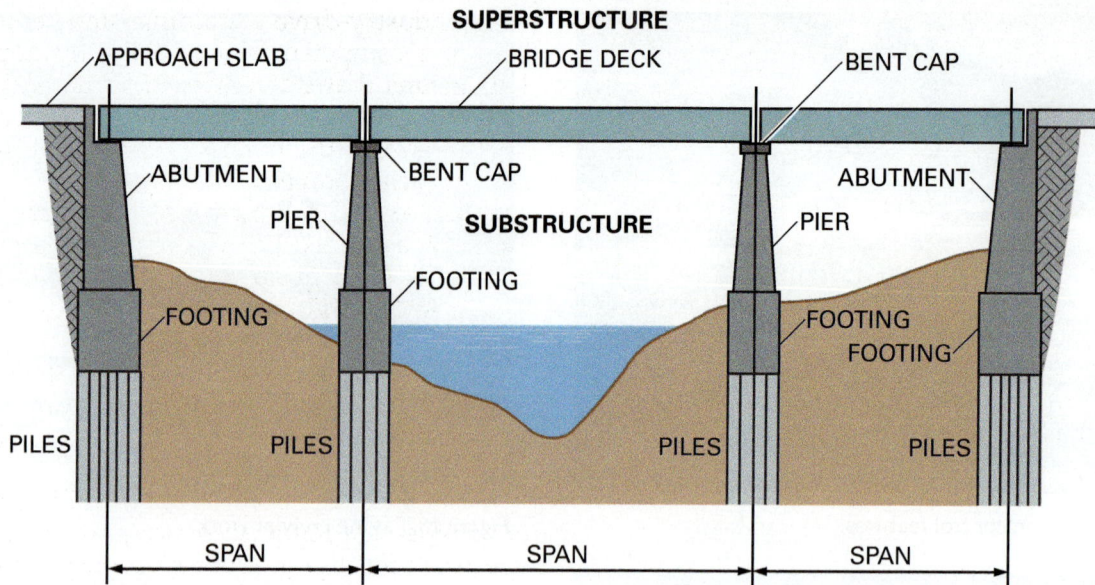

Figure 19 Bridge substructure and superstructure.

2.4.0 Inspection and Delivery

The following final steps are taken when a high-way or bridge construction project is complete:

Step 1 *Final inspection* – Once the roads and bridges are in place, they will be inspected by the proper authorities to verify compliance with all applicable regulations.

Step 2 *Create a punch list* – The owner inspects the project and creates a list of issues to be corrected to bring the project within the specifications of the contract.

Step 3 *Deliver to owner* – When the project has been completed to the owner's satisfaction, it is formally delivered to the owner as the final step in the construction process.

Additional Resources

Scranton Gillette Communications. "Roads and Bridges." **www.roadsbridges.com**

2.0.0 Section Review

1. Which of the following steps in the highway construction process takes place along with finish grading?

   a. Compacting the site
   b. Clearing the site
   c. Bringing in fill soil
   d. Rodbusting

2. Which of these statements about bridge building is true?

   a. The deck must be grooved before rails are added.
   b. The bridge is completed and accepted by the owner before work can start on the road.
   c. The road must be completed before work can start on the bridge.
   d. The substructure must be built before the superstructure.

3.0.0 TRAINING AND APPRENTICESHIP PROGRAMS

Objective

Describe apprenticeship training.
  a. Explain how apprenticeship programs are structured.
  b. Describe the basic principles of apprenticeship training.
  c. Explain licensing requirements for apprenticeship training.

There are several ways to gain entry into the highway/bridge industry, and multiple ways to advance your career in the construction trades. One of the most potentially rewarding is the apprenticeship program, which can not only make you a better worker, but can also help to increase your income and gain credentials that follow you to any job.

3.1.0 Apprenticeship Structuring

The US Department of Labor (DOL) Office of Apprenticeship sets the minimum standards for apprenticeship programs across the country. These programs rely on mandatory classroom instruction (termed related technical instruction) and on-the-job learning (OJL). They require at least 144 hours of related instruction per year and 2,000 hours of OJL per year. In a typical highway construction apprenticeship program, trainees spend a total of at least 288 hours in classroom instruction and 4,000 hours in OJL before receiving journey level certificates issued by registered apprenticeship programs.

To address the training needs of highway and bridge construction workers, NCCER has developed a two-level highway and bridge construction program that meets the requirements for a registered apprenticeship. This program (*Heavy Highway Construction Level One* and *Heavy Highway Construction Level Two*) provides for many possible career paths.

NCCER uses the minimum Department of Labor standards as a foundation for a comprehensive curricula that provides trainees with in-depth classroom and OJL experience. The standardized NCCER curriculum provides trainees with industry-driven training and education. It uses a competency-based learning approach. This means that trainees must demonstrate that they possess the knowledge and skills needed to safely perform the hands-on tasks that are covered in each module.

When an NCCER-certified instructor is satisfied that a trainee has the required knowledge and skills for a given module, that information is sent to NCCER and kept in the Registry system. NCCER's Registry can verify satisfactorily completed training for workers as they move from state to state, company to company, or even within a company. See the *Appendix* for examples of credentials issued by NCCER.

Whether you enroll in an NCCER program or another apprenticeship program, make sure you work for an employer or sponsor who supports a nationally standardized training program that includes credentials to confirm your skill development.

Did You Know?

George Washington (surveyor), Benjamin Franklin (printer), and Paul Revere (silversmith) were apprentices. According to the US Department of Labor, there were 375,000 apprentices in training in 19,000 apprenticeship programs in 2014. More than 113,000 of these were active construction industry apprentices in 21 different construction crafts.

3.2.0 Apprenticeship Training

Apprenticeship training goes back thousands of years, and its basic principles have not changed. First, it is a means for a person entering the craft to learn from those who have mastered the craft. Second, it focuses on learning by doing, applying theory in context with real-world skills. Theory is presented in the classroom in a way that helps trainees understand the purpose behind the skill that is to be learned.

Did You Know?

Apprenticeship in the United States

Every year, more than half a million apprentices receive registered apprenticeship training toward their journey-level credentials.

Equipment Simulators

Equipment simulators, like the one shown here, can be used to train new operators, assess the skills of experienced operators, and screen potential new employees. The simulators duplicate the cab environment and controls. The simulation software causes the screen images to respond to the controls as the real machine would. Trainees using the simulator learn to perform tasks the safest and most efficient way. There are significant savings in terms of machine ownership or rental, cost of fuel, and reduced wear and tear on machines, and most importantly there are also significant safety benefits. Simulators are available for most types of heavy equipment.

CM LABS SIMULATIONS INC. – WWW.CM-LABS.COM

3.2.1 Apprenticeship Standards

All apprenticeship standards prescribe certain work-related or on-the-job learning. This OJL is broken down into specified tasks in which the apprentice receives hands-on training. In addition, a specified number of hours is required in each task. The total number of OJL hours for an apprenticeship program traditionally ranges from 4,000 to 8,000. In a competency-based program, it may be possible to shorten this time by testing out of specific tasks through a series of performance exams. The apprenticeship program for highway and bridge construction workers has been established as a two-year program.

In a traditional apprenticeship program, the required OJL may be acquired in increments of 2,000 hours per year.

The apprentice must log all work time and turn it in to the apprenticeship committee so that accurate time control can be maintained. After each 1,000 hours of related work, the apprentice may receive a pay increase if so prescribed by the apprenticeship standards.

For those entering an apprenticeship program, a high school or technical school education is desirable. Courses in shop, mechanical drawing, and general mathematics are helpful. Manual dexterity, good physical conditioning, and quick reflexes are important. The ability to solve problems quickly and accurately and to work closely with others is essential. You must also have a high concern for safety.

The prospective apprentice must submit certain information to the apprenticeship committee. This may include the following:

- Aptitude test (General Aptitude Test Battery [GATB] Form Test) results usually administered by the local Employment Security Commission
- Proof of educational background (candidate should have high school transcripts sent to the committee)
- Letters of reference from past employers and friends
- Proof of age
- If a candidate is a veteran, a copy of Form DD214
- A record of technical training received that relates to the construction industry and/or a record of any pre-apprenticeship training

The apprentice must do the following:

- Wear proper safety equipment on the job
- Purchase and maintain tools of the trade as needed and required by the contractor
- Submit a monthly on-the-job training report to the committee
- Report to the committee if a change in employment status occurs
- Attend classroom-related instruction and adhere to all classroom regulations such as attendance requirements

3.2.2 Youth Apprenticeship Program

Youth apprenticeship programs are also available that allows students to begin their apprentice training while still in high school. In some programs, a student entering the program in eleventh grade may complete as much as two years of the NCCER standardized four-year curriculum by high school graduation. In addition, the program, in cooperation with local craft employers, allows students to work in the trade and earn money while still in school. Upon graduation, the student can enter the industry at a higher level and with more pay than someone just starting the apprenticeship program.

The training program is similar to the one used by NCCER learning centers, contractors, and colleges across the country. Students are recognized through official transcripts and can enter the next year of the program whenever it is offered. They may also have the option of applying the credits at a two-year or four-year college that offers degree or certification programs in the construction trades.

3.3.0 Licensing

After you complete your training, you may be required to take your state or local licensing exam. Licensing is used by states to ensure that contractors have the required knowledge to successfully perform the work. The purpose of licensing is to provide assurance that you are qualified to perform skilled tasks in bridge and road construction. Not all crafts require a license. Being licensed, you will be able to work independently and earn a higher income. With your credentials, you are not only responsible for your own work, but you are liable for that work as well. If someone is working for you, then you are also responsible for that person's work.

Electricians are licensed in just about every state; plumbers generally require licenses as well. Licensing requirements vary from state to state, and may also vary by municipality. Contact your local building department for the requirements in your area. After you receive your license, your state or locality may require continuing education in order to renew your license.

Additional Resources

29 *CFR* 29, **www.ecfr.gov**

3.0.0 Section Review

1. What is the minimum number of annual classroom instruction hours required for a Department of Labor-registered apprenticeship program?

   a. 100
   b. 144
   c. 150
   d. 188

2. Apprentices may receive a prescribed pay increase after _____.

   a. every 1,000 hours of OJL
   b. completing an apprenticeship level
   c. passing a test of trade knowledge
   d. registering for an apprenticeship program

3. Which of these statements about licensing requirements is true?

   a. Licensing of construction workers is required by all states.
   b. Licensing requirements are standardized by the federal government.
   c. Licensing requirements are the same for all crafts.
   d. Some crafts do not require licensing.

SUMMARY

Highway and bridge construction is complex, demanding work. It calls for knowledge of several fields, and precision implementation. While this may seem daunting, hard workers can move up through the ranks to positions of greater responsibility and bigger rewards.

Part of the complexity in bridge and highway construction is the number of processes required to make the road safe and long-lasting. While some of these processes can be performed out of the sequence provided here, in general the sequence presented is common among construction crews.

One option to consider when starting in highway or bridge construction is a registered apprenticeship program if one is available to you. While an apprenticeship requires extra effort from you in the form of classroom education in your off hours, the benefits of an apprenticeship can be substantial. These benefits include higher pay earlier in your career; a greater understanding of the trade; and credentials that follow you wherever you go, proving your knowledge to prospective employers.

Review Questions

1. The craft professional who prepares loads to be lifted by a crane is a _____.
   - a. heavy equipment operator
   - b. carpenter
   - c. rigger
   - d. surveyor

2. Pile driving is normally done with _____.
   - a. excavators
   - b. cranes
   - c. motor graders
   - d. bulldozers

3. The person in charge of heavy equipment maintenance is called the _____.
   - a. superintendent
   - b. foreman
   - c. crew leader
   - d. maintenance supervisor

4. Protecting soil from being washed away by rainwater runoff is called _____.
   - a. erosion control
   - b. washout control
   - c. soil conservation
   - d. water shielding

5. Which of the following is the correct order for the steps involved in highway construction?
   - a. Compaction, rough grade, finish grade
   - b. Rough grade, finish grade, compaction
   - c. Rough grade, compaction, finish grade
   - d. Finish grade, rough grade, compaction

6. After paving is added to road surface of a bridge it is _____.
   - a. compacted
   - b. grooved
   - c. hoisted
   - d. surveyed

7. A bridge's substructure supports its _____.
   - a. right-of-way
   - b. foundation
   - c. piles
   - d. superstructure

8. An apprenticeship education is composed of classroom learning combined with _____.
   - a. on-the-job learning
   - b. Department of Labor credentials
   - c. supervisor mentoring
   - d. personal study

9. What is the established duration of a highway/bridge construction apprenticeship program?
   - a. 2 years
   - a. 3 years
   - b. 4 years
   - c. 5 years

10. Trade licensing is typically regulated by states, but may be regulated by _____.
    - a. a Civil Engineering Board
    - b. employer representatives
    - c. the Department of Commerce
    - d. municipalities

Trade Terms Quiz

Fill in the blank with the correct term that you learned from your study of this module.

1. The _____ of a bridge includes the vertical support structures.

2. The land where a highway will be built is the _____.

3. The upper elements of a bridge are called the _____.

4. A program designed to mentor new construction workers is called a(n) _____.

5. _____ is used to reinforce concrete structures.

Trade Terms

Apprenticeship
Rebar
Right-of-way
Substructure
Superstructure

Rick L. Wilson

BATC Heavy Equipment Operator Instructor,
Bridgerland Applied Technology College

Working summers on a family ranch in Canada, Rick was no stranger to hard work. It was only after going to work as a laborer for a local contractor for a few seasons, that he discovered his real love for running heavy equipment. He now applies his passion for heavy equipment to training and being able to give back to the industry that gave so much to him.

How did you get started in the construction industry?

After high school I spent a few seasons working on a crusher and a paving crew for a local contractor, where I discovered my great love of running heavy equipment by building many miles of freeway. After some mining and hazardous waste containment, I found myself in an open pit mine in southern Idaho, learning to identify and mine phosphate ore.

Who or what inspired you to enter the construction industry?

The combination of having a hard-working father as a role-model who put himself through school working in the construction industry and later as an estimator, and my own experience with running equipment on the farm earning a really great wage, both inspired me to pursue a career path in construction.

What do you enjoy most about your career?

I especially enjoy training new people and seeing that light come on when they "get it," and being able to watch and help the next generation learn and grow.

What types of training have you completed?

Over the years I have participated in weekly safety meetings, annual MSHA safety refreshers, hazardous waste recognition and handling, hydraulic shovel and other equipment operation training, phosphate ore identification, mining procedures, welding, operating forklifts, and most recently as an NCCER certified craft trainer.

How would you define craftsmanship?

Applying yourself to do a task or job with a high level of skill gives you a finished product and something to look back on and be proud of.

Why do you think credentials are important in the construction industry?

Training gives you a good way to show your interest in the field, as well increasing an awareness and responsibility for your safety and that of those around you. Credentials and certificates can help in advancements and promotions in your field of work, showing discipline and dedication to your chosen field.

What advice would you give to someone who is new to the construction industry?

Pay attention to the details. It's the details that go into a job well done that can make the difference between an entry level employee and a good operator. Always be 15 minutes early. If you are not 15 minutes early, you're late! Challenge yourself to do better and have fun.

Would you recommend construction as a career to others?

Heavy equipment is a very rewarding career when you can look back 30 years to a highway you helped build, and it's still there to prove it. Working with heavy equipment in construction is a great career where you're not stuck in an office or warehouse all day, and have all the comforts of radio, air conditioning, and ride control. But the best part is, your office window has a view and it's always changing.

Trade Terms Introduced in This Module

Apprenticeship: A system of providing training and mentoring for entry-level trade workers.

Rebar: A contraction of reinforcing bars. Used to reinforce concrete structures.

Right-of-way: A type of easement which designates land as reserved for transportation or other restricted use.

Substructure: The part of a bridge below the top of the bearings. The substructure of a bridge supports the superstructure.

Superstructure: The part of a bridge above the top of the bearings. The superstructure of a bridge supports the traffic load.

SAMPLES OF NCCER TRAINING CREDENTIALS

NCCER

Board of Trustees confers upon

Sample Student

this certificate of completion for

Heavy Highway Construction Level One

in the Standardized Craft Training program
on this Twenty-seventh day of February, 2016

Donald E. Whyte
Donald E. Whyte
President, NCCER

Student 7 Sample
Certified Plus
4671784

Sample Student
2781481

Additional Resources

This module is intended as a thorough resource for task training. The following reference works are suggested for further study.

29 CFR 29, **www.ecfr.gov**

The following websites offer resources for products and training:

Build Your Future (initiative of NCCER), **www.byf.org**

Occupational Information Network (O*NET), sponsored by the US Department of Labor / Employment and Training Administration. "My Next Move." **www.mynextmove.org**

Scranton Gillette Communications. "Roads and Bridges." **www.roadsbridges.com**

Figure Credits

Section Review Answer Key

Answer	Section Reference	Objective
Section One		
1. b	1.1.0	1a
2. c	1.2.0	1b
Section Two		
1. a	2.2.0	2b
2. d	2.3.0	2c
Section Three		
1. b	3.1.0	3a
2. a	3.2.1	3b
3. d	3.3.0	3c

NCCER CURRICULA — USER UPDATE

NCCER makes every effort to keep its textbooks up-to-date and free of technical errors. We appreciate your help in this process. If you find an error, a typographical mistake, or an inaccuracy in NCCER's curricula, please fill out this form (or a photocopy), or complete the online form at **www.nccer.org/olf**. Be sure to include the exact module ID number, page number, a detailed description, and your recommended correction. Your input will be brought to the attention of the Authoring Team. Thank you for your assistance.

Instructors – If you have an idea for improving this textbook, or have found that additional materials were necessary to teach this module effectively, please let us know so that we may present your suggestions to the Authoring Team.

NCCER Product Development and Revision
13614 Progress Blvd., Alachua, FL 32615

Email: curriculum@nccer.org
Online: www.nccer.org/olf

❏ Trainee Guide   ❏ Lesson Plans   ❏ Exam   ❏ PowerPoints   Other _____

Craft / Level: _____   Copyright Date: _____

Module ID Number / Title: _____

Section Number(s): _____

Description: _____

_____

_____

_____

_____

Recommended Correction: _____

_____

_____

_____

_____

Your Name: _____

Address: _____

_____

Email: _____   Phone: _____

_____

_____

_____

_____

_____

_____

_____

_____

_____

36111-17

Heavy Highway Construction Equipment

OVERVIEW

Highway and bridge construction requires the use of many different types of equipment. While most pieces of equipment are designed for a particular function, many can be used to perform other tasks. In addition, attachments are commonly used to expand the capabilities of the equipment. In order to work safely and effectively, and to have career advancement opportunities, construction workers need to be able to recognize the equipment and machines commonly used on highway and bridge projects and understand their uses.

Module Two

Trainees with successful module completions may be eligible for credentialing through the NCCER Registry. To learn more, go to **www.nccer.org** or contact us at 1.888.622.3720. Our website has information on the latest product releases and training, as well as online versions of our *Cornerstone* magazine and Pearson's product catalog.

Your feedback is welcome. You may email your comments to **curriculum@nccer.org**, send general comments and inquiries to **info@nccer.org**, or fill in the User Update form at the back of this module.

This information is general in nature and intended for training purposes only. Actual performance of activities described in this manual requires compliance with all applicable operating, service, maintenance, and safety procedures under the direction of qualified personnel. References in this manual to patented or proprietary devices do not constitute a recommendation of their use.

Objectives

When you have completed this module, you will be able to do the following:

1. Identify the heavy equipment used in highway and bridge construction and describe their uses.
   a. Identify heavy equipment used in excavation and grading.
   b. Identify the types of trucks used in highway and bridge construction.
   c. Identify paving equipment used in highway construction.
2. Identify utility equipment used on the job site and describe their uses.
   a. Identify and describe compaction equipment.
   b. Identify and describe electrical generators.
   c. Identify and describe air compressors.
3. Identify the mobile cranes used in highway and bridge construction and describe their uses.
   a. Identify and describe crawler cranes.
   b. Identify and describe truck-mounted cranes.

Performance Tasks

This is a knowledge-based module; there are no Performance Tasks.

Trade Terms

Apron
Articulated
Auger
Bowl
Dipper stick
Haul roads
Haul truck
Hydraulic breaker
Knuckle boom
Lattice boom
Outriggers

Pneumatic
Power takeoff (PTO)
Ripper
Scarifying
Screed
Shooting-boom excavator
Slipform paver
Squirt boom
Tamping roller
Undercarriage

Industry Recognized Credentials

If you are training through an NCCER-accredited sponsor, you may be eligible for credentials from NCCER's Registry. The ID number for this module is 36111-17. Note that this module may have been used in other NCCER curricula and may apply to other level completions. Contact NCCER's Registry at 888.622.3720 or go to **www.nccer.org** for more information.

Contents

Figures

SECTION ONE

1.0.0 EXCAVATION, GRADING, AND PAVING EQUIPMENT

Objective

Identify the heavy equipment used in highway and bridge construction and describe their uses.

a. Identify heavy equipment used in excavation and grading.
b. Identify the types of trucks used in highway and bridge construction.
c. Identify paving equipment used in highway construction.

Trade Terms

Apron: A movable section on the forward wall of the bowl on a scraper.

Articulated: Two parts connected by a joint so as to move independently.

Auger: A screw conveyor that is used to move bulk material such as asphalt.

Bowl: The area on a scraper where soil is stored when it is scraped from the surface.

Dipper stick: A pivoting section that connects the bucket to the boom on a hydraulic excavator.

Haul roads: Compacted dirt roads used to move material and equipment on and off the site.

Haul truck: A name that is sometimes used to describe a rigid-frame dump truck or a mining truck.

Hydraulic breaker: A hydraulic attachment for an excavator that is used for breaking boulders and other solid objects.

Knuckle boom: A term sometimes used for a boom and stick combination that resembles a knuckle at the pivot point of the boom and stick.

Outriggers: Stabilizer legs that can be extended to widen the stance of a piece of equipment to keep it from tipping or rolling.

Power takeoff (PTO): A system found on construction tractors that uses a shaft to transfer power from the tractor to an attachment.

Ripper: A towed attachment with teeth used on dozers, motor graders, and other machines to loosen heavily compacted soil and soft rock.

Scarifying: Using an attachment with teeth on a motor grader to loosen soil in front of the moldboard.

Screed: A blade-like component on a paver that levels and smoothes asphalt or concrete as it is applied.

Shooting-boom excavator: A term sometimes used to describe a telescoping-boom excavator.

Slipform paver: A type of concrete paver that evenly spreads bulk concrete that is dumped in front of it.

Squirt boom: A component on some telescoping-boom forklifts that allows the fork carriage to be moved in the horizontal plane while the boom remains stationary.

Undercarriage: The lower frame of an excavator that supports the turntable and has the tracks or wheels attached.

Highway and bridge construction involves several fundamental phases, including excavation, grading, and paving. In a typical project, there might be a dozen or more pieces of heavy equipment involved at one time or another. This section identifies common excavation and grading equipment, different types of trucks, and paving equipment.

1.1.0 Excavation and Grading Equipment

Construction workers on highway and bridge projects are certain to encounter many different types of excavation and grading equipment. Machines like loaders, excavators, dozers, and graders are a common sight at virtually every job site. Workers must be familiar with these pieces of equipment and understand each the purpose of each type.

1.1.1 Utility Tractors

The types of tractors normally used in agricultural work are sometimes used as utility vehicles on job sites. The smaller tractors typically fall into the 60 to 100 horsepower range, but some of the larger ones have engines that range between 300 and 600 horsepower. They are usually equipped with a hitch, which allows them to pull trailers and other attachments that do not require any kind of drive power to be supplied by the tractor. These tractors are also equipped with power takeoff (PTO) and hydraulic systems that allow them to power a variety of construction-related implements that may be used for grading, sweeping, and even post-hole digging. *Figure 1* shows examples of tractors that might be used on construction sites.

(A) UTILITY TRACTOR

(B) QUAD-TRACK TRACTOR

(C) WHEELED ARTICULATED TRACTOR WITH SCRAPERS

Figure 1 Tractors used in construction.

Large, articulated tractors like the one shown in *Figure 1* are used for heavy-duty tasks such as pulling scrapers, oversized rollers, and water tanks. These tractors are in the 375-horsepower and up range. Some companies may prefer to use rubber track vehicles for the heavy work.

1.1.2 Backhoe Loaders

A backhoe loader (*Figure 2*), often called a *backhoe*, is a common piece of equipment used in highway and bridge construction. In a typical configuration, a backhoe consists of three separate parts: a tractor, a loader attachment, and a backhoe attachment. The loader attachment is mounted on the front of the tractor, while the backhoe attachment is mounted at the rear of the tractor. Stabilizer legs, or outriggers,

are used to distribute the lateral loads created by the bucket as it digs. They widen the stance of the backhoe to keep it from tipping or rolling.

Site Safety

During the site development phase of a project, it is common to find many machines operating at the same time, as well as people who may be working near the machines. Operators must remain alert to their surroundings at all times in order to avoid accidents that could kill or injure personnel or damage expensive equipment. Likewise, the workers on the ground must remain alert to the movement of equipment.

DIPPER HYDRAULIC CYLINDER   BOOM   ROLLOVER PROTECTIVE STRUCTURE (ROPS)   OPERATOR'S CAB

BUCKET/DIPPER   DIPPER ARM (STICK)   STABILIZER/ OUTRIGGER   LOADER BOOM   FRONT LOADER BUCKET

Figure 2 Backhoe loader.

The loader attachment on a backhoe is used primarily to move soil, gravel, stone, and similar materials. It is also used to fill in ditches after pipes or other items have been installed. Like its larger cousin the loader, the loader attachment on a backhoe can lift material from the ground and dump it into the bed of a dump truck (*Figure 3*). The operator faces the loader end when operating the loader or moving the machine.

The backhoe attachment is used to dig and clean ditches, excavate hard to reach areas, and hoist light loads. One common use is digging trenches for pipes and other utilities. Many backhoes, including the one shown in *Figure 4*, have a boom that can slide left and right, which adds versatility to the machine. The backhoe attachment has separate controls and the operator simply turns the seat around to operate that attachment.

Numerous attachments can be fitted onto a backhoe loader, including a ripper, an auger, and a hydraulic breaker (*Figure 5*). Other attachments are available to replace the loader bucket, such as a tool/material carrier. With all of these options, the backhoe loader becomes a very versatile piece of equipment.

Figure 3 Backhoe loader loading a truck.

Figure 4 Backhoe trenching and loading a truck.

Figure 5 Backhoe with hydraulic breaker for demolition work.

There are many sizes and models of backhoe loaders. The backhoe attachment is usually mounted on a rubber-tired, two- or four-wheel-drive tractor. It may also be fitted onto a crawler tractor or a truck. The tractor usually has permanent mountings, such as a three-point hitch, to which the backhoe can be attached with pins. Most models, however, are a combination backhoe and loader.

Other backhoe configurations include the following:

- Crawler-type hydraulic
- Multi-purpose hydraulic
- Trenching machine with backhoe accessory

Tractor-mounted backhoes are smaller and less powerful than most other hydraulic excavating equipment. Nevertheless, their compact size, mobility, and maneuverability make them an attractive option for working in areas that are hard to reach for most excavators. In addition, the rubber tires allow it to be driven a short distance to and from jobs on public roads.

1.1.3 Loaders

Loaders are machines that have a large bucket on the front for scooping up material and moving it to another area. The bucket is mounted in such a way that it can be lifted high enough and tilted to dump material onto piles or into trucks, conveyers, or augers.

Loaders are used for a variety of purposes on highway and bridge construction sites, including the following:

- Loading trucks
- Stockpiling soil, gravel, and other material
- Rough grading
- Digging
- Hauling material and equipment

Like many other pieces of construction equipment, loaders come in crawler and wheel configurations. The wheel version of the loader, shown in *Figure 6* (*A*), is more common and has a wider range of models and sizes than the track version shown in *Figure 6* (*B*).

Skid steer loaders, shown in *Figure 6* (*C*) and *Figure 6* (*D*), are a compact version of the loaders. They are equipped with either wheels or tracks. Tracked skid steer loaders have better traction. Regardless of how a skid steer is equipped (wheels or tracks), it is steered by pushing or pulling on two levers or joysticks in the cab. The ability to turn in small spaces is where the term *skid steer* came from. Like the backhoe loader, the skid steer is available with a range of attachments, such as pallet forks, augers, brooms, and grapple buckets.

1.1.4 Excavators

Excavators are found on many highway and bridge construction sites. An excavator is a motor-driven piece of heavy equipment used primarily for digging. It has a boom and a bucket much like a backhoe, and an operator's cab. Both the boom and the cab are mounted onto a rotating platform that enables the excavator to pivot 360 degrees. The rotating platform sits atop an undercarriage, which has tracks or wheels so that the excavator can move under its own power.

(A) WHEEL LOADER

(B) CRAWLER LOADER

(C) WHEELED SKID STEER WITH BUCKET

(D) TRACKED SKID STEER WITH AUGER

Figure 6 Loaders.

Attachments Used with a Skid Steer

A skid steer can be used to perform the functions of many other machines when equipped with attachments available from the manufacturer. This skid steer is equipped with a dozer blade. Many different attachments are made by the manufacturer for use with this machine, including backhoe, auger, trenching attachments, and pallet forks.

Like most pieces of construction equipment, excavators are available in a vast array of sizes and configurations, and most of these machines can be fitted with a variety of attachments that make the equipment useful for earthmoving, trenching, ditching, material handling, riprap placement, and many other applications.

Generally speaking, the most common types of excavators used in highway and bridge construction can be divided into two basic categories: so-called standard hydraulic excavators and telescoping-boom excavators.

Standard hydraulic excavators, such as the one shown in *Figure 7*, are usually referred to simply as *excavators*. These excavators are available in various sizes and capacities from mini to very large. This type of excavator has a boom, a pivoting dipper stick, and a bucket. Some manufacturers refer to the dipper stick as a *dipper arm*. In addition, the boom with its pivoting dipper stick is sometimes called a knuckle boom or *hoe*.

A less common and more specialized type of hydraulic excavator is a telescoping-boom excavator, sometimes called a shooting-boom excavator. As shown in *Figure 8*, this type of excavator has a straight boom equipped with a pivoting bucket. The boom can be extended, retracted, and rotated. Rotation is used to tilt the bucket from side to side. This type of excavator is generally used for restricted-clearance digging as well as grading and finishing work. They range in sizes and capacities from small to large.

Both basic types of excavators are available as track-mounted, wheel-mounted, and truck-mounted machines. The specific type of excavator that is used on a job will depend on the features required. The excavators shown in *Figure 7* and *Figure 8* are both track-mounted machines. They can move under their own power using the tracks mounted on the undercarriage. This type of excavator is a common choice at construction sites where the machine does not have to travel to, from, or around the site for any great distance.

A wheel-mounted excavator, such as the one shown in *Figure 9*, can perform the same type of operations as the track-mounted excavator. It sits on basically the same type of undercarriage. This undercarriage, however, is attached to axles with rubber tires that make it suitable for driving on the highway and around the job site without damaging pavement.

Truck-mounted or wheel-mounted telescoping-boom excavators can be driven on public highways and do not have to be moved on a trailer. *Figure 10* shows a typical truck-mounted telescoping-boom excavator. The excavator portion of the machine is mounted on the rear part of the truck frame.

Figure 7 Standard hydraulic excavator.

Figure 8 Telescoping-boom excavator.

There are separate engines and operator cabs for the operation of the undercarriage truck and the excavator.

1.1.5 Trenchers

Trenchers are machines used to dig trenches. They use either a gasoline or diesel engine and have a digging chain and boom that are very similar to the blade and chain of a chainsaw. When activated, the chain moves around the boom. When the boom is lowered onto the ground, the chain begins to dig into the ground.

While there are different types of trenchers available, depending on the scale of the work to be performed (*Figure 11*), most fall within three basic styles. The lighter-weight one is called a *pedestrian trencher* because the operator walks behind it. A heavier, but still small one is called a *compact trencher* and the operator rides along on the back end of it. A pedestrian trencher is shown in *Figure 11 (A)*.

Figure 9 Wheel-mounted excavator.

BUCKET

TELESCOPING BOOM

Figure 10 Truck-mounted telescoping-boom excavator.

The third style is more of a heavy construction model used to dig trenches for pipelines on large construction jobs. With the large trenchers, the operator sits in an operator cab position on top of the machine. These large construction trenchers ride on tracks, similar to the tracks of a dozer, so that they have much more traction. Some companies make trencher attachments that mount on the front of a skid steer machine, like the one shown in *Figure 11* (*B*). Such skid steer attachments operate like the compact trenchers in that the operator sits inside the skid steer and operates the trenching attachment from additional controls installed in the skid steer cab.

Hydraulic Excavator Choices

As with most pieces of heavy equipment, there are pros and cons that apply to different types of hydraulic excavators. The specific types of excavators found at a construction site are based on factors such as size and capacity, traction capabilities, the means of locomotion, maneuverability, stability, and work site conditions. The following characteristics often apply to the particular type of hydraulic excavator used on site.

Track-Mounted Excavator	Wheel-Mounted Excavator
Flotation	Mobility and speed
Traction	Less pavement damage
Maneuverability	Better stability with outriggers
Severe soil conditions	Leveling machine with outriggers
Faster machine	Usable for backfilling repositioning trenches

(A) PEDESTRIAN TRENCHER

(B) SKID STEER WITH TRENCHER ATTACHMENT

Figure 11 Examples of trenchers.

1.1.6 Scrapers

A scraper (*Figure 12*) is used to remove dirt and other material from a site by scraping it from the surface into a bowl, as it moves along. Scrapers can be used to remove soil in rough terrain, and to finish large horizontal areas where tolerances are not as close as those requiring the use of a motor grader. With its power, weight, and a blade, it can neatly trim horizontal surfaces and pick up excess material without disturbing the compacted surface underneath. Scrapers are also used to spread dirt and other material. There are four basic types of scrapers:

- Standard or self-propelled scrapers
- Elevating scrapers

(A) ELEVATING SCRAPER

(B) PULL-TYPE SCRAPER TOWED BY A TRACTOR

Figure 12 Scrapers.

- Tandem-powered scrapers
- Pull-type scrapers

A standard scraper is a self-propelled unit consisting of a bowl pulled by a single tractor. The cutting edge at the bottom of the bowl can be levered to dig into the earth by two feet (0.6 meters) or more. The bowls of some units can hold more than 40 cubic yards (31 cubic meters) of earth. The width of the cut can range up to 13 feet (4

Utility Concerns when Trenching and Excavating

Damage to underground utility lines is a major concern during trenching operations. Always call the One-Call System (dial 811) or other designated authority to arrange in advance for buried utilities to be located and marked. Trenching operators are required to walk the area to be trenched and verify that all underground utility lines are well marked and easily seen from the operator's position. Nevertheless, any construction workers in the area of a trenching operation should remain aware of the hazards associated with underground utility lines.

meters), also depending on the model in use. The bowl contains a mechanism for ejecting the material. The forward wall of the bowl is a movable section called the apron. Standard scrapers are often pushed by a dozer when additional power is needed to load the bowl.

The bowl of the elevating scraper, also called a *paddle wheel scraper*, contains a conveyor system that moves the material toward the rear of the bowl as it is picked up from the ground. The capacity of these units is not as great as that of the standard scraper because of the space taken by the elevator system. This type of scraper is used for fine soils and sand. Elevating scrapers are used primarily for fine grading work.

Tandem-powered (push-pull) scrapers have two engines, one in the tractor and a second at the rear of the scraper unit. The advantage offered by these units is that they can work on steeper grades and rougher terrain. Tandem-powered scrapers are also available in a push-pull model that is equipped with a cushioned plate on the front and a hook at the rear of the bowl. When operating in a cut, the second unit pushes the first unit. When the first unit is full, it then uses its engines to pull the second scraper through the cut.

The pulled, or towed, scraper is pulled along by a heavy-duty tractor. These types of scrapers have large capacities and can be towed in tandem to pick up twice the material in a single run.

1.1.7 Dozers

A dozer is a heavy-duty tractor with a pusher blade mounted on the front. Depending on the manufacturer, these machines may be called *tractors, track-type tractors, crawler tractors, crawlers, dozers, crawler dozers,* or *bulldozers*. While most dozers are not very fast, their traction, stability, and low center of gravity make them especially suited to work on irregular terrain and steep slopes. Because dozers are available in many different models and can accept numerous attachments, they can be used for a wide variety of excavation and grading jobs.

Dozers are used for a variety of purposes including the following:

- Clearing land, including removing tree stumps
- Moving soil, rock, and other material
- Spreading material dumped on the site
- Rough and finish grading
- Maintaining haul roads and temporary roads through rocky or rough terrain

Dozers can either be wheeled or tracked. Most are tracked. Two basic types of tracked (crawler) dozers are low-track dozers (*Figure 13*) and high-track dozers. The low-track machine is used primarily for grading. The high-track type is used primarily for pushing. Some manufacturers also make heavy-duty four-wheel drive dozers. Wheel dozers have excellent traction, and a steering wheel makes it easy to operate and maneuver. High-speed dozers are capable of operating at up to 16 miles per hour (26 kilometers per hour). The high-speed dozer in *Figure 13* (*B*) has an articulated frame to enhance maneuverability and rubber tracks that enable it to move across concrete or paved material without damaging the surface.

Many dozers are equipped with a powered winch or a ripper for breaking up hard soil, rock, and other materials such as concrete and asphalt (*Figure 14*).

(A) TRACK DOZER

(B) HIGH-SPEED DOZER

Figure 13 Dozers.

Figure 14 Dozer with a ripper attachment.

1.1.8 Forklifts

Forklifts are machines that have two forks that travel up and down on a mast. They are used to lift and move equipment and materials. They are also used extensively to unload materials from trucks and trailers. The forklifts used in construction work are generally diesel-powered, rough-terrain forklifts. These forklifts have a higher ground clearance and larger tires than warehouse forklifts.

Three basic types of forklifts are fixed-mast, telescoping boom, and articulating boom. On a fixed-mast forklift (*Figure 15*), the forks can be raised as high as the limits of the mast will allow. The mast may be able to be tilted forward 15 to 20 degrees and backward by a few degrees. These forklifts must be able to get very close to the pickup or landing point because of the limited movement range of the mast.

A telescoping-boom forklift (*Figure 16*) is able to lift material much higher than fixed-mast forklifts and has much more flexibility in positioning a load. Some telescoping-boom forklifts have a level-reach fork carriage, which is often called a squirt boom. A squirt boom allows the fork carriage to be moved in the horizontal plane while the boom remains stationary.

Articulating forklifts (*Figure 17*) are built with a fork carriage that is hinged so that it will pivot left or right of center line. The operator uses hydraulic controls and cylinders to move the fork carriage. Most articulating forklifts are designed for operations within narrow, confined spaces such as inside truck trailers.

Figure 15 Fixed-mast forklifts.

Laser Guidance and GPS

Modern graders and dozers, as well as other excavating and grading equipment, use laser guidance systems or a global positioning system (GPS) to precisely control the position of the dozer blade. In a GPS system, antennas (one or two) are mounted near the outer tips of the blade. A GPS base station (surveying system) is set up near where the dozer is working. The base station works with the GPS devices on the dozer to control the positioning of the blade.

Figure 16 Telescoping-boom forklift.

1.1.9 Motor Graders

A motor grader is a rubber-tired, hydraulically operated, single-engine machine used to shape and finish materials. It is one of the most used pieces of equipment on a construction site. Its main purpose is to mix, place, and smooth material on the ground or on other surfaces such as embankments or ditches.

The basic design of a grader has not changed much from the earliest days when they were pulled by horses or steam-powered tractors. The adjustable blade, or moldboard, is carried on a

Figure 17 Articulating forklift.

long, narrow machine. Today, power is supplied by a large diesel engine and the controls are either hydraulic or electronic.

Historically, there have been two basic types of motor graders: rigid-frame motor graders and articulated-frame motor graders. A rigid-frame motor grader has a single metal frame that extends from the front to the back of the motor grader. Rigid-frame motor graders are typically smaller and less common than articulated-frame graders.

Most modern graders are articulated-frame graders. An articulated-frame motor grader (*Figure 18*) has a frame that is hinged at the center. The hinge point is either in front of the cab or behind the cab, depending on the manufacturer's design. This allows the front of the motor grader to work at an offset angle to the back tandem wheels for such work as cleaning out ditches. It also allows for tighter turning and greater maneuverability.

Figure 18 Articulated-frame motor grader.

There are many types, sizes, and manufacturers of motor graders, but their basic features and operations are essentially the same. Regardless of the type of motor grader being used, there are seven primary operations that a grader performs using its blade, including the following:

- Rough grading
- Mixing
- Spreading new material
- Finish grading
- Ditch cutting and clearing
- Snow plowing

Other operations such as ripping and scarifying can also be performed with a motor grader by using attachments. *Figure 19* shows motor graders with scarifier and ripper attachments. A scarifier is used to break up and mix soft soil, while a ripper is needed to break up hard soil.

1.2.0 Trucks

Highway and bridge construction jobs often require large amounts of material to be moved from one area to another. One of the most common pieces of equipment used for this purpose is a dump truck. Dump trucks are built on a heavy-duty chassis and have a large bed that is used to move loose material such as sand, soil, gravel, and asphalt. The beds can be raised and tilted back with a hoist that is mounted on the truck, so the driver can quickly unload the cargo without help. There are two basic types of dump trucks: on-road dump trucks and off-road dump trucks.

(A) SCARIFIER ATTACHMENT

(B) RIPPER ATTACHMENT

Figure 19 Motor grader attachments.

1.2.1 On-Road Dump Trucks

On-road dump trucks (*Figure 20*) are designed for use on highways, and must be registered for that purpose. Anyone operating these trucks on the highway must have a commercial driver's license (CDL).

The key difference in dump trucks of different sizes is the number of axles (*Figure 21*). Increasing the number of axles increases the amount of weight that a truck can safely carry. However, the truck is not permitted to exceed the maximum load allowed on a road or bridge, regardless of the number of axles.

Standard dump trucks include the following rear axle configurations:

- *Single-axle* – These trucks have one rear axle and are used to haul light loads. They normally have six wheels: two on the front axle and four on the rear axle.
- *Dual-axle* – These trucks with two rear axles and are used to haul heavy loads. They normally have ten wheels: two on the front axle and eight on the rear axles. (Some states refer to any truck with dual-axle or above

(A)   **(B)**

Figure 20 On-road dump trucks.

as tandem-axle trucks, while some refer to a tandem-axle as a dual-axle only.)

- *Tri-axle* – These trucks have three rear axles and are used to haul heavier and larger loads than dual-axle vehicles. They normally have 12 wheels: two on the front axle and ten on the rear axles.
- *Quad-axle* – These are trucks with four rear axles that are used to haul the heaviest and largest loads.
- *Auxiliary-axle* – These axles, which are placed in front of or behind the drive wheels, are used only when the truck is loaded. They are lowered into place to provide additional load capacity. Auxiliary axles are located in front of the drive wheels. If the auxiliary axle is located behind the drive wheels, it is called a tag axle. The truck shown in *Figure 21* has an auxiliary axle.

Figure 21 Dump truck axles.

In addition to standard dump trucks, there are dump trucks and trailers made for special hauling situations. One example is a transfer dump truck (*Figure 22*). Transfer dump trucks are standard dump trucks that tow a second dump bed on a trailer. When the bed of the standard dump truck is empty, the dump bed on the trailer is slid into the bed of the truck and then emptied using the truck's hoist mechanism.

> **WARNING!**
>
> Never operate an overweight dump truck. Not only can it damage the vehicle, but it is also a safety hazard. Carrying a load that is too heavy decreases the maneuverability of the truck and stresses its brakes, suspension, axles, and tires. If the weight limits for a vehicle are not known, check the user manual.

Size Matters

In 1982, the federal government increased the maximum width of vehicles traveling on public roads from 96" (2.4 meters) to 102" (2.6 meters) or (8' to 8'-6"). Some states still limit the legal width of on-road vehicles to 96" (2.4 meters). The legal length of on-road vehicles varies greatly from state to state.

Figure 22 Dump truck with transfer trailer.

Pup trailers are similar to the trailers used in transfer dump trucks (*Figure 23*). The chief difference is that pup trailers have their own hoist mechanisms, so they can be emptied without using a standard dump truck.

Bottom-dump trucks (*Figure 24*), also called *belly dumpers*, are trailers that are towed behind a truck tractor. As the name implies, the load is dumped through doors in the bottom of the trailer. These trailers are often used to dump materials like asphalt, gravel, and dirt in a row on a roadbed.

Side-dump trucks are trailers that are towed behind a truck tractor. This type of trailer dumps its load by tilting the trailer body on its side. A side-dumper can only be used when the dumpsite is long enough to permit the trailer access to it.

1.2.2 Off-Road Dump Trucks

Off-road, or off-highway, trucks barely resemble on-road dump trucks. In fact, they are seldom referred to as dump trucks at all, even though they

Figure 23 Pup trailer.

have bodies that carry large amounts of material and are lifted for dumping. These are rugged vehicles built for the demanding conditions that exist at many large road-building sites. They are usually much larger than on-road vehicles and are

Weigh Stations

Anyone who travels on the Interstate Highway System no doubt has seen the weigh stations at intervals along the highway. Commercial vehicles using the highways are required to pass through these stations. More modern weigh stations can record the weight while the vehicle is moving through the station. Others require the vehicle to stop on a scale. The primary purpose of these stations is to perform safety inspections. Another purpose is to determine if any vehicle exceeds the maximum weight specified by the state for the type of vehicle. In addition to these permanent weigh stations, police and Department of Transportation (DOT) agents also have portable scales that can be set up at any location desired.

Figure 24 Bottom-dump truck.

prohibited by law from routine operation on public roads because of their great size and weight.

Off-road dump trucks serve the same basic purpose as on-road dump trucks: to safely and efficiently move large volumes of material from one area of a work site to another. They are designed to be loaded on site with loaders, excavators, or other equipment (*Figure 25*). The loaded truck is then driven to a dump site, where hoist cylinders raise the body and dump the load. The efficiency of the loading, driving, and dumping process depends in large part on an operator's ability to safely maneuver the truck.

Figure 25 Off-road dump truck being loaded by a shovel excavator.

Roll-Off Trucks

Some trucks are designed to leave their body at the job site and return to pick it up when it has been filled. Once reattached to the truck, the bed can be raised and dumped like that of any other dump truck. Trucks like this are convenient for demolition sites.

While off-road trucks vary a great deal in size and capacity, they can be divided into two basic categories: rigid-frame dump trucks and articulated-frame dump trucks.

A rigid-frame dump truck (also known as a haul truck) has a rigid, or non-pivoting, frame that is similar to what is used on a standard on-road dump truck. The cab and the body are mounted on a common frame and operate together. The rigid frame is strong enough to handle the heavy loads it is designed to carry. *Figure 26* shows a typical rigid-frame dump truck.

Rigid-frame dump trucks usually have two axles. The front axle has some type of conventional steering mechanism for steering the vehicle, while the rear axle has the necessary drivetrain gears to drive the vehicle. Most rigid-frame dump trucks have dual wheels on each side of the rear axle for strength, traction, and stability.

All rigid-frame dump trucks have a large diesel engine, but many of them also use electric motors in combination with the diesel engine. In this diesel/electric powertrain configuration (*Figure 27*), the diesel engine drives an AC alternator or DC generator, which provides power to electric motors located inside the axle at the rear wheels. This arrangement increases efficiency by providing greater power to each drive wheel and better braking, but with less weight.

Rigid-frame dump trucks have huge capacities. Even smaller trucks can haul 30 tons (27 tonnes) of material. Larger trucks can haul up to 400 tons (363 tonnes). Any rigid-frame dump truck that is

Figure 26 Rigid-frame dump truck.

capable of hauling 272 tons (247 tonnes) or more is categorized as an ultra-class truck.

Rigid-frame dump trucks can be used in most heavy construction applications, as long as the work-site terrain is amenable. Because of their massive size and weight, these trucks are best suited for heavy construction applications where the terrain is relatively stable and/or haul roads have been established (*Figure 28*). In addition, these trucks require a lot of space to operate. It is difficult for operators to see personnel, obstacles, or other vehicles unless they are a good distance from the truck itself.

An articulated-frame dump truck (*Figure 29*) has a pivoting point in the frame between the cab and the dump body. This permanent hinge

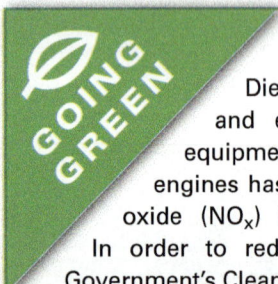

Diesel Engine Emission Controls

GOING GREEN

Diesel engines provide the power and efficiency needed to drive heavy equipment. A major drawback of past diesel engines has been the level of soot and nitrous oxide (NO_x) they emit through their exhaust. In order to reduce these emissions, the Federal Government's Clean Air Act of 1990 included regulations for reducing diesel engine emissions. The regulations were phased-in over time.

One early element of these regulations was the requirement for diesel engines to use ultra-low sulfur diesel fuel (USDF). On-road vehicles manufactured since 2010 are required to meet Tier IV exhaust emission standards by using and by implementing a system for treating the engine exhaust.

Most engine manufacturers adopted the selective catalytic reduction (SCR) exhaust treatment system. This

DEF TANK

system injects a solution known as diesel emission fluid (DEF) into the engine exhaust system. DEF is an ammonia-based solution known as *urea*. The urea is mixed with air in a mixing valve before being injected and is stored in a tank with a capacity of 10 to 30 gallons. The photo below shows a DEF tank on a Tier IV-compliant dump truck.

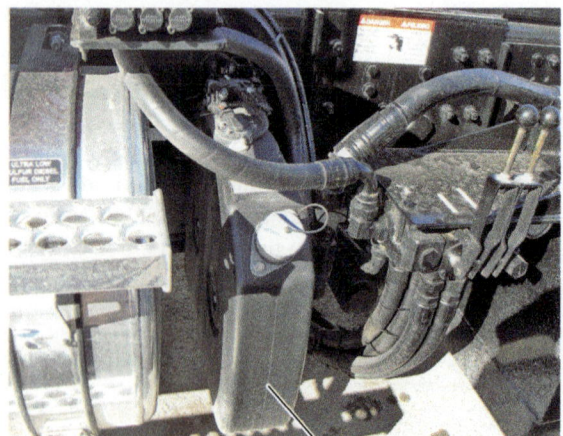

ELECTRIC GENERATING EQUIPMENT
FOR AC DRIVETRAIN MOTORS

Figure 27 Rigid-frame dump truck with diesel/electric drive system.

enables the front of the truck to turn in a way that all of the wheels follow the same path. However, an articulated dump truck should not be confused with a tractor trailer arrangement in which a separate tractor pulls a trailer that is connected to it. An articulated dump truck is permanently connected in much the same way as an articulated loader or an articulated motor grader.

Articulated-frame dump trucks, which are sometimes called *articulated haulers*, are typically all-wheel drive vehicles. Most models have one front axle and two rear axles. Hydraulic cylinders are used to provide the steering by turning the entire front part of the vehicle. As *Figure 29* shows, articulated trucks are well adapted to rough terrain.

Articulated-frame trucks are capable of handling loads from about 25 tons (23 tonnes) up to about 46 tons (42 tonnes). They can be used in basically any heavy hauling application. However, because of their all-wheel drive and maneuverability, they are most likely to be used in rugged applications that may require lots of turning. Generally, they do not haul as much material as rigid-frame dump trucks, but they do a better job of moving material across muddy terrain and sites that may not have well-maintained haul roads.

On many highway construction sites, there are numerous side roads and sloping areas where dump trucks need to travel. An articulated truck is able to maneuver around these areas with little

Figure 28 Rigid-frame dump trucks traveling over haul road.

World's Largest Haul Truck

The title for the world's largest haul truck changes periodically. Some of the largest trucks stand over 25 feet (8 meters) tall, weigh over 250 tons (227 tonnes) empty, can haul 400 tons (363 tonnes), and have V-24 diesel engines capable of producing over 3,700 horsepower.

Figure 29 Articulated-frame dump truck.

Figure 30 Crawler dump truck.

problem. In recent years, crawler dump trucks on rubber tracks (*Figure 30*) have become popular for use on wet construction sites.

1.3.0 Paving Equipment

Paving equipment is used during the final stages of road building, after all the excavating and grading has been completed. Pavers are typically used during this stage, but road reclaimers and chip spreaders might also be used during repaving operations.

1.3.1 Pavers

Paving machines are used to lay down asphalt, concrete, or base material on prepared roadways. As with other highway construction equipment, pavers vary in size and operation. As a general rule, pavers are either track-mounted machines or wheel-mounted machines (*Figure 31*).

> **WARNING!**
>
> Paving machine operators must be especially careful when working near power lines. When the paver's bed is raised, it can be an electrical hazard.

Asphalt pavers have a powerful diesel engine, an operator area, a hopper that is usually fed from a dump truck, and conveyors and augers that move the asphalt from the hopper to a screed at the front of the machine. The screed spreads the asphalt evenly over the entire width of the paving area, which can be 50 feet (15 meters) or more. The paver also provides some compaction during operation, but not enough to negate the need for further compaction by a compactor.

(A) TRACK-MOUNTED PAVER

(B) WHEEL-MOUNTED PAVER

Figure 31 Asphalt pavers.

Figure 32 Slipform paver spreading concrete.

Some asphalt pavers can also be used to lay down concrete in much the same way as they lay down asphalt. Dump trucks feed mixed concrete into the paver's hopper, and the paver spreads the concrete evenly over the width of the paving area. Other pavers are designed to spread only concrete or base material.

One type of concrete paver is called a slipform paver. With this type of paver, bulk concrete is dumped in front of the paver. As the paver moves forward the screed spreads the concrete evenly over the width of the paving area. *Figure 32* shows a slipform paver being used to create a concrete roadway.

1.3.2 Road Reclaimers and Mills

Road reclaiming equipment goes by many names, including road *reclaimers,* road recyclers, *asphalt mills, asphalt pavement grinders,* and *cold planers.* These machines grind down the upper surface of a road and direct the material into a dump truck. The ground material is then hauled away to be recycled into new asphalt. In some cases, the ground asphalt is mixed with the material underneath the asphalt and deposited immediately back onto the roadway as a foundation layer prior to paving.

(A) TRACK-MOUNTED RECLAIMER

(B) WHEEL-MOUNTED RECLAIMER

Figure 33 Milling equipment (reclaimers).

Large pieces of milling equipment are commonly mounted onto tracks for better traction and stability. Some smaller pieces of milling equipment are mounted on rubber tires. *Figure 33* shows a track-mounted reclaimer and a wheel-mounted reclaimer.

1.3.3 Chip Spreaders

Chip spreaders are machines used to pave roads and apply surface treatments to extend the life of paved roads. Most large chip spreaders are self-propelled machines with a hopper for storing aggregate, and rubber tires. There are also chip spreader attachments that can be mounted onto trucks or other types of equipment. Either way, these machines are commonly used in rural areas to resurface existing roads.

The process performed by a chip spreader is called chip sealing, which is sometimes referred to as *tar and chip paving*. Once a road has been properly prepared, hot liquid asphalt (sometimes incorrectly referred to as *tar*) is applied to the road surface, often by a spray truck. Next, the chip spreader applies an even layer of aggregate, typically crushed gravel. A compactor is then used to roll the surface to ensure that the aggregate is properly embedded in the liquid asphalt. *Figure 34* shows a chip-sealing machine being used.

Figure 34 Chip sealing in process.

Additional Resources

The Earthmover Encyclopedia, Keith Haddock. 2007. St. Paul, MN: MBI Publishing.

Construction Equipment Guide, David A. Day and Neal B.H. Benjamin. New York, NY: John Wiley & Sons.

Heavy Equipment Operations Level One, NCCER. Third Edition. 2012. New York, NY: Pearson Education, Inc.

Heavy Equipment Operations Level Two, NCCER. Third Edition. 2013. New York, NY: Pearson Education, Inc.

Heavy Equipment Operations Level Three, NCCER. Third Edition. 2014. New York, NY: Pearson Education, Inc.

Machinery's Handbook, Erik Oberg, Franklin D. Jones, Holbrook L. Horton, and Henry H. Ryffel. Latest Edition. New York, NY: Industrial Press Inc.

1.0.0 Section Review

1. A piece of heavy equipment that has a long, narrow articulating frame and an adjustable blade or moldboard is a(n) _____.

   a. loader
   b. grader
   c. dozer
   d. excavator

2. Which of the following types of trucks have the largest hauling capacity?

   a. On-road dump trucks
   b. Articulated-frame dump trucks
   c. Tandem-axle dump trucks
   d. Rigid-frame dump trucks

3. The part of a slipform paver that spreads concrete evenly over the entire width of the paving area is called the _____.

   a. screed
   b. conveyor
   c. mill
   d. auger

2.0.0 UTILITY EQUIPMENT

Objective

Identify utility equipment used on the job site and describe their uses.

   a. Identify and describe compaction equipment.
   b. Identify and describe electrical generators.
   c. Identify and describe air compressors.

Trade Terms

Pneumatic: Inflated with compressed air.

Tamping roller: A type of compaction roller that has projecting pads on a steel drum. Also called a *sheepsfoot roller* or a *segmented-pad roller*.

(A) UPRIGHT TAMPER   (B) FLAT PLATE

Figure 35 Upright tamper and flat plate compactor.

In addition to the heavy equipment used for grading, excavating, and paving on highway and bridge construction sites, there are numerous other pieces of equipment that construction workers are likely to encounter. Some of this equipment is used to compact soil before any paving takes place, or to compact pavement that has just been laid. Other equipment is used to provide electrical power or compressed air at the site.

2.1.0 Compaction Equipment

The surfaces on which structures and roads are built must be able to carry the weight of the structure and its contents or the road and the traffic that will travel on it. To do this, each structure and road is designed with certain foundation specifications. Specialized compaction equipment is used to make the underlying soil, asphalt, or gravel as smooth, level, and solid as possible. These machines roll, tamp, vibrate, and/or press the soil to make it denser. The denser the soil, the more likely it is to support the weight of the roadway. If the soil gives way under the weight, the road is likely to crack.

Compactors are available in a variety of styles and sizes, including upright tampers, plate tampers, walk-behind rollers, and ride-on rollers. Each type of compactor is recommended for different soil conditions and activities. Compactors can also be installed on other types of equipment as attachments. *Figure 35* shows two types of compactors in which the operator stands and

holds the compactor during operation. These types of compactors are used for small compacting applications.

Ride-on compacting equipment is used for larger applications. Several companies manufacture ride-on rolling equipment in different configurations and sizes. Although there are many different models and makes of equipment, currently there are four basic designs. The four basic types of rollers include the following:

- **Pneumatic** tire
- Steel-wheel
- Vibratory
- Sheepsfoot

2.1.1 Pneumatic Tire Roller

Pneumatic tire rollers (*Figure 36*) achieve excellent results in almost every type of compaction. However, some types of wet soils may stick to the tires, limiting their effectiveness. Scrapers are often attached to scrape mud from the tires. The primary design is a weight-filled box mounted on two rows of tires. Some models are self-propelled, while others are towed.

The tires are the compaction devices on a pneumatic tire roller. Typically, there are two axles with a set of tires on each axle. The tires are smooth with no treads. There are four to five tires on one axle and five or six on the other. Compaction is varied by changing the air pressure in the tires. The normal range is between 60 and 120 psi (414 and 827 kPa).

Rubber-tired rollers are used to compact all types of surfaces. In most cases, the rubber-tired roller is followed by finish steel-wheel rolling. When used as an intermediate roller, the rubber-tired roller contributes to compaction and particle alignment.

Figure 36 Pneumatic tire roller.

> **NOTE**
>
> Pneumatic tire rollers are used in asphalt road construction. The tendency of the asphalt binder to stick to the rubber tires increases as the temperature of the tires decreases. Heavy mats made of rubber or plywood can be mounted around the tire area to prevent heat loss and keep the tires at or near mat temperature. This will reduce the tendency of the binders to stick to the tires and increase rolling efficiency.

2.1.2 Steel-Wheel Roller

Steel-wheel rollers (*Figure 37*) are rollers as well as compactors. Their rolls are machined to provide a smooth, concentric surface. They are designed primarily for pavement course rolling, but are also used for rolling gravel roads, road bases, and some subgrades.

There are three basic types of steel-wheel rollers: three-wheel, two-wheel, and single-wheel. The three-wheel roller has one wide guide roll and two narrow drive rolls. The two-wheel (tandem) roller has two rolls of equal width. The single-wheel roller has one steel rolling wheel and rubber drive tires. The drive tires are typically mounted on the rear of the machine. There is also a unique version of the tandem roller, called the *three-axle tandem*, which is similar to the tandem in appearance, but has two guide rolls on a walking beam. The walking beam allows the forces exerted by the rolls to be varied. The three-axle tandem is generally the heaviest of all steel-wheel rollers, but there are different size categories in each style.

2.1.3 Vibratory Compactor

Vibratory compactors combine the static weight of the machine with a generated, cyclic force. The cyclic, or dynamic, force is created by rotating a series of off-balance weights within a steel compacting drum. This rotation develops a centrifugal force that is sufficient to lift and drop the steel drum as it turns. The lifting and dropping action (the vibration) moves soil particles while the static pressure of the machine forces them into a compact structure. *Figure 38* shows a vibratory steel-wheel roller.

> **CAUTION**
>
> The vibratory compactor must be in motion when the vibrator is on.

2.1.4 Sheepsfoot Roller

The sheepsfoot roller, or tamping roller (*Figure 39*), offers fast compaction and high production. Self-propelled units have good maneuverability and pulling power. In addition, a blade can be mounted on the front of a self-propelled compactor to provide grading capabilities.

Figure 37 Steel-wheel roller.

Figure 38 Vibratory steel-wheel roller.

Figure 39 Sheepsfoot roller.

Figure 40 Portable generator.

On a sheepsfoot roller, the steel drum has projecting pads or feet. These compactors are sometimes called *segmented-pad rollers*. The feet are normally 7 to 9 inches (18 to 23 centimeters) long. Sheepsfoot rollers compact a little at the surface, but provide greater compaction under the feet. Cleaner bars can be added to the back of the machine to remove dirt caught between the feet.

2.2.0 Electrical Generators

Generators are used to provide electrical power for power tools and/or lighting at the job site. The power is also used for electric engine heaters in cold climates. Generators are available in many different sizes and configurations. They range from small portable generators (*Figure 40*) to large installed backup power systems for emergency power generation.

Construction site requirements for power vary depending on the scale of the work being performed. One of the most common types of generators used at highway and bridge construction sites are tow-behind generators. *Figure 41* shows a typical tow-behind generator and a tow-behind generator with a light plant. Another type of generator commonly used in highway construction is the light plant. Lights are part of the light

plant, which provides lighting, as well as electrical power for other purposes.

> **WARNING!**
>
> High voltages are present when a generator is operating. Exercise caution and follow the established safety practices and manufacturer's instructions to avoid electric shock. The generator may need to be grounded per the manufacturer's instructions. Make sure circuits are equipped with ground fault circuit interrupters (GFCIs).

2.3.0 Air Compressors

Compressors are used to provide compressed air for pneumatic tools at the job site. Compressors are available in many different sizes and configurations. They range from portable units, such as the one shown in *Figure 42*, to large installed units for industrial applications. Construction site requirements for compressed air vary depending on the scale of the work being performed.

One type of air compressor commonly found on highway and bridge construction projects is a tow-behind compressor (*Figure 43*).

Combining Compaction Equipment

Some types of compaction equipment are designed to be used in conjunction with other equipment. For instance, a tow-type sheepsfoot roller can be towed behind a dozer.

(A) TOW-BEHIND GENERATOR

(B) TOW-BEHIND GENERATOR WITH LIGHT PLANT

Figure 41 Electrical generators.

Figure 42 Small, portable air compressor.

Figure 43 Tow-behind air compressor.

Additional Resources

The Earthmover Encyclopedia, Keith Haddock. 2007. St. Paul, MN: MBI Publishing.

Construction Equipment Guide, David A. Day and Neal B.H. Benjamin. New York, NY: John Wiley & Sons.

Heavy Equipment Operations Level One, NCCER. Third Edition. 2012. New York, NY: Pearson Education, Inc.

Heavy Equipment Operations Level Two, NCCER. Third Edition. 2013. New York, NY: Pearson Education, Inc.

Heavy Equipment Operations Level Three, NCCER. Third Edition. 2014. New York, NY: Pearson Education, Inc.

Machinery's Handbook, Erik Oberg, Franklin D. Jones, Holbrook L. Horton, and Henry H. Ryffel. Latest Edition. New York, NY: Industrial Press Inc.

2.0.0 Section Review

1. On a pneumatic tire roller, the compaction is accomplished by the _____.

   a. sheepsfoot pads
   b. steel rollers
   c. vibrating drum
   d. rubber tires

2. A common piece of equipment used to provide electrical power and lighting at highway and bridge construction sites is a _____.

   a. pup trailer
   b. portable compressor
   c. tow-behind generator
   d. transfer trailer

3. Construction site requirements for compressed air vary depending on _____.

   a. the scale of the work being performed
   b. ambient temperature at the tools
   c. altitude of the work site
   d. the types of compressors available

SECTION THREE

3.0.0 MOBILE CRANES

Objective

Identify the mobile cranes used in highway and bridge construction and describe their uses.
 a. Identify and describe crawler cranes.
 b. Identify and describe truck-mounted cranes.

Trade Terms

Lattice boom: A type of boom used on cranes and excavators that has a crisscross pattern of braces that enable the machine to lift heavy loads.

Cranes are machines that are used for lifting and moving heavy loads. Two common categories of cranes are stationary cranes and mobile cranes. Stationary cranes are used in applications where it is not necessary to move the crane on a regular basis. Mobile cranes, on the other hand, are suited for highway and bridge construction because they often need to be moved from one area to another. Mobile cranes consist primarily of crawler cranes and truck-mounted cranes.

3.1.0 Crawler Cranes

Crawler cranes (*Figure 44*) get their name from the fact that they have tracks much like those used on a dozer or excavator. The tracks enable the crane to move across rough terrain where a rubber-tired machine cannot. They also allow the crane to move while carrying a load.

Common uses of crawler cranes in highway and bridge construction include loading and unloading trucks and lifting and moving objects such as culvert and drainage piping, steel and concrete bridge components, and utility equipment needed on site. Cranes are also used for pile driving. They can drive sheet piles into place to prevent erosion or falling rocks, and they can drive foundation parts into place for bridge abutments.

Figure 44 Crawler crane.

3.2.0 Truck-Mounted Cranes

Truck-mounted cranes have rubber tires, so they can travel on highways and other fairly level, compact surfaces. These types of cranes can have a lattice boom or a telescoping hydraulic boom. *Figure 45* shows a typical truck-mounted crane with a telescoping boom.

Some truck-mounted cranes are made for rough terrain. The crane shown in *Figure 46* has larger tires and enhanced steering to enable it to operate on rough terrain. While it can be driven on the worksite, its speed is limited compared to that of a truck-mounted crane. This type of crane is typically transported by truck to the work site.

No matter what type of terrain a truck-mounted crane is designed for, it must be level when it is being operated. For that reason, truck-mounted cranes have outriggers that can be deployed to stabilize the crane during operation.

Figure 46 Rough-terrain crane.

Figure 45 Truck-mounted crane.

Additional Resources

The Earthmover Encyclopedia, Keith Haddock. 2007. St. Paul, MN: MBI Publishing.

Construction Equipment Guide, David A. Day and Neal B.H. Benjamin. New York, NY: John Wiley & Sons.

Heavy Equipment Operations Level One, NCCER. Third Edition. 2012. New York, NY: Pearson Education, Inc.

Heavy Equipment Operations Level Two, NCCER. Third Edition. 2013. New York, NY: Pearson Education, Inc.

Heavy Equipment Operations Level Three, NCCER. Third Edition. 2014. New York, NY: Pearson Education, Inc.

Machinery's Handbook, Erik Oberg, Franklin D. Jones, Holbrook L. Horton, and Henry H. Ryffel. Latest Edition. New York, NY: Industrial Press Inc.

3.0.0 Section Review

1. The tracks on a crawler crane enable the crane to move across rough terrain and to move while _____.

   a. towing a compactor
   b. carrying a load
   c. on the highway
   d. on a steep incline

2. In order to level and stabilize the machine during operation, a truck-mounted crane is equipped with _____.

   a. augers
   b. a power takeoff
   c. outriggers
   d. hydraulic breakers

SUMMARY

Highway and bridge construction require the use of many different types of heavy equipment. From excavation and grading to moving material and equipment to paving and finishing, heavy equipment is a vital part of the entire construction process. A typical construction site has numerous pieces of heavy equipment operating at the same time. In order to work safely and effectively in these conditions, and to acquire the knowledge needed to advance in the construction business, workers must be able to recognize the equipment and machines on site and thoroughly understand their uses.

1. Which of the following tasks can a backhoe loader be used for?

    a. compacting soil
    b. finish grading
    c. loading a truck
    d. paving

Figure RQ01

2. The machine shown in *Figure RQ01* is a(n) _____.

    a. skid-steer loader
    b. scraper loader
    c. motor grader
    d. mini excavator

Figure RQ02

3. The piece of equipment shown in *Figure RQ02* is a(n) _____

    a. compactor
    b. dozer
    c. motor grader
    d. scraper

4. The amount of weight that an on-road dump truck can legally carry varies with the number of _____.

    a. outriggers
    b. bowls
    c. axles
    d. tracks

5. A slipform paver is typically used to _____.
   a. recycle asphalt from the upper surface of a road
   b. spread concrete that has been dumped in front of it
   c. apply surface treatments to new asphalt pavement
   d. form concrete shoulders and sidewalks as it moves

6. What feature of a vibratory roller creates the force used to help compact newly laid asphalt?
   a. The front and rear sheepsfoot rollers
   b. The pneumatic force inside the wheels
   c. The heat that is applied to the scraper
   d. The off-balance weights inside the drum

7. The main purpose of a generator on a job site is to _____.
   a. create sine waves for soil compaction
   b. provide electrical power
   c. power pneumatic vibratory rollers
   d. add energy to compact hoisting equipment

8. Pneumatic tools used on a construction site are typically powered by a(n) _____.
   a. electrical generator
   b. steam turbine
   c. solar array
   d. air compressor

Figure RQ03

9. The piece of heavy equipment shown in *Figure RQ03* is called a _____.
   a. crawler crane
   b. telescoping loader
   c. rough-terrain crane
   d. stationary crane

10. Which of the following is an accurate statement about truck-mounted cranes that are made for rough terrain?
    a. They can be driven on the highway and the work site.
    b. They do not require the use of outriggers for stability.
    c. They are typically transported by truck to the work site.
    d. They use crawler tracks on the rear axles for better traction.

Trade Terms Quiz

Fill in the blank with the correct term that you learned from your study of this module.

1. A hydraulic attachment for an excavator that is used for breaking boulders and other solid objects is called a(n) _____.

2. A boom and stick combination that resembles a knuckle at the pivot point of the boom and stick is called a(n) _____.

3. A screw conveyor that is used to move bulk material such as asphalt is called a(n) _____.

4. The area on a scraper, sometimes called a hopper, where soil is stored when it is scraped from the surface is the _____.

5. A pivoting section that connects the bucket to the boom on a hydraulic excavator is called a(n) _____.

6. Compacted dirt roads used to move material and equipment on and off a site are called _____.

7. A name that is sometimes used to describe a rigid-frame dump truck or a mining truck is _____.

8. A movable section on the forward wall of the bowl on a scraper is called a(n) _____.

9. When two parts are connected by a joint so that they move independently, the joint is said to be _____.

10. A type of boom used on cranes and excavators that has a crisscross pattern of braces that enable the machine to lift heavy loads is called a(n) _____.

11. Something that is inflated or powered with compressed air is said to be _____.

12. A system found on construction tractors that uses a shaft to transfer power from the tractor to an attachment is known as a(n) _____.

13. Stabilizer legs that can be extended to widen the stance of a piece of equipment to keep it from tipping or rolling are called _____.

14. A telescoping-boom excavator is sometimes referred to as a(n) _____.

15. A towed attachment with teeth used on dozers, motor graders, and other machines to loosen heavily compacted soil and soft rock is a(n) _____.

16. Using an attachment with teeth on a motor grader to loosen soil in front of the moldboard is called _____.

17. A blade-like component on a paver that levels and smoothes asphalt or concrete as it is applied is a(n) _____.

18. A type of concrete paver that evenly spreads bulk concrete that is dumped in front of it is a(n) _____.

19. A component on some telescoping-boom forklifts that allows the fork carriage to be moved in the horizontal plane while the boom remains stationary is a(n) _____.

20. A sheepsfoot roller is sometimes referred to as a(n) _____.

21. The lower frame of an excavator that supports the turntable and has the tracks or wheels attached to it is the _____.

Trade Terms

Apron	Haul truck	Power takeoff (PTO)	Slipform paver
Articulated	Hydraulic breaker	Outriggers	Squirt boom
Auger	Knuckle boom	Ripper	Tamping roller
Bowl	Lattice boom	Scarifying	Undercarriage
Dipper stick	Pneumatic	Screed	
Haul roads		Shooting-boom excavator	

Mark S. Jones
Utility Relations, Charah, Inc.

Mark Jones is one of the lucky few who were able to parlay a summer job into a lifelong successful career. His is the classic story of a man who started at the bottom in a company and worked his way up to a senior management position.

How did you get started in the construction industry?

When I was at the University of Alabama studying corporate finance, I was fortunate to be able to spend summers working for a construction company based in Birmingham, AL as a heavy equipment operator. When I graduated, I was offered a full-time job as operations coordinator. Since then, I have worked as a project manager, estimator, and operations manager.

What inspired you to enter the construction industry?

While in college, I was offered a summer job to operate heavy equipment, by my mentor Donald Stansberry. I hadn't considered a career in construction, but his influence convinced me to give it a try. That was 25 years ago and I have never regretted the decision.

What kinds of work have you done during your career?

During my career, I have managed the civil work for power plants, landfills, subdivisions, shopping centers, quarries, and automobile manufacturing plants. Currently I am working for one of the largest providers of coal combustion product (CCP) management and power plant support services for the coal-fired electric utility industry.

What do you enjoy most about your job?

I enjoy working in the outdoors and the challenge of pulling a complex project together. There are so many things happening at once on a construction site and so many different skills making them happen. It takes a great deal of teamwork and it's exciting to me to be one of the people who make it all come together. I also get a great deal of satisfaction when I drive by a finished project that I worked on, especially one that I was responsible for. My work as a subject matter expert in the development of the Heavy Equipment training program has been very rewarding because it gives me the opportunity to bring my knowledge of the industry to people who are just starting out.

How do you feel about training in general and NCCER training specifically?

Training, especially safety training, is an important part of a successful career in construction. Anyone who wants to succeed in the construction industry needs to make a lifelong commitment to learning because new equipment, methods, and safety regulations are constantly evolving. Anyone who doesn't make an effort to keep up will fall behind. NCCER training is the industry standard. Their training courses are reviewed and developed by industry professionals and their credentials are accepted by companies all over the United States.

What kinds of training have you had?

In addition to my college education, I have completed MSHA instructor training, the OSHA 10-hour course, the NCCER Supervisor course, and the NCCER Project Management course. An in-depth knowledge of OSHA safety requirements is critical for anyone working in construction. In addition, anyone working around heavy equipment, especially in mining or quarrying work, must also know the MSHA safety regulations.

Would you recommend a career in construction to others?

Yes. There is a great demand for people who are skilled at operating heavy equipment. The pay is excellent and the work is challenging.

What advice would you give to someone just entering the heavy equipment field?

Have patience and always keep your mind on your work because distractions are the number one reason why people have accidents or are injured in the construction industry. Listen to the people who have made their careers in the industry.

How would you define craftsmanship?

Craftsmanship is having a unique skill that enables you to perform a craft with a consistently high standard of quality.

Trade Terms Introduced in This Module

Apron: A movable section on the forward wall of the bowl on a scraper.

Articulated: Two parts connected by a joint so as to move independently.

Auger: A screw conveyor that is used to move bulk material such as asphalt.

Bowl: The area on a scraper where soil is stored when it is scraped from the surface.

Dipper stick: A pivoting section that connects the bucket to the boom on a hydraulic excavator.

Haul roads: Compacted dirt roads used to move material and equipment on and off the site.

Haul truck: A name that is sometimes used to describe a rigid-frame dump truck or a mining truck.

Hydraulic breaker: A hydraulic attachment for an excavator that is used for breaking boulders and other solid objects.

Knuckle boom: A term sometimes used for a boom and stick combination that resembles a knuckle at the pivot point of the boom and stick.

Lattice boom: A type of boom used on cranes and excavators that has a crisscross pattern of braces that enable the machine to lift heavy loads.

Pneumatic: Inflated with compressed air.

Power takeoff (PTO): A system found on construction tractors that uses a shaft to transfer power from the tractor to an attachment.

Outriggers: Stabilizer legs that can be extended to widen the stance of a piece of equipment to keep it from tipping or rolling.

Ripper: A towed attachment with teeth used on dozers, motor graders, and other machines to loosen heavily compacted soil and soft rock.

Scarifying: Using an attachment with teeth on a motor grader to loosen soil in front of the moldboard.

Screed: A blade-like component on a paver that levels and smoothes asphalt or concrete as it is applied.

Shooting-boom excavator: A term sometimes used to describe a telescoping-boom excavator.

Slipform paver: A type of concrete paver that evenly spreads bulk concrete that is dumped in front of it.

Squirt boom: A component on some telescoping-boom forklifts that allows the fork carriage to be moved in the horizontal plane while the boom remains stationary.

Tamping roller: A type of compaction roller that has projecting pads on a steel drum. Also called a *sheepsfoot roller* or a *segmented-pad roller*.

Undercarriage: The lower frame of an excavator that supports the turntable and has the tracks or wheels attached.

Additional Resources

This module presents thorough resources for task training. The following reference material is recommended for further study.

The Earthmover Encyclopedia, Keith Haddock. 2007. St. Paul, MN: MBI Publishing.

Construction Equipment Guide, David A. Day and Neal B.H. Benjamin. New York, NY: John Wiley & Sons.

Heavy Equipment Operations Level One, NCCER. Third Edition. 2012. New York, NY: Pearson Education, Inc.

Heavy Equipment Operations Level Two, NCCER. Third Edition. 2013. New York, NY: Pearson Education, Inc.

Heavy Equipment Operations Level Three, NCCER. Third Edition. 2014. New York, NY: Pearson Education, Inc.

Machinery's Handbook, Erik Oberg, Franklin D. Jones, Holbrook L. Horton, and Henry H. Ryffel. Latest Edition. New York, NY: Industrial Press Inc.

Figure Credits

Deere & Company, Module Opener, Figures 2, 4, 6, 7, 11B, SA02, 13, 14, 19, 27, 29, RQ01, Exam Figures 4, 5

New Holland, Figure 1A

Images courtesy of Case IH, Figure 1B-C, 12B, RQ02

Volvo Construction Equipment, Figures 3, 5, SA01, 31, 33, 37, 39, Exam Figures 1, 2

Gradall Industries, Inc., Figures 8–10, Exam Figure 3

Topaz Publications, Inc., Figures 11A, 21, SA04, 41B

Courtesy of **www.GeneralContractor.com**, Figure 12A

Courtesy of Sellick Equipment Limited, Figure 15

Courtesy of JLG Industries, Inc., Figure 16

Landoll Corporation, Figure 17

Komatsu America Corp., Figures 18, 25, 26, 28

Images courtesy of Ray Walker Trucking, Figure 20A

Navistar Inc., Figure 20B

Courtesy of Wardlaw Trucking, Figure 22

Courtesy of CBI Manufacturing, Figure 23

Courtesy of Robert Lafrenière, Figure 24

Courtesy of Trucks.com, SA03

Images courtesy of Dominion Equipment Parts, LLC, Figure 30

Crowder Construction Company, Figure 32

West Contracting, Figure 34

Courtesy of BOMAG Americas, Inc., Figures 36, 38

Courtesy of Multiquip Inc., Figure 40

©iStockphoto.com/Andyqwe, Figure 41A

Campbell Hausfeld, Figure 42

Courtesy of Sullair, LLC, Figure 43

Carolina Bridge Co., Figure 44

Manitowoc Cranes, Figures 45, 46, RQ03

Section Review Answer Key

Answer	Section Reference	Objective
Section One		
1. b	1.1.9	1a
2. d	1.2.2	1b
3. a	1.3.1	1c
Section Two		
1. d	2.1.1	2a
2. c	2.2.0	2b
3. a	2.3.0	2c
Section Three		
1. b	3.1.0	3a
2. c	3.2.0	3b

NCCER CURRICULA — USER UPDATE

NCCER makes every effort to keep its textbooks up-to-date and free of technical errors. We appreciate your help in this process. If you find an error, a typographical mistake, or an inaccuracy in NCCER's curricula, please fill out this form (or a photocopy), or complete the online form at **www.nccer.org/olf**. Be sure to include the exact module ID number, page number, a detailed description, and your recommended correction. Your input will be brought to the attention of the Authoring Team. Thank you for your assistance.

Instructors – If you have an idea for improving this textbook, or have found that additional materials were necessary to teach this module effectively, please let us know so that we may present your suggestions to the Authoring Team.

NCCER Product Development and Revision
13614 Progress Blvd., Alachua, FL 32615

Email: curriculum@nccer.org
Online: www.nccer.org/olf

❏ Trainee Guide ❏ Lesson Plans ❏ Exam ❏ PowerPoints Other _____

Craft / Level: _____ Copyright Date: _____

Module ID Number / Title: _____

Section Number(s): _____

Description: _____

_____

_____

_____

_____

Recommended Correction: _____

_____

_____

_____

_____

Your Name: _____

Address: _____

_____

Email: _____ Phone: _____

36110-17

Highway and Bridge Safety

Overview

Highway and bridge construction takes place around large motorized equipment and power tools, in trenches and excavations, and in severe weather. Although health and safety hazards are everywhere, injuries can easily be prevented by taking proper precautions. In fact, most accidents and injuries occur when workers stop paying attention to safety rules and the hazards around them. Be aware of and alert to all the dangers around highway construction to keep yourself and your co-workers safe.

Module Three

Trainees with successful module completions may be eligible for credentialing through the NCCER Registry. To learn more, go to **www.nccer.org** or contact us at 1.888.622.3720. Our website has information on the latest product releases and training, as well as online versions of our *Cornerstone* magazine and Pearson's product catalog.

Your feedback is welcome. You may email your comments to **curriculum@nccer.org**, send general comments and inquiries to **info@nccer.org**, or fill in the User Update form at the back of this module.

This information is general in nature and intended for training purposes only. Actual performance of activities described in this manual requires compliance with all applicable operating, service, maintenance, and safety procedures under the direction of qualified personnel. References in this manual to patented or proprietary devices do not constitute a recommendation of their use.

Objectives

When you have completed this module, you will be able to do the following:

1. Identify the common hazards found in highway and bridge construction and explain how to minimize the risks associated with those hazards.
   a. Identify the hazards commonly associated with equipment used in highway and bridge construction.
   b. Identify common job site hazards.
   c. Describe methods used to identify hazards on a job site.
   d. Describe hazards commonly associated with confined spaces.
   e. Describe the purpose of a job safety analysis.
2. Identify the hazards associated with working in hot and cold weather and the precautions for minimizing these hazards.
   a. Describe heat-related hazards.
   b. Describe cold-related hazards.
3. Identify the four main causes of accidents on the job site.
   a. Explain the importance of fall protection.
   b. Explain pinch points and caught-between hazards.
   c. Explain struck-by hazards.
   d. Explain the hazards associated with electricity.
4. Explain traffic zones and the importance of flagging.
   a. Explain traffic zones.
   b. Explain flagging.
5. Describe the safety hazards associated with trenches.
   a. Identify trench-related hazards and the safety requirements associated with trenches.
   b. Explain how to work safely in and around trenches.

Performance Task

Under the supervision of your instructor, you will be able to do the following:

1. Develop a job safety analysis.

Trade Terms

Caissons
Cofferdam
Combustible
Competent person
Cribbing
Culverts

Flaggers
Flammable
Heat index
Locked out
Machine guarding
Personal fall arrest system (PFAS)

Personal flotation device (PFD)
Shoring
Spoil
Temporary traffic control (TTC)
Trench box

Industry Recognized Credentials

If you are training through an NCCER-accredited sponsor, you may be eligible for credentials from NCCER's Registry. The ID number for this module is 36110-17. Note that this module may have been used in other NCCER curricula and may apply to other level completions. Contact NCCER's Registry at 888.622.3720 or go to **www.nccer.org** for more information.

Contents

Figures and Tables

SECTION ONE

1.0.0 HIGHWAY AND BRIDGE CONSTRUCTION HAZARDS

Objective

Identify the common hazards found in highway and bridge construction and explain how to minimize the risks associated with those hazards.

   a. Identify the hazards commonly associated with equipment used in highway and bridge construction.
   b. Identify common job site hazards.
   c. Describe methods used to identify hazards on a job site.
   d. Describe hazards commonly associated with confined spaces.
   e. Describe the purpose of a job safety analysis.

Performance Task

1. Develop a job safety analysis.

Trade Terms

Caissons: Watertight chambers commonly used in building or repairing bridge pilings.

Cofferdam: A structure built to keep water away from a construction area.

Combustible: A substance that readily ignites and burns.

Competent person: As defined by OSHA, an individual who is capable of identifying existing and predictable hazards in the surroundings or working conditions which are unsanitary, hazardous, or dangerous to employees, and who has the authorization to take prompt corrective measures to eliminate such hazards. In the context of trenches, a person with knowledge of soil types and soil stability who will inspect the trench whenever required.

Cribbing: Alternately stacked timbers used to support heavy loads.

Culverts: Drains or channels that cross under a road.

Flammable: Capable of burning.

Personal flotation device (PFD): A sleeveless jacket composed of a buoyant or inflatable material and used to prevent drowning; commonly called a *life vest*.

Construction of highways and bridges is a more efficient process today than in the past. However, there are still many hazards involved that require the awareness and consideration of every worker.

1.1.0 Equipment-Related Hazards

Construction equipment is large, heavy, powerful, and difficult to maneuver. In addition, some equipment doesn't allow for a good view from the operator's cab. While equipment operators are responsible for managing their equipment safely and responsibly, you must also take responsibility for your own safety by behaving conscientiously when working around heavy equipment. This includes making operators aware of your presence. Common types of heavy construction equipment you may encounter include the following:

- Mobile cranes
- Utility tractors
- Dump trucks
- Compactors
- Scrapers
- Backhoe loaders
- Excavators
- Dozers
- Loaders
- Forklifts
- Motor graders
- Trenchers

Figure 1 shows an articulated tractor. Tractors like this may be used on a construction site to pull equipment such as scrapers. Note that, even though the operator's cab is situated high on the vehicle with an almost unobstructed view, there are still areas the operator cannot easily see,

Figure 1 Articulated tractor with some limited sight lines.

especially low and near the vehicle. What would happen if you knelt to tie a boot string just as the operator reversed? The operator might look around to make sure his path was safe, but still might not see you.

1.1.1 Equipment Capacity and Stability

There are specific hazards associated with mobile cranes and other construction equipment used to hoist loads. They can experience a loss of load control by hoisting in excess of their capacity, and they can experience a loss of stability.

While this section refers primarily to cranes, there are also other types of equipment used for hoisting, such as backhoes and excavators. Lifts are made to hoist loads, and other equipment such as backhoes and excavators are also frequently used for that purpose. The information discussed in this section can be applied to any piece of heavy equipment used to lift material.

Each crane configuration will have a capacity chart (load chart), which should be readily available to the operator. The baseline capacity is a function of factors such as the hoisting cable and the rigging strategy. There are several other factors that affect both the capacity and the stability of the equipment, including the following:

- *Ground conditions* – Ground conditions at a work site can vary widely. Some areas may have been graded to hardpan; these areas will be very stable. Other areas may consist of recent fill that has yet to be compacted. These areas will be very unstable.
- *Crane base* – The parts of the crane that rest on the ground and which supports the crane and the load. The crane base can be tires and wheels, rubber or metal tracks, or outriggers.
- *Bearing surface* – The ground surface that supports the hoisting equipment. This can be earth, or a man-made support such as barges, mats, concrete, or cribbing. The bearing surface must be as solid and level as possible. Any instability or tilt will reduce the crane's hoisting capacity.
- *Load center of gravity* – The lower the center of gravity, the more stable the hoisting equipment. Hoisting a load will raise the center of gravity.
- *Quadrant of operation* – The quadrant of operation refers to the direction the boom is pointing relative to the supporting wheels, tracks, or outriggers. In general, hoisting in a fore or aft direction is more stable than hoisting to the side.

- *Boom length, angle, operating radius, and boom point elevation* – All of these factors share one thing in common; they change the effective distance of the load from the center of gravity. The further the load is from the center of gravity, the more unstable the load and the hoisting equipment are.
- *Swing out, side loading, and dynamic loading* – These are are three different types of load movement that can occur while the load is suspended, which can cause potentially dangerous stress on the boom. Each movement also affects the stability of the crane.

Even if you are not operating hoisting equipment, you should be aware of hoisting operations occurring near you. Pay attention to the factors that affect the stability of hoisting equipment to help maintain the safety of the project and to keep yourself safe during hoisting operations. Be specifically aware of the possibility of a crane becoming unbalanced as it swings its load to the side. Also, never get under a load while it is being lifted or moved.

1.2.0 Common Job Site Hazards

Workers involved in highway and bridge construction face a number of hazards, including those related to working above water, working around concrete reinforcing materials, and working in confined spaces. This section deals with practices used to recognize and avoid these hazards.

1.2.1 Water Safety

Bridges are built to cross a given terrain more easily. Frequently, the terrain being crossed is water. All workers must be aware of potential hazards. If working over or near water, or on floating vessels such as support barges, the following precautions apply:

- You must wear a personal flotation device (PFD), either a life jacket or buoyant work vest (*Figure 2*).
- US Coast Guard-approved 30-inch (76-cm) life rings (*Figure 3*), with at least 90 ft. (27 m) of line, must be placed at intervals of no less than 200 ft. (61 m).
- All workers must know the established man-overboard procedures. The procedures should be practiced regularly.
- A ladder (permanent or portable) must be available to help workers who have fallen into the water.
- A rescue vessel must be available.

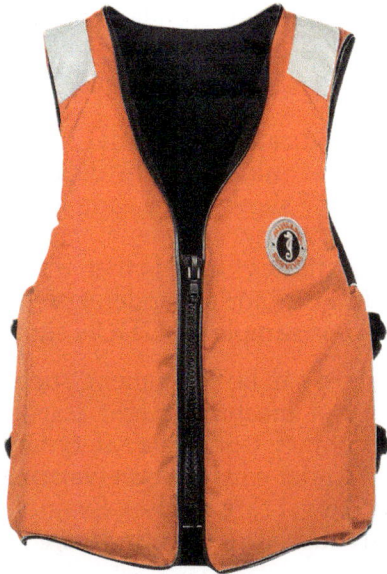

Figure 2 Personal flotation device.

Figure 3 Life ring.

A cofferdam (Figure 4) is a closed-cell retaining structure designed to temporarily hold back water while construction is underway. The following special precautions are required when working in a cofferdam:

- A cofferdam should be treated as a confined space. This includes monitoring oxygen levels and the buildup of potentially toxic gases, and providing for adequate ventilation. Requirements for working in confined spaces are covered in 29 *CFR* 1910.146.
- Gasoline-powered equipment should not be operated in a cofferdam.
- Interior areas of a cofferdam must be equipped with walkways, ladders, stairs, and railings to provide access to all work areas.

Figure 4 Example of a cofferdam.

- Fall protection is required for personnel working 6 ft. (2 m) or more from the bottom of the cofferdam.

1.2.2 Rebar-Related Hazards

Concrete is reinforced to help it stand up to tension, shear forces, and vibration. One method of reinforcing concrete involves embedding steel reinforcing bars, or rebar, in the concrete. Rebar must be set up before the concrete is poured. This prepared rebar can be hazardous.

Rebar that is placed horizontally in preparation for concrete is a trip and fall hazard. Furthermore, many rebar placements are vertical (*Figure 5*). Vertically positioned rebar must be capped to prevent anyone who might fall on the rebar from being impaled.

Figure 5 Rebar with safety caps.

1.3.0 Hazard Identification

Hazard identification involves the marking of hazards that exist during the construction process. These can be divided into two categories: hazards to workers and hazards to civilians.

1.3.1 Personal Protective Equipment

One type of markings for workers consists of the markings that you wear. Personal protective equipment (PPE) helps make you visible to equipment operators. A high-visibility vest is standard equipment for anyone working around heavy equipment. You will also need a hard hat. A high visibility hard hat helps make sure that you are seen.

1.3.2 Safety Labels and Signs

The labeling of equipment and materials is another important form of safety markings for workers. Some signs have a safety alert symbol, which is an exclamation point inside a triangle. This is an international symbol signaling a threat. Safety alerts are broken into the following three categories:

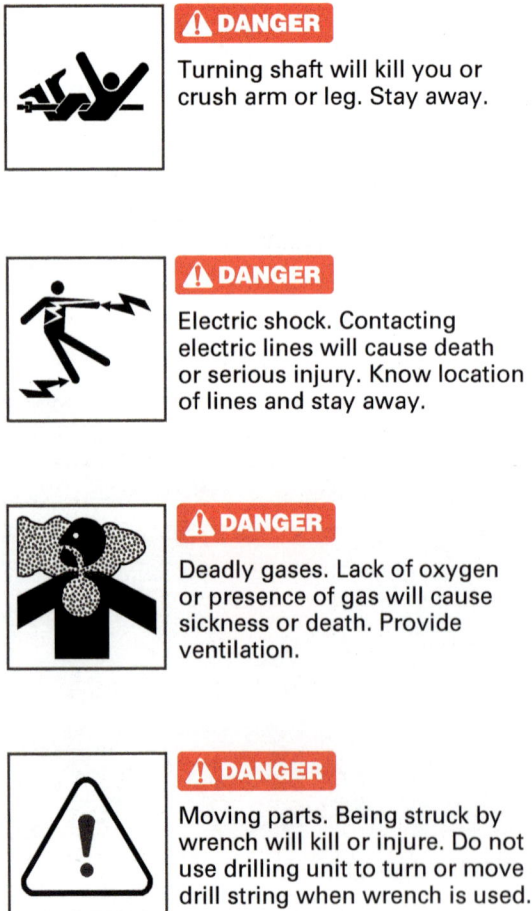

⚠ DANGER
Turning shaft will kill you or crush arm or leg. Stay away.

⚠ DANGER
Electric shock. Contacting electric lines will cause death or serious injury. Know location of lines and stay away.

⚠ DANGER
Deadly gases. Lack of oxygen or presence of gas will cause sickness or death. Provide ventilation.

⚠ DANGER
Moving parts. Being struck by wrench will kill or injure. Do not use drilling unit to turn or move drill string when wrench is used.

Figure 6 Danger signs.

- *Danger* – Indicates an immediate hazard that, if not avoided, will result in death or serious injury. *Figure 6* shows some sample DANGER signs.
- *Warning* – Indicates a potential hazard that, if not avoided, could result in death or serious injury. *Figure 7* shows some sample WARNING signs.
- *Caution* – Indicates hazards that, if not avoided, may result in minor or moderate injury. *Figure 8* shows some sample CAUTION signs.

Hazard signs can be permanent (such as a label warning that a machine might start automatically) or temporary (such as a safety tag attached to a scaffold under construction warning that the structure is not yet safe for use). Examples of temporary safety tags are shown in *Figure 9*.

1.3.3 Material Labeling

Materials should be labeled according to any safety threat they pose. Two common systems are the National Fire Protection Association (NFPA) system (*Figure 10*) and the Hazard Material Identification System (*Figure 11*). Each system divides the potential safety hazard into categories: flammability, health, and reactivity. The NFPA system (sometimes referred to as the *diamond*) also adds a rating of the potential hazard from 0 (no danger) to 4.

1.3.4 Traffic Control

Construction projects frequently occur on operating highways, or near to operating highways and connected by temporary ramps. In bridge construction, a new bridge is often built right next to an existing bridge. In both instances, workers are at risk from traffic and civilians are at risk from the hazards commonly associated with construction work.

Hazard markings for civilians help make the construction area safer by alerting them to the dangers they pose to workers, as well as helping to protect the civilians themselves. They also act as access control, inhibiting travel into the construction area. *Figure 12* shows some typical traffic control devices in use. *Figure 13* and *Figure 14* show some common traffic control signage.

1.4.0 Confined Spaces

Spaces on a job site are considered confined when their size and shape restrict the movement of anyone who must enter, work in, and exit the space. Confined spaces are often poorly ventilated and difficult to enter and exit. For example, employ-

⚠ WARNING	⚠ WARNING
Explosion possible. Serious injury or equipment damage could occur. Follow directions carefully.	Looking into fiber-optic cable could result in permanent vision damage. Do not look into ends of fiber-optic or unidentified cable.
⚠ WARNING	⚠ WARNING
Fluid or air pressure could pierce skin and cause injury or death. Stay away.	Fire or explosion possible. Fumes could ignite and cause burns. No smoking, no flame, no spark.
⚠ WARNING	⚠ WARNING
Job-site hazards could cause death or serious injury. Use correct equipment and work methods. Use andmaintain proper safety equipment.	Moving traffic - hazardous situation. Death or serious injury could result. Avoid moving vehicles, wear high visibility clothing, post appropriate warning signs.
⚠ WARNING	⚠ WARNING
Crushing weight could cause death or serious injury. Use proper procedures and equipment or stay away.	Hot pressurized cooling system fluid could cause serious burns. Allow to cool before servicing.
⚠ WARNING	⚠ WARNING
Moving parts could cut off hand or foot. Stay away.	Improper control function could cause death or serious injury. If control does not work as described in instructions, stop machine and have it serviced.

Figure 7 Warning signs.

ees who work in manholes have to squeeze in and out through narrow openings and perform their tasks while in a cramped or awkward position. Deep excavations, even though they may be open at the top, may also be classified as confined spaces because toxic or flammable gases can accumulate in the lower part of the excavation. In some cases, confinement itself creates a hazard. Examples of confined spaces found in highway and bridge construction include the following:

- Cofferdams
- Caissons
- Manholes
- Drains
- Culverts

In a confined space, hazards such as poor air quality, toxins, explosions, fire, and moving machinery parts tend to be far more deadly. According to the US Department of Labor (DOL), approximately two-thirds of these fatalities are caused by atmospheric issues, such as reduced oxygen or the presence of toxins in the air.

Case History

Culvert Checks Save Lives

A worker was about to enter a culvert, and decided to check it out first. When he looked inside, he found a nest of rattlesnakes.

The Bottom Line: Always be sure to look inside a confined space before entering.

CAUTION

Flying objects may cause injury. Wear hard hat and safety glasses.

CAUTION

Hot parts may cause burns. Do not touch until cool.

CAUTION

Exposure to high noise levels may cause hearing loss. Wear hearing protection.

CAUTION

Fall possible. Slips or trips may result in injury. Keep area clean.

CAUTION

Battery acid may cause burns. Avoid contact.

CAUTION

Improper handling or use of chemicals may result in illness, injury, or equipment damage. Follow instructions on labels and in material safety data sheets (MSDSs).

Figure 8 Caution signs.

Confined spaces may contain unknown hazards. In one instance, a worker was lowered into a 21 ft. (7 m) deep manhole on a looped chain seat. Twenty seconds after entering the manhole, he started gasping for air and fell. An autopsy determined that he died from lack of oxygen.

Most confined spaces have restricted entrances and exits. Workers are often injured as they enter or exit through small doors and hatches. It can also be difficult to move around in a confined space and workers can be struck by moving equipment. Escapes and rescues are much more difficult in confined spaces than they are elsewhere.

Confined spaces are entered for inspection, equipment testing, repair, cleaning, or emergencies. They should only be entered for short periods.

A written confined-space entry program can protect you. It will identify the hazards and specify the equipment or support that is needed to avoid injury. All industrial and some construction sites have written confined-space entry programs. You need to know and follow your company's policy.

1.4.1 Confined-Space Classifications

Confined spaces must be inspected before work can begin. This inspection helps to identify possible hazards. After an inspection by a company-authorized person, the confined space is classified based on any hazards that are present. The two classifications are *non-permit* and *permit-required*. A permit-required confined space is shown in *Figure 15*. These two classifications are defined as follows:

> **WARNING!**
>
> No one may enter a confined space unless they have been properly trained and have a valid entry permit. An attendant must be stationed outside the space while it is occupied in order to monitor the work and assist with exits.

- *Non-Permit Confined Space (NPCS)* – A non-permit confined space (NPCS) is a work space free of any mechanical, physical, electrical, and atmospheric hazards that can cause death or injury. After a space has been classified as a non-permit space, workers can enter using the appropriate PPE for the type of work to be performed. Always check with your supervisor if it is unclear what PPE is required. OSHA defines a confined space as any space that has the following characteristics:

  - It is large enough and configured in such a way that an employee can enter and perform assigned work
  - Has a limited or restricted means of entry or exit (such as tanks, vessels, silos, storage bins, hoppers, vaults, and pits)
  - Is not designed for continuous employee occupancy

- *Permit-Required Confined Space (PRCS)* – A permit-required confined space (PRCS) is a confined space that has real or possible hazards. These hazards can be toxic, flammable, physical, electrical, or mechanical. OSHA defines a permit-required confined space as a confined space that has one or more of the following characteristics:

(A)

START-UP INSTRUCTIONS

EQUIPMENT ID_____
LOCATION_____

Step No.	Instructions

(B)

DO NOT OPERATE

Signed by_____

Date_____

(C)

OUT OF ORDER

Signed by_____

Date_____

(D)

AUTHORIZED PERSONNEL ONLY

OPERATION OF THIS EQUIPMENT IS RESTRICTED TO THE FOLLOWING PERSONS

NAME_____
_____
_____

Signed by_____

Date_____

(E)

DANGER LIVE WIRE

SIGNED BY_____

DATE_____

(F)

DANGER DO NOT REMOVE THIS TAG

REMARKS_____
_____
_____

SEE OTHER SIDE

Figure 9 Temporary safety tags.

- Contains or has the potential to contain a hazardous atmosphere
- Contains a material that has the potential for engulfing an entrant
- Has an internal configuration such that an entrant could be trapped or asphyxiated by inwardly converging walls or by a floor that slopes downward and tapers to a small cross-section

- Contains any other recognized serious safety or health hazard

WARNING!

All permit-required confined spaces must be identified, and the associated permit must be posted. A permit is required unless it has been determined that no atmospheric hazard exists in the space.

HEALTH	
4	DEADLY
3	EXTREME DANGER
2	DANGEROUS
1	SLIGHT HAZARD
0	NO HAZARD
REACTIVITY	
4	MAY DETONATE
3	EXPLOSIVE
2	UNSTABLE
1	NORMALLY STABLE
0	STABLE
SPECIFIC HAZARD	
OXY	MAY DETONATE
ACID	EXPLOSIVE
ALK	UNSTABLE
COR	NORMALLY STABLE
W	STABLE
☢	RADIATION

Figure 10 NFPA hazard labeling.

An entry permit must be issued by a compe-tent person and signed by the job-site or entry supervisor before the confined space is entered. No one is allowed to enter a confined space unless there is a valid entry permit. The permit is to be kept at the confined space while work is being performed. Always check with your supervisor if it is unclear whether or not you need a permit to enter a confined space.

1.4.2 Confined-Space Entry Permits

Confined spaces can be extremely dangerous. Entry into the space begins when any part of your body passes the entrance or opening of a confined space. Before entering a permit-required confined space, you must have an entry permit. An example entry permit is shown in *Figure 16A* and *Figure 16B*.

NAME OF MATERIAL		
3	HEALTH	
4	FLAMMATORY	
2	PHYSICAL HAZARD	
B	PROTECTIVE EQUIPMENT	

4	SEVERE HAZARD	
3	SERIOUS HAZARD	
2	MODERATE HAZARD	
1	SLIGHT HAZARD	
0	MINIMAL HAZARD	
*	CHRONIC HAZARD	
1	SLIGHT HAZARD	
0	MINIMAL HAZARD	

Figure 11 HMIS hazard labeling.

Figure 12 Traffic barricades in place.

An entry permit is a job checklist that verifies that the space has been inspected. It also lets everyone on the site know about the hazards of the job. All entry permits must be filled out and signed by the supervisor before anyone enters the space. The permit must also be posted at the entrance to the site and be available for workers to review. Entry permits must include the following information:

- A description of the space and the type of work that will be done

- The date the permit is valid and how long it lasts
- Test results for all atmospheric testing including oxygen, toxin, and flammable material levels
- The name and signature of the person who did the tests
- The name and signature of the entry supervisor
- A list of all workers, including supervisors, who are authorized to enter the site
- The means by which workers and supervisors will communicate with each other
- Special equipment and procedures that are to be used during the job
- Other permits needed for work done in the space, such as welding
- The contact information for the emergency response rescue team

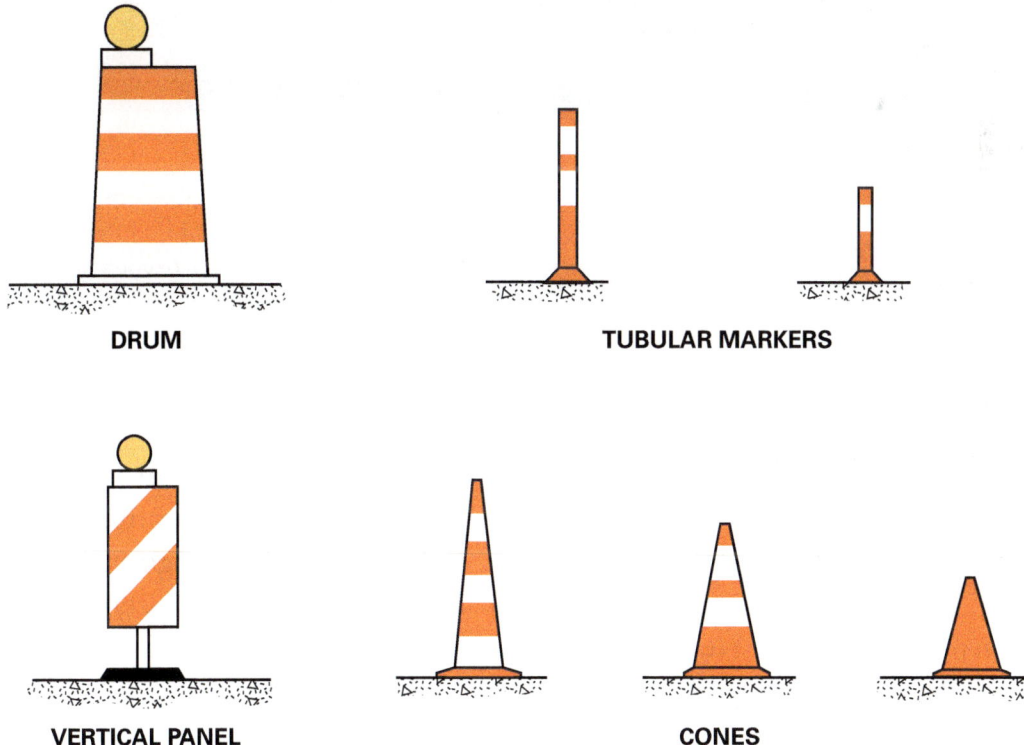

Figure 13 Traffic control signs.

NOTE

Some confined spaces require both a permit to enter and a permit to work in the space.

1.4.3 Confined-Space Hazards

Confined spaces are dangerous. The main hazards include poor airflow and restricted movement. Poor ventilation can allow toxic gases to build up. Physical hazards are more dangerous because escape is limited.

Atmospheric hazards are the most common hazards in a confined space. In a hazardous atmosphere, the air can have either too little or too much oxygen, be explosive or flammable, or contain toxic gases. Special meters are used to detect these atmospheric hazards (*Figure 17*).

A confined space that does not have enough oxygen is called an oxygen-deficient atmosphere. A confined space that has too much oxygen is called an oxygen-enriched atmosphere. For safe working conditions, the oxygen level in a confined space must range between 19.5 and 23.5 percent by volume, with 21 percent being considered the normal level. Oxygen concentrations below 19.5 percent by volume are considered oxygen-deficient, and those above 23.5 percent by volume are considered oxygen-enriched.

DRUM   **TUBULAR MARKERS**

VERTICAL PANEL   **CONES**

Figure 14 Traffic channeling devices.

Figure 15 Permit-required confined space.

Many of the processes that occur in a confined space use oxygen and may reduce the percentage of oxygen to an unsafe level. These processes include the following:

- Cutting/welding
- Rusting of metal
- Infiltration of external exhaust fumes
- Breaking down of plants or garbage
- Oxygen mixing with other gases

When the oxygen in a confined space is reduced, it becomes harder to breathe. The symptoms of insufficient oxygen happen in the following order:

1. Fast breathing and heartbeat
2. Impaired mental judgment
3. Extreme emotional reaction
4. Unusual fatigue
5. Nausea and vomiting
6. Inability to move your body freely
7. Loss of consciousness
8. Death

Master Card No._____

1. Work Description

Area_____   Equipment Location_____

Work to be done:

2. Gas Test		Results	Recheck	Recheck
Required	☐ Instrument Check			
☐ Yes	☐ Oxygen % 20.8 Min.			
☐ No	☐ Combustible % LFL			
		Date/Time/Sig.	Date/Time/Sig.	Date/Time/Sig.

3. Special Instructions: ☐ Check issuer before beginning work   ☐ None

4. Hazardous Materials: ☐ None   What did the line / equipment last contain?

5. Special Protection Required: ☐ None   ☑ Forced Air Ventilation

☐ Avoid Skin Contact   ☐ Gloves _____   ☐ Suit _____

☐ Goggles or Face Shield   ☐ Respirator _____   ☐ Safety Harness

☐ Self-Contained Breathing Equipment   ☐ Hoseline Breathing Equipment

☐ Other, Specify _____   ☐ Standby - Name: _____

6. Fire Protection Required: ☐ None   ☐ Portable Fire Extinguisher   ☐ Fire Hose and Nozzle

☐ Fire Watch   ☐ Other, specify:

7. Condition of Area and Equipment

Required   THESE KEY POINTS MUST BE CHECKED
Yes   No

Yes	No		
		a.	Lines disconnected & blinded or where disconnecting is not possible, blinds installed? (Includes drains, vents and instrument leads) and appropriate valves locked out?
		b.	Equipment cleaned, washed, purged, ventilated?
		c.	Low voltage or GFCI-protected electrical equipment provided?
		d.	Explosion-proof electrical equipment provided?
		e.	Life lines required to be attached to safety harnesses?

Comments

rev.6/10/91

Figure 16A Entry permit (1 of 2).

8. Approval	Date	Time	Permit Authorization Area Supervisor	Permit Acceptance Maint./ Contractor Supervisor/Engineer	Date	Time
Issued by						
Endorsed by						
Endorsed by						
Endorsed by						

9. Individual Review

I have been instructed on proper Safety Procedures and proper Confined Space Entry Procedures. I have signed in on the appropriate Master Card and have affixed personal locks on energy isolation devices as appropriate.

_____   _____

_____   _____

_____   _____

_____   _____

_____   _____

_____   _____

_____   _____

_____   _____

_____   _____

_____   _____

_____   _____

Signature of all personnel covered by this permit.

Forward to Production Superintendent 7 days after completion of work.

rev.6/10/91

Figure 16B Entry permit (2 of 2).

Too much oxygen in a confined space is a fire hazard and can cause an explosion. Materials like clothing and hair are highly flammable and will burn rapidly in oxygen-enriched atmospheres. Fires can start easily in a confined space that contains oxygen-enriched air.

> **WARNING!**
> When opening a manhole, always step back away from the opening in case any toxic vapors are present in the space.

Combustible Atmospheres – Air in a confined space becomes combustible when chemicals or gases reach a certain concentration. Flammable gases can be trapped in confined spaces. These gases include acetylene, butane, propane, methane, and others. Dust and work byproducts from spray painting or welding can also form a combustible atmosphere. A spark or flame will cause an explosion in a combustible atmosphere.

Some flammable gases are lighter than air and have a higher concentration at the top of a confined space. Vapors from fuels are generally heavier than air and will form a greater concentration at the bottom of the space.

> **WARNING!**
> Explosion-proof lights, motors, exhaust fans, and other equipment must be used to prevent fires and explosions in any combustible area.

Toxic Atmospheres – Toxic gases and vapors come from many sources. They can be deadly when they are inhaled or absorbed through the skin above certain concentration levels. In spaces with no ventilation, high concentrations can gather and quickly become toxic. Even in lower doses, some chemicals can seriously affect your breathing and brain function.

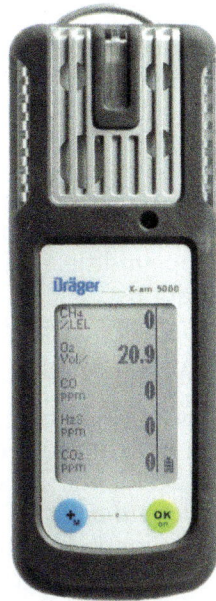

(A)

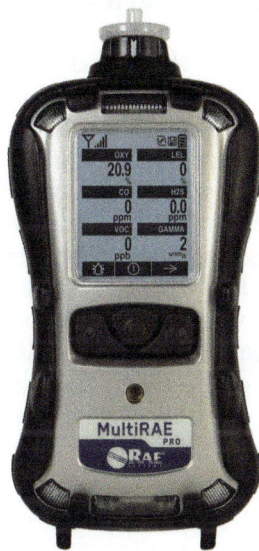

(B)

Figure 17 Detection meters.

Case History

Added Oxygen Causes Lost Life

A welder entered a 24-in. (61-cm) diameter steel pipe to grind a bad weld on a valve about 30 ft. (9 m) from the entry point. Before he entered, other crew members decided to add oxygen to the pipe near the bad weld to make sure the air was safe. The welder had been grinding off and on for about five minutes when a fire broke out. The fire covered his clothing. He was pulled from the pipe and the fire was put out. The burns were so serious that the welder died the next day.

The Bottom Line: This accident could have been avoided. It happened because of poor communication between workers and because of unsafe work practices.

Source: The Occupational Safety and Health Administration (OSHA)

The harmful effects of toxic gases and vapors vary. Many toxic gases, such as carbon monoxide, cannot be detected by sight or smell. Some toxic gases have harmful effects that may not show up until years after contact. Others, such as nitric oxide, can kill quickly.

> **NOTE**
>
> The safety data sheet (SDS) attached to the entry permit will have information about any toxins you may encounter.

Monitoring the Atmosphere in Confined Spaces – The air inside a permit-required confined space must be tested before anyone is allowed to enter. Atmospheric tests must be done in the following order:

1. Oxygen content
2. Flammable gases and vapors
3. Potential toxic contaminants

The test for oxygen must be performed first because most combustible gas meters are oxygen-dependent and will not provide reliable readings in an oxygen-deficient atmosphere. The test for combustible gases is performed next because the threat of fire or explosion is usually more urgent and more life-threatening than exposure to toxic gases and vapors. Both tests are performed at multiple levels. In addition, any time the conditions change, the atmosphere must be retested. An industry best practice is to retest any time the excavation has been evacuated for more than 30 minutes. Under conditions in which the atmosphere is likely to change, a best practice is to provide continuous real-time monitoring. Always refer to your company policy.

Various instruments are used to test and monitor confined-space atmospheres. Portable, battery-operated gas detection meters can measure oxygen levels by changing the sensor in the detection meter. Gas detection meters must be calibrated and operated according to the manufacturer's instructions. The meter must be able to detect oxygen levels, as well as toxic and combustible gases at the top, middle, and bottom levels as specified by regulations.

1.5.0 Job Safety Analysis

A job safety analysis (JSA) is a process that focuses on job tasks in order to recognize hazards before they cause accidents. This practice is also referred to as *job hazard analysis*. The process focuses on the connection between the worker, the project or task, the tools, and the environment. After potential safety hazards are identified and brought to light, workers can take steps to eliminate or reduce the risk to themselves and their co-workers.

The JSA is one tool used to help employers and workers manage hazards that are specific to tasks and work environments. JSA documentation is generally completed during the course of a meeting between involved workers and supervisors, but management may initiate the forms and thought process prior to the meeting. Tasks to be completed are first identified, then they are broken down into smaller steps so potential hazards can be more easily identified. Although the primary function of the JSA is to reduce or eliminate potential hazards and risks, it can also serve as a planning tool for the coming task. An example of a safety planning form used by a construction contractor is shown in *Figure 18A* and *Figure 18B*.

Case History

Four Die in Confined Space

A project involved the upgrade/replacement of a sewer pumping station and the contractor prepared a confined-space entry permit for the work. One employee was disconnecting a sewer bypass connection in a manhole while three others were at the manhole entrance. The manhole filled with sewage and gases from the sewer line, and the employee was overcome by a lack of oxygen. The other three employees tried to help. Each entered the manhole one at a time, apparently to attempt rescue. Each was overcome by the sewer gases and died.

The Bottom Line: Asphyxiation (lack of oxygen) is far more dangerous than it is portrayed to be in movies. You can't go in unprotected to save someone, and then exit coughing but unharmed. In reality, you pass out upon your first breath. If a retrieval system is not in place, anyone who enters the space to rescue you will die. Many workers have died simply by putting their heads in manholes to assess a situation. Even without fully entering the confined space, these workers were immediately incapacitated.

Source: The Occupational Safety and Health Administration (OSHA)

PTP (Pre-Task Plan)

Date: | Day of the Week: | Division: | Job #: | Foreman / Supervisor: | Crew Size:

Description of Activity:

Location of Activity:

PLAN YOUR WORK...WORK YOUR PLAN

With the end in mind, what is the specific goal for the task or day?

FOREMAN DAILY PLANNING LOG

Name / Schedule of Crew Members Assigned to Activity	7AM	8AM	9AM	10AM	11AM	12PM	1PM	2PM	3PM	4PM	5PM

EVALUATE YOUR WORK AREA

Have you walked your area?.. Yes.....No
Do you have the PPE needed for this task?.. Yes.....No

Are you working around live systems?.. Yes.....No
Are the required materials and tools provided?.. Yes.....No

Does this task require special training?.. Yes.....No
Have all tools/equipment been inspected before use?...................................... Yes.....No

Is an MSDS review necessary for this task?.. Yes.....No
Does every crew member know how to use assigned tools?................................ Yes.....No

Is air monitoring required?... Yes.....No
Does this task involve a confined space?.. Yes.....No

Are work permits required for this task?.. Yes.....No
Will weather conditions affect the safety or quality of the work?.................... Yes.....No

Are you familiar with evacuation routes?.. Yes.....No
Is there a safety issue that has not been addressed?...................................... Yes.....No

Emergency equipment located (fire extinguishers, eyewash stations, safety showers, phones)?.... Yes.....No
If yes, describe the safety issue that needs to be addressed:

If the work area is congested, has the work plan been coordinated with other crafts?.... Yes.....No

IDENTIFY POTENTIAL HAZARDS (Place a Checkmark if applicable)

___ Pinch Points
___ Thermal Burns
___ Particles in Eyes
___ Elevated Work
___ Poor Housekeeping
___ Electrical Shock
___ Chemical Burns
___ Fire/Explosion

___ Inadequate Access
___ High Noise Levels
___ Falling Objects
___ Manual Lifting
___ Chemical Spill
___ Plant Operations
___ Scaffolding
___ Mobile Equipment

___ Hazardous Chemicals
___ Heat Exhaustion/Stress
___ Sharp Objects or Tools
___ Radiation
___ Excavations
___ Lockout/Tagout
___ Ladders
___ Rigging

___ Falls from Elevations
___ Confined Spaces
___ Line Breaking
___ Inhalation Hazard
___ Critical Lift
___ Other:

PERSONAL PROTECTION PROTECTION EQUIPMENT?

All Crowder Construction Company Jobsites require Hard Hats, Safety Glasses and Safety Vests.

List any other PPE Required for this PTP:

Figure 18A Pre-task safety plan (1 of 2).

Based on task to be performed and potential hazards, prepare your Pre-Task Plan below:

List Steps of Work to be Performed (in the sequence it is to be completed for maximum efficiency)	Identify Potential Hazards Associated with Each Step	Specify Actions to Eliminate or Control the Hazard

Ask for input from the crew, and add any additional items to consider based on the crew's discussion:

The tasks have been reviewed in the work area where they will be performed, and this plan has been reviewed by all workers on this crew:

Foreman Name: _____   Foreman Signature: _____

Work Crew Sign-off (Signatures)

VISITOR SIGN-OFF:

IF WORK CONDITIONS OR TASK CHANGES, WORK MUST STOP AND A NEW PTP MUST BE PREPARED AND DISCUSSED WITH THE CREW

PTP FORM–Revision 5/15/09

Figure 18B Pre-task safety plan (2 of 2).

The use of a checklist helps ensure that areas of concern are not overlooked. Checklists assist all personnel involved to think clearly and bring up any potential hazard that comes to mind as the items are read aloud. Workers must not hesitate to contribute to this process and raise any concerns they may have.

JSA forms will differ from site-to-site to fit the needs of the employer, the team, and the specific site. JSA forms should be considered living documents, in that they can and should be modified any time a deficiency is found in them or a new potential hazard is identified.

The analysis process is best considered a discovery activity, where supervisors and workers alike consider five important questions:

- What can go wrong?
- What are the consequences or results when it does go wrong?
- How could it happen?
- What other factors contribute to the possibility of it happening?
- How likely is it that it will occur?

As the activity progresses, answers to this series of questions should be documented in an organized and consistent manner. As a hazard scenario is developed, try to answer the questions in the following sequence:

- Where is it happening (in what environment or workstation)?
- Who or what is exposed to the potential for injury or damage?
- What happens that sets the hazard in motion?
- What will be the result or consequence if it happens?
- What other factors contribute to the sequence of events?

Additional Resources

Building Safer Highway Work Zones: Measures to Prevent Worker Injuries from Vehicles and Equipment (PDF), US Department of Health and Human Services. 2001. Available at **www.cdc.gov/niosh/docs/2001-128**

29 *CFR* 1926.200, **www.ecfr.gov**

29 *CFR* 1926.146, **www.ecfr.gov**

Work-Related Roadway Crashes: Challenges and Opportunities for Prevention, US Department of Health and Human Services. 2003. Available at **www.cdc.gov/niosh/docs/2003-119**

OSHA 3071, *Job Hazard Analysis Guide*, OSHA. Available at **www.osha.gov**

1.0.0 Section Review

1. Cribbing is a type of _____.
   a. warning sign
   b. bearing surface
   c. crane
   d. storage enclosure

2. Which of the following is a correct statement regarding cofferdams?
   a. Cofferdams are considered confined spaces.
   b. Fall protection is not required in a cofferdam.
   c. Gasoline engines can be used in a cofferdam.
   d. It is not necessary to monitor oxygen levels.

3. A sign that indicates an immediate threat to health and safety is a _____.
   a. danger sign
   b. warning sign
   c. caution sign
   d. safety sign

4. A confined space with 25 percent oxygen by volume would be classified as _____.
   a. oxygen-deficient
   b. oxygen-enriched
   c. normal
   d. oxygen-deprived

5. The primary purpose of a JSA is to _____.
   a. identify PPE required
   b. prevent workers from entering a confined space
   c. reduce or eliminate hazards
   d. report injuries that occur on the job

2.0.0 WEATHER-RELATED HEALTH HAZARDS

Objective

Identify the hazards associated with working in hot and cold weather and the precautions for minimizing these hazards.

    a. Describe heat-related hazards.
    b. Describe cold-related hazards.

Trade Terms

Heat index: A value that combines temperature and humidity to establish an equivalent value that represents the effect of humidity on perceived temperature.

Large mobile equipment, potentially dangerous power tools, and drivers speeding through a construction zone are all obvious hazards. Most people would recognize these hazards as threats to personal safety. However, most people do not think of weather as a threat except in the most extreme circumstances (such as tornadoes, hurricanes, or blizzards). The reality is that constant exposure to unusually high or low temperatures can be as much of a threat as walking in front of moving equipment.

2.1.0 Heat-Related Illnesses

Working in the heat is uncomfortable and tiring. Working in extreme heat without taking the proper precautions can lead to heat exhaustion or heat stroke, both of which can be deadly.

Heat exhaustion is caused by losing water and sodium from the body through sweating. A victim of heat exhaustion can exhibit pale, clammy skin, heavy sweating, nausea (and possibly vomiting), a fast, weak pulse, and possibly fainting.

Heat exhaustion can lead to heat cramps. A victim of heat cramps can exhibit abdominal pain, nausea and dizziness, heavy sweating, muscular twitching, and even severe muscle cramps.

Heat stroke is the more serious of the two heat-related illnesses presented here. A victim of heat stroke can exhibit a very high body temperature (as high as 106°F, or 41°C), as well as the following:

- Hot, dry, flushed skin

- Slow, deep breathing
- Convulsions
- Loss of consciousness

Heat stroke is a life-threatening emergency that requires immediate medical care for the victim.

2.1.1 Preventing and Treating Heat-Related Illnesses

Heat-related illnesses can creep up on you suddenly. In addition, you may not realize that you are suffering from a heat-related illness, even after the effects begin to show themselves. The best response to the problem of heat-related illness is to prevent the problem before it starts by doing the following:

- *Drink plenty of water* – OSHA recommends that you drink 4 cups of water (32 ounces, or about 1 liter) every hour if the heat index is over 103°F (40°C). If the heat index is higher, adjust your water intake upward.
- *Avoid over-exertion* – You can avoid pushing yourself to limits that are unsafe during periods of high heat.
- *Wear light-colored, lightweight clothing* – You may have a work uniform or other clothing that you are required to wear, but whenever you have a choice, choose light-colored clothing during periods of high heat. Avoid heavy cloth, which will soak up perspiration and block cooling breezes.
- *Keep your head covered and your face in the shade* – You will be required to wear a hard hat, but you can choose one with a built-in sun visor,
- *Take frequent, short work breaks* – Instead of taking long breaks, take several shorter breaks. Whenever possible, take your breaks in the shade.

You and your co-workers should watch one another for signs of heat-related illness. Co-workers usually will be the first to notice a problem. If any of your co-workers appear to be suffering from a heat-related illness, immediately call 911 / seek medical assistance, and perform the following first-aid measures while waiting for help:

- Get the worker out of direct sunlight, and if possible, out of the heat.
- If the worker is alert and can drink, administer cool water. Avoid cold water and beverages with caffeine and alcohol. Don't let the worker gulp water.
- Remove tight or unnecessary clothing.

- Apply a compress soaked with cool water. The forehead, neck, armpits, wrists, and groin are particularly effective areas to apply compresses. Do NOT wrap the worker in wet towels; these will act as an insulator and keep the worker's body from bleeding off heat.
- Use a fan to create a flow of cooling air over the worker.

2.2.0 Cold-Related Illnesses

Extremely cold temperatures can lead to cold-related illnesses, but even relatively mild temperatures can cause cold-related illnesses if accompanied by wind and/or moisture such as rain. Cold weather, possibly accompanied by winds and rain, can lead to hypothermia and frostbite.

Hypothermia is a condition resulting from an abnormally low body temperature. Hypothermia cases typically begin with the victim shivering, and in some cases, a state of confusion. As the condition worsens, the victim may begin to lose muscle control or begin to be confused. These conditions can be accompanied by numbness and/or drowsiness. Symptoms continue to worsen as the victim's core temperature drops: heart rate, blood pressure, and respiratory rate slow, and the victim may have difficulty speaking or remembering. Death can follow shortly thereafter.

While hypothermia can lead to death, its effects on the body are relatively small when compared to the effects of frostbite. Frostbite is caused by water in tissues freezing, damaging the surrounding tissues. While a mild case of frostbite (sometimes called *frostnip*) make not cause long-lasting damage, completely developed frostbite can require amputation of limbs. Some of the symptoms of frostbite include the following:

- Tingling in the skin, followed by numbness
- Pain in the frostbitten area, which then disappears
- Blisters in the affected area
- Swelling and hardness in the affected area
- White, gray, or waxy yellow skin, indicating deep tissue damage

The nerve endings in a frostbitten area are not functioning, so victims may not be aware of the damage.

Did You Know?

Hypothermia is a relatively well-known term for describing a dangerously low body temperature. The opposite condition, a dangerously high body temperature such a heat exhaustion or heat stroke, is called *hyperthermia*.

2.2.1 Preventing and Treating Cold-Related Illnesses

Cold-weather illnesses are connected to the prevailing outside temperature, and possibly worsened by moisture such as high humidity, rain, or heavy perspiration. Follow these tips to help prevent these injuries:

- *Dress in layers* – Layers allow you to easily remove excess clothing if you start to build body heat as you work. Layer up again if your body temperature starts to drop.
- *Wear thermal underwear* – Thermal underwear is not only warmer, it also covers more of the body.
- *Wear wind- and water-repelling outer layers* – These layers will help reduce wind chill and the added danger of dampness against your body, which reduces core body temperature much faster.
- *Cover the face, ears, and hands* – The face, ears, and hands are particularly sensitive to frostbite. Keep them covered, dry, and warm. Lined, water- and wind-repellent gloves can keep your hands warm and still allow you to work.
- *Wear warm boots* – Keep your feet warm and dry as well. If working in wet or snowy conditions, carry an extra pair of socks. Wicking socks, worn under the regular socks against your skin, can help draw moisture away from your feet if they get wet.

As with heat-related illness, you and your co-workers should keep an eye on each other for signs of cold-related illness. If you see someone who appears to be suffering from hypothermia, call 911 / seek medical assistance as soon as possible, and perform the following first-aid measures while waiting for help:

- Get the worker to a warm area as soon as possible.

- Remove wet, frozen, or restrictive clothing, and anything that is binding, such as a watch, a ring, or a belt.
- Dress the worker in warm, dry clothing, or wrap them in a blanket.
- Administer warm (but not hot) liquids. Avoid alcohol.

- Check for signs of frostbite. If frostbite is found, seek immediate medical help.
- Keep the frostbitten area from becoming re-frozen. Do not apply heat to the frostbitten area, or massage the area.

Additional Resources

Heat Stress Quick Card™, *Protecting Workers from Heat Stress*, OSHA. Available at **www.osha.gov/pls/publications**

Cold Stress Quick Card™, *Protecting Workers from Cold Temperature Stress*, OSHA. Available at **www.osha.gov/pls/publications**

2.0.0 Section Review

1. A symptom of heat exhaustion is _____.
   a. heat stroke
   b. heat cramps
   c. hyperthermia syndrome
   d. hypothermia syndrome

2. If a worker has frostbite, you should help them by _____.
   a. removing wet clothing
   b. massaging the frostbitten area
   c. applying heat to the frostbitten area
   d. giving the victim hot liquids to drink

3.0.0 PRIMARY JOB SITE HAZARDS

Objective

Identify the four main causes of accidents on the job site.
  a. Explain the importance of fall protection.
  b. Explain pinch points and caught-between hazards.
  c. Explain struck-by hazards.
  d. Explain the hazards associated with electricity.

Trade Terms

Locked out: Of machinery or equipment, de-energized and removed from service during periods of maintenance or repair, rendering the equipment safe.

Machine guarding: The process of protecting an operator from the moving parts of a machine.

Personal fall arrest system (PFAS): A system designed to prevent fall injuries by securing a worker to a solid structure.

All hazards should be taken seriously. However, there are four categories of hazards that account for the majority of workplace injuries every year. These hazards, referred to by OSHA as the *Fatal Four*, are the following:

- Fall hazards
- Pinch points and caught-between hazards
- Struck-by hazards
- Electrocution hazards

3.1.0 Fall Hazards and Fall Protection

A fall hazard is defined as a potential fall of 6 ft. (2 m) or greater. The most common causes of fall hazards are the following:

- Unprotected sides, edges, and holes
- Improperly constructed walking/working surface
- Improper use of access equipment
- Failure to properly use a personal fall arrest system (PFAS)

There are several types of fall protection. Each of the following protective methods is meant to provide protection from one or more particular types of fall hazards:

- *Guard rails* – Guard rails must be constructed with a top rail that is 39 to 45 in. (99 to 114 cm) from the supporting surface. They must also have a mid-rail that is 18 to 24 in. (44 to 61 cm) from the supporting surface. Guard rails must have a smooth surface with no projecting ends and must be able to support 200 pounds (91 kg) in any direction. If personnel are working below, there must also be a toeboard installed, with a minimum 4 in. (10 cm) height. *Figure 19* shows a scaffold with guard rails and a toeboard.
- *Personal fall arrest system* – A personal fall arrest system (PFAS) is composed of a harness and a lanyard, along with a lifeline or anchorage (*Figure 20*). All the parts of the PFAS must be rated for a minimum of 5,000 lbs (2,268 kg). A PFAS must be designed so that there will not be a free fall of more than 6 ft. (2 m) or to allow you to contact a lower level. A PFAS must be inspected prior to each use.
- *Controlled access zones* – Controlled access zones must be protected by a rope, cable, or rail that is 42 in. (1.1 m) high and which must have a minimum breaking strength of 200 lbs (91 kg). It must also be flagged at 6 ft. (2 m) intervals, be located between 6 and 25 ft. (2 and 7.6 m) from the leading edge, and extend the entire length of the leading edge, tying into a guard-rail or wall on each end.
- *Housekeeping* – All work areas should be as clean and clear of any trip or slip hazards as possible.
- *Protection from falling objects* – Employees must wear hard hats at all times. Toeboards or other protective measures must be used when other employees are working below.

Fall protection is required whenever you will be working 6 ft. (2 m) or more above the ground. When working in an area where you might encounter fall hazards, you should do the following:

- Inspect your PFAS before each use, and always stay tied off.
- Make sure that holes are securely covered.
- Make sure that required guard rails are properly configured.
- Keep passageways clear from tools, equipment, and debris. When completing a work segment, clean the area as much as possible.

3.2.0 Pinch Points and Caught-Between Hazards

Pinch point hazards are also referred to as *caught-in hazards* or *caught-between hazards*. OSHA defines these hazards as the result of parts of the body be-

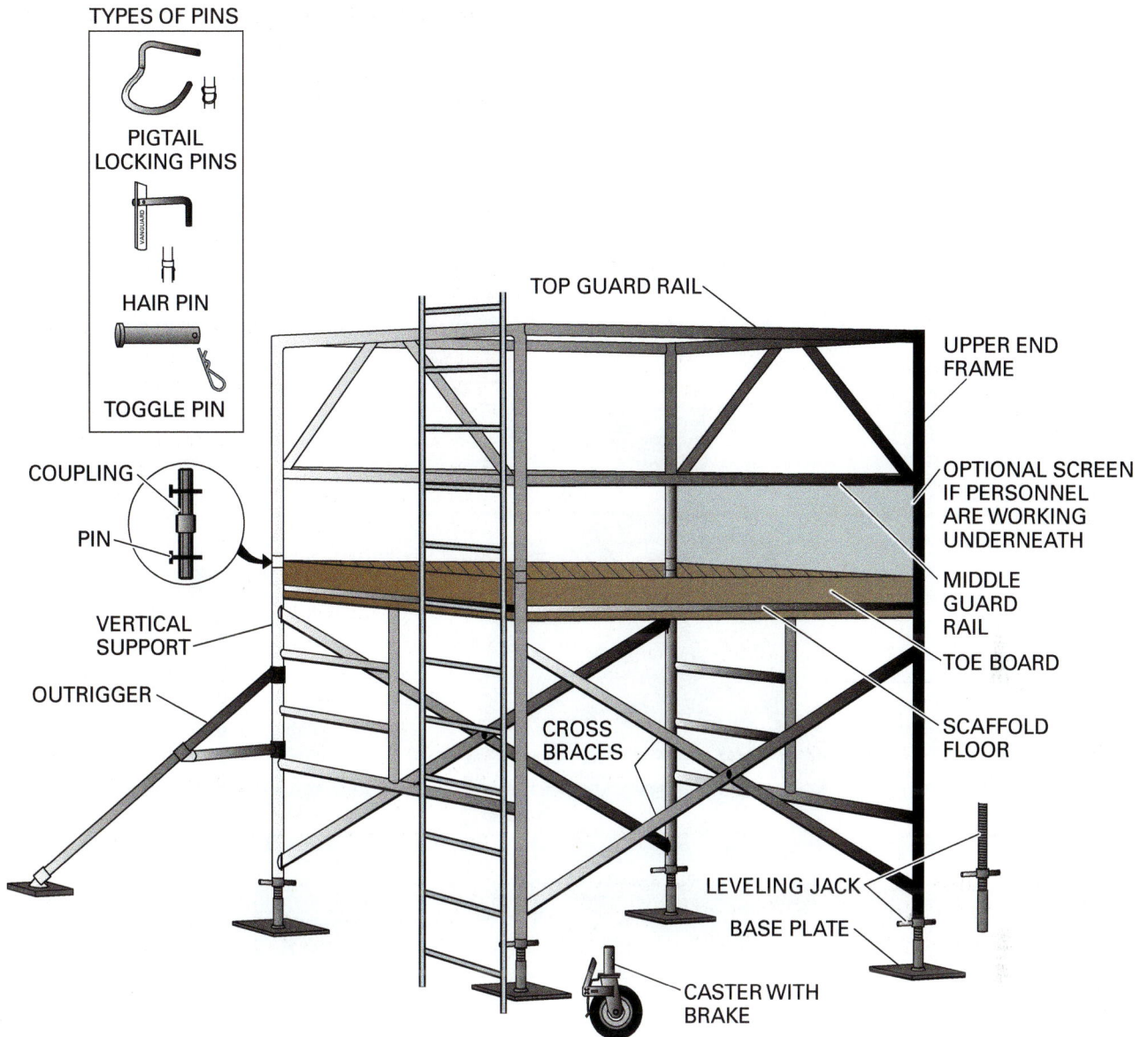

TYPES OF PINS

PIGTAIL LOCKING PINS

HAIR PIN

TOGGLE PIN

COUPLING

PIN

VERTICAL SUPPORT

OUTRIGGER

TOP GUARD RAIL

UPPER END FRAME

OPTIONAL SCREEN IF PERSONNEL ARE WORKING UNDERNEATH

MIDDLE GUARD RAIL

TOE BOARD

SCAFFOLD FLOOR

CROSS BRACES

LEVELING JACK

BASE PLATE

CASTER WITH BRAKE

Figure 19 A scaffold with guard rails and toeboard.

ing caught between two objects, either machines or parts of machines, or large stationary objects. Common causes of these hazards are the following:

- Rotating equipment
- Improper machine guarding
- Equipment rollovers
- Improper maintenance procedures
- Trench collapse

Almost all sites use machinery that has moving or rotating parts or that requires maintenance or repair at some point during construction. If machinery is not properly guarded, or is not de-energized during maintenance or repair, injuries from caught-in or caught-between hazards may result, ranging from amputations and fractures to death. When machines or power tools are not properly guarded, workers can get their clothing or parts of their bodies caught in the machines. If machines are not de-energized and secured (locked out) when they are being repaired, they may cycle or otherwise start up and catch a worker's body part or clothing, causing injury or death. Finally, workers can be trapped and crushed under equipment that tips.

To avoid pinch point hazards, observe the following safety precautions:

- Use only properly guarded machines. Never remove a guard for any reason.
- Avoid wearing loose clothing and jewelry.
- When performing maintenance, make sure the equipment is de-energized and cannot be
- started accidently. Follow lockout/tagout procedures.

PERSONAL FALL ARREST SYSTEM

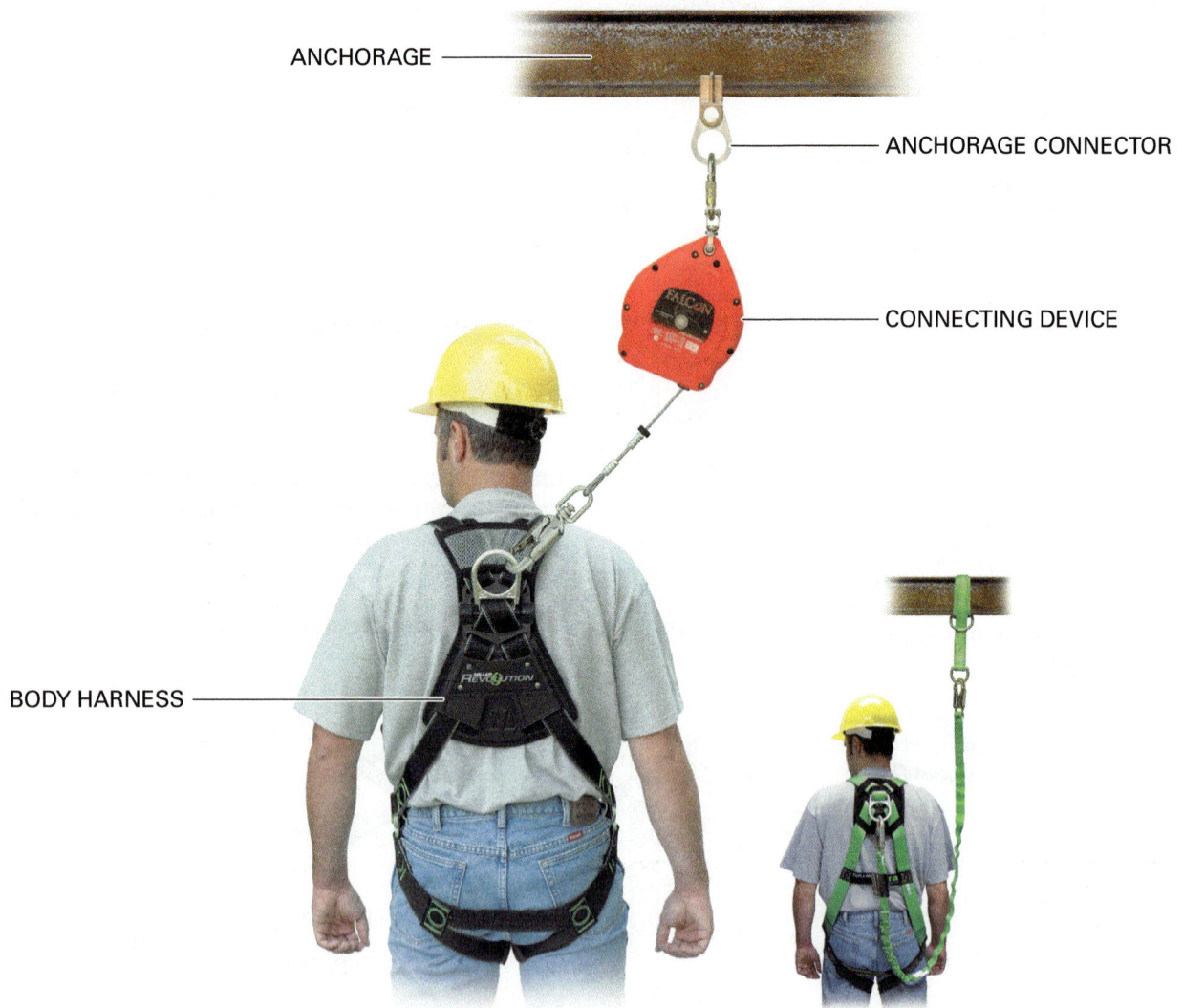

ANCHORAGE

ANCHORAGE CONNECTOR

CONNECTING DEVICE

BODY HARNESS

Figure 20 A PFAS.

- Be aware at all times of the equipment around you and stay a safe distance from it.
- Never place yourself between moving materials and an immovable structure, vehicle, equipment, or stacked materials.
- Stay out of the swing radius of cranes and other equipment.
- Wear a seatbelt.
- Do not work in an unprotected trench that is 5 ft. (1.5 m) deep or more.

3.3.0 Struck-By Hazards

Struck-by hazards are the result of an impact on the victim by an object or a piece of equipment. Struck-by hazards can be caused by the following:

- *Flying objects* – An object propelled across a space, either purposefully or accidentally. Note that an object propelled across a space because of contact with a rapidly moving machine or machine part is considered a flying object.
- *Falling objects* – An object that falls from a higher level and strikes a victim at a lower level. Tool lanyards should be used when working at heights.
- *Swinging objects* – An impact by a mechanically lifted load that is not properly controlled.
- *Rolling objects* – Impact on a victim by a moving vehicle, or by a sliding or rolling object or equipment.

Struck-by hazards have the potential to be deadly. To avoid struck-by incidents, follow these guidelines:

- Stay clear of lifted loads. Never work under a suspended load.
- Beware of unbalanced loads.

- Make sure that all workers and other personnel are in the clear before using dumping or lifting devices.
- Confirm and receive acknowledgement from nearby heavy equipment operators that you are visible to them.
- If people will be working below you, use toeboards or screens, as well as tool lanyards, and secure all tools and materials.
- Wear safety glasses all the time and face shields when required.

> **NOTE**
>
> Additional safety procedures may be required when working with compressed air or hydraulic equipment.

3.4.0 Electrocution

Electrocution hazards are caused by contact with improperly handled electricity. The acronym BE SAFE stands for the types of hazards associated with electricity.

- *Burns* – The most common electrical-related injury, caused by the heat of current flow.
- *Electrocution* – Death resulting from current flow passing through the body. The most common result of electrocution is immediate heart failure.
- *Shock* – Caused by the entry of electricity at one point on the body and the exit through another point. In other words, the body completes an electrical circuit.
- *Arc flash* – The sudden release of electrical energy through the air at a high-voltage gap between conductors. Temperatures of arc flashes have been recorded as high as 35,000°F (19,427°C). High-voltage arcs can also produce considerable pressure waves by rapidly heating the air and creating a blast (*Figure 21*).
- *Fire* – The heat generated by providing an unintended current path can cause fires.
- *Explosions* – Explosive mixtures in the air can be ignited by an arc.

Electrocution hazards can have several causes. Some of the most common are the following:

- Contact with overhead power lines
- Contact with energized sources such as damaged or bare wires, or defective equipment
- Improper use of extension cords
- Contact with underground electrical lines

To avoid electrocution and electrical accidents, use the following safety precautions:

- Follow safe procedures for working around power lines.
    - Do not work closer than 10 ft. (3 m) from overhead power lines carrying 50kV.
    - Call 811 to locate buried power lines before digging.

- Use ground fault circuit interrupters (GFCIs) and test them before each use.
- Inspect tool electrical cords and extension cables before each use. Replace damaged equipment.
- Follow site lockout/tagout procedures when working around electrical equipment. Lockout/tagout procedures are designed to prevent the unwanted release of energy.
- Do not use electrical tools in wet, damp, or explosive environments. Store electrical tools in a dry place when not in use.

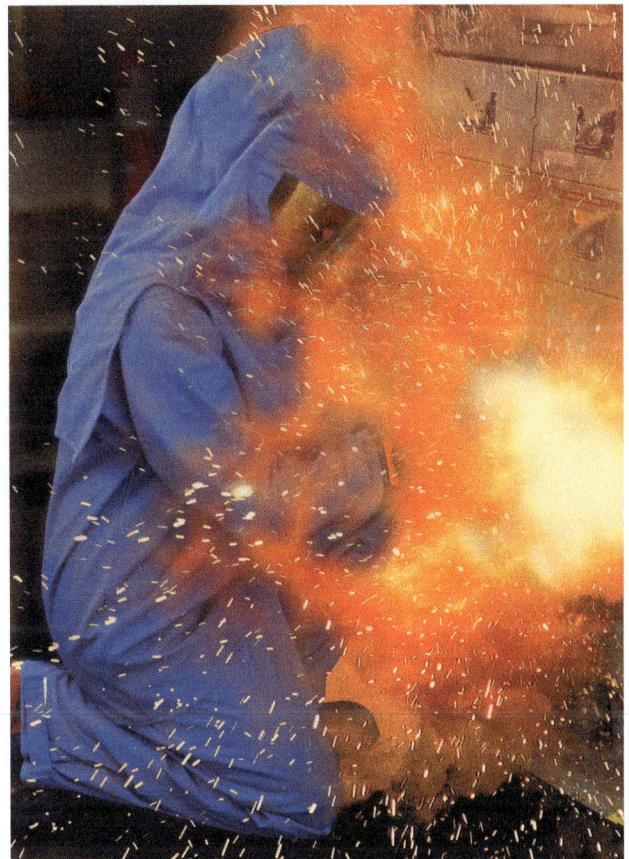

Figure 21 An arc flash.

Additional Resources

29 *CFR* 1910.212, **www.ecfr.gov**

OSHA 3416-05R 2015, *Fall Protection in Construction* (PDF), OSHA. Available at **www.osha.gov/ Publications/OSHA3146.pdf**

3.0.0 Section Review

1. What is the minimum height of a toeboard?
   a. 1 in. (2.5 cm)
   b. 2 in. (5 cm)
   c. 4 in. (10 cm)
   d. 6 in. (15 cm)

2. If you are working around a machine with a pinch hazard, which of the following should you do?
   a. Make sure the machine is correctly guarded.
   b. Remove any clothing with a loose fit.
   c. Use scrap material to block the pinch point.
   d. Keep one hand on the power switch.

3. A mechanically lifted load that is not controlled would be considered a(n) _____.
   a. falling hazard
   b. flying hazard
   c. swinging hazard
   d. unstable hazard

4. If a power cable is determined to be faulty, it must be _____.
   a. tagged
   b. tested
   c. repaired
   d. replaced

SECTION FOUR

4.0.0 TRAFFIC CONTROL AND FLAGGING

Objective

Explain traffic zones and the importance of flagging.

   a. Explain traffic zones.
   b. Explain flagging.

Trade Terms

Flaggers: Workers who are specially trained to direct traffic through and around a work zone.

Temporary traffic control (TTC): Traffic controls enacted when the normal use of a road is disrupted by construction work.

Construction sites on or near public roads present additional hazards to both construction workers and motorists using the roads. Dangers to construction workers working near moving traffic are so great that all states have enacted laws that provide for additional fines for speeders and other violators of traffic laws in work zones.

4.1.0 Temporary Traffic Control

When the normal use of a road is disrupted, a temporary traffic control (TTC) zone must be established. A traffic control plan is a key element of this process. Specially trained workers design a temporary traffic flow to help ensure the safety of both construction workers and motorists. Using TTC aids such as barricades, cones, and signs, they design a new traffic flow with provisions to maintain safety.

A TTC must be designed to allow for expected changes in traffic and changes in work. Different lanes might be closed on different days, for instance, resulting in different traffic patterns. Flaggers may need to stop traffic for trucks and equipment entering and leaving the highway.

In addition to providing safety zones and helping to protect workers and motorists, a TTC must also allow workers to complete their work quickly and efficiently, including moving necessary construction equipment around the construction zone. Finally, TTCs must be designed to allow accident responders such as police and rescue workers to maneuver to areas where they are needed.

4.1.1 Elements of a TTC

While each TTC is unique, there are elements of TTC design that are common and are required for the TTC to function properly. You may not be involved in designing a TTC, but understanding the function of each TTC element will help you stay safe as well as help you support the operation of the TTC.

TTCs are composed of an advance warning area, a transition area, a longitudinal buffer zone, a lateral buffer zone, the construction workspace, and a termination zone. *Figure 22* shows the elements of the TTC, which are described as follows:

* *Advance warning area* – The advance warning area is where warning signs prepare motorists for the upcoming TTC.
* *Transition area* – The transition area is used to gradually move traffic from its normal flow and into the desired path.
* *Longitudinal buffer zone* – The longitudinal buffer zone is the beginning of the actual protection for workers and motorists. The longitudinal buffer must have some type of barrier to mark its border. If there is no room for any other type of barrier, the longitudinal transition zone can be marked with one or more truck-mounted attenuators (*Figure 23*). The longitudinal transition zone is divided into a buffer area and the workspace.
* *Lateral buffer zone* – The lateral buffer is next to the border. It is designed to protect workers from closely passing vehicles. Buffer zones must be clear areas to give workers more reaction time in case a vehicle breaks through a barrier.
* *Workspace* – The workspace is for workers, equipment, and material only. You should not leave the workspace, and motorists should not be in the workspace.
* *Termination area* – The final area of the TTC is the termination area. This is where construction ends. Motorists are channeled back to the regular traffic flow, and signage lets them know that they have left the construction area.

4.2.0 Construction Flagging

The term *flagging* applies to both the equipment used to signal motorists within a construction zone and the personnel who are trained to perform the signaling.

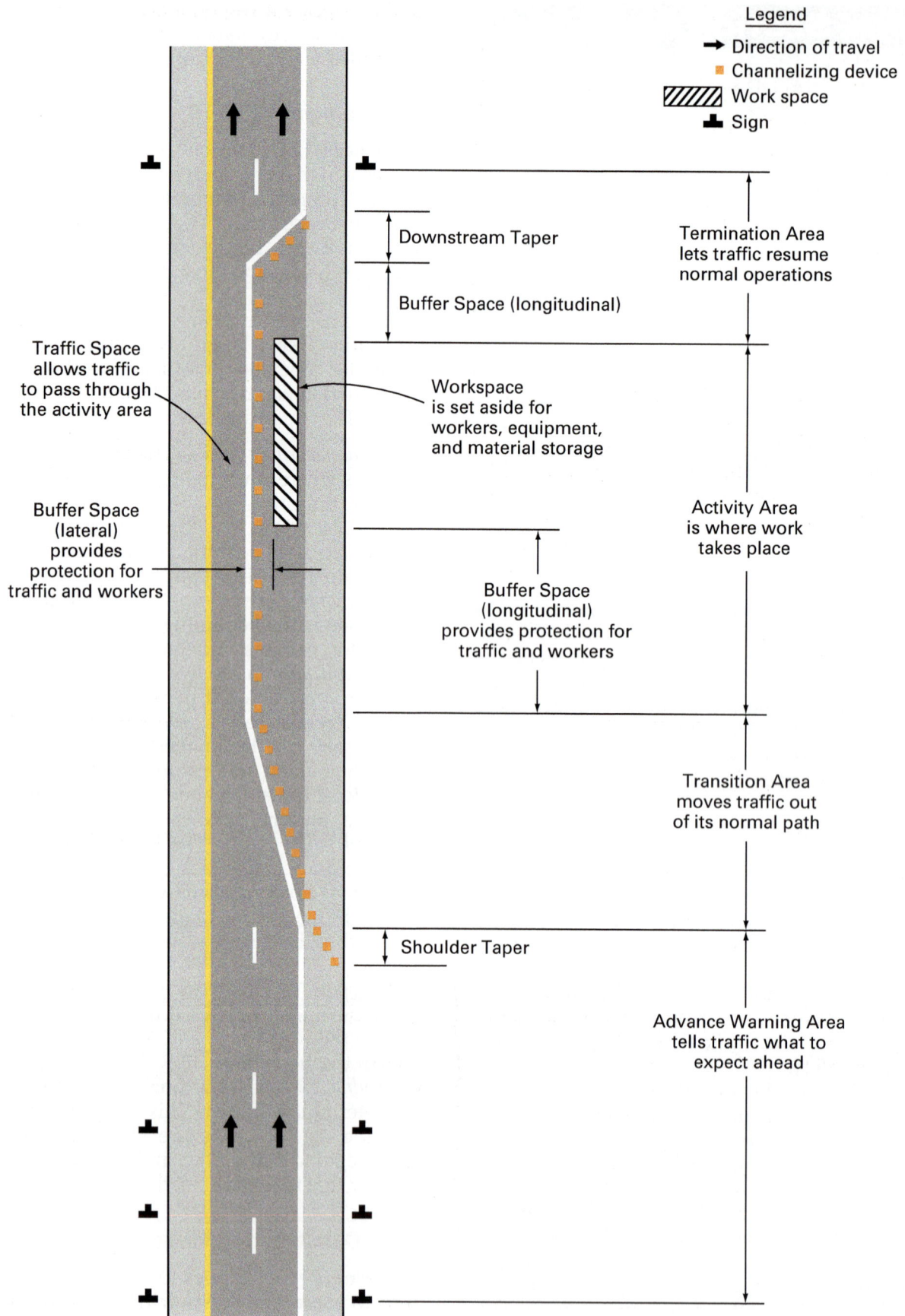

Legend
→ Direction of travel
■ Channelizing device
▨ Work space
⊥ Sign

Downstream Taper

Buffer Space (longitudinal)

Termination Area lets traffic resume normal operations

Traffic Space allows traffic to pass through the activity area

Workspace is set aside for workers, equipment, and material storage

Activity Area is where work takes place

Buffer Space (lateral) provides protection for traffic and workers

Buffer Space (longitudinal) provides protection for traffic and workers

Transition Area moves traffic out of its normal path

Shoulder Taper

Advance Warning Area tells traffic what to expect ahead

Figure 22 A typical TTC zone.

NCCER – *Heavy Highway Construction Level One* 36110-17

Figure 23 Truck-mounted traffic attenuator.

4.2.1 Flaggers

Flaggers are part of the construction team. However, because flaggers are responsible for public safety, they must receive special training in traffic control practices. In some jurisdictions, flaggers must be certified. On most jobs, flaggers will be dedicated, but it is not uncommon for other team members to perform the flagger role, especially in an emergency. A flagger must be able to do the following:

- Give specific instructions clearly, firmly, and politely
- Move quickly to avoid danger
- Use signaling devices, such as paddles and flags, to provide clear direction to drivers
- Apply safe traffic control practices, sometimes in stressful and/or emergency conditions
- Identify unsafe traffic conditions and warn workers in enough time to avoid injury

> **NOTE**
>
> Flaggers must wear high-visibility safety apparel. According to the US Department of Transportations's *Manual on Uniform Traffic Control Devices (MUTCD)*, Section 6E.02, Class 2 material is required for daytime work, and Class 3 is needed at night. Class 3 can be used for daytime work as well. The background must be fluorescent orange-red, fluorescent yellow-green, or a combination of the two as defined by the American National Standards Institute (ANSI). The retroreflective material must be orange, yellow, white, silver, yellow-green, or a fluorescent version of these colors. The material must be visible from a distance of 1,000 ft. (305 m).

4.2.2 Traffic Control Devices

Traffic control devices are the components of a TTC, such as signs and barricades, which are used to direct traffic safely through the work zone. Traffic control devices must meet the following five basic requirements in order to be useful in a TTC:

- Fulfill a need
- Gain motorists' attention
- Have a clear and simple meaning
- Command respect from motorists
- Give motorists enough time to respond

Motorists will likely encounter traffic control devices before flaggers in a TTC. There are a variety of these devices, including the following:

- Warning signs
- Merge signs (static or animated)
- Channeling devices

  - Drums
  - Barricades
  - Vertical panels
  - Tubular markers
  - Cones

Figure 24 shows channeling devices in use.

When motorists encounter a flagger, the flagger will typically be using either a Stop/Slow paddle or a flag for additional traffic control. Paddles are mounted high on poles and allow the flagger to rapidly flip the paddle from stop to slow. Flags are red cloth warning signs, usually without text and made of nylon. Flags are acceptable traffic controls, but paddles are preferable.

Figure 24 Examples of channeling devices.

Stop/Slow paddles are octagonal. The stop side looks just like a regular stop sign. The slow side has an orange background and black letters. When used at night, the paddle must be made of material that reflects light. A Stop/Slow paddle can also have flashing lights, in either red or white on the stop side and in either yellow or white on the slow side.

Flags must be at least 24 in. (155 cm) square and made of a durable red material. The free edge of a flag must be weighted so the material will hang vertically, even in heavy winds. Flags used at night must be made of red retro-reflective material.

When using a paddle, it should be held still, with the flagger's arm extended horizontally away from the body. The flagger's free hand is used for additional traffic direction:

- Above the shoulder with palm facing oncoming traffic to signal the traffic to stop
- Using a beckoning motion to communicate that traffic may proceed
- Moving slowly up and down with the palm facing the ground to signal traffic to proceed slowly

When using a flag, the flagger must face oncoming traffic, with the flag in the hand closest to the traffic. To stop traffic, the flagger extends the flag horizontally across the motorists' lane so that the full area of the flag is visibly hanging below the staff, with the free arm making the same motion as with a Stop/Slow paddle. To release stopped traffic, the flag is hung down at the flagger's side, with the free arm motion also the same as for a Stop/Slow paddle. To slow traffic with a flag, the flagger stands with the free hand down and slowly waves the flag up and down without raising the flag above shoulder level. Note that motorists often mistake the direction to slow down as a direction to stop. Be prepared to correct them with additional signals.

Figure 25 shows typical uses of a Stop/Slow paddle, a flag, and the flagger's free arm and hand when directing traffic.

NOTE

Traffic is generally lighter at night, so highway construction and repaving work is frequently performed during nighttime hours. It is common to close one lane of a road for work activities, while motorists use the other lane. Work sites must be well lighted so that flaggers and workers are clearly visible to drivers. While providing the necessary light, the lighting must not interfere with the vision of drivers on the highway.

18"
MIN

STOP

36"

24"

24"

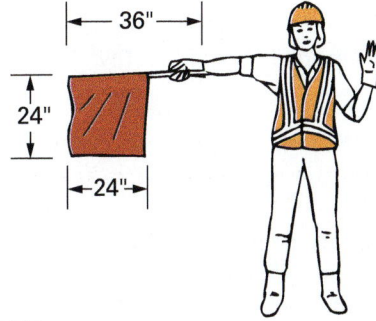

TO STOP TRAFFIC

SLOW

TO LET TRAFFIC PROCEED

SLOW

TO ALERT AND SLOW TRAFFIC

Figure 25 Flagger signals.

Additional Resources

Defensive Flagging: A Survivor's Guide, Texas Engineering Extension Service. 2011. College Station, TX: Texas A&M University.

Manual on Uniform Traffic Control Devices (MUTCD), US Department of Transportation Federal Highway Administration. Available at **www.mutcd.fhwa.dot.gov**

Flagger Illumination during Nighttime Construction and Maintenance Operations, John A. Gambatese. 2012. Reston, VA: American Society of Civil Engineers. Available at **www.ascelibrary.org**

4.0.0 Section Review

1. TTCs must allow _____.
   a. space for an emergency vehicle to maneuver around traffic inside the workspace
   b. space for a disabled vehicle inside the work area
   c. traffic to be rerouted quickly if a task is completed
   d. for workers to park their own cars

2. Which of the following is a primary traffic control tool used by a flagger?
   a. Go/Don't Go paddle
   b. Stop/Go paddle
   c. Stop/Slow paddle
   d. Double-sided stop sign

SECTION FIVE

5.0.0 TRENCH SAFETY

Objective

Describe the safety hazards associated with trenches.

  a. Identify trench-related hazards and the safety requirements associated with trenches.

  b. Explain how to work safely in and around trenches.

Trade Terms

Spoil: Soil removed from a trench.

Shoring: Structures used to brace the sides of a trench.

Trench box: A structure placed into a trench to brace the trench walls and prevent cave-ins.

Trenches are typically created to create footings, lay pipe, or place manholes. In an average year, more than 30 construction workers die in trenches. Hazards around trenches include the following:

- Collapse from unstable soil
- Falls
- Struck-by hazards from falling objects or equipment

> **WARNING!**
> Workers sometimes die while trying to rescue victims of a trench collapse. Be sure to follow site safety practices to help victims of trench-related injuries.

5.1.0 Trench Hazards

In the case of trenches, the primary danger is collapse, or cave-in, of the trench walls. There is also a danger from other gases displacing oxygen, and a danger of striking utilities buried below the surface. No mechanical digging is permitted within 24 in. of buried utilities.

Before digging a trench, be sure to notify the One-Call System by calling 811, or contact the designated private utility locator and make sure that all utility lines in the area have been marked. The 811 number is a national dig safely service that contractors can contact for help in locating and marking buried utilities. The One-Call System will contact the appropriate utility companies. It is usually necessary to give at least two days notice.

A minimum distance of 10 ft. (3 m) from 50 kilowatt power lines must be maintained, with an additional 4 in. (10 cm) for every additional 10,000 kilowatts.

A trench can be in stable rock, or in type A, type B, or type C soil, in increasing order of hazard. Stable rock is the safest, followed by Type A soil. Most soils are Type B. Type C soils are sandy, or moist. Clays can be Type A, B, or C, depending on the moisture content.

> **WARNING!**
> Many trench cave-ins happen in clay because the clay looks solid. Always make sure that the trench is verified safe by the responsible competent person. When a trench in clay soil fails, it is likely to collapse in large slabs.

A trench that is more than 5 ft. (1.5 m) deep must have some form of protection such as a **trench box** (*Figure 26*) to protect workers against

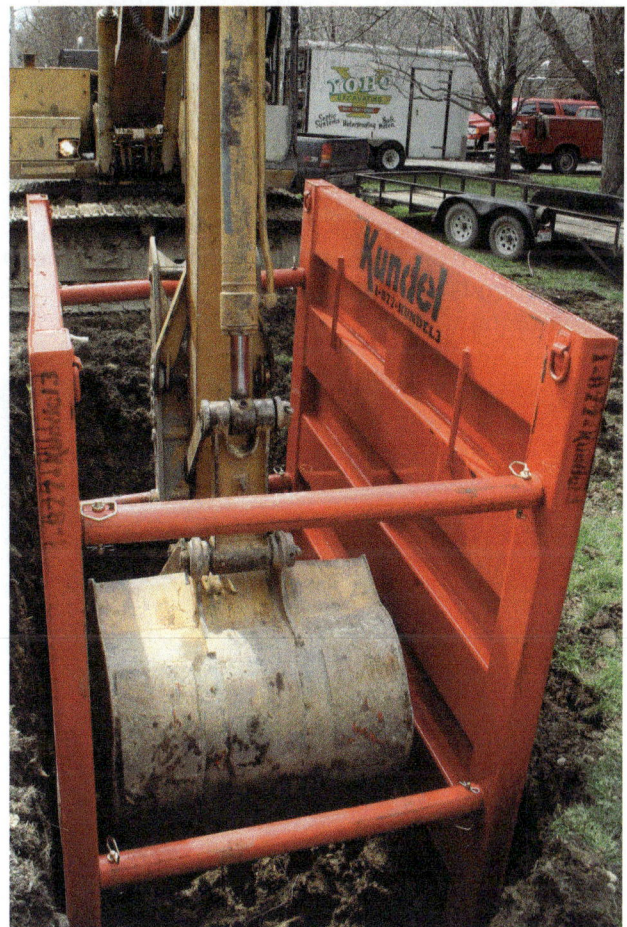

Figure 26 Trench box.

---

cave-ins. A trench box is a box-like structure placed in an excavation and backfilled. If a trench box is not used, the trench must be sloped, benched, or shored to help prevent injury and death. Shoring (*Figure 27*) is created by supporting the walls of an excavation to prevent their movement and collapse.

5.2.0 Trench Safety Precautions

OSHA standards require that any employer digging a trench must designate a competent person who will be responsible for safety in that trench. A competent person is trained or otherwise knowledgeable about soil types and stability, as well as safety practices and equipment to be used in and around trenches. The competent person must inspect the trench:

- After initial construction.
- Before every shift.
- Whenever trench conditions have changed, including after a rainfall or after the trench is subjected to significant vibration, such as a passing train or the operation of nearby heavy equipment.
- After any vehicle near the trench that hauls material has been loaded, and thus becomes heavier.
- After anything occurs that can increase the risk of being in the trench, including changes to the trench wall such as cracking, scaling, or bulging.

A trench in solid rock does not require inspection if the trench is less than 5 ft. (1.5 m) deep. Inspection is only required after the initial trench construction.

Figure 27 Trench shoring.

OSHA requires that your employer identify the competent person. If you have any doubt who the competent person is, ask your employer. Before working in a trench it is recommended that a JSA be developed containing the following information:

- Verify that the competent person has inspected the trench and approved it before each shift.
- Verify that all equipment necessary for working in the trench, such as water pumps and ventilators, is in good working condition.
- Verify that there is a way to exit the trench. This is typically a ladder. If the trench is 4 ft. (1.2 m) or more, there must be one ladder for every 25 ft. (7.6 m) of trench. Ladders must extend out of the trench at least 3 ft. (1 m).
- Have a competent person test the air in the trench if bad air is expected.
- Have a competent person verify that the trench is safe if water has accumulated in the trench.
- Verify that you are using the worker-protection system(s) that have been identified by the competent person.

Working in a trench requires caution. After the trench is approved for work, you should keep the spoil at least 2 ft. (0.6 m) from the edge of the trench. Additionally, all other materials (such as rocks or other soils) should be separated from the trench, using barriers if necessary. Avoid working above or below co-workers on sloped or benched trenches. Tools and equipment must be kept at least 4' (1.2 m) away from the edge of the trench.

5.2.1 Emergency Response to Trench Accidents

If a trench collapses, evacuate the area and call 911 or emergency services. Do not attempt any rescues without first consulting a competent person.

If a trench has partially collapsed, do not enter the trench. Wait until the trench has been rebuilt if necessary and inspected by a competent person.

If a trench has bad air, do not enter it, even to help co-workers. You can be quickly overcome, adding to the severity of the accident.

5.0.0 Section Review

1. Before digging a trench, utility companies must be contacted so the utility lines in the trench can be _____.

   a. located
   b. shut down
   c. evaluated
   d. replaced

2. Ladders must extend out of a trench at least _____.

   a. 1 ft. (0.3 m)
   b. 2 ft. (0.6 m)
   c. 3 ft. (1 m)
   d. 4 ft. (1.3 m)

SUMMARY

Highway construction presents several threats to life and health. Heavy equipment is nearly everywhere on a job site, and the equipment operators sometimes have trouble seeing workers near their vehicles. Workers perform their jobs near heavy loads being moved overhead, or in trenches dug for subterranean access. Even the weather can be a threat, especially extremes of heat and cold. Highway and bridge workers may encounter confined space hazards in trenches, cofferdams, manholes, and caissons and must know how to avoid these hazards.

According to OSHA rules, job site hazards must be controlled. Controls can be as simple as a warning sign, or more complex, such as lock-out and machine guarding systems. While signs, labels, and guards can help protect you, ultimately you must be aware of the threats around you and take steps to protect yourself. The main root cause of accidents at a construction site is a lapse of attention to safety. There are four main categories that represent the majority of fatal accidents on construction job sites: falls, struck-by hazards, caught-between hazards, and electrocution. Keep those categories in mind when you approach each new site and each new work area. Workers must be especially cautions when working in trenches, as the sides can collapse into the trench if proper protection is not used.

1. When working around heavy equipment you should _____.
   a. verify that the operator is aware of your presence
   b. verify that the vehicle is not operating within 6 ft. (2 m) of you
   c. verify that hoisted loads will be more than 6 ft. (2 m) overhead
   d. wear hearing protection

2. Which of the following is a safety concern about some heavy equipment?
   a. Forklifts are not designed to lift over 10 in. (25 cm).
   b. Metal tracks are prone to slipping.
   c. Larger tires lead to instability.
   d. Ground conditions can cause instability.

3. A higher center of gravity leads to _____.
   e. a less stable vehicle
   f. a more stable vehicle
   g. faster load movement
   h. greater load capacity

4. When working around water, which of the following must be available?
   a. A 30-inch (76 cm) life ring with 90 ft. (27 m) of line
   b. A 30-inch (76 cm) life ring with 120 ft. (36 m) of line
   c. A 36-inch (91 cm) life ring with 90 ft. (27 m) of line
   d. A 36-inch (91 cm) life ring with 120 ft. (36 m) of line

5. A common type of hazard in a confined space is _____.
   a. atmospheric
   b. electric
   c. temperature
   d. noise

6. The purpose of a job safety analysis is to _____.
   a. determine the cause of an accident
   b. report an accident
   c. support an insurance claim
   d. recognize potential hazards

7. Which of the following temperature-related illnesses requires immediate emergency medical care?
   a. Frostnip
   b. Heat exhaustion
   c. Heat stroke
   d. Heat failure

8. The term hypothermia describes _____.
   a. abnormally high body temperature
   b. abnormally low body temperature
   c. a type of cold-weather clothing
   d. dehydration caused by excessive sweating

9. Which of the following methods is a way to avoid cold-related illness?
   a. Wear light-weight clothing and stay active.
   b. Wear heavy clothing.
   c. Dress in layers.
   d. Drink hot liquids.

10. Guard rails must be able to support _____.
    a. 100 pounds (45 kg) horizontally, 200 pounds (90 kg) vertically
    b. 200 pounds (90 kg) pressure, 100 pounds (45 kg) tension
    c. 200 pounds (90 kg) horizontally, 100 pounds (45 kg) vertically
    d. 200 pounds (90 kg) in any direction

11. A personal fall arrest system should be _____.
    a. fluorescent red or fluorescent green
    b. inspected before each use
    c. custom-fitted to the user
    d. stored in a warm, dry area

12. Workers must not enter an unprotected trench that is _____.
    a. cut into hardpack
    b. more than 5 feet deep
    c. freshly cut
    d. unmarked

13. Which of these would be classified as a caught-between hazard?

    a. Being struck by a falling object.
    b. Being crushed by rolling machinery.
    c. Being pinched by an unguarded machine.
    d. Contact with a live power line.

14. When working around electrical lines carrying 50kV, you should maintain a distance of at least _____.

    a. 1 ft. (0.3 m)
    b. 5 ft. (1.5 m)
    c. 8 ft. (2.4 m)
    d. 10 ft. (3 m)

15. Temporary traffic control zones are used when _____.

    a. the normal use of a road is disrupted
    b. traffic speeds exceed 55 mph
    c. traffic exceeds 1,000 vehicles per hour
    d. construction will last more than 30 days

16. When working at night, flaggers must wear high-visibility safety apparel that is Class _____.

    a. 1
    b. 2
    c. 3
    d. 4

17. In order to be effective, a traffic control device must be able to _____.

    a. be rapidly located
    b. be broken down easily
    c. be used day and night
    d. attract the attention of motorists

18. Before trenching work is started, utility lines in the vicinity of the trench must be _____.

    a. removed
    b. marked
    c. measured
    d. shut down

19. A competent person must re-inspect a trench _____.

    a. after the temperature drops below 32°F (0°C)
    b. after every shift
    c. before every shift
    d. after the trench has been re-filled

20. How far from the edge of a trench must the spoil be kept?

    a. At least 1 ft. (0.3 m)
    b. At least 2 ft. (0.6 m)
    c. At least 5 ft. (1.5 m)
    d. At least 10 ft. (3 m)

Trade Terms Quiz

Fill in the blank with the correct term that you learned from your study of this module.

1. _____ control traffic in a construction zone.

2. A(n) _____ must inspect a trench before you are allowed to work in it.

3. If you will be working more than 6 ft. above the ground you will need a(n) _____.

4. A(n) _____ zone is a systematically designed area of traffic re-routed due to construction.

5. When a trench is being dug, the _____ must be kept at least 2 ft. (0.6 m) away from the trench.

6. When building a bridge piling in a river, you will need a(n) _____ to keep the work area dry.

7. Some soils require a(n) _____ be placed in a trench before you can work in it.

8. Never perform maintenance on a machine before it has been _____.

9. If you will be working over water you will need a(n) _____.

10. The principles of _____ require pinch points to be covered so that workers do not accidentally become caught up in a machine.

11. Watertight chambers commonly used in building or repairing bridge pilings are called _____.

12. A substance that readily ignites and burns is considered _____.

13. A substance that is capable of burning is classified as _____.

14. Alternately stacked timbers used to support heavy loads is called _____.

15. Drains or channels that cross under a road are known as _____.

16. Structures used to brace the sides of a trench are called _____.

17. A value that represents the combination of temperature and humidity is known as the _____.

Trade Terms

Caissons
Cofferdam
Combustible
Competent person
Cribbing
Culverts

Flaggers
Flammable
Heat index
Locked out
Machine guarding
Personal fall arrest system (PFAS)

Personal flotation device (PFD)
Shoring
Spoil
Temporary traffic control (TTC)
Trench box

Trade Terms Introduced in This Module

Caissons: Watertight chambers commonly used in building or repairing bridge pilings.

Cofferdam: A structure built to keep water away from a construction area.

Combustible: A substance that readily ignites and burns.

Competent person: As defined by OSHA, an individual who is capable of identifying existing and predictable hazards in the surroundings or working conditions which are unsanitary, hazardous, or dangerous to employees, and who has the authorization to take prompt corrective measures to eliminate such hazards. In the context of trenches, a person with knowledge of soil types and soil stability who will inspect the trench whenever required.

Cribbing: Alternately stacked timbers used to support heavy loads.

Culverts: Drains or channels that cross under a road.

Flaggers: Workers who are specially trained to direct traffic through and around a work zone.

Flammable: Capable of burning.

Heat index: A value that combines temperature and humidity to establish an equivalent value that represents the effect of humidity on perceived temperature.

Locked out: Of machinery or equipment, de-energized and removed from service during periods of maintenance or repair, rendering the equipment safe.

Machine guarding: The process of protecting an operator from the moving parts of a machine.

Personal fall arrest system (PFAS): A system designed to prevent fall injuries by securing a worker to a solid structure.

Personal flotation device (PFD): A sleeveless jacket composed of a buoyant or inflatable material and used to prevent drowning; commonly called a *life vest*.

Shoring: Structures used to brace the sides of a trench.

Spoil: Soil removed from a trench.

Temporary traffic control (TTC): Traffic controls enacted when the normal use of a road is disrupted by construction work.

Trench box: A structure placed into a trench to brace the trench walls and prevent cave-ins.

Additional Resources

This module presents thorough resources for task training. The following reference material is recommended for further study.

Building Safer Highway Work Zones: Measures to Prevent Worker Injuries from Vehicles and Equipment (PDF), US Department of Health and Human Services. 2001. Available at **www.cdc.gov/niosh/docs/2001-128**

29 *CFR* 1926.146, **www.ecfr.gov**

29 *CFR* 1926.200, **www.ecfr.gov**

29 *CFR* 1910.212, **www.ecfr.gov**

OSHA 3071, *Job Hazard Analysis Guide*, OSHA. Available at **www.osha.gov**

OSHA 3416-05R 2015, *Fall Protection in Construction* (PDF), OSHA. Available at **www.osha.gov/Publications/OSHA3146.pdf**

OSHA Fact Sheet, *Trenching and Excavation Safety*, OSHA. Available at **www.osha.gov/Publications/trench_excavation_fs.html**

Heat Stress Quick Card™, *Protecting Workers from Heat Stress*, OSHA. Available at **www.osha.gov/pls/publications**

Cold Stress Quick Card™, *Protecting Workers from Cold Temperature Stress*, OSHA. Available at **www.osha.gov/pls/publications**

Work-Related Roadway Crashes: Challenges and Opportunities for Prevention, US Department of Health and Human Services. 2003. Available at **www.cdc.gov/niosh/docs/2003-119**

Defensive Flagging: A Survivor's Guide, Texas Engineering Extension Service. 2011. College Station, TX: Texas A&M University.

Manual on Uniform Traffic Control Devices (MUTCD), US Department of Transportation Federal Highway Administration. Available at **www.mutcd.fhwa.dot.gov**

Flagger Illumination during Nighttime Construction and Maintenance Operations, John A. Gambatese. 2012. Reston, VA: American Society of Civil Engineers. Available at **www.ascelibrary.org**

Figure Credits

Section Review Answer Key

Answer	Section Reference	Objective
Section One		
1. b	1.1.1	1a
2. a	1.2.1	1b
3. a	1.3.2	1c
4. b	1.4.3	1d
5. c	1.5.0	1e
Section Two		
1. b	2.1.0	2a
2. a	2.2.1	2b
Section Three		
1. c	3.1.0	3a
2. a	3.2.0	3b
3. c	3.3.0	3c
4. d	3.4.0	3d
Section Four		
1. d	4.1.0	4a
2. c	4.2.2	4b
Section Five		
1. a	5.1.0	5a
2. c	5.2.0	5b

NCCER CURRICULA — USER UPDATE

NCCER makes every effort to keep its textbooks up-to-date and free of technical errors. We appreciate your help in this process. If you find an error, a typographical mistake, or an inaccuracy in NCCER's curricula, please fill out this form (or a photocopy), or complete the online form at **www.nccer.org/olf**. Be sure to include the exact module ID number, page number, a detailed description, and your recommended correction. Your input will be brought to the attention of the Authoring Team. Thank you for your assistance.

Instructors – If you have an idea for improving this textbook, or have found that additional materials were necessary to teach this module effectively, please let us know so that we may present your suggestions to the Authoring Team.

NCCER Product Development and Revision

13614 Progress Blvd., Alachua, FL 32615

Email: curriculum@nccer.org
Online: www.nccer.org/olf

❏ Trainee Guide ❏ Lesson Plans ❏ Exam ❏ PowerPoints Other _____

Craft / Level: _____ Copyright Date: _____

Module ID Number / Title: _____

Section Number(s): _____

Description: _____

_____

_____

_____

Recommended Correction: _____

_____

_____

_____

_____

Your Name: _____

Address: _____

Email: _____ Phone: _____

75104-13

Work-Zone Safety

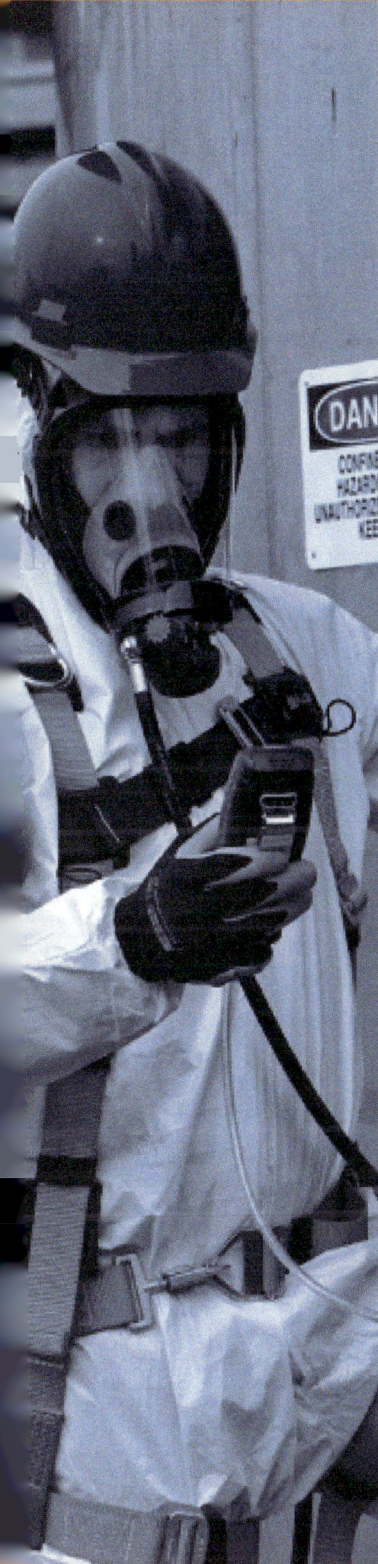

OVERVIEW

All construction workers will be exposed to numerous signs, signals, and barricades on the job. This module explains how to identify various types of signs, as well as audible and hand signals. It also covers traffic control and the safe movement of heavy equipment on a job site.

Module Four

Trainees with successful module completions may be eligible for credentialing through NCCER's National Registry. To learn more, go to **www.nccer.org** or contact us at **1.888.622.3720**. Our website has information on the latest product releases and training, as well as online versions of our *Cornerstone* newsletter and Pearson's product catalog.

Your feedback is welcome. You may email your comments to **curriculum@nccer.org**, send general comments and inquiries to **info@nccer.org**, or fill in the User Update form at the back of this module.

This information is general in nature and intended for training purposes only. Actual performance of activities described in this manual requires compliance with all applicable operating, service, maintenance, and safety procedures under the direction of qualified personnel. References in this manual to patented or proprietary devices do not constitute a recommendation of their use.

Objectives

When you have completed this module, you will be able to do the following:

1. Identify signs, signals, and barricades used on a job site.
   a. Identify the meaning of various signs.
   b. Identify the meaning of audible signals.
   c. Identify the meaning of barricades.
2. Identify highway work-zone safety requirements.
   a. Describe the use of temporary traffic control.
   b. Identify the responsibilities of a flagger.
   c. List the requirements for moving equipment safely.

Performance Task

Under the supervision of your instructor, you should be able to do the following:

1. Demonstrate how to properly use traffic control devices.

Trade Terms

Center of gravity
Conspicuous location
Flagger

Pinch points
Temporary traffic control (TTC)

Industry-Recognized Credentials

If you are training through an NCCER-accredited sponsor, you may be eligible for credentials from NCCER's Registry. The ID number for this module is 75104-13. Note that this module may have been used in other NCCER curricula and may apply to other level completions. Contact NCCER's Registry at 888.622.3720 or go to **www.nccer.org** for more information.

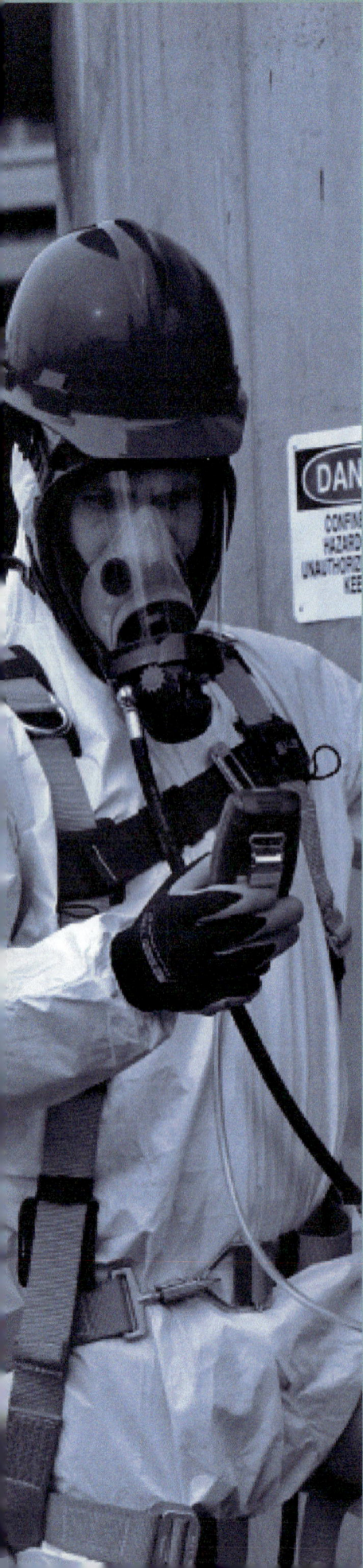

Contents

Topics to be presented in this module include:

Figures ─────────────────────────────────────

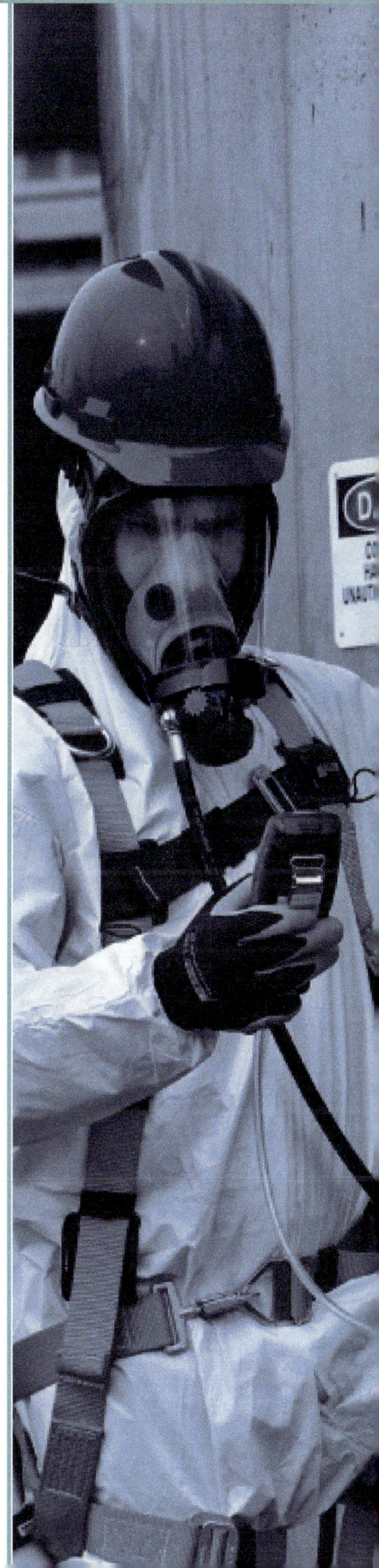

SECTION ONE

1.0.0 IDENTIFYING SIGNS, SIGNALS, AND BARRICADES

Objectives

Identify signs, signals, and barricades used on a job site.
   a. Identify the meaning of various signs.
   b. Identify the meaning of audible signals.
   c. Identify the meaning of barricades.

Trade Terms

Conspicuous location: A particularly noticeable spot, as would be appropriate for posting an important sign or tag to ensure it is seen.

One of the ways accidents can be avoided is the use of signs, safety tags, barricades and barriers, and other warning devices. When used properly, these devices not only prevent accidents, but also assist in the smooth flow of work and permit the job to be completed on schedule.

Construction work zones are often located near public areas and sometimes even on public roads (*Figure 1*). This creates the additional danger of pedestrian and motor traffic in the surrounding area. Work on public roads is so dangerous that every state in the United States has enacted laws that allow added fines and penalties for motorists who speed and commit other traffic violations in roadway construction zones. Specific standard procedures must be used without fail in these types of work zones to ensure everyone's safety.

Did you know?
Safety Training

Safe working conditions on mine sites fall under the Mine Safety and Health Administration (MSHA), and every other job site is regulated by the Occupational Safety and Health Administration (OSHA). However, except for a few industry-specific requirements, the regulations are the same. Safety training is required for all activities. Never operate tools, machinery, or equipment without prior training.

There are also many signs, signals, and barricades on a typical job site. Remember, these safety devices are just tools. They cannot prevent accidents unless used correctly. Stay alert for information and conditions that can affect everyone's safety (*Figure 2*). Do not rely on someone else to keep you safe. Think a job through before starting it. Look around for potential hazards and stay alert for changes in the work area. Remember the following:

- Accidents can kill and disable.
- Accidents cost you and your company money.
- Accidents can be prevented.

1.1.0 Identifying Signs

There are many types of warning signs on a typical job site. Common signs include danger signs, caution signs, informational signs, safety signs, and safety tags. It is important to understand the meaning of and respond appropriately to each type of sign.

1.1.1 Danger Signs

Danger signs are usually red, black, and white. They are used to inform workers that an immediate hazard exists and specific precautions must be observed to avoid an accident (*Figure 3*). Examples include the following:

- Danger – High Voltage
- Danger – No Smoking, Matches, or Open Lights
- Danger – Keep Away

Some of the dangers that may be found on a site are radiation and biological hazards. Be on the lookout for these types of signs as well.

Radiation Signs – Radiation signs are used in the workplace to alert workers to radiation hazards (*Figure 4*). The radiation hazard sign contains the word *Radiation* as well as the radiation symbol. The sign's background is ususally yellow and the panel is a reddish-purple color with yellow letters. Any additional lettering used against the yellow background is black.

Biological Hazard Signs – Biological hazard signs are used to warn workers of the actual or potential presence of a biological hazard (*Figure 5*). Biological hazards, or biohazards, can be any infectious agents that create a real or potential health risk. Biological hazard signs are commonly used to identify contaminated equipment, containers, rooms, materials, and areas housing experimental animals.

75104-13_F01.EPS

Figure 1 Traffic control.

DANGER

DO NOT ENTER

CAUTION

OPEN DOOR SLOWLY

POISON GAS

75104-13_F02.EPS

Figure 2 Stay alert for safety signs.

The biohazard symbol used on the sign is fluorescent orange or orange-red in color. Background colors may vary, but must contrast enough for the symbol to be easily identified. Wording is used on the sign to indicate the nature of the hazard or to identify the specific hazard.

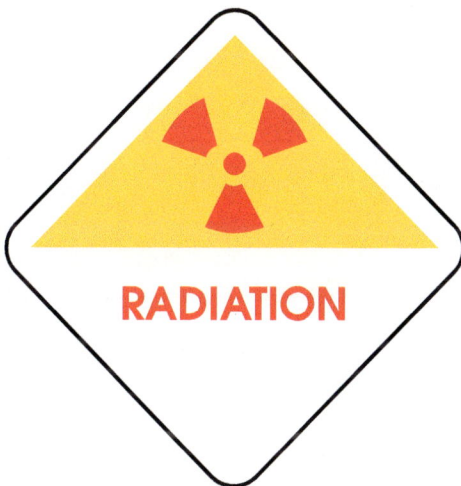

Figure 3 Typical danger sign.

Figure 4 Example of a radiation sign.

1.1.2 Caution Signs

Caution signs are used to inform workers about potential hazards or unsafe practices (*Figure 6*). When you see a caution sign, take action to protect yourself. Caution signs are yellow with black letters. Examples include the following:

- Caution – Do Not Operate
- Caution – Keep Aisles Clear
- Caution – Electric Fence

Figure 5 Biological hazard sign.

Figure 6 Typical caution sign.

Case History

Be Alert

A 23-year-old apprentice lineman died of injuries he received after being run over by the tandem dual rear tires of a digger derrick truck. He was part of a five-person crew that had been working on the job for three days. At the time of the incident, the victim and two other crewmembers had just finished setting and back-filling around a utility pole. They proceeded to the next pole requiring framing and setting, walking 30 feet ahead of the digger derrick in one lane of the road. The digger derrick moved slowly in reverse to the same pole. At a point approximately midway between the two poles, the victim knelt with his back to the truck to apparently inscribe a word or initials into some seal coating on the roadway. He was hit and run over by the backing truck's passenger-side tandem dual rear tires. The digger derrick truck did not have a back-up alarm system.

The Bottom Line: Keep your mind on the job. Communication with other workers can save your life. Check equipment for warning signals before you use it. If it does not have warning signals, establish hand or radio signals. Always check the surrounding area for potential hazards.

Source: The National Institute for Occupational Safety and Health (NIOSH)

Yellow is the basic color used for caution. It is used to identify places where physical hazards may be caused by striking against objects, stumbling, falling, tripping, and being caught between obstacles. Solid yellow, yellow and black stripes, or yellow and black checkers caution workers against these hazards.

The caution signs on piping systems that contain dangerous materials are yellow. Yellow is also used to warn workers against starting machinery under repair. Painted barriers and flags should be located at the starting point or power source. They should be displayed so that workers will notice them easily on such things as electrical controls, ladders, scaffolds, vaults, valves, dryers, boilers, elevators, and tanks.

1.1.3 Informational Signs

Informational signs are used where it is necessary to provide general information that is not related to safety (*Figure 7*). The standard color used is blue. The background, the entire sign, or just a panel may be blue. Examples include the following:

- No Admittance
- No Trespassing
- For Employees Only

Traffic and Housekeeping Signs – Black and white informational signs are used as traffic and house-keeping markings. These signs identify things such as the following:

- Dead ends of aisles or passageways
- Location of trash cans
- Location and width of aisle-ways
- Rooms or passageways
- Stairways (risers, direction, borders)
- Drinking fountains and food-dispensing machines

Directional Signs – Directional signs help workers and visitors find locations such as restrooms, stairways, and locker rooms. Standard colors for directional signs are black and white.

Exit and fire extinguisher signs are specialized directional signs that provide vital safety information and must be distinctive in color. They are red and white, as shown in *Figure 8*.

On-Site Traffic Signs – On-site traffic signs help in the safe movement of vehicles and pedestrians. Just as traffic signs must be obeyed on public highways and streets, they must also be observed in the workplace.

A slow-moving vehicle emblem is used on vehicles that move at speeds of 25 miles per hour (mph) or less. The emblem is a fluorescent yellow-orange triangle with a dark red reflective border (*Figure 9*). It is frequently seen on construction and farm equipment.

75104-13_F07.EPS

Figure 7 Common informational sign.

75104-13_F08.EPS

Figure 8 Directional signs.

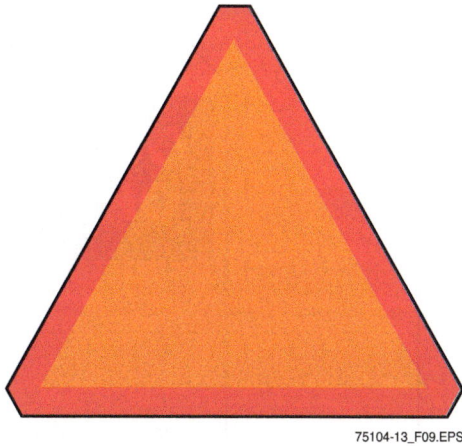

Figure 9 Slow-moving vehicle sign.

1.1.4 Safety Signs

Safety instruction signs are used when there is a need for general instructions and suggestions related to safety measures (*Figure 10*). The background and lettering on these signs are white and green, but can vary depending on the message and the location of the sign. Any letters used against the white background are black. Examples include the following:

- Report All Unsafe Conditions to Your Supervisor
- Walk, Don't Run
- Help Keep This Plant Safe and Clean

1.1.5 Accident-Prevention Tags

Accident-prevention tags, also known as safety tags, are used as a temporary warning to workers about immediate and potential hazards (*Figure 11*). They are similar to signs; however, they are not designed to be used in place of signs or as a permanent means of protection. For example, an Out of Order tag may be used on damaged equipment until it can be disposed of or repaired. A Do Not Start tag may be placed on machinery during lockout procedures. Tags can be an effective

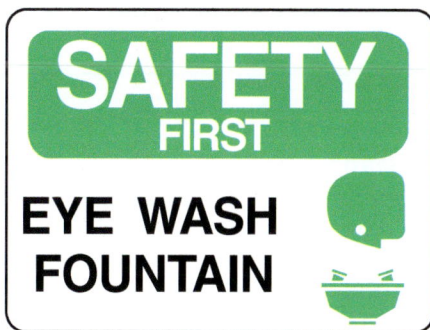

Figure 10 Common safety sign.

means of protecting workers and property. Tags and the devices used to attach them must meet specific physical requirements to ensure their durability and effectiveness.

Do Not Start Tags – Do Not Start tags are used to prevent the unexpected energizing of equipment that could result in injury, equipment damage, or both. The tags must be placed in a conspicuous location that effectively blocks the starting mechanism. The tag is white. The square panel is red and the letters may be white, grey, or visibly etched on the tag.

Danger Tags – Danger tags are used where an immediate hazard exists. They tell workers that specific precautions must be observed. Danger tags are white. White letters appear on a red oval in a black square. A danger tag may read Danger – Unsafe – Do Not Use.

Caution Tags – Caution tags are used to warn workers of potential hazards and to caution workers against unsafe practices. These tags inform workers that proper precautions should be taken. For example, a tag may read Caution—Stop Machinery To Clean, Oil, Or Repair. Caution tags are yellow. They contain a black square panel with yellow letters.

Out of Order Tags – Out of Order tags should only be used to notify workers that a particular machine or tool is out of order and that using it may present a hazard. Out of Order tags are usually white with black letters.

Radiation Tags – Radiation tags are yellow with a reddish-purple panel. Any letters used against the yellow background are black. The radiation symbol should also appear on the tag. Radiation tags notify workers of actual or potential radiation hazards.

Biological Hazard Tags – Biological hazard tags signify the presence of an actual or potential biological hazard. Biological hazards can include any infectious agents that present a risk or potential risk to workers. Biological hazard tags are usually white or fluorescent orange with black letters. The tag also shows the biological hazard or biohazard symbol.

1.2.0 Identifying Audible Signals

Audible signals such as alarms, bells, buzzers, whistles, and horns can be used to communicate hazards to workers. For example, back-up alarms are used on forklifts, trucks, and other heavy equipment. Fire alarms are used to clear work areas. Conveyer belt lines have buzzers, bells, and/or whistles to let workers know that they are about to start. However, when heavy equipment

START-UP INSTRUCTIONS

EQUIPMENT ID _____
LOCATION _____

Step No.	Instructions

(A)

DO NOT OPERATE

Signed by _____

Date _____

(B)

OUT OF ORDER

Signed by _____

Date _____

(C)

AUTHORIZED PERSONNEL ONLY

OPERATION OF THIS EQUIPMENT IS RESTRICTED TO THE FOLLOWING PERSONS

NAME _____
_____
_____

Signed by _____

Date _____

(D)

DANGER LIVE WIRE

SIGNED BY _____

DATE _____

(E)

DANGER DO NOT REMOVE THIS TAG

REMARKS _____
_____
_____

SEE OTHER SIDE

(F)

75104-13_F11.EPS

Figure 11 Examples of safety tags.

is present on a work site, there may be so much background noise that these types of alarms cannot be heard.

Hand signals can be safely used over long distances, but those using hand signals must never be out of each other's sight. Always use slow and exaggerated gestures to be sure that you are understood. Heavy equipment operators who lose sight of their co-worker must immediately bring their vehicle to a safe stop. Always make eye contact with nearby personnel to be sure they are aware of your presence.

NOTE

Standards may vary from state to state. Be sure to use hand signals for your local jurisdiction.

1.3.0 Identifying Barricades and Barriers

Barricades and barriers are used to alert workers to potential hazards and to help prevent accidents. Each work zone may have different policies and procedures for how and when to use them, so learn and follow the rules at the job site. Sometimes barriers will be placed in an area to prevent injury to workers, and other times they will be placed to alert heavy equipment operators of dangers. For example, a barrier may be erected to keep a bulldozer off steeply graded ground. *Figure 12* shows the various uses for barricades and barriers.

Barricades and barriers help to prevent accidents and injuries, but they are not foolproof. When you are assigned to a job site, carefully examine the area that you will be working in. Before moving heavy equipment, walk the area to ensure that there are no hidden hazards. This is especially important when there are obstructions, such as snow, mud, leaves, and other debris in the area. Never move a barricade, even in an emergency, unless certain that it is safe to enter the barricaded area.

Any opening in a wall, floor, or the ground is a safety hazard. There are two types of protection for these openings: they can be guarded or they can be covered. Cover any hole whenever possible. When it is not practical to cover a hole, use barricades. If the bottom edge of a wall opening is less than 3' above the floor and would allow someone to fall 4' or more, place guards around the opening. There are several different guard methods, including:

- Railings are used across wall openings or as barriers around floor openings to prevent falls. See *Figure 12(A)*.

- Warning barricades alert workers to hazards but provide no real protection. See *Figure 12(B)*. Typical warning barricades are made of plastic tape or rope strung from wire or between posts. The tape or rope is color-coded:
  - Red means danger. No one can enter an area with a red warning barricade. A red barricade is used when there is danger from falling objects or when a load is suspended over an area.
  - Yellow means caution. You can enter an area with a yellow barricade, but you must know what the hazard is, and be careful. Yellow barricades are used around wet areas or areas containing loose dust. Yellow with black lettering warns of physical hazards such as bumping into something, stumbling, or falling.
  - Yellow and reddish-purple indicates a radiation warning. These barricades are often used where piping welds are being X-rayed.

- Protective barricades provide both a visual warning and protection from injury. See *Figure 12(C)*. They can be wooden posts and rails, posts and chain, or steel cable. People cannot get past protective barricades.
- Blinking lights are placed on barricades so they can be seen at night. See *Figure 12(D)*.

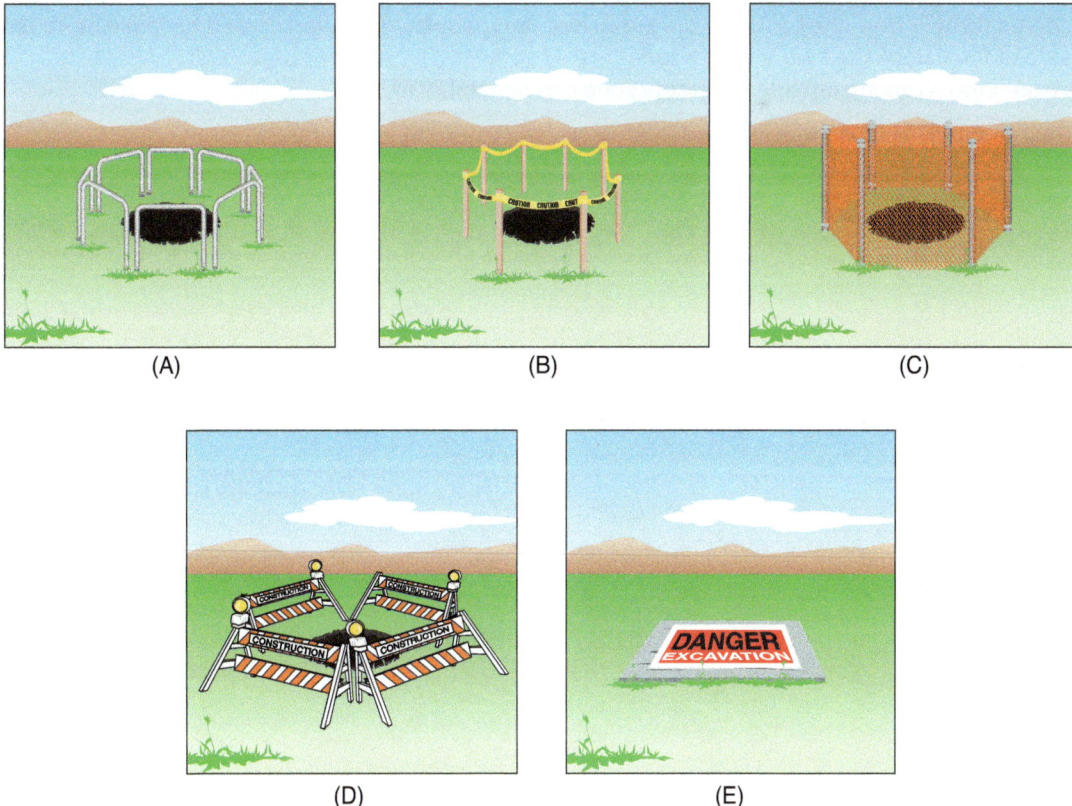

(A)   (B)   (C)

(D)   (E)

75104-13_F12.EPS

Figure 12 Common types of barricades and barriers.

Stay Alert

Many sidewalks and roadways use grates as barriers to cover vaults or pits. During a snowstorm in 2003, a skid-steer loader that was being used to clear snow fell into a vault when the grate system failed under the weight of the loader. The operator died.

The Bottom Line: Obstacles can make an unsafe area appear to be safe. Do not take chances. Examine the work area before moving any equipment.

- Hole covers are used to cover open holes in the floor or ground. See *Figure 12(E)*. They must be secured and labeled. They must be strong enough to support twice the weight of anything that may be placed on top of them.

The types of barriers and barricades used will vary between job sites. There may also be different procedures for when and how barricades are put up. Learn and follow the policies at your job site.

Work Inside the Zone

A 41-year-old electrician died after he fell from the basket of an aerial lift truck. He and two co-workers were installing electrical wiring for a sign on the side of a highway bridge. The two left-most traffic lanes were closed to traffic. The workers arrived at the scene with a pickup truck and a truck equipped with a basket-type work platform attached to a hydraulic boom. At this time the basket was positioned above the closed traffic lanes. While the co-workers cut the pipe, the victim moved the basket to a location above the nearest open traffic lane. A tractor trailer approached the site and observed the warning cones and signs directing traffic to merge to the right. The driver proceeded toward the work site in the open traffic lane nearest the closed lanes. When the truck neared the work area, the driver noticed the basket over the traffic lane in which he was driving. Before the driver could slow down or safely move to the next lane, the tractor trailer struck the basket. The collision caused the victim to be thrown from the basket to the surface of the roadway. This worker died because he chose to work outside of the established work safety zone.

The Bottom Line: Working outside the safety zone can be fatal.

Source: The National Institute for Occupational Safety and Health (NIOSH)

Additional Resources

Additional information on *OSHA Standard 29 CFR, Part 1926, Subpart G, Accident Prevention Signs and Tags*, can be found at **www.osha.gov**.

1.0.0 Section Review

1. A demolition area in which there is an immediate hazard from falling materials should be marked with a(n) _____.

    a. danger sign
    b. caution sign
    c. informational sign
    d. safety sign

2. One of the risks of using audible signals on a construction site is that the background noise may be too loud for them to be heard.

    a. True
    b. False

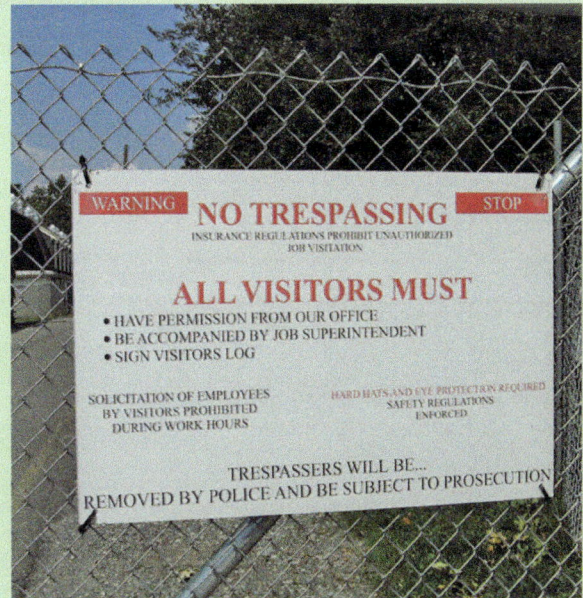

Figure 1   75104-13_SR01.EPS

3. The type of barricade shown in the *Section Review Figure 1* is a _____.

    a. work-zone barrier
    b. protective barricade
    c. warning barricade
    d. zone marker

2.0.0 HIGHWAY WORK-ZONE SAFETY REQUIREMENTS

Objectives

Identify highway work-zone safety requirements.
  a. Describe the use of temporary traffic control.
  b. Identify the responsibilities of a flagger.
  c. List the requirements for moving equipment safely.

Performance Task 1

Demonstrate how to properly use traffic control devices.

Trade Terms

Center of gravity: The point where all of an object's weight is evenly distributed.

Flagger: A person who is specially trained to direct traffic through and around a work zone.

Pinch points: The area in which two moving equipment parts come together.

Temporary traffic control (TTC): Device and plan used to safely divert traffic when the normal use of the road is disrupted due to construction.

When construction sites are near or in public roads, moving traffic presents an additional hazard to construction workers. Dangers to construction personnel working near moving traffic are so great that all states have laws that provide for additional fines for drivers who speed or commit other traffic violations in a traffic work zone. When the work site is located near traffic, supervisors study the area and make plans to ensure that the job can be completed safely and efficiently, but safety is ultimately your responsibility. Stay alert.

2.1.0 Using Temporary Traffic Control

When the normal use of a road is disrupted, temporary traffic control (TTC) measures must be used (*Figure 13*). TTC measures help to ensure that motorists, pedestrians, and bicyclists can use the road safely while construction is going on. Planning traffic flow is key to keeping traffic moving smoothly and managing traffic incidents. Specially

trained workers create traffic flow plans following direction from the US Department of Transportation (US DOT). These workers also decide on the need for and position of TTC devices such as barricades, cones, and signs, but you may be called on to help place some of these devices. In addition, it may be necessary to temporarily change the TTC during an accident or other emergency.

The US DOT Federal Highway Administration publishes a manual that states the basic principles for changing the flow of traffic at a work site. It also explains how to design and use traffic control devices. This manual is called the *Manual on Uniform Traffic Control Devices (MUTCD) for Streets and Highways*. It is used for streets and highways open to the public, regardless of type or class or the public agency having control of the road.

For a traffic control device to be useful, it must meet five basic requirements:

- Fulfill a need
- Attract attention
- Have a clear and simple meaning
- Command respect from road users
- Give enough time for proper response

TTC measures and devices are used to provide for safe and orderly movement of traffic through or around work zones and to protect workers, responders to traffic accidents, and equipment. At the same time, the TTC zone must permit workers to complete the project quickly and well.

Conditions in TTC zones are always changing. These conditions include lane closures and merges, speed changes, flaggers, trucks and other equipment entering and leaving the roadway, and uneven pavement. This can be very confusing to motorists and dangerous for construction workers and heavy equipment operators.

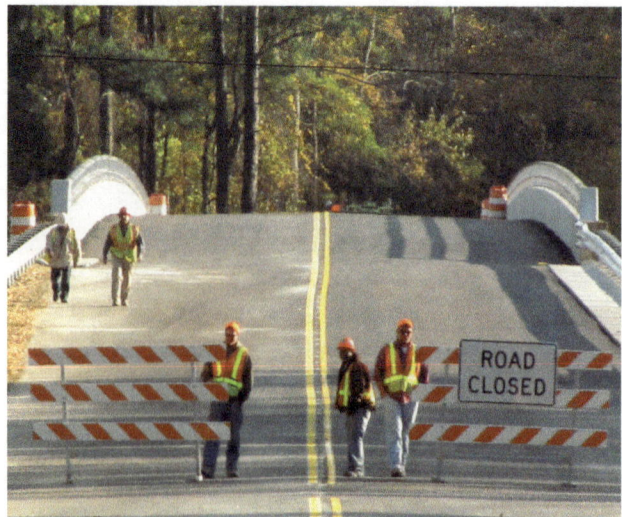

75104-13_F13.EPS

Figure 13 Typical use of barricades.

Various methods and devices are used to mark the TTC zone. It is important to understand the how the TTC zone is set up and how TTC devices are used to keep workers safe. It is also important to know the purpose and use of flaggers in the TTC zone, and to be aware of the signals that flaggers use, just in case you need to control traffic in an emergency.

2.2.0 Flagger Responsibilities

Flaggers are part of the construction team. They are responsible for public safety, so they must be trained in safe traffic-control practices. In some states, flaggers must be certified. Flaggers must keep themselves and the construction team safe, as well as direct traffic using signs, paddles, or flags. On most jobs, it is unlikely that an equipment operator will need to be a flagger, but in the case of an emergency, the operator may need to perform this task for a short period. A flagger must be able to do the following:

- Clearly, firmly, and politely give specific instructions.
- Move quickly to avoid danger.
- Use signaling devices such as paddles and flags to provide clear direction to drivers.
- Apply safe traffic-control practices, sometimes in stressful or emergency conditions.
- Identify unsafe traffic situations and warn workers in time to avoid injury.

Case History

Stay Alert

An 18-year-old flagger, outfitted in full reflective vest, pants, and hard hat, was directing traffic at one end of a bridge approach during a night milling operation. The work zone was correctly marked with cones and signs, and the entire bridge was lit with streetlights. The flagger was standing under portable floodlights in the opposing traffic lane close to the centerline, facing oncoming traffic. A pickup truck traveling in the wrong lane at an estimated speed of 55 to 60 miles per hour struck the flagger head on and carried him approximately 200'. He died of multiple traumatic injuries at the scene.

The Bottom Line: Highway construction is very dangerous. Even if you are following all of the safety rules, you are still at risk. Stay alert.

Source: The National Institute for Occupational Safety and Health (NIOSH)

One of the ways that flaggers keep themselves safe is by wearing highly visible safety apparel, usually a vest that is either fluorescent orange-red or fluorescent yellow-green. The vest must be approved by the American National Standards Institute (ANSI) and needs to be visible at a minimum of 1,000 feet.

Hand-signaling devices, such as Stop/Slow paddles, and lights are used to direct motorists through TTC zones. The Stop/Slow paddle is the best device to use because it is easy for drivers to understand. Red flags may be used in emergencies.

The Stop/Slow paddle is octagonal in shape, and the Stop side looks like the Stop signs used at intersections. The Slow side has an orange background with black letters and border. When used at night, the paddle must be made of material that reflects light. The Stop/Slow paddle can also have either white or red flashing lights on the Stop side and either white or yellow flashing lights on the Slow side.

When flags are used they must be at least 24 inches square and made of a good grade of red material. The free edge of a flag must be weighted so the flag hangs vertically, even in heavy winds. When used at night, flags must be made of red retro reflective material.

> **WARNING!**
>
> Never step in front of a vehicle in order to stop it.

The paddle must be held still, with the flagger's arm extended horizontally away from the body (*Figure 14*). The other arm is used to give motorists additional directions. To stop traffic, the flagger's free hand is held above shoulder level with the palm facing oncoming traffic. To direct stopped traffic to continue, the Slow side of the paddle must face oncoming traffic, and the flagger's free arm is used to motion drivers to continue. To slow moving traffic, the flagger can slowly motion up and down with the free hand, palm down.

When a red flag is used in an emergency situation, the flagger must stand facing oncoming traffic with the red flag in the hand closest to traffic. To stop traffic, the flagger extends the flag horizontally across the road users' lane so that the full area of the flag is visibly hanging below the staff. The free arm is used in the same way as described when using the Stop/Slow paddle. To release stopped traffic, the flag is hung down at the flagger's side and the free arm is used to wave traffic on. To slow traffic, the flagger must stand

18"
MIN

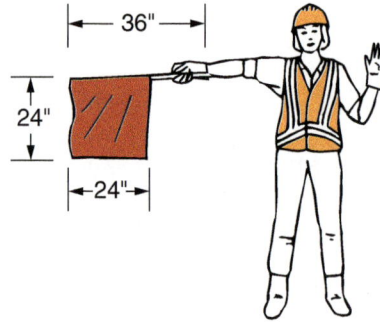

TO STOP TRAFFIC

36"

24"

24"

TO LET TRAFFIC PROCEED

TO ALERT AND SLOW TRAFFIC

75104-13_F14.EPS

Figure 14 Flagger signals.

with the free hand down and slowly wave the flag up and down without raising the arm above shoulder level. Motorists often mistake this gesture as a stop signal.

2.3.0 Moving Equipment Safely

All heavy equipment is moved during the course of a job. It is moved to the site, on the site, and away from the site. Many injuries and deaths happen during the movement of heavy equipment. It is important to be especially safety conscious whenever equipment is moving. When operating heavy equipment, be responsible. A responsible operator must be a qualified and authorized operator. To be qualified, you must understand the instruction manual supplied by the manufacturer of the equipment, have training in the actual operation of the equipment, and know the safety rules and regulations for the work site. Never operate any equipment if under the influence of alcohol or drugs. If taking prescription or over-the-counter drugs, seek medical advice about whether you can safely operate machinery. Never knowingly allow anyone to operate heavy equipment when he or she is impaired.

Always observe the following guidelines when driving equipment on public roads:

- Drive slowly and never speed.
- Know the distance it takes to stop safely.
- Allow extra time to enter traffic.
- Know and follow the traffic control pattern.
- Travel with lights on.
- Use proper warning signs and flags.
- Secure all attachments and loose gear.
- Turn cautiously; allow for extensions or attachments and for structural clearances. Some equipment is top-heavy and will tip over if a turn is made too fast.
- Know the turning radius of the equipment.
- Know the swing radius of the equipment and make sure the swing path is clear of people and objects (*Figure 15*).
- Know and obey all state and local laws.

Always follow these guidelines for driving equipment on the job site:

- Never drive a machine in a congested area, or around people, without a spotter or flagger. The spotter or flagger is responsible for determining and directing the driver's speed.
- Be sure everyone is in the clear while backing up, connecting, or moving attachments.
- Never move buckets or shovels over the heads of other workers.
- If you cannot see the surrounding area clearly from the operator's seat, get a spotter or do not operate the equipment.
- Wait for an all-clear signal before moving.
- Signal a forward move with two blasts of the horn.
- Signal a reverse move with three blasts of the horn.
- Yield the right-of-way to loaded equipment on construction sites.
- Maintain a safe distance from all other vehicles.
- When moving, keep the equipment in gear at all times; never coast.
- Maintain a ground speed consistent with ground conditions and posted speed limits.
- Know where overhead electrical power lines are located.

75104-13_F15.EPS

Figure 15 Warning tape used to indicate swing radius.

- Ensure that buried pipes and power and gas lines have been located and marked.

Each piece of heavy equipment has its own unique set of hazards. Be aware of its pinch points (*Figure 16*), also known as crush points. Pinch points occur when there is motion between equipment parts. Serious injury or death can result from getting part of your body or your clothing caught in a pinch point.

Another danger that heavy equipment operators must face is tipping over. Improper loading, unsafe driving, or unstable terrain can cause tipping. These factors affect the center of gravity, which is the point at which the machine and load are in balance. To understand how to prevent tipping, you must first understand center of gravity. The center of gravity is the point where all of an object's weight is evenly distributed. The addition of a load causes a machine's center of gravity to move toward the load. This effect is very noticeable on a machine like a telescoping boom forklift.

Ergonomics

Ergonomics is the science that deals with identifying and reducing stress on equipment users. In the past, jarring movements, uncomfortable seats, tight compartments, deafening noise, and hard-to-use controls were common. In the long term, operators were faced with a variety of physical problems, including bad backs and hearing loss, because of the harshness of their working environment.

In recent years, equipment designers have focused on eliminating these problems. Their efforts have resulted in such improvements as the following:

- Adjustable suspension seats
- Roomier cabs with better visibility
- More leg and foot room
- Easy-to-operate joystick controls
- Quieter operation
- Easier-to-reach service components, such as oil dipsticks and fuel tank nozzles

75104-13_F16.EPS

Figure 16 Pinch points.

Case History
Double Check Your Equipment

A 49-year old bulldozer operator was crushed to death while compacting earthen fill during the construction of an oil exploration island. The victim was operating the bulldozer on level ground, alternately in forward and reverse at half throttle. With the blade completely down, the victim shifted the transmission control lever toward neutral. The transmission was partially disengaged, but not fully in neutral.

He assumed, without checking, that the bulldozer transmission was in a stable neutral position. He exited the cab on the right side. When he did this, the transmission slipped back into the first gear of reverse, causing the bulldozer to suddenly move. As a result, the victim was pulled between the underside of the fender and the top of the track cleats. As the bulldozer continued in reverse, the victim was fatally crushed beneath the track cleats.

The Bottom Line: Always use parking brakes, and double-check settings before exiting the equipment. Never leave equipment when it is still running.

Source: The National Institute for Occupational Safety and Health (NIOSH)

Additional Resources

Extensive information on the National Highway Work Zone Safety Program can be found at **safety.fhwa.dot.gov**.

2.0.0 Section Review

1. TTC zones can be designed by anyone with a license to operate heavy equipment.

   a. True
   b. False

2. Flagger vests must be fluorescent and visible at a minimum of _____.

   a. 250 feet
   b. 500 feet
   c. 750 feet
   d. 1,000 feet

3. When moving equipment on a job site, loaded equipment has the right-of-way.

   a. True
   b. False

SUMMARY

Work-zone safety is an important part of overall job safety. Many accidents happen because of a lack of communication. Signs, signals, and barricades make communication clear on a work site and in public areas where work is being done. It's important to know the hazards of your job and be able to recognize the signs, signals, and barricades that could save your life.

You must set up a clear work zone with tape, fencing, cones, or barriers. This is particularly important when working on a busy street or highway. Make sure that your work zone includes all work areas. Stay alert for traffic.

1. The sign used to inform workers that an immediate hazard exists is a(n) _____.
   a. caution sign
   b. danger sign
   c. informational sign
   d. warning sign

2. Red is the standard color used for caution.
   a. True
   b. False

3. A barricade that alerts workers to hazards but provides no real protection is called a(n) _____
   a. hole cover
   b. temporary barricade
   c. railing
   d. warning barricade

4. Where it is necessary to tell workers about general information not related to safety, you would post a(n) _____.
   a. caution sign
   b. danger sign
   c. informational sign
   d. accident-prevention tag

5. To inform workers about potential hazards or unsafe practices, you would see a(n) _____.
   a. caution sign or tag
   b. informational sign or tag
   c. red flag
   d. Stop/Slow paddle

6. A danger sign warns workers that _____.
   a. an immediate hazard exists and specific precautions must be observed
   b. equipment is out of order
   c. radiation is present
   d. general safety rules apply

7. When there is a need for general instructions and suggestions related to safety measures, you would find a(n) _____.
   a. informational sign or tag
   b. safety sign or tag
   c. accident-prevention tag
   d. flagger

8. Accident-prevention tags are used _____.
   a. where it is necessary to tell workers about information not related to safety
   b. where it is necessary to inform workers about general instructions related to safety measures
   c. on landmarks where traffic must slow or stop completely.
   d. as a temporary way of warning workers about immediate and potential hazards.

9. Safety tags can be used in place of safety warning signs.
   a. True
   b. False

10. To use the Stop/Slow paddle, the flagger faces road users and _____.
    a. waves the flag rapidly across the lane of oncoming traffic
    b. steps into the lane of oncoming traffic and raises the flag
    c. repeatedly raises and lowers the flag to shoulder level
    d. extends the arm holding the paddle horizontally away from the body

Trade Terms Introduced in This Module

Center of gravity: The point where all of an object's weight is evenly distributed.

Conspicuous location: A particularly noticeable spot, as would be appropriate for posting an important sign or tag to ensure it is seen.

Flagger: A person who is specially trained to direct traffic through and around a work zone.

Pinch points: The area in which two moving equipment parts come together.

Temporary traffic control (TTC): Device and plan used to safely divert traffic when the normal use of the road is disrupted due to construction.

Additional Resources

This module presents thorough resources for task training. The following resource material is suggested for further study.

Extensive information on the National Highway Work Zone Safety Program can be found at **safety.fhwa.dot.gov**.

Additional information on *OSHA Standard 29 CFR, Part 1926, Subpart G, Accident Prevention Signs and Tags*, can be found at **www.osha.gov**.

Figure Credits

Carolina Bridge Co., Module opener, Figures 1, 13

Accuform Signs, Figures 7, 10

Topaz Publications, Inc., SR01

Manual on Uniform Traffic Control Devices (MUTCD), 2009 Edition, published by FHWA at **http://mutcd.fhwa.dot.gov/pdfs/2009/pdf_index.htm**, Figure 14

LPR Construction, Figure 15

Courtesy of Atlas Copco, Figure 16

Answer	Section Reference	Objective
Section One		
1. a	1.1.1	1a
2. a	1.2.0	1b
3. b	1.3.0	1c
Section Two		
1. b	2.1.0	2a
2. d	2.2.0	2b
3. a	2.3.0	2c

NCCER CURRICULA — USER UPDATE

NCCER makes every effort to keep its textbooks up-to-date and free of technical errors. We appreciate your help in this process. If you find an error, a typographical mistake, or an inaccuracy in NCCER's curricula, please fill out this form (or a photocopy), or complete the online form at **www.nccer.org/olf**. Be sure to include the exact module ID number, page number, a detailed description, and your recommended correction. Your input will be brought to the attention of the Authoring Team. Thank you for your assistance.

Instructors – If you have an idea for improving this textbook, or have found that additional materials were necessary to teach this module effectively, please let us know so that we may present your suggestions to the Authoring Team.

NCCER Product Development and Revision

13614 Progress Blvd., Alachua, FL 32615

Email: curriculum@nccer.org
Online: www.nccer.org/olf

❏ Trainee Guide   ❏ Lesson Plans   ❏ Exam   ❏ PowerPoints   Other _____

Craft / Level: _____ Copyright Date: _____

Module ID Number / Title: _____

Section Number(s): _____

Description: _____

_____

_____

_____

_____

Recommended Correction: _____

_____

_____

_____

_____

Your Name: _____

Address: _____

_____

Email: _____ Phone: _____

22308-13

Soils

OVERVIEW

Soil conditions vary widely from site to site. In order to effectively dig, place, and finish earthwork, a heavy equipment operator must be able to identify soil types and understand how these soils will react to various conditions.

Module Five

Trainees with successful module completions may be eligible for credentialing through NCCER's National Registry. To learn more, go to **www.nccer.org** or contact us at **1.888.622.3720**. Our website has information on the latest product releases and training, as well as online versions of our *Cornerstone* newsletter and Pearson's product catalog.

Your feedback is welcome. You may email your comments to **curriculum@nccer.org**, send general comments and inquiries to **info@nccer.org**, or fill in the User Update form at the back of this module.

This information is general in nature and intended for training purposes only. Actual performance of activities described in this manual requires compliance with all applicable operating, service, maintenance, and safety procedures under the direction of qualified personnel. References in this manual to patented or proprietary devices do not constitute a recommendation of their use.

Objectives

When you have completed this module, you will be able to do the following:

1. Describe the different types and characteristics of soils.
   a. Identify the types of soils.
   b. Describe the properties of soils.
   c. Explain how soil density is determined.
   d. Explain how moisture affects soil.
2. Describe the factors that affect soil excavation.
   a. Explain what the swell factor is and how to calculate the swell factor of soils.
   b. Explain what the shrink factor is and how to calculate the shrink factor of soils.
   c. Describe how swell and shrink factors affect cycle times and equipment selection.
3. Describe working in various soil conditions.
   a. Describe the weight-bearing and flotation properties of different soils.
   b. Explain how soil characteristics affect machine performance.
   c. Describe how soil conditions can affect trenching safety.

Performance Tasks

Under the supervision of the instructor, you should be able to do the following:

1. Identify five basic types of soils, and summarize their characteristics.
2. Read results from a field density test and explain what additional compaction effort is needed.
3. Compute shrinkage and relative compaction for two different types of soils.

Trade Terms

American Association of State Highway and Transportation Officials (AASHTO)
American Society of Testing Materials (ASTM)
Banked
Bedrock
Capillary action
Cohesive
Consolidation
Density

Elasticity
Expansive soil
Fines
Friable
Horizon
Humus
In situ
Inorganic
Loading
Liquid limit
Optimum moisture

Organic
Peat
Plastic limit
Plasticity
Settlement
Shrinkage
Swell
Swell factor
Voids
Water table
Well-graded

Industry Recognized Credentials

If you are training through an NCCER-accredited sponsor, you may be eligible for credentials from NCCER's Registry. The ID number for this module is 22308-13. Note that this module may have been used in other NCCER curricula and may apply to other level completions. Contact NCCER's Registry at 888.622.3720 or go to **www.nccer.org** for more information.

Contents

Topics to be presented in this module include:

Figures and Tables —————————————

SECTION ONE

1.0.0 TYPES AND CHARACTERISTICS OF SOIL

Objective 1

Describe the different types and characteristics of soils.

a. Identify the types of soils.
b. Describe the properties of soils.
c. Explain how soil density is determined.
d. Explain how moisture affects soil.

Performance Task 1

Identify five basic types of soils, and summarize their characteristics.

Trade Terms

American Association of State Highway and Transportation Officials (AASHTO): An organization representing the interest of all state government highway and transportation agencies throughout the United States. This organization establishes design standards, materials-testing requirements, and other technical specifications concerning highway planning, design, construction, and maintenance.

American Society of Testing Materials (ASTM): A national organization that establishes standards for testing and evaluation of manufactured and raw materials.

Bedrock: The solid layer of rock under Earth's surface. Its solid-rock state distinguishes it from boulders.

Capillary action: The tendency of water to move into free space or between soil particles, regardless of gravity.

Cohesive: The ability to bond together in a permanent or semipermanent state. To stick together.

Density: Ratio of the weight of material to its volume.

Elasticity: The property of a soil that allows it to return to its original shape after a force is removed.

Expansive soil: A soil that expands and shrinks with moisture. Clay is an expansive soil.

Fines: Very small particles of soil. Usually particles that pass the No. 200 sieve.

Friable: Crumbles easily.

Horizon: Layers of soil that develop over time.

Humus: Dark swamp soil or decaying organic matter. Also called peat.

In situ: In the natural or original place on site.

Inorganic: Derived from other than living organisms, such as rock.

Liquid limit: The amount of moisture that causes a soil to become a fluid.

Loading: Applying a force to soil. A building can be a permanent load at a site, and a truck can be a passing load on a roadway.

Optimum moisture: The percent of moisture at which the greatest density of a particular soil can be obtained through compaction.

Organic: Derived from living organisms, such as plants and animals.

Peat: Dark swamp soil or decaying organic matter. Also called humus.

Plastic limit: The amount of water that causes a soil to become plastic (easily shaped without crumbling).

Plasticity: The range of water content in which a soil remains plastic or is easily shaped without crumbling.

Voids: Open space between soil or aggregate particles. A reference to voids usually means that there are air pockets or open spaces between particles.

Water table: The depth below the ground's surface at which the soil is saturated with water.

Well-graded: Soil that contains enough small particles to fill the voids between larger ones.

The stability of any building relies on the stability of the soil on which it is built. There are various types of soils. Each type has its own characteristics and requires different preparations for construction. The most important property for any soil for construction purposes is its strength, which is its ability to support a building or road without compressing or otherwise deforming.

Soil strength is measured by density, or how tightly the soil is packed. Soil that is densely packed and has few air pockets, called voids, weigh more per cubic unit than lightly packed soil, which has many air pockets. See *Figure 1*. Density is measured in weight per volume and can be expressed as pounds of wet or dry soil per cubic foot. The greater the weight per unit, the more tightly the soil is packed and the greater its density and strength. Densely compacted soil is usually made of a mixture of different sizes and types of soil particles.

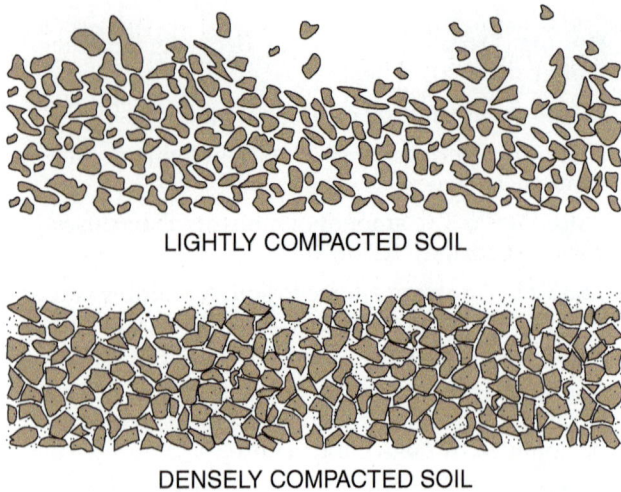

LIGHTLY COMPACTED SOIL

DENSELY COMPACTED SOIL

22308-12_F01.EPS

Figure 1 Soil density.

In construction, soil is often compacted to improve its density and strength. Compaction is the deliberate application of pressure or vibration to eliminate air pockets and increase the density of soil. It is one of the most important tasks in construction, and it helps ensure the durability and safety of a building or roadway. Poorly compacted soil naturally settles over time. The resulting movement of the ground can cause a cracked foundation (see *Figure 2*) or road surface failure.

Soils that have poor structural strength, even with compaction, can sometimes be improved by using additional materials or chemicals that stabilize the soil. Different geographic regions have predominant soil types, but most areas have a mixture of soils. It is possible to find many different kinds of soils on the same project.

1.1.0 Types of Soils

Soil is the loose material on the surface of Earth that is laid on the area's bedrock. Soil develops

The Importance of Recognizing Soil Types

Engineers, architects, and specially trained workers identify soils on a site and make technical decisions about how to prepare the site for construction. A heavy equipment operator must implement those decisions in order to safely operate the equipment in various soils and environmental conditions. To do this, the operator must be able to identify and understand the general behavioral characteristics of the different types of soils, as well as the various soil factors that affect equipment operation.

22308-12_F02.EPS

Figure 2 Cracks caused by foundation settling.

over time. When a hole is dug, definite layers of soil are noticeable. These layers, called horizons, represent the various times in which the soil developed. *Figure 3* shows the soil layers from an excavated lime rock bed in Florida.

There are several soil classification systems used, but the most common are the Unified Soil Classification System, which is also known as the American Society for Testing and Materials (ASTM)

22308-12_F03.EPS

Figure 3 Soil layers.

Unified Soil Classification System, and the **American Association of State Highway and Transportation Officials (AASHTO)** Soil Classification System. Other classification systems are used in agriculture and geology; however, these names occasionally show up on construction documents.

For construction purposes, soil is classified into two types: granular or fine-grained. Fine-grained soil is often referred to as **fines**. Soils with particles that can be seen with the naked eye are granular. They include sand, gravel, and rock. Soils with particles that cannot be seen with the naked eye are fines. They include clays and silts.

Table 1 shows soil characteristics for the Unified Soil Classification System. This system classifies soils as coarse-grained or fine-grained, and gives each subcategory a two-letter designation. For example, gravel is designated with the letter G and clay with the letter C, so gravel and clay mixtures are designated as GC. The table shows that silts and clays are not measured by particle size, but by **liquid limit**. Liquid limit refers to the percent of moisture content in the soil at the point where it turns into a liquid. Silts and clays have very small soil particles and are referred to as fine-grained soils. These soils have almost no strength, but small amounts of these soils mixed with coarser soils give the mixture other desirable characteristics for building.

> **NOTE**
>
> Soils are usually classified by particle size, but keep in mind that the various soil classification organizations may use the same name for a different particle size. The important thing to remember is that the particle size greatly influences soil behavior. The tables shown in this section are for reference. It is unlikely that you will need this kind of information often. If you do need the information, don't rely on memory; look it up!

Table 1 Unified Soil Classification and Symbol Chart

Unified Soil Classification and Symbol Chart		
Coarse-grained soils (more than 50% of material is larger than No. 200 sieve size.)		
Gravel More than 50% of the coarse material is larger than No. 4 sieve size.	GW GP	Clean Gravels (less than 5% fines): Well-graded gravels, gravel-sand mixtures, little or no fines Poorly graded sands, gravelly sands, little or no fines
	GM GC	Gravels With Fines (more than 12% fines): Silty gravels, gravel-sand-silt mixtures Clayey gravels, gravels-sand-clay mixtures
Sands More than 50% of the coarse material is smaller No. 4 sieve size.	SW SP	Clean Sands (less than 5% fines): Well-graded sands, gravelly sands, little or no fines Poorly graded sand, gravelly sands, little or no fines
	SM SC	Sands With Fines (more than 12% fines): Silty sands, sand-silt mixtures Clayey sands, sand-clay mixtures
Fine-grained soils (50% or more of the material is smaller than No. 200 sieve size.)		
Silts and Clays Liquid limit less than 50%.	ML	Inorganic silts and very fine sands, rock flour, silty or clayey fine sands or clayey silts with slight plasticity.
	CL	Inorganic clays of low to medium plasticity, gravelly clays, sandy clays, silty clays, lean clays
	OL	Organic silts and organic silty clays of low plasticity.
Silts and Clays Liquid limit 50% more.	MH	Inorganic silts, micaceous or diatomaceous fine sandy or silty soils, elastic silts.
	CH	Inorganic clays of high plasticity, fat clays
	OH	Organic clays of medium to high plasticity, organic silt
Highly Organic Soils	PT	Peat and other highly organic soils.

22308-12_T01.EPS

Table 2 shows soil characteristics for the AASHTO Soil Classification System. This system divides soils into granular and silt-clay materials, and then further by the sizes of the soil particles in a mixture. Granular material includes gravel and sand. The sieve number designates the size of the particles that pass the sieve. No. 10 passes big particles and No. 200 passes very tiny particles. At the bottom of the table, it can be seen that well-mixed material that is mostly granular gets the best rating for construction. Sand and clay are often blended in specified proportions to increase the strength of the soil at a site.

The Unified Soil Classification System and the AASHTO Soil Classification System define the main classes of soils as gravels, sands, silt and clay, and organics. In general, soils are divided as follows for construction purposes:

- *Gravel* – Gravel is any rock-like material above 0.125 inches (⅛ inch) in diameter. Larger particles are called cobbles or stones. Particles larger than 10 inches are called boulders. Gravel occurs naturally or it can be made by crushing rock. Natural gravel is usually rounded from the effects of water, while crushed rock is usually angular (*Figure 4*). Gravel is very strong.
- *Sand (coarse and fine)* – Sand is made of mineral grains measuring 0.002 to 0.125 inches. Sand comes from grinding or decaying rock. It usually contains a high amount of quartz, which is a very hard mineral. It is called granular material because it separates easily, giving it almost no cohesive strength. Coarse sand is frequently

rounded like gravel. It is often found mixed with gravel, but fine sand is usually more angular. See *Figure 5*.

- *Inorganic silt* – **Inorganic** silt is very fine sand with particles that are 0.002 inches or less. Silt is sand that has been ground very fine. Silt is often called rock flour or rock dust because it has a dusty appearance and powdery texture when dry. When wet, it sticks together, but silt has almost no cohesive strength. Dried lumps are easily crushed. Silt has a tendency to absorb moisture by **capillary action**, which means that moisture wicks up through the soil (see *Figure 6*), making silts problematic in areas where the **water table** is shallow.

22308-12_F04.EPS

Figure 4 Gravel.

Table 2 AASHTO Soil Classification System

General Classification	Granular Material (35% or less passing No. 200 Sieve size)							Silt-Clay Materials (More than 35% passing No. 200 sieve size)			
Group Classification	A-1		A-3	A-2				A-4	A-5	A-6	A-7 A-7-5 A-7-6
	A-1-a	A-1-b		A-2-4	A-2-5	A-2-6	A-2-7				
Sieve Analysis (percent passing)											
No. 10	50 max	–	–	–	–	–	–	–	–	–	–
No. 40	30 max	51 max	–	–	–	–	–	–	–	–	–
No. 200	15 max	25 max	10 max	35 max	35 max	35 max	35 max	36 min	36 min	36 min	36 min
Characteristics of fraction passing No. 40											
Liquid Limit	–		–	40 max	41 min	40 max	41 min	40 max	41 min	40 max	41 min
Plasticity Index	6 max		NP	10 max	10 max	11 min	11 min	10 max	10 max	11 min	11 min
Usual types of significant constituent materials	Stone fragments gravel and sand		Fine Sand	Silty or clayey gravel and sand				Silty soils		Clayey soils	
General rating as subgrade	Excellent to good							Excellent to good			

22308-12_T02.EPS

Figure 5 Fine sand.

- *Clay* – Clay is the finest size of soil particles. Clay is very cohesive. When wet, clayey soils feel like putty and can be easily molded and made into long ribbons (see *Figure 7*). When dry, clay is very strong and clumps are difficult to crush. Clay is an expansive soil. It swells and shrinks with moisture changes, so pure clay is not suitable for building. However, a small amount mixed uniformly with granular material is desirable.

- *Organic matter (top soil) and colloid clays* – Organic matter is partly decomposed vegetable and animal material. Organic matter is sometimes called humus or peat and is usually soft, fibrous, and may have an offensive odor when warm. This material is not suitable for building or as fill because as it decays it loses volume, which may cause air pockets that make the ground unstable. Colloidal clays are very fine clay particles that remain suspended in water for long periods of time and don't settle easily under the force of

Figure 7 Clay ribbon.

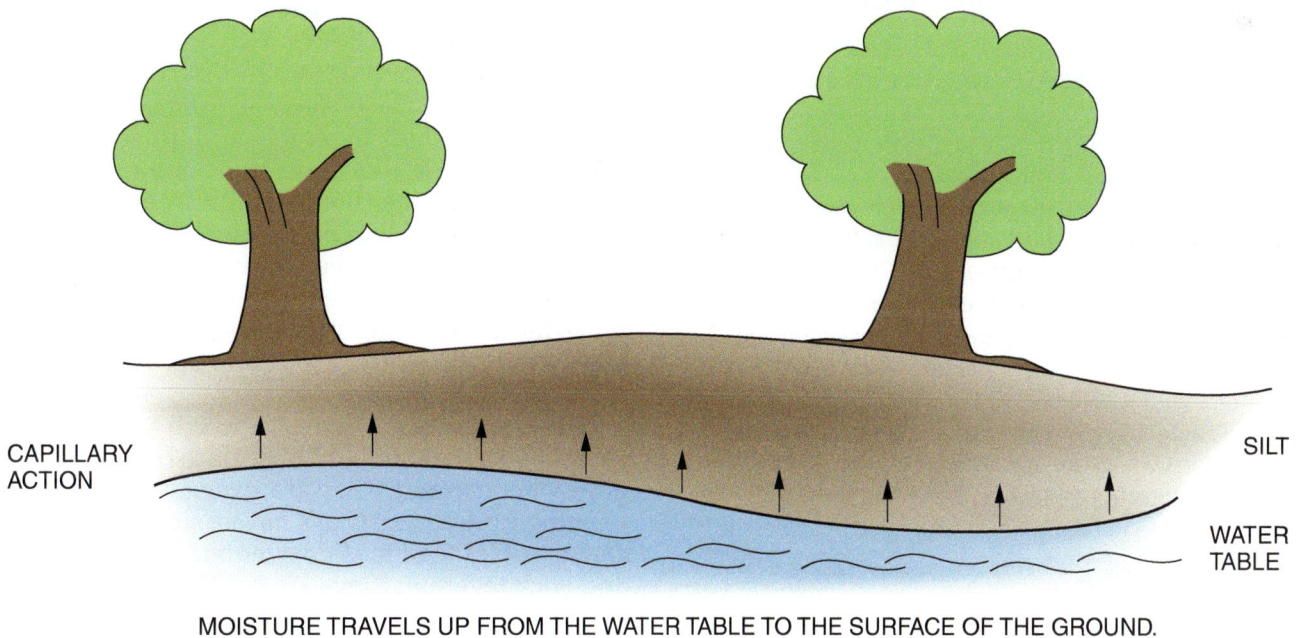

CAPILLARY ACTION

SILT

WATER TABLE

MOISTURE TRAVELS UP FROM THE WATER TABLE TO THE SURFACE OF THE GROUND.

Figure 6 Capillary action.

gravity. Individual particles cannot be seen with the naked eye. Colloidal clays are very susceptible to swelling and shrinking, so they are not suitable for construction.

1.2.0 Properties of Soils

There are many different types of soils and each type reacts differently to environmental conditions such as moisture and temperature. This reaction helps determine the strength and suitability of a soil for a construction project. The most useful characteristics in predicting the behavior of a soil are grain size, shape, surface texture, and soil composition. *Table 3* summarizes the properties of various soils that may be encountered on a job site.

1.2.1 Grain Size

Sieves are used to measure the grain size of soil particles (*Figure 8*). The number of the sieve tells you how many holes there are per linear inch of screen. A No. 10 sieve has 10 holes per inch, so particles smaller than a tenth of an inch can pass. A No. 200 sieve has 200 holes per inch, so it passes only very small particles. The size and distribution of soil particles throughout the soil mixture are important factors that determine a soil's behavior under load. *Table 4* shows the range in particle sizes of different soil materials based on the AASHTO classification.

Soil-Related Safety

The US Occupational Health and Safety Administration (OSHA) views soils from a safety perspective. OSHA issued its first Excavation and Trenching Standard in 1971 for the purpose of protecting workers from excavation hazards. OSHA developed the standard in order to prevent cave-ins that were injuring and killing workers. The purpose of the standard is to specify sloping, benching, and mechanical protection measures for specific types of soils. The standard requires that excavations be inspected by a competent person, which is defined as "one who is capable of identifying existing and predictable hazards in the surroundings or working conditions which are unsanitary, hazardous, or dangerous to employees, and who has authorization to take prompt corrective measures to eliminate them."

Notice in *Table 4* that silts and clays are not measured by sieve. These soils are measured using a hydrometer, which measures liquid content. For most construction projects, having soil that consists of predominately coarse-grain soil, such as sand, is preferable to fine-grain soil, such as clay. Sands are very permeable, so they quickly drain water. Clay particles attract water, so they swell and shrink with changes in moisture, causing movement in the ground. Permanent structures should never be built on purely clay soils. The best soil mixture for construction is one that contains enough small particles to fill the voids between large particles. This is called well-graded soil.

1.2.2 Grain Shape

Soil mixtures that contain a lot of gravel and sand are good for highway and building projects. Crushed rock and gravel have angular surfaces that resist sliding. This increases the stability of the soil under load. Small soil particles, such as fine sand, can easily slide into the voids between the larger and more angular gravel, increasing the density of the soil mass (*Figure 9*). Building specifications often call for a soil mixture that contains certain percentages of crushed rock particles or gravel and fine sands. This mixture of angular particles with rough edges is desirable because it can be compacted to a high density.

1.2.3 Surface Texture

The strength of a soil mixture is partly determined by the surface texture of soil particles. Friction is the resistance two materials have to sliding over each other. Soil particles with a smooth, slick surface (such as quartz sand that has been polished by water action) slip easily over one another; they have a low amount of friction. Soil mixtures that are high in quartz content have little resistance to deformation when under pressure. However, a small amount of this type of material can help the soil mixture compress more easily, because the smooth, slick particles easily slip into air pockets between larger, more angular particles.

Gravels and sands that have rough angular surfaces have a high amount of surface friction and a grainy appearance. They are strong and resist deforming under load. Clays have a smooth texture; clay particles tend to slide over one another easily and stick together. Clays are easily deformed.

Table 3 Soil Properties

Soil Texture	Visual Detection of Particle Size and General Appearance of Soil	Squeezed in Hand	Soil Ribboned Between Thumb and Finger
Sand	Soil has a granular appearance in which the individual grain sizes can be detected. It is free-flowing when in a dry condition.	*When air dry*: Will not form a cast and will fall apart when pressure is released. *When moist*: Forms a cast which will crumble when lightly touched.	Cannot be ribboned.
Sandy Loam	Essentially a granular soil with sufficient silt and clay to make it somewhat coherent. Sand characteristics dominate.	*When air dry*: Forms a cast which readily falls apart when lightlytouched. *When moist*: Forms a cast which will bear careful handling without breaking.	Cannot be ribboned.
Loam	A uniform mixture of sand, silt, and clay. Grading of sand fraction quite uniform from coarse to fine. It is mellow, has a somewhat gritty feel, yet is fairly smooth and slightly plastic.	*When air dry*: Forms a cast which will bear careful handling without breaking. *When moist*: Forms a cast which can be handled freely without breaking.	Cannot be ribboned.
Silt Loam	Contains a moderate amount of the finer grades of sand and only a small amount of clay. Over half of the particles are silt. When dry it may appear quite cloddy, which can be readily broken and pulverized to a powder.	*When air dry*: Forms a cast which can be freely handled. Pulverized, it has a soft, flour-like feel. *When moist*: Forms a cast which can be freely handled. When wet, soil runs together and puddles.	It will not ribbon, but has a broken appearance, feels smooth, and may be slightly plastic.
Silt	Contains over 80 percent silt particles, with very little fine sand and and clay. When dry, it may be cloddy, and readily pulverizes to powder with a soft flour-like feel.	*When air dry*: Forms a cast which can be handled without breaking. *When moist*: Forms a cast which can be freely handled. When wet, it readily puddles.	It has a tendency to ribbon with a broken appearance, feels smooth.
Clay Loam	Fine textured soil breaks into very hard lumps when dry. Contains more clay than silt loam. Resembles clay in a dry condition; identification is made on the physical behavior of moist soil.	*When air dry*: Forms a cast which can be handled freely without breaking. *When moist*: Forms a cast which can be freely handled without breaking. It can be worked into a dense mass.	Forms a thin ribbon which readily breaks, barely sustaining its own weight.
Clay	Fine textured soil breaks into very hard lumps when dry. Difficult to pulverize into a soft flour-like powder when dry. Identification is based on cohesive properties of the moist soil.	*When air dry*: Forms a cast which can be freely handled without breaking. *When moist*: Forms a cast which can be freely handled without breaking.	Forms long, thin, flexible ribbons. Can be worked into a dense, compact mass. Considerable plasticity.
Organic Soils	Identification based on the high organic content. Muck consists of thoroughly decomposed organic material with considerable amount of mineral soil finely divided with some fibrous remains. When considerable fibrous material is present, it may be classified as peat. The plant remains or sometimes the woody structure can easily be recognized. Soil color ranges from brown to black. They occur in lowlands, swamps, or swales. They have high shrinkage upon drying.		

Figure 8 Soil sieves.

1.2.4 Soil Composition

What is normally considered to be soil is usually a mixture of two or more soil types, organic matter, and multiple chemicals. No single soil type is acceptable for construction. Gravel and sand have great strength but no significant elasticity, plasticity, or cohesiveness. Once gravel or sand is deformed under load, it stays deformed. Silt and clay are very plastic and cohesive, but have almost no strength. These soils easily compress and break apart under load. Soil that is easy to crush, crumble, or break apart is said to be friable.

The best soil for construction is a mixture that contains a high amount of large-particle soils, such as gravel and sand, and a small amount of small particle soil, such as silt and clay. Soils that contain moderate to large amounts of organic

matter are not suitable for construction because as the matter decays, it will decrease in mass and cause settling to occur in the soil. This may damage building foundations and road surfaces. Looking back at *Table 1* and *Table 2*, it can be seen that soils with excellent to good subgrade ratings contain a high amount of gravel and sand, and a small amount of silt or clay.

Although it is difficult to determine the chemical components of a soil, they can affect the performance of the soil under load. In particular, chemicals may affect the amount of water the soil can absorb, which changes the strength of the soil.

1.2.5 Engineering Properties of Soil

For the soil at a site to be strong enough for construction, it must have characteristics that allow it to be stable under a variety of environmental conditions. These characteristics are called the engineering properties of soil and include the following:

- Permeability
- Elasticity
- Plasticity
- Cohesion
- Shearing strength
- Shrinkage and swelling
- Frost susceptibility

Permeability is the ability of a soil to allow water to flow through it. Soils that hold water swell and cause building foundations and road surfaces to move. Coarse-grained soil, such as sand, is more permeable than fine-grained soil, such as clay, so

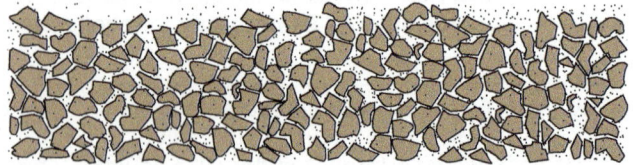

Figure 9 Different sized particles make soil easy to compact.

Table 4 Soil Particle Size Using US Sieve Sizes

| Soil Type | Particle Diameter | | Passing | Retained |
	millimeters	inches		
Gravel	76.2 to 2.0	3.0 to 0.08	3-inches	No. 10
Coarse Sand	2.0 to 0.42	0.08 to 0.017	No. 10	No. 40
Fine Sand	0.42 to 0.074	0.017 to 0.003	No. 40	No. 200
Silt	0.074 to 0.005	0.003 to 0.0002	No. 200	—
Clay	0.005 to 0.001	Less than 0.0002		
Colloidal Clay	Less than 0.001	Less than 0.00004		

22308-12_T04.EPS

untreated clayey sites are poor choices for construction. *Table 5* lists the coefficients of permeability for various soil mixtures. The coefficient is measured in the amount of water that can drain from the soil in a period of time. In this case, it is the number of feet of water that drain in a minute.

Elasticity is the ability of the soil to return to its original shape after having been deformed by some temporary force, such as vehicular traffic. In most soils, this deformation is small, but it is important in highway engineering. Coarse-grained soils have almost no elasticity, so mixing clay with sand gives the resulting mixture some elasticity.

Plasticity is the ability of a soil to be molded into a different shape without breaking apart. Plasticity is used to classify fine-grained soils and is directly related to their moisture content. As moisture increases from 0 percent, plastic soils go from solids to liquids (*Figure 10*). The moisture content at which the soil goes from semisolid to plastic is called the **plastic limit**, and the moisture content at which the soil goes from plastic to liquid is called the liquid limit. These are referred to as the Atterberg limits, named after the Swedish scientist who defined them Albert Atterberg.

Clay is a plastic soil. When clay is saturated with water, it is a very thick liquid, but it dries into solid clumps. At some point between the liquid and solid states, the clay can be molded and is plastic.

The shear strength of a soil is its ability to withstand pressure without breaking apart. A soil's shear strength is determined by its cohesiveness and friction. Cohesiveness is the ability of the soil particles to stick together, while friction is the resistance the soil particles have to sliding over one another. Clay is a cohesive soil, while sand has a high amount of friction. Cohesion and friction, and thus shear strength, vary with moisture, **loading**, and other factors. Clay is extremely cohesive when dry, but less cohesive when wet.

Table 5 Coefficient of Permeability for Various Soil Mixtures

Type of Sand (Unified Soil Classification System)	Coefficient of Permeablility ft/min
Sandy Silt	0.001 to 0.004
Silty Sand	0.004 to 0.01
Very Fine Sand	0.01 to 0.04
Fine Sand	0.04 to 0.1
Fine to Medium Sand	0.1 to 0.2
Medium Sand	0.2 to 0.3
Medium to Coarse Sand	0.3 to 0.4
Coarse Sand and Gravel	0.4 to 10

22308-12_T05.EPS

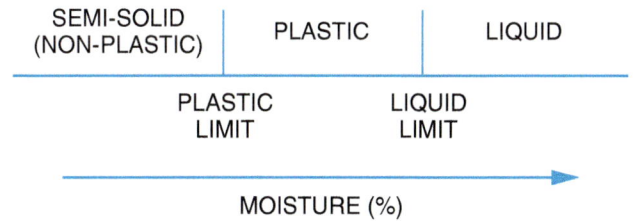

SEMI-SOLID (NON-PLASTIC)	PLASTIC	LIQUID
	PLASTIC LIMIT	LIQUID LIMIT

MOISTURE (%)

22308-12_F10.EPS

Figure 10 Atterberg limits.

Shrinkage and swelling of soils due to moisture changes are problematic in construction because the resulting movement of the ground may damage building foundations, underground utilities, and road surfaces (*Figure 11*). Clay soils are particularly prone to shrinkage and swelling with moisture changes.

Frost is troublesome for soil in cold climates. In freezing weather, moisture in the soil freezes and causes the soil to expand; the movement of the ground may damage unprotected building foundations and road surfaces. The moisture can come from above the ground, such as melting snow, or below the ground, such as a shallow water table. Where freezing temperatures are prolonged and the frost line penetrates deep into the soil, ice layers cause the soil to heave at the surface.

1.3.0 Soil Density

The most important characteristic to a building site is its **in situ** soil density. Soil density is a ratio of a soil's weight to its volume (usually per cubic foot). Each site will have a natural density that depends on the engineering properties of the soil, as well as the stressors to which the site is subjected. Keep in mind that no single soil type is ideal for construction, but rather it is a mixture of types that make a good building site.

22308-12_F11.EPS

Figure 11 Cracked road surface.

A soil's origin determines its engineering properties and can be classified as organic, inorganic, and rock. Organic soils are made from living matter such as plants or animals, while inorganic soils are made from rocks that have been broken down due to weathering (rain, wind, freezing temperatures, or sun). Rock is any solid or packed material that cannot be easily dug or loosened by machinery. The soils that you work with will most likely be a mixture of these three categories.

Stressors that affect a site's soil density can be natural or man-made. Weather is the primary natural stressor; soil that is subject to heavy rains and freeze/thaw cycles have a greater density than the same soil that is not exposed to such conditions. Man-made stressors are usually determined by land use. Soils at a site subjected to a great deal of traffic or vibration from nearby roadways have a higher density than if the site were located far away from heavy traffic.

Engineers measure the soil density to determine whether the site needs additional compacting or stabilization. It is doubtful that an equipment operator will be required to do soil density tests, so they are discussed very briefly in the following paragraphs to familiarize you with the technology. There are three types of tests primarily in use today: sand cone testing, penetrometer testing, and nuclear density testing.

1.3.1 Sand Cone Test

Many engineers consider the sand cone test to be the most accurate testing method. To perform this test, dig a round hole with a volume of one-tenth of a cubic foot. Weigh the soil taken out of the hole. Send the soil removed from the hole to a laboratory. At the laboratory, the soil is dried and weighed again to determine how much of the total weight is water. An additional sample is taken

and analyzed the same way so the lab can plot a moisture density curve of the site.

The specific volume of the hole is determined by using a calibrated jar of dry sand. The sand is calibrated so that the volume can be easily determined from the weight. Weigh the jar full of dry sand. Fill the hole with sand using the jar and cone device shown in *Figure 12*. Weigh the container and remaining sand again. Subtract that from the original weight to determine the weight of sand in the hole. Use the conversion factor to convert the weight of the sand into the volume of the hole.

The dry weight of the soil removed from the hole is divided by the volume of sand needed to fill the hole. This gives the density of the compacted soil in pounds per cubic foot. This result is compared to the theoretical maximum density, which gives the relative density of the soil that was just compacted.

1.3.2 Penetrometer Testing

The cone penetrometer is used to determine the bearing capacity of soil. *Figure 13* shows a portable penetrometer. There are also truck-mounted penetromers that are used for extensive soil testing. The penetrometer density test is also called a cone test because of the conical attachment that is screwed onto the end of the probe. The cone and the friction sensor behind it contain instrumentation. The cone reacts to the pressure of the soil, while the friction sleeve detects the resistance (friction) of the soil.

1.3.3 Soil Density Testing

Soil density testing involves either placing the testing machine on the ground to get a reading or inserting a probe attached to the machine into a small hole drilled into the soil. Either method sends impulses into the soil that are reflected back to the device and recorded. Denser soils absorb more impulses. The more a soil is compacted, the fewer impulses are returned. The lab technician then creates a moisture density curve of the site in the same way as the sand cone test. The test can be performed with nuclear instruments such as the one shown in *Figure 14A*, or non-nuclear instruments.

Soil Density

In the 1930s, experiments conducted by American architect R.R. Proctor established the relationship between soil density and moisture content. Proctor determined that varying the water content of soil had a direct bearing on the amount of compacting required. Soil with no moisture (such as sand) is impossible to compact. As water is added, the soil becomes easier to compact. The conclusion to be drawn from Proctor's studies is that in order to be compacted, soil must contain the correct amount of water.

WARNING!

Nuclear equipment should only be used by trained personnel. Exposure to nuclear materials may cause serious illness.

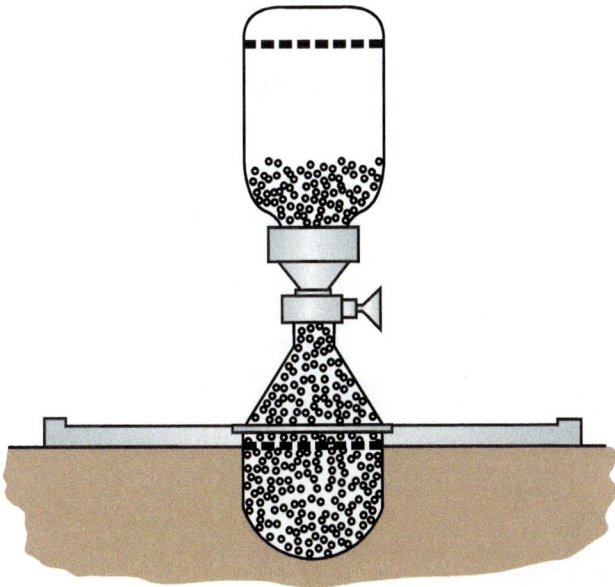

Figure 12 Sand cone test.

Figure 13 Portable penetrometer.

The electrical density gauge (EDG) shown in Figure 14B is a non-nuclear instrument that measures the physical properties of compacted soils used in roadbeds and foundations. This device is battery-operated and can determine the wet and dry density, gravimetric moisture content, and percent of compaction. The kit includes a console, four electrodes, a hammer, soil sensor and cables, template, temperature probe, and battery charger.

1.4.0 Effects of Moisture on Soils

The single greatest factor affecting a soil's behavior is its moisture content. Clay attracts and holds moisture, causing the clay to swell and increase in volume. Sand is very permeable to water, allowing quick drainage, so its volume changes very little with increased moisture.

Moisture also affects the compaction of soil. Moisture surrounds rough-surfaced particles of sand and allows them to slide across one another more easily than with dry particles, making compaction easier. Soil mixtures that are dry have a high degree of surface friction. This makes them difficult to compact, tending to leave air pockets and lower soil density. Soil mixtures that contain too much moisture can make the soil rubbery and resistant to compacting. They may even approach the point where they are fluid. This makes them impossible to compact until some of the moisture is removed. When a soil has just the right amount of moisture to achieve maximum density, it is said to have optimum moisture content.

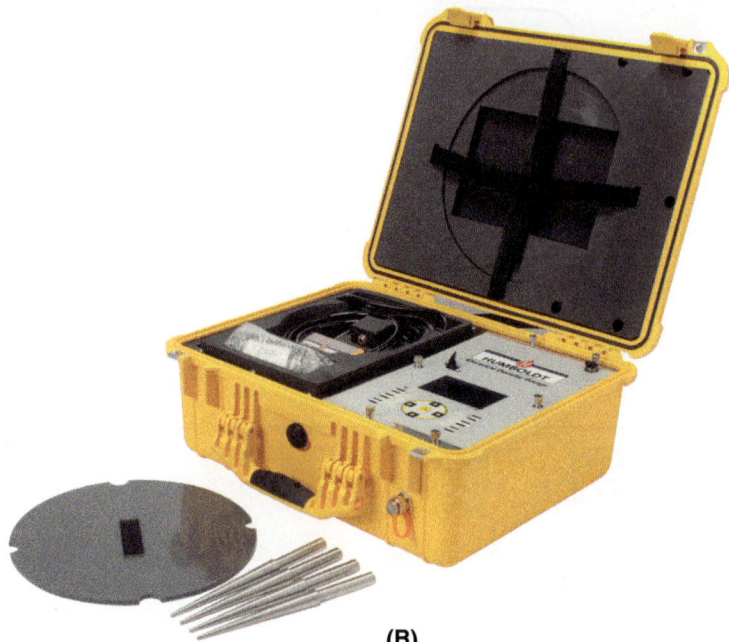

(A)   (B)

22308-12_F14.EPS

Figure 14 Soil density testers.

Figure 15 compares the moisture content and density of various soils. The highest point on the curve is maximum density. Soil density is a ratio of a soil's weight to its volume, so it is measured in pounds per cubic foot. When there are more soil particles in a cubic foot of space, the weight and density both increase. *Figure 15* shows that the soil with the highest density and lowest water content is well-graded sand, which is a mixture of predominantly sand and small amounts of clay and silt.

On the Beach

Think about a beach. Near the water's edge, the sand is tightly compacted. When you walk on it, the sand feels hard. As you move away from the water, the sand becomes softer, and your feet will sink into it. This condition demonstrates the relationship between moisture and compaction.

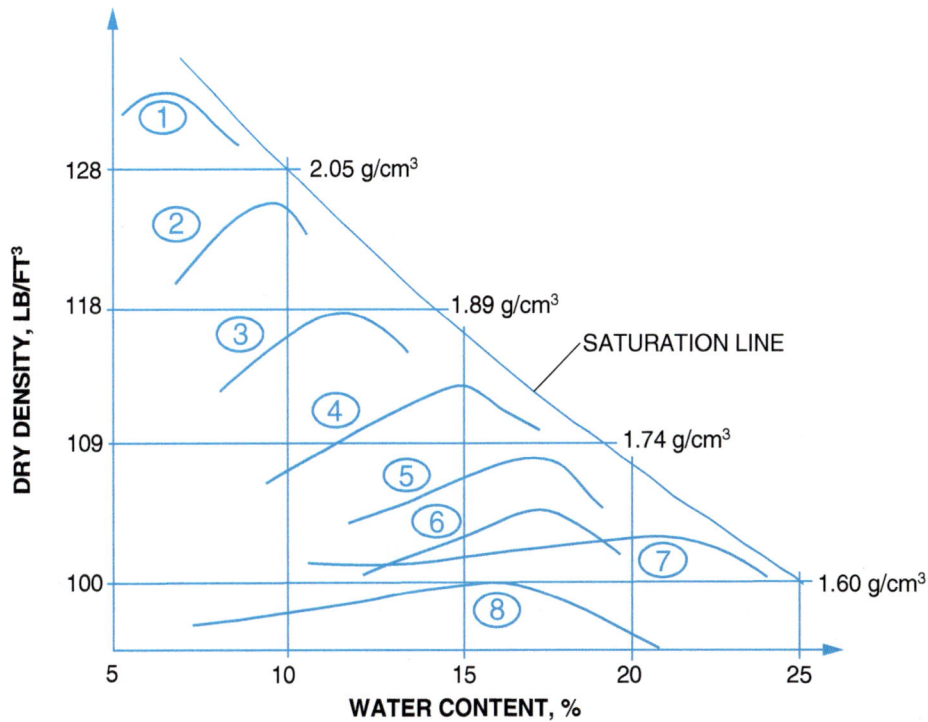

Figure 15 How moisture affects soil density.

No.	Description	Sand %	Silt %	Clay %	w_L	I_P
1	Well-Graded Sand	88	10	2	16	–
2	Well-Graded Sandy Marl	72	15	13	16	–
3	Medium Sandy Marl	73	9	18	22	4
4	Sandy Clay	32	33	35	28	9
5	Silty Clay	5	64	31	36	15
6	Loess Silt	5	85	10	26	2
7	Clay	6	22	72	67	40
8	Poorly Graded Sand	94	6	–	–	–

22308-12_F15.EPS

1.0.0 Section Review

1. Which of these soils has a tendency to ribbon?

   a. Sand
   b. Sandy loam
   c. Silt loam
   d. Clay

2. The grain size of soil is measured using a _____.

   a. sieve
   b. scale
   c. micrometer
   d. ruler

3. Soil density is measured as the _____.

   a. ratio of weight to volume
   b. ratio of moisture to volume
   c. pressure per cubic inch
   d. ratio of rock to sand

4. Which of these soils has the lowest percentage of clay content?

   a. Medium sandy marl
   b. Loess silt
   c. Sandy clay
   d. Silty clay

2.0.0 EXCAVATING SOIL

Objective 2

Describe the factors that affect soil excavation.
a. Explain what the swell factor is and how to calculate the swell factor of soils.
b. Explain what the shrink factor is and how to calculate the shrink factor of soils.
c. Describe how swell and shrink factors affect cycle times and equipment selection.

Performance Tasks 2 and 3

Read results from a field density test and explain what additional compaction effort is needed.
Compute shrinkage and relative compaction for two different types of soil.

Trade Terms

Banked: Any soil mass that is to be excavated from its natural position.

Consolidation: To become firm by compacting the particles so they will be closer together.

Settlement: To become firm by compacting the particles so they will be closer together.

Shrinkage: Decrease in volume when soil is compacted.

Swell: Increase in volume when soil is excavated.

Swell factor: The ratio of the banked weight of a soil to the loose weight of a soil.

As soil is excavated from its natural position, it is mixed with air and gives the appearance of swelling in volume. When the soil is used as fill, it is compacted, pushing air pockets out, and gives the appearance of shrinking in volume. This is important to know because heavy equipment is rated for maximum volumes and weights, so the operator needs to be able to select the proper equipment for the job.

Soil is measured in cubic units, with cubic yards being the most common measure. In the construction industry, a cubic yard is casually referred to as a yard. Undisturbed soil is called banked, while excavated soil is called loose. *Table 6* shows the typical loose and bank volumes and weights of many common soil mixtures.

2.1.0 Swell Factor

In *Table 6* it can be seen that the loose weight of the soil is always less than the bank weight for the same volume. This is because air, which weighs almost nothing, is mixed with the soil when it is excavated, increasing the volume and decreasing the weight per cubic yard. This increase in volume is called **swell**. To select the correct equipment for an excavation job, the operator needs to figure out how much the volume of compacted soil will increase when it is excavated.

Table 6 shows that wet clay has a banked weight of 3,500 and a loose weight of 2,800 pounds per cubic yard. The **swell factor** of a soil is found by dividing its banked weight by its loose weight. For this example, 3,500 pounds divided by 2,800 pounds equals 1.25 (3,500 ÷ 2,800 = 1.25). One cubic yard of banked wet clay equals 1.25 (1¼) cubic yards when it is excavated. See *Figure 16*.

Refer to *Figure 17* when reading the following example. A material has a banked weight of 1,000 pounds per cubic yard and a loose weight of 800 pounds per cubic yard. This means 800 pounds is used to make one loose cubic yard with 200 pounds of material left over. Calculate the swell factor by dividing the banked weight by the loose weight (1,000 ÷ 800 = 1¼ or 1.25). The calculation shows that one banked cubic yard equals 1¼ or 1.25 loose cubic yards.

Returning to the original problem (see *Figure 18*), wet clay weighs 3,500 pounds in a banked cubic yard and 2,800 pounds in a loose cubic yard, so after the first loose cubic yard is made, 700 pounds of material is left over. Those 700 pounds are enough

Effect of Soil Compaction

Water from such sources as rain and melting snow is absorbed by the soil. However, areas occupied by buildings, roads, and parking lots do not absorb water. Such areas are said to be impervious to water. The problem is that the water still needs to go somewhere. Populated areas need to have stormwater-runoff systems to compensate for the impervious area. For that reason, municipalities maintain stormwater-processing facilities to filter the water and return it to lakes, rivers, etc. As the amount of impervious area grows, the need to expand stormwater management grows with it. Many municipalities charge builders and developers an impact fee to cover the cost of stormwater management. Property owners may also be charged an annual fee for this purpose.

Table 6 Material Weights

Material	Loose Weight		Bank Weight	
	kg/m³	lb/yd³	kg/m³	lb/yd³
Clay – Natural Bed	1,660	2,800	2,020	3,400
Dry	1,480	2,500	1,840	3,100
Wet	1,660	2,800	2,080	3,500
Clay and Gravel – Dry	1,420	2,400	1,660	2,800
Wet	1,540	2,600	1,840	3,100
Decomposed Rock				
75% Rock, 25% Earth	1,960	3,300	2,790	4,700
50% Rock, 50% Earth	1,720	2,900	2,280	3,850
25% Rock, 75% Earth	1,570	2,650	1,960	3,300
Earth – Dry packed	1,510	2,550	1,900	3,200
Wet Excavated	1,600	2,700	2,020	3,400
Loam	1,250	2,100	1,540	2,600
Gravel – Dry (¼ to 2 inches)	1,690	2,850	1,900	3,200
Wet (¼ to 2 inches)	2,020	3,400	2,260	3,800
Sand – Dry, loose	1,420	2,400	1,600	2,700
Damp	1,690	2,850	1,900	3,200
Wet	1,840	3,100	2,080	3,500
Sand and Clay – Loose	1,600	2,700	2,020	3,400
Sand and Gravel – Dry	1,720	2,900	1,930	3,250
Wet	2,020	3,400	2,230	3,750

22308-12_T06.EPS

to make ¼ of a cubic yard more of loose material (700 ÷ 2,800). This material has a swell factor of 1.25 because 1 banked cubic yard makes 1.25 loose cubic yards. This is calculated by dividing the banked weight by the loose weight (3,500 ÷ 2,800).

2.2.0 Shrink Factor

Fill is brought onto the job site to bring the existing ground to the desired grade. When a fill site compacts naturally over time due to rain, freeze/thaw cycles, traffic, or the weight of a building, it is called settlement or consolidation; that is, the soil particles rearrange themselves so that they fit together more tightly. This is not desirable because the ground movement can damage foundations and road surfaces. To avoid settlement, the soil at construction sites is uniformly compacted before building begins.

When soil is compacted, its volume shrinks. The amount of shrinkage depends on the characteristics of the soil, the soil's moisture content, and the degree of compaction. The shrink factor can be calculated by dividing the compacted volume by the fill volume. The tricky part is determining the fill volume. Fill volume can be divided into the following three categories:

- *In situ* – There is no cut and no fill. The natural ground is cleared, grubbed, and then compacted.
- *Loose fill* – Fill is ordered and trucked in by the loose cubic yard, and then compacted.
- *Banked fill* – Fill is cut from another location, hauled to the building site, and compacted.

Fortunately, the method of calculating shrinkage is the same regardless of how the fill is obtained. The formula is: compacted volume divided by the loose volume (compacted volume/loose volume). For example, when 100 cubic yards of fill is compacted to 80 cubic yards, then the shrink factor is 0.8 (80 ÷ 100 = 0.8). See *Figure 19*.

The shrink factor can also be calculated by dividing the weight of a cubic yard of loose fill by the weight of a cubic yard of compacted fill (loose weight / compacted weight). For example, if there are 2,800 pounds in a cubic yard of loose fill, and it takes 3,500 pounds of loose fill to make one cubic yard of compacted fill, the shrink factor is 0.8, because 2,800 ÷ 3,500 equals 0.8. *Figure 20* shows that it takes 1¼ (1.25) cubic yards of loose fill to make a single cubic yard of compacted fill.

Now, how is the shrink factor used to determine how much fill is needed? When a material has a shrink factor of 0.8, then a compacted cubic

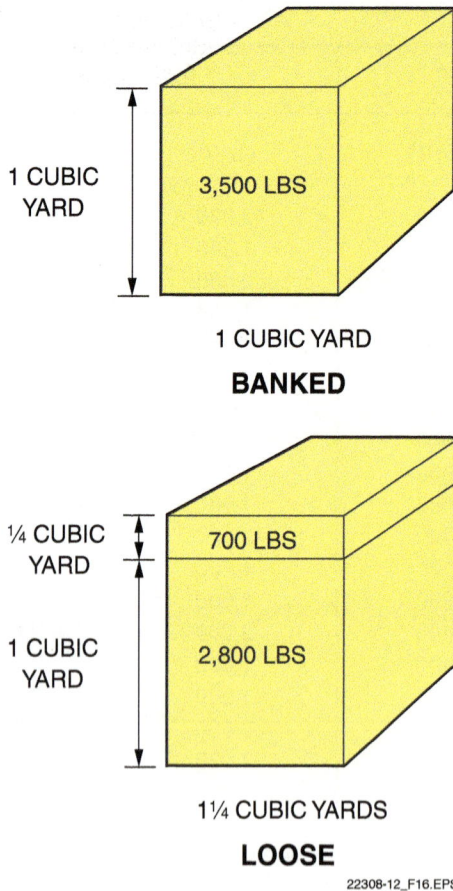

Figure 16 Swell factor.

yard equals 1¼ loose cubic yards. The amount of fill can be calculated by dividing the compacted volume by the shrink factor (1 ÷ 0.8 = 1¼ or 1.25). If an area needs to be filled with the volume of 10 cubic yards, using fill that has a shrink factor of 0.80, 12.5 cubic yards of fill (10 ÷ 0.8 = 12½ or 12.5) would need to be ordered.

These numbers should sound familiar, since they were used previously in the swell factor section. Shrinkage can be thought of as the opposite of swell. But keep in mind that when soil is being excavated, it may not be compacted to the density required at the building site. Therefore, assuming that a cubic yard of in situ banked soil will equal a cubic yard of compacted fill may get you in trouble.

2.3.0 Using Shrink and Swell Factors

In this section, you will learn to apply shrink and swell factors. You will also learn how these factors affect cycle times.

Overloading equipment can make it hard to control and places the operator at risk for an accident. Further, it can damage the equipment, resulting in costly down time and repairs. To ensure that equipment is not overloaded, the operator must apply the load and swell factors. Information about the equipment's weight and volume capacities can be found in the manufacturer's instructions.

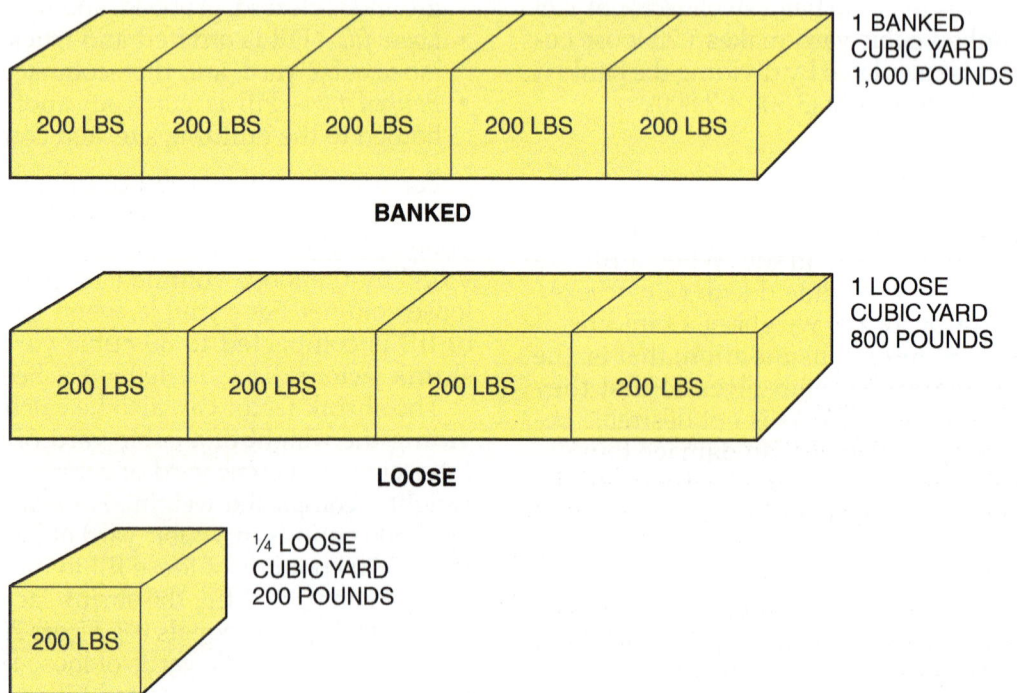

Figure 17 Swell factor example.

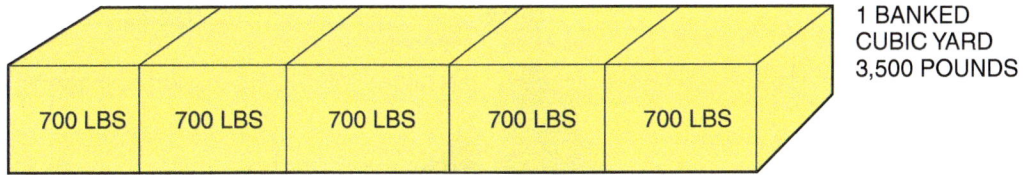

1 BANKED
CUBIC YARD
3,500 POUNDS

BANKED

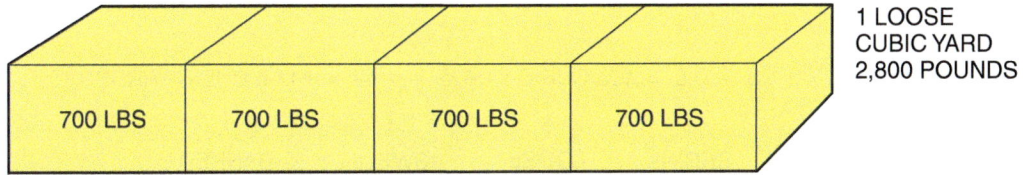

1 LOOSE
CUBIC YARD
2,800 POUNDS

LOOSE

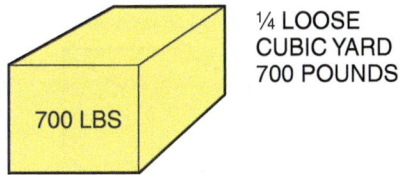

¼ LOOSE
CUBIC YARD
700 POUNDS

$$\text{SWELL FACTOR} = \frac{3,500}{2,800} = 1.25$$

22308-12_F18.EPS

Figure 18 Wet clay swell factor.

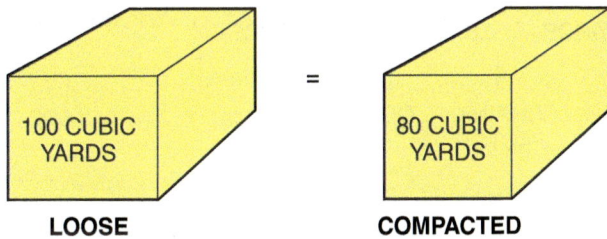

LOOSE   **COMPACTED**

$$\text{SHRINK FACTOR} = \frac{80}{100} = 0.8$$

22308-12_F19.EPS

Figure 19 Shrink factor using volume.

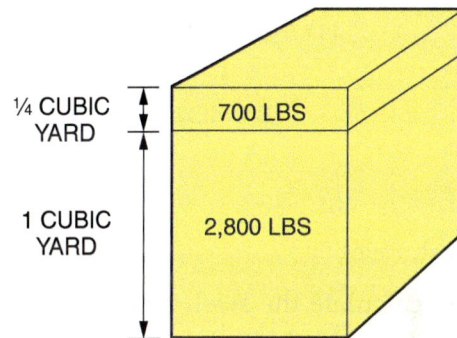

¼ CUBIC YARD — 700 LBS

1 CUBIC YARD — 2,800 LBS

LOOSE FILL
1¼ CUBIC YARDS

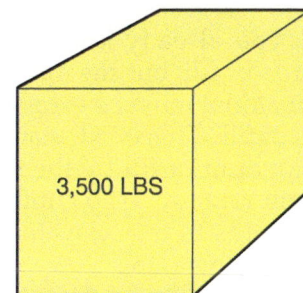

3,500 LBS

COMPACTED
1 CUBIC YARD

22308-12_F20.EPS

Figure 20 Shrink factor using weight.

2.3.1 Scenario

Let's assume that you need to excavate some damp sand from a borrow pit and take it to a building site. The sand has a banked weight of 3,200 pounds per cubic yard and a loose weight of 2,850 pounds per cubic yard (this information is from *Table 6*). You need to fill an area with a volume of 100 cubic yards. This material has a shrink factor of 0.8 when thoroughly compacted. This may be confusing, so refer to *Figure 21A* while carefully reading the following paragraph.

First, determine how much fill is needed. Since the compacted volume is 100 cubic yards and the shrink factor is 0.8, you need 125 cubic yards of loose fill (100 ÷ 0.8 = 125).

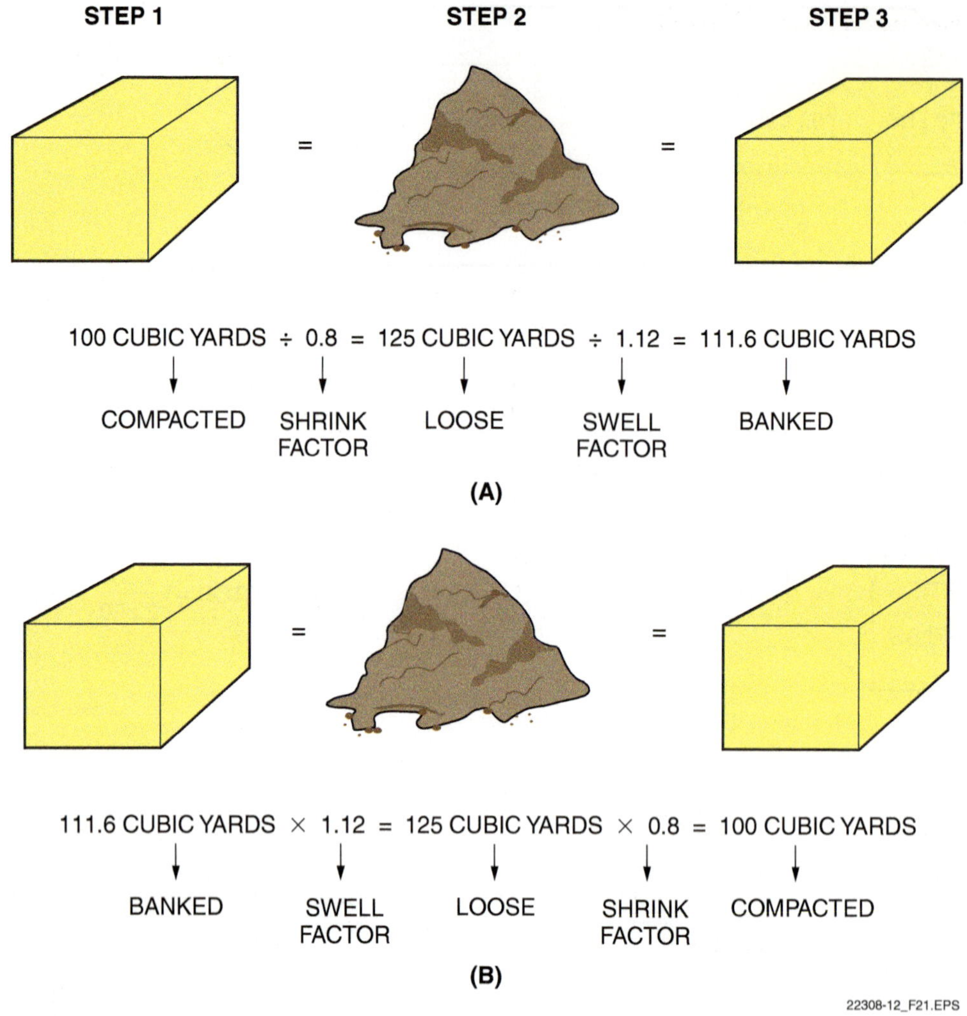

STEP 1 **STEP 2** **STEP 3**

100 CUBIC YARDS ÷ 0.8 = 125 CUBIC YARDS ÷ 1.12 = 111.6 CUBIC YARDS

COMPACTED   SHRINK   LOOSE   SWELL   BANKED
FACTOR   FACTOR

(A)

111.6 CUBIC YARDS × 1.12 = 125 CUBIC YARDS × 0.8 = 100 CUBIC YARDS

BANKED   SWELL   LOOSE   SHRINK   COMPACTED
FACTOR   FACTOR

(B)

22308-12_F21.EPS

Figure 21 Calculating fill.

Second, calculate the swell factor of the excavated material. Since the banked weight of the material is 3,200 pounds per cubic yard and the loose weight is 2,850 pounds per cubic yard, the swell factor is 1.12 (3,200 ÷ 2,850 = 1.1228, rounded to 1.12).

Third, you already know that you need 125 cubic yards of fill, but the question is how much banked material must be excavated to get 125 cubic yards of fill? This is calculated by dividing the desired amount of fill by the swell factor. 125 ÷ 1.12 equals 111.6, so 111.6 cubic yards of banked material needs to be excavated to get 125 cubic yards of loose material.

To prove that this calculation is correct, go through the problem backwards (see *Figure 21B*). When 111.6 cubic yards of fill is excavated, it swells to 125 cubic yards (111.6 multiplied by the 1.12 swell factor equals 124.9, rounded to 125). When this material is compacted, it shrinks to 100 cubic yards because it has a shrink factor of 0.8 (125 multiplied by the shrink factor of 0.8, equals 100).

Select the truck that best suits the needs of this excavation based on weight and volume. *Table 7* shows the volume and weight capacity for three articulated trucks. Your instinct is probably telling you to select the truck with the largest payload, so you can get your work done quickly. This is a good instinct, but you need to check that the weight capacity is acceptable for the material you are going to haul, because the cost of operating a large truck is greater than that of using a smaller truck.

You already know that the weight of the loose material is 2,850 pounds per cubic yard, so you need to calculate the loaded weight of the material.

> **NOTE**
>
> The term *heaped* refers to material that has been heaped on the equipment. It is commonly assumed that material will heap in a two to one slope. The term *struck* refers to material level with the sides of the equipment.

Table 7 Articulated Trucks Payload Capacities

Truck Model	Volume Capacity (heaped SAE 2:1*, cubic yards)	Volume Capacity (struck, cubic yards)	Rated Payload (calculated, pounds)
TRUCK A	18.8	14.5	52,000
TRUCK B	22.1	17.1	62,000
TRUCK C	30	22.8	84,000

*SAE 2:1 refers to the angle of repose for the loaded material. For every 2 feet of horizontal run, there is 1 foot of rise. Source: Caterpillar® website.

22308-12_T07.EPS

Truck A:

- Heaped weight is 18.8 multiplied by 2,850 or 53,580 pounds. This exceeds the weight capacity of the vehicle.
- Struck weight is 14.5 multiplied by 2,850 or 41,325 pounds. This is within the weight capacity of the vehicle.

Truck B:

- Heaped weight is 22.1 multiplied by 2,850 or 62,985 pounds. This exceeds the weight capacity of the vehicle.
- Struck weight is 17.1 multiplied by 2,850 or 48,735 pounds. This is within the weight capacity of the vehicle.

Truck C:

- Heaped weight is 30 multiplied by 2,850 or 85,500 pounds. This exceeds the weight capacity of the vehicle.

- Struck weight is 22.8 multiplied by 2,850 or 64,980 pounds. This is within the weight capacity of the vehicle.

All of the available trucks can carry the load when it is struck. To get the job done as quickly as possible, you would select the largest truck because it can do the job in six trips, whereas Truck A would require nine trips and Truck B eight trips. (Divide the total quantity of 125 yards by the struck volume capacity.)

> **NOTE**
> This exercise demonstrated equipment selection based on speed. Large vehicles cost more to purchase, maintain, and operate. To decide which truck is the most cost efficient, you would need to compare operating costs.

2.0.0 Section Review

1. The banked weight of clay is less than its loose weight.
   a. True
   b. False

2. Given soil with a loose weight of 2,600 pounds per cubic yard and a bank weight of 3,100 pounds per cubic yard, the swell factor is _____.
   a. 1.19
   b. 1.25
   c. 1.30
   d. 1.33

3. The shrink factor of soil is found by _____.
   a. observing it for 1 hour after it is moistened
   b. dividing its compacted volume by its loose volume
   c. dividing its banked weight by its loose weight
   d. multiplying its weight by its compacted volume

SECTION THREE

3.0.0 WORKING IN VARIOUS SOIL CONDITIONS

Objective 3

Describe working in various soil conditions.
  a. Describe the weight bearing and flotation properties of different soils.
  b. Explain how soil characteristics affect machine performance.
  c. Describe how soil conditions can affect trenching safety.

Different soils have different characteristics that affect the operation of heavy equipment. An operator must operate the equipment safely in different soil types and environmental conditions. Operators must know how to use the different types of heavy equipment to overcome problems caused by different soil characteristics. Three important characteristics that you need to keep in mind are the weight-bearing capability of a soil; the amount of traction a piece of equipment can get on a soil; and the resistance of the material to excavation.

> **NOTE**
> This section contains a number of tables that show technical information. Use the information in the tables to compare the characteristics of different soil types rather than trying to memorize the information.

3.1.0 Weight-Bearing and Flotation Properties

It is important to understand the relationship between the equipment and the weight bearing capacity of the soil on which the equipment is used. This section explains that relationship.

3.1.1 Weight Bearing

The weight-bearing capability of soil is how much weight it can safely support. It is related to soil density, so when a soil density is low, meaning it is not well compacted, its weight-bearing capacity is low. Compaction increases a soil's weight-bearing capability and makes it more stable. *Table 8* shows the approximate weight bearing capability of different types of soils, assuming a high degree of compaction. Weight bearing capability is usually measured in terms of pounds per square inch (psi), but it can be measured in any terms of weight per unit of measure, such as pounds per square foot. The abbreviation psi means the number of pounds that the soil can support on one square inch without sliding, excessively deforming, or collapsing.

Figure 22 illustrates the concept of pounds per square inch of pressure. In *Figure 22A*, a 200-pound person stands on one foot on a block of wood that is a 1-inch cube. The surface area in contact with the ground is 1 square inch, so the ground under the block is supporting 200 psi.

In *Figure 22B* the same person is standing with each foot on a 1-inch cube, so the ground under each cube is supporting 100 psi (200 pounds ÷ 2). This same concept may be applied to heavy equipment, because the total weight of the machine needs to be supported on the area of the tires or tracks in contact with the ground. When the weight on the tires exceeds the weight-bearing ability of the soil, the vehicle sinks into the ground until the pressure between the tires and the ground is equal.

3.1.2 Flotation

The ability of a tire or track to support the machine's weight on the ground is called flotation. To determine flotation capacity, consider the total weight of the machine, including its payload, the contact area (in square inches) of the tires or tracks on the ground, and the weight bearing ability of the ground. The total weight of the vehicle is divided by the total contact area of the tires or tracks with the ground to determine the pressure of the machine on the ground. If this pressure is greater than the load-carrying capability of the ground, the equipment will sink. Larger tires improve flotation because more tire surface is in contact with the ground. Reducing tire pressure can also increase flotation because more tire tread is in contact with the ground (*Figure 23*).

Table 8 Weight-Bearing Capability

Material	Weight-Bearing Capability in lb per in² (psi)
Rock (semi-shattered)	70
Rock (solid)	350
Clay (dry)	55
Clay (medium dry)	27
Clay (soft)	14
Gravel (cemented)	110
Sand (compact dry)	55
Sand (clean dry)	27
Quicksand	7

22308-12_T08.EPS

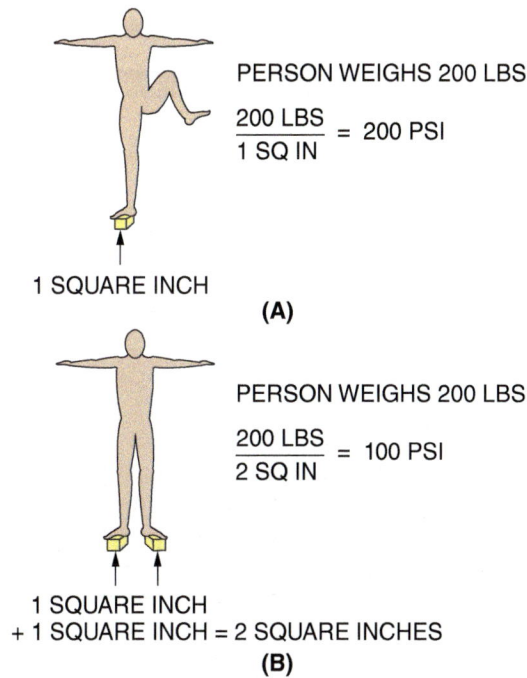

PERSON WEIGHS 200 LBS

$$\frac{200 \text{ LBS}}{1 \text{ SQ IN}} = 200 \text{ PSI}$$

1 SQUARE INCH

(A)

PERSON WEIGHS 200 LBS

$$\frac{200 \text{ LBS}}{2 \text{ SQ IN}} = 100 \text{ PSI}$$

1 SQUARE INCH
+ 1 SQUARE INCH = 2 SQUARE INCHES

(B)

22308-12_F22.EPS

Figure 22 Determination of psi.

> **WARNING!**
>
> Always keep tire inflation pressure within the manufacturer's recommendation. Under-inflation may damage tires and cause a sudden tire failure, which may result in an accident.

3.2.0 Effects of Soil Conditions on Machine Performance

The condition of the soil on which a machine is working can have a significant effect on the performance of the machine. Soil conditions that affect machine performance include rolling resistance, digging resistance, and traction.

3.2.1 Rolling Resistance

Rolling resistance is the resistance of the tires to movement. You have probably experienced having your personal vehicle sink in mud or snow. In this situation, more power is needed to move the vehicle. The deeper the tires sink, the more power is needed to move the vehicle. This increases rolling resistance, and it seems as if the vehicle is always going uphill. Tires do not need to sink for the vehicle to experience rolling resistance. When the road surface flexes as the vehicle moves over it, as shown in *Figure 24*, the effect is the same as sinking. Rolling resistance is measured in terms of percentage of the machine's total weight. The effect of a rolling resistance of 20 percent is as if the vehicle is towing an additional 20 percent of

its weight. *Table 9* shows some typical rolling resistances based on the type of tires or tracks.

3.2.2 Digging Resistance

All material has some degree of resistance to excavation. Some materials have more resistance than others. The type of material, along with its hardness, cohesiveness, weight, and degree of compaction, determine its resistance. Naturally, more resistant material will requires heavy-duty

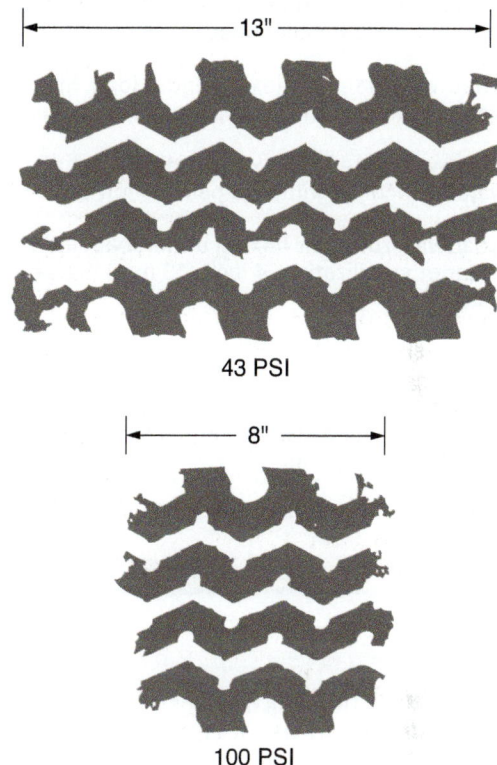

13"

43 PSI

8"

100 PSI

22308-12_F23.EPS

Figure 23 Reducing tire inflation improves flotation.

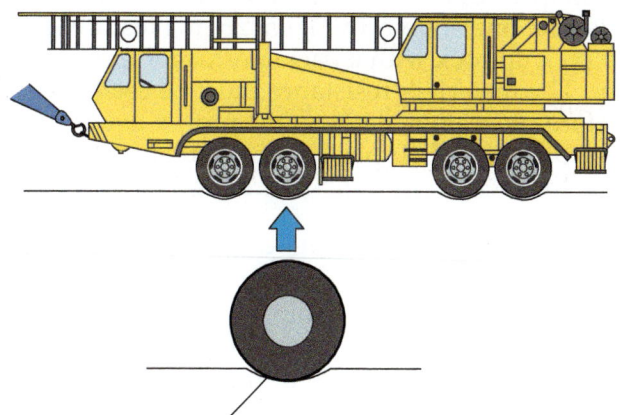

ROAD SURFACE FLEXES, SO THE EFFECT IS THAT THE VEHICLE IS ALWAYS GOING UPHILL.

22308-12_F24.EPS

Figure 24 Flexing of road surface increases rolling resistance.

Table 9 Rolling Resistance

Surface Conditions	Rolling Resistance Percentage			
	Tires		Track	Track and Tires
	Bias	Radial		
Very hard, smooth concrete, cold asphalt, or dirt surface with no penetration or flexing.	1.5%	1.2%	0.0%	1.0%
Hard, smooth stabilized surface that is maintained and watered with no penetration under load.	2.0%	1.7%	0.0%	1.2%
Firm, smooth, rolling roadway with dirt or light surfacing, flexing slightly under load, maintained fairly regularly, watered.	3.0%	2.5%	0.0%	1.8%
Dirt roadway, rutted or flexing under load, little maintenance, no water, 1 inch (25 mm) tire penetration or flexing.	4.0%	4.0%	0.0%	2.4%
Dirt roadway, rutted or flexing under load, little maintenance, no water, 2 inch (50 mm) tire penetration or flexing.	5.0%	5.0%	0.0%	3.0%
Dirt roadway, rutted or flexing under load, little maintenance, no water or stabilization, 4 inch (10 mm) tire penetration or flexing.	8.0%	8.0%	0.0%	4.8%
Loose sand or gravel.	10.0%	10.0%	2.0%	7.0%
Dirt roadway, rutted or flexing under load, little maintenance, soft under travel, no stabilization, 8 inch (200 mm) tire penetration and flexing.	14.0%	14.0%	5.0%	10.0%
Very soft, muddy, rutted roadway, 12 inch (300 mm) penetration, no flexing.	20.0%	20.0%	8.0%	15.0%

SOURCE: CATERPILLAR

22308-12_T09.EPS

equipment and has higher excavation costs. It may be necessary to use a ripper attachment to help with the excavation. A ripper (*Figure 25*) is pulled behind a piece of equipment and has large teeth that comb through the excavation material to help loosen it. A bucket with similar teeth, like the one shown in *Figure 26*, makes it easier to pick up rock with a backhoe/loader or excavator. The material's resistance must be established before excavation starts so the proper equipment can be selected. Typically, excavation is divided into three categories: rock, hard digging, and easy digging.

Rock – Rock is generally solid, although some material classified as rock can have small voids. Rock needs to be broken into small pieces for excavation, but it is often resistant to heavy machinery and rippers. The surface of the rock, which has been exposed to the elements, is often easy to rip. But lower layers, which have been shielded from the weather, are more resistant. Sometimes rocks have been formed with fissures.. It can be broken along the fissures by approaching the rock mass from a different angle with the equipment. Often, rock can be broken by excavating beneath it and allowing the weight to shear off pieces of the rock, although this practice can be dangerous.

22308-12_F25.EPS

Figure 25 Ripper attachment.

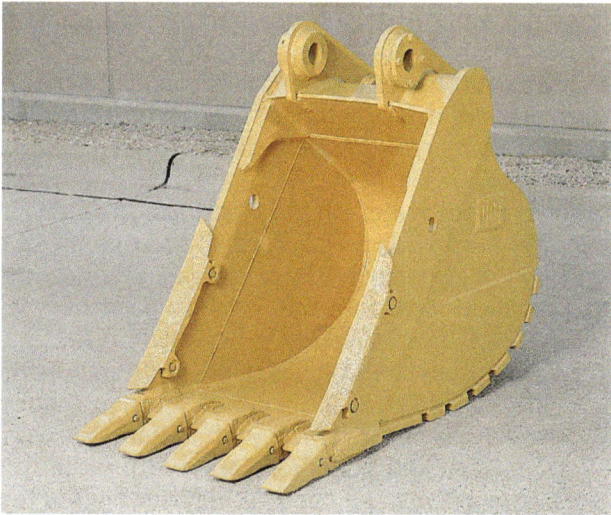

Figure 26 Heavy-duty rock bucket.

Some material can be broken mechanically with a pneumatic jackhammer or a pneumatic drill (*Figure 27*), but sometimes it requires blasting. Blasting involves drilling a hole into the rock, packing it with explosives, and detonating them. Excavation of solid rock is the most expensive and dangerous type of earthwork.

> **WARNING!**
>
> Blasting requires advanced training and specialized knowledge of rock formations and explosives. A heavy equipment operator should never set explosives.

Hard digging – Hard digging doesn't require drilling or blasting, but the material is still very strong and resistant to digging. Materials that require hard digging include heavy material, such as crushed stone and gravel; or cohesive material, such as soft moist clay. These materials can be dug with heavy-duty machinery, such as front-end loaders, power buckets, bulldozers, and rippers.

Easy digging – Easy digging materials are usually light and dry, such as soft or fine loose earth or sand. Although these materials offer little resistance to digging, they may make it difficult for heavy equipment to move because they may lack traction.

3.2.3 Traction

Traction is the friction between the road surface and the tire or tracks of a vehicle. Low friction between the road surface and vehicle tires allows the tires to spin without moving the vehicle, forcing some of the engine power to be used to overcome the spinning of the tires rather than to move the vehicle. Some of the things that affect vehicle traction are the surface material coefficient of traction, the weight of the vehicle, and the type of tires or tracks.

Equipment weight distribution – Equipment weight distribution refers to the force of an engine's power on the drive wheels of a vehicle moves the vehicle across a surface. Ideally, all of the engine's power would be used to move the vehicle, but this is not possible because of traction and rolling resistance. The coefficient of traction and the amount of weight on the vehicle's drive wheels determine the amount of force a vehicle can deliver before the tires start to slip.

Another point to keep in mind is that when climbing a hill, gravity shifts some of the vehicle weight to the rear wheels, thus increasing traction for rear drive vehicles. Unfortunately, when driving down a steep grade, gravity shifts vehicle weight to the front wheels, thus decreasing traction of rear drive vehicles, which can make a vehicle difficult to handle.

Tire treads and traction – Tire treads are the working surface of a tire. The type of tread helps determine the flotation capacity of a vehicle. Many tires are designed to work effectively in many different environmental conditions. Generally, tires with a deep tread provide good traction under most conditions, but shallower treads are best on dry sand and ice, because they have better flotation. For tires to operate correctly, the operator needs to be sure that they are inflated according to manufacturer's specification and that the treads are free of foreign matter, such as mud.

Figure 27 Pneumatic rock drill.

Tracks – Tracked equipment is often used in muddy work because it has better flotation than wheeled equipment. Some tracks have especially wide track shoes to further increase flotation. All tracks are not equal, however; it is important to match the track shoe to the job that needs to be done.

Smooth or worn tracks are not suitable for mud work because they offer little traction on wet, slippery surfaces. It may be possible to improve the traction of worn tracks by replacing some of the shoes with new shoes, but this is not a good choice when the equipment must also be used on hard ground.

Grouser shoes (*Figure 28*) are often used to improve equipment footing on muddy ground, but they are not without problems. Grousers dig into loose soil very quickly and sometimes mud gets packed between them, giving the effect of a single, smooth surface with no traction.

3.2.4 Working in Mud

Mud is particularly troublesome for heavy equipment operators. Mud occurs when enough water has been mixed with a soil and to bring it to a liquid consistency. Mud has a very poor weight-bearing capability, so equipment sinks until it finds support. This increases rolling resistance on the tires and forces the engine to use extra power to overcome the resistance in order to move the vehicle. Some soils are more likely to turn muddy than others. Coarse-grained soils, such as sand, tend to drain water quickly and are unlikely to become muddy. Fine-grained soils, such as clay and silt, hold water for long periods and become muddy. Soil mixtures that contain a lot of fine-grained soil are also likely to become muddy.

Mud is often shallow. The vehicle tires can sink a small amount to reach solid ground, but when the mud is deeper, a vehicle with better floatation, such as a tracked vehicle or one with wide, soft tires, may need to be used. When the mud is too deep to support any vehicle, platforms, timbers, or mats will need to be laid on its surface in order to support the vehicle (*Figure 29*).

Even under the best of circumstances, driving on timbers or platforms is risky. Whatever is used must be wider than the vehicle tire or track and strong enough to support the weight of the vehicle. Both sides of the vehicle must be supported equally to avoid having one side sink into soft ground, causing the vehicle to tip over. Also, the timbers or platform must be moved as the vehicle is moved.

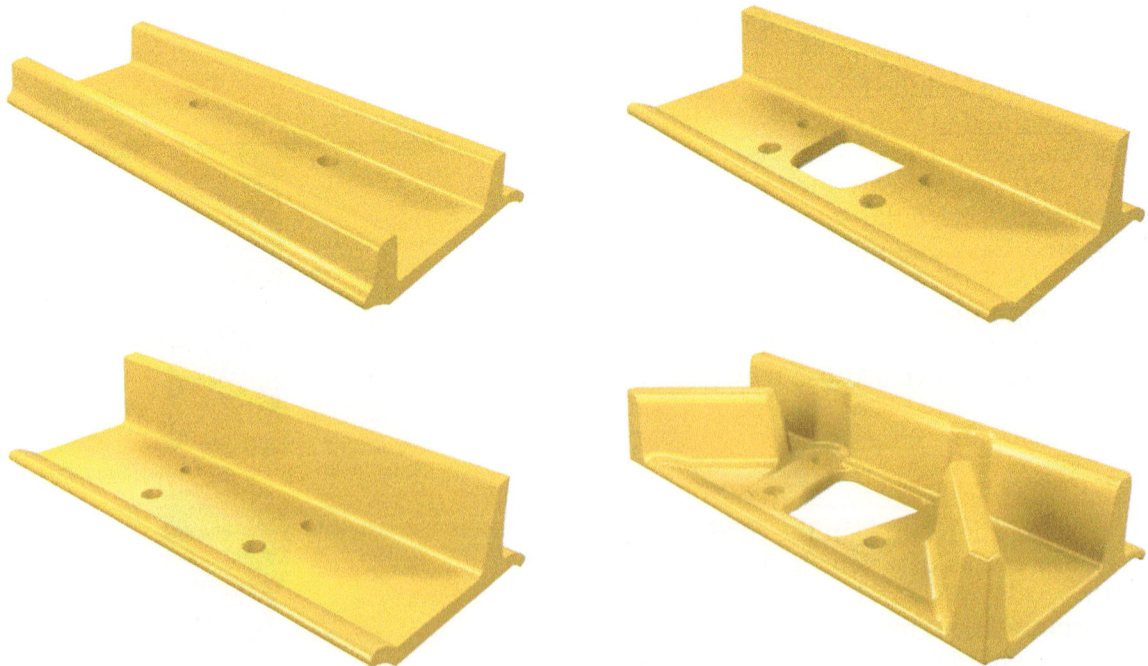

22308-12_F28.EPS

Figure 28 Grouser shoes.

22308-12_F29.EPS

Figure 29 Mud mat.

3.3.0 Trenching Hazards

The type of soil in and around a trench can contribute to the collapse of trench walls. Soil type is a major factor that must be considered in trenching operations. Only a competent person who has additional education, experience, and authority can decide if the soil in and around a trench is safe and stable. However, it is still the operator's responsibility to know some of the basics about soil and its hazards.

The soil found on most construction sites is a mixture of many types, including sand, loam, clay, and silt. It is the type of mixture that gives the soil its properties. For example, sand with a small amount of silt and clay may compact well and permit excavation of a stable trench.

Each of the various soil types, depending on their conditions at the time of excavation, behave differently. Sandy soils tend to collapse straight down, and wet clay and loams tend to slab off the sides of the trench (*Figure 30*). Firm, dry clays and loams tend to crack, and wet sand and gravel tend to slide.

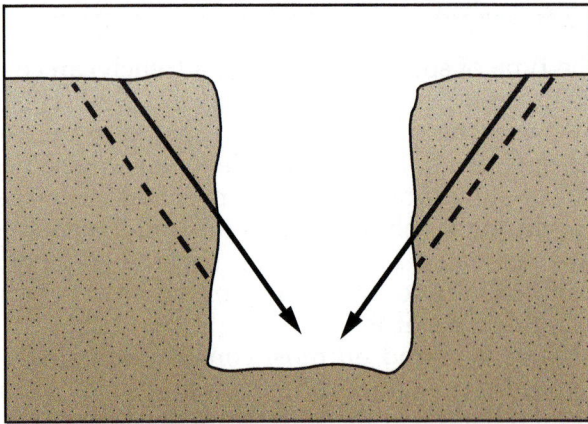

SANDY SOIL COLLAPSES

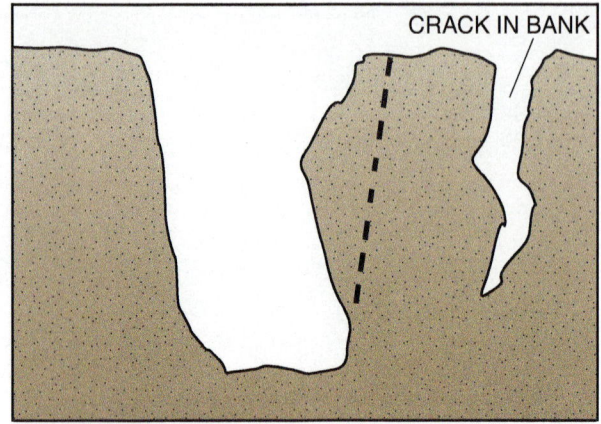

FIRM DRY CLAY AND LOAMS CRACK

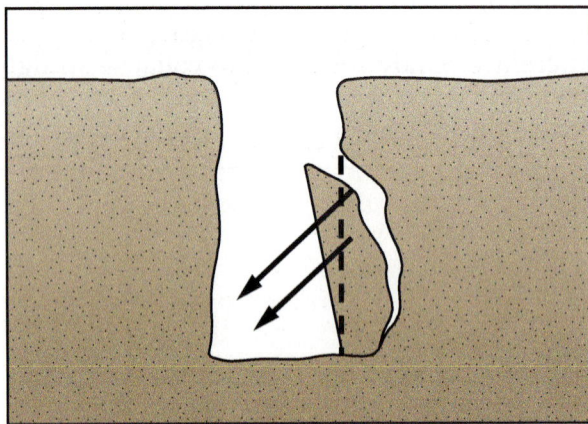

CRACK IN BANK

WET CLAY AND LOAMS SLAB OFF

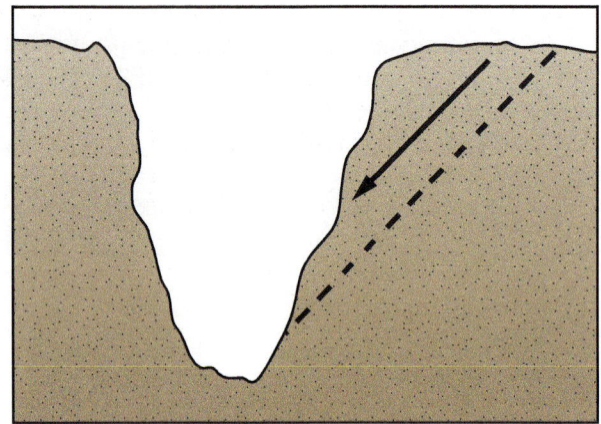

WET SANDS AND GRAVELS SLIDE

22308-12_F30.EPS

Figure 30 Behavior of different soils in trenches.

3.0.0 Section Review

1. Which of these soils has the greatest weight-bearing capacity?

   a. Dry clay
   b. Cemented gravel
   c. Compacted dry sand
   d. Medium dry clay

2. A ripper is used to _____.

   a. pick rocks out of soil
   b. loosen heavily compacted soil
   c. improve traction in damp soil
   d. break up rocks

3. A grouser shoe is used to _____.

   a. improve equipment footing on muddy ground
   b. compact soil that has been loosened by a ripper
   c. improve the traction of the tires on wheeled vehicles
   d. make it easier to walk on muddy ground

4. Which of these soil conditions is likely to cause a trench to fail by slabbing off?

   a. Sandy soil
   b. Firm, dry clay
   c. Wet sand
   d. Wet clay

SUMMARY

The term *soil* means any type of earthen material, including rock. It does not include bedrock, which is the base of Earth's surface. The soil forms on the bedrock in layers, which are called horizons. In general, there are three categories of soil: organic, inorganic, and rock. Organic material comes from living matter such as plants or animals. Inorganic material comes from rocks that have been broken down due to natural forces. Rock is any material that cannot be dug or loosened by available machinery.

The two soil classification systems that are used in construction are the ASTM Unified Soil Classification System and the AASHTO Soil Classification System. Soils are generally classified by particle size as granular and fine-grained. Sand, gravel, and rock are considered granular, while clay and silt are fine-grained. No single type of soil is ideal for construction; rather, it is a mixture of several types that will yield the ideal construction site.

Construction workers are mainly concerned with a soil's strength, which is related to its density. The density of soil is measured as a ratio of its weight to its volume, usually in pounds of wet soil or dry soil per cubic foot. Soil is compacted to improve its density. Compaction refers to the act of artificially increasing the density of the soil using vibration or weight. It is one of the most important activities in construction. To increase the density of soil, more soil particles must be moved into the same space by expelling air or water. A large amount of water makes it impossible to compact soil, but a small amount aids in compaction because it acts like a lubricant to rough-surfaced particles, allowing them to slip into voids more easily.

When working in various soil conditions, it is necessary to know how certain types of equipment will react to different soil conditions. The conditions that must be watched are the ability to rip rock, traction, and the flotation capacity of the equipment. Rocks have different levels of hardness and therefore cannot be broken up in the same way as soil.

It is important for heavy equipment operators to be able to identify different soil types and to understand how different environmental factors—especially rain—affect soil conditions. The equipment will behave differently in wet soil than it will in dry soil, so it is important to stay alert to soil changes to complete the job safely.

1. Air space or open pockets between soil particles are called _____.

   a. fines
   b. voids
   c. rakes
   d. spoils

2. Which of these factors has the most effect on the classification of soil?

   a. Size of particles
   b. Hardness of particles
   c. Behavior when moisture content varies
   d. Static weight of particles

3. Which of the following soil types has the finest grain size?

   a. Silt
   b. Gravel
   c. Clay
   d. Fine sand

4. The grain size of soil particles is measured by _____.

   a. compacting with a ram
   b. sieving
   c. measuring each particle and taking the average
   d. performing the sand cone test

5. When soil has enough small particles to fill the voids between large ones it is _____.

   a. well-graded
   b. sieved
   c. gradated
   d. poorly differentiated

6. Which of the following is *not* an engineering property of soil?

   a. Elasticity
   b. Conductivity
   c. Cohesion
   d. Plasticity

7. Soil density tests are performed to _____.

   a. see if a site needs additional compaction
   b. set a standard for measuring field tests
   c. test material for uniform gradation
   d. compare soils from different parts of the country

8. The ratio of a weight of a material to its volume is called _____.

   a. density
   b. lift
   c. mass
   d. specific gravity

9. The soil with the highest density and lowest water content is _____.

   a. poorly graded sand
   b. clay
   c. well-graded sand
   d. silty clay

10. The single most important factor in the compaction of soil is _____.

    a. surface texture
    b. chemical composition
    c. moisture content
    d. weight of the compaction equipment

11. When soil is removed from its natural (or banked) state, what typically happens to it?

    a. It shrinks.
    b. The chemical composition changes.
    c. It swells.
    d. Nothing happens.

12. The swell factor is the _____.

    a. number of loose yards in a banked yard
    b. amount of consolidation
    c. size of the soil particles
    d. small drainage slope between buildings

13. A material has a shrink factor of 0.8. In order to fill a 10 cubic yard trench, how many cubic yards of fill will be needed?

    a. 8
    b. 10
    c. 12.5
    d. 80

14. The purpose of compacting soil is to _____.
    a. produce a more stable surface
    b. make the surface smooth
    c. keep the excavation from failing
    d. squeeze out all of the moisture

15. Which of the following soils is difficult to dig because of strong, uniform cohesion?
    a. Soft clay
    b. Wet sand
    c. Gravel
    d. Boulders

Trade Terms Introduced in This Module

American Association of State Highway and Transportation Officials (AASHTO): An organization representing the interest of all state government highway and transportation agencies throughout the United States. This organization establishes design standards, materials-testing requirements, and other technical specifications concerning highway planning, design, construction, and maintenance.

American Society of Testing Materials (ASTM): A national organization that establishes standards for testing and evaluation of manufactured and raw materials.

Banked: Any soil mass that is to be excavated from its natural position.

Bedrock: The solid layer of rock under Earth's surface. Its solid-rock state distinguishes it from boulders.

Capillary action: The tendency of water to move into free space or between soil particles, regardless of gravity.

Cohesive: The ability to bond together in a permanent or semipermanent state. To stick together.

Consolidation: To become firm by compacting the particles so they will be closer together.

Density: Ratio of the weight of material to its volume.

Elasticity: The property of a soil that allows it to return to its original shape after a force is removed.

Expansive soil: A soil that expands and shrinks with moisture. Clay is an expansive soil.

Fines: Very small particles of soil. Usually particles that pass the No. 200 sieve.

Friable: Crumbles easily.

Horizon: Layers of soil that develop over time.

Humus: Dark swamp soil or decaying organic matter. Also called peat.

In situ: In the natural or original place on site.

Inorganic: Derived from other than living organisms, such as rock.

Liquid limit: The amount of moisture that causes a soil to become a fluid.

Loading: Applying a force to soil. A building can be a permanent load at a site, and a truck can be a passing load on a roadway.

Optimum moisture: The percent of moisture at which the greatest density of a particular soil can be obtained through compaction.

Organic: Derived from living organisms, such as plants and animals.

Peat: Dark swamp soil or decaying organic matter. Also called humus.

Plastic limit: The amount of water that causes a soil to become plastic (easily shaped without crumbling).

Plasticity: The range of water content in which a soil remains plastic or is easily shaped without crumbling.

Settlement: To become firm by compacting the particles so they will be closer together.

Shrinkage: Decrease in volume when soil is compacted.

Swell: Increase in volume when soil is excavated.

Swell factor: The ratio of the banked weight of a soil to the loose weight of a soil.

Voids: Open space between soil or aggregate particles. A reference to voids usually means that there are air pockets or open spaces between particles.

Water table: The depth below the ground's surface at which the soil is saturated with water.

Well-graded: Soil that contains enough small particles to fill the voids between larger ones.

Additional Resources

This module presents thorough resources for task training. The following resource material is suggested for further study.

"Arrest that Fugitive Dust," 2002. Roberta Baxter, *Erosion Control* magazine, Forester Communications, Inc.

Basic Equipment Operator, NAVEDTRA 14081, 1994 Edition. Morris, John T. (preparer), Naval Education and Training Professional Development and Technology Center.

Caterpillar Performance Handbook, Edition 33. A CAT® Publication. Peoria, IL: Caterpillar, Inc.

Moving the Earth, 3rd Edition, 1988. H.L. Nichols. Greenwich, CT: North Castle Books.

"Optimizing Soil Compaction and Other Strategies," 2004. Donald H. Gray, *Grading and Excavation Contractor* magazine. Forester Communications, Inc.

Soil Stabilization for Pavements Mobilization Construction Engineer Manual (EM 1110-3-137), 1984. Department of the Army, Corps of Engineers.

Temporary Stream and Wetland Crossing Options for Forest Management, 1998. Charles R. Blinn, Rick Dahlman, Lola Hislop, and Michael A. Thompson, USDA Forest Service, North Central Research Station, General Technical Report NC-202.

Using Lime for Soil Stabilization and Modification, 2001. National Lime Association website (**www.lime.org**).

Figure Credits

Reprinted courtesy of Caterpillar Inc., Module opener, Figures 25, 26, and 28

USDA Natural Resources Conservation Service, Figure 3

Topaz Publications, Inc., Figures 4, 5, 7, 11

Courtesy of American Educational Products, Figure 8

Courtesy of Durham Geo Slope Indicator, Figure 12B

Courtesy of Humboldt Mfg. Co., Figures 13 and 14

Courtesy of American Pneumatic Tools, Figure 27

Courtesy of SVE Portable Roadway Systems, Inc., Figure 29

Section Review Answers

Answer	Section Reference	Objective
Section One		
1 d	1.1.0, Table 3	1a
2 a	1.2.1	1b
3 a	1.3.0	1c
4 b	1.4.0, Figure 15	1d
Section Two		
1 b	2.1.0	2a
2 a*	2.1.0	2b
3 b	2.2.0	2b
Section Three		
1 b	3.1.0, Table 8	3a
2 b	3.2.2	3b
3 a	3.2.3	3b
4 d	3.3.0	3c

*$3,100 \div 2,600 = 1.192$

NCCER CURRICULA — USER UPDATE

NCCER makes every effort to keep its textbooks up-to-date and free of technical errors. We appreciate your help in this process. If you find an error, a typographical mistake, or an inaccuracy in NCCER's curricula, please fill out this form (or a photocopy), or complete the online form at **www.nccer.org/olf**. Be sure to include the exact module ID number, page number, a detailed description, and your recommended correction. Your input will be brought to the attention of the Authoring Team. Thank you for your assistance.

Instructors – If you have an idea for improving this textbook, or have found that additional materials were necessary to teach this module effectively, please let us know so that we may present your suggestions to the Authoring Team.

NCCER Product Development and Revision

13614 Progress Blvd., Alachua, FL 32615

Email: curriculum@nccer.org
Online: www.nccer.org/olf

❏ Trainee Guide   ❏ Lesson Plans   ❏ Exam   ❏ PowerPoints   Other _____

Craft / Level: _____   Copyright Date: _____

Module ID Number / Title: _____

Section Number(s): _____

Description: _____

_____

_____

_____

_____

Recommended Correction: _____

_____

_____

_____

_____

Your Name: _____

Address: _____

_____

Email: _____   Phone: _____

22210-13

Site Work

OVERVIEW

Understanding how to operate various pieces of heavy equipment is only the beginning for a heavy equipment operator. In addition, an operator must understand safety issues, equipment transportation, groundwater control, operational costs, and advanced grading methods.

Module Six

Trainees with successful module completions may be eligible for credentialing through NCCER's National Registry. To learn more, go to **www.nccer.org** or contact us at **1.888.622.3720**. Our website has information on the latest product releases and training, as well as online versions of our *Cornerstone* newsletter and Pearson's product catalog.

Your feedback is welcome. You may email your comments to **curriculum@nccer.org**, send general comments and inquiries to **info@nccer.org**, or fill in the User Update form at the back of this module.

This information is general in nature and intended for training purposes only. Actual performance of activities described in this manual requires compliance with all applicable operating, service, maintenance, and safety procedures under the direction of qualified personnel. References in this manual to patented or proprietary devices do not constitute a recommendation of their use.

Objectives

When you have completed this module, you will be able to do the following:

1. Describe the safety practices associated with site grading work.
   a. Explain the purpose of a site safety program.
   b. Describe why safety inspections and investigations are important.
   c. Explain how hazardous materials are controlled on a job site.
   d. Describe safety practices associated with trenching and excavations.
   e. Describe how to prepare heavy equipment for transporting.
2. Describe the methods used to control water on job sites.
   a. Explain the importance of maintaining proper drainage on a job site.
   b. Describe the methods used to control ground water and surface water.
   c. Describe the safety practices and construction methods used when working around bodies of water.
3. Explain how grades are established on a job site.
   a. Describe how to set grades from a benchmark.
   b. Describe how grades are set for highway construction.
   c. Describe how grades are set for building construction.
   d. Explain how grading operations are performed.
   e. Describe the use of stakeless and stringless grading systems.
4. Describe grading and installation practices for pipe-laying operations.
   a. Explain how grades are established for pipe-laying operations.
   b. Describe the equipment and methods used to lay pipe.

Performance Tasks

Under the supervision of the instructor, you should be able to do the following:

1. Interpret layout and marking methods to determine grading requirements and operation.
2. Set up a level and determine the elevations at three different points, as directed by the instructor.

Trade Terms

Aquifer	Dewater	String line
Balance point	Erosion	Subsidence
Bedding material	Groundwater	Sump
Berm	Sedimentation	Swale
Boot	Shielding	Topographic survey
Competent person	Stations	Uprights
Cross braces	Stormwater	Walers

Industry Recognized Credentials

If you are training through an NCCER-accredited sponsor, you may be eligible for credentials from NCCER's Registry. The ID number for this module is 22210-13. Note that this module may have been used in other NCCER curricula and may apply to other level completions. Contact NCCER's Registry at 888.622.3720 or go to **www.nccer.org** for more information.

Contents ——————————————

Topics to be presented in this module include:

Figures

Figures

1.0.0 JOB SITE SAFETY

Objective 1

Describe the safety practices associated with site grading work.

a. Explain how to a site safety program is set up.
b. Describe safety inspection and investigation practices.
c. Explain how hazardous materials are controlled on a job site.
d. Describe safety practices associated with trenching and excavations.
e. Describe how to prepare heavy equipment for transporting.

Trade Terms

Competent person: A person who is capable of identifying existing and predictable hazards in the area or working conditions that are unsanitary, hazardous, or dangerous to employees, and who has the authority to take prompt corrective measures to fix the problem.

Cross braces: The horizontal members of a shoring system installed perpendicular to the sides of the excavation, the ends of which bear against either uprights or walers.

Shielding: A structure that is able to withstand the forces imposed on it by a cave-in and thereby protect employees within the structure.

Subsidence: Pressure created by the weight of the soil pushing on the walls of the excavation. It stresses the excavation walls and can cause them to bulge.

Uprights: The vertical members of a trench shoring system placed in contact with the earth and usually positioned so that individual members do not contact each other. Uprights placed so that individual members are closely spaced, in contact with, or interconnected to each other, are often called sheeting.

Walers: Horizontal members of a shoring system or coffer dam placed parallel to the excavation face whose sides bear against the vertical members of the shoring system or the earth. Also, supports for piles in a coffer dam.

Safety is everyone's responsibility. You and your co-workers have an obligation to your employer to operate your equipment safely. You are also obligated to make sure that anyone you supervise works safely. Your employer has an obligation to you to provide you with safe machinery and equipment, and to maintain a safe workplace for all employees.

The Occupational Health and Safety Administration (OSHA) sets forth a number of safety regulations that you, your co-workers, and your employer must follow. The *Code of Federal Regulation (CFR) 29* has to do with labor standards, while *CFR 29, Part 1926* has to do specifically with standards for the construction industry. Your employer's safety program is based on these standards.

Safety begins with the proper training and orientation of employees. New operators should not begin work until they have read and demonstrated an understanding of the company's safety program. In addition, operators should not perform work in hazardous areas or operate new types of equipment until they have received a safety orientation and checkout from their supervisor or safety officer.

It is important to take the time to learn as much about safe operating practices as possible. Do not be put off by safety requirements or regulations that seem unnecessary. They are established for a reason. Follow them for your own protection, as well as for the protection of workers around you.

The basic rule to follow every working day is to report everything that you consider unsafe. These situations cannot be corrected if they are not reported. If you are a lead operator or a first-line supervisor, it is your responsibility to follow up on these reports and see that the situations are corrected.

The person who has the greatest effect on safety is the worker. Without safety-conscious employees, an employer cannot have a good safety record. Part of a worker's responsibility for safety is to follow good safety practices while working, and to report any unsafe conditions to the employer. Workers have an obligation to themselves, their families, and their co-workers to report any unsafe conditions. Safe workers take steps to remain alert while working. This can be difficult when performing repetitive tasks. The repetition can be monotonous and a person can become distracted. To combat this, it is important to get adequate rest. Caffeine and other stimulants are no substitute for sleep. In addition, eat on a regular schedule so that your body has the fuel it needs to work and think. Avoid any prescription or over-the-counter medications that can cause drowsiness. Recreational drug use and excessive alcohol use must be avoided during off-duty hours. Never use drugs or alcohol on the job.

Some hazards have an immediate effect on health. For example, exposure to excessive levels of carbon monoxide or an oxygen-deficient environment can be deadly in a matter of minutes. Other hazards may not have any noticeable effect until years after the exposure. For example, repeated exposure to noise has a cumulative effect rather than an immediate effect.

In addition to protecting yourself for your family's sake, you need to protect your family's health. Dust particles that cling to your clothing can contain materials that are hazardous to the health of people—especially children—with whom you are in close contact. If you are working with any potentially hazardous materials, you can help protect your family by showering immediately after work. If workplace showers are available, take advantage of them. Change into clean clothing before socializing with your family or others. If possible, immediately launder your work clothing after work. If this is not possible, keep work clothing separate from all other clothing and launder it separately.

1.1.0 Site Safety Program

Although there is no specific requirement for a company to have a safety officer, it makes sense to designate someone who can oversee the safety program and act as a point of contact for all safety-related matters. The safety officer also coordinates safety programs and activities. Most companies have an appointed person responsible for dealing with safety issues. Larger job sites often have a site safety officer. Larger companies generally have a full-time safety officer or engineer. The safety officer is responsible for training employees in safety and for the development of the company's safety program. This safety officer is an important resource for safety concerns.

The general activities of a safety officer are as follows:

- Ensuring that all safety rules are communicated and followed. The safety rules are enforced by making periodic inspections of all job sites and offices. If the rules are not enforced consistently and uniformly, their importance will be diminished.
- Initiating training programs to ensure that every employee is aware of any safety and health hazards associated with their assigned tasks or areas.
- Investigating accidents. When an accident occurs, the safety officer should review the accident report to help supervisory personnel determine the cause of the accident and recommend methods to prevent a recurrence.
- Conducting safety meetings. The safety officer is often the person best suited to conduct safety meetings to ensure that employees receive the most from the meeting and get involved in the safety program.
- Performing inspections of offices and job sites upon request to detect unsafe conditions or safety rule violations.
- Analyzing work practices to detect unsafe conditions, and to develop procedures to eliminate accidents or injuries.

Depending upon the size and the type of the company, the safety officer may have several other related responsibilities in the areas of project planning, equipment maintenance, and operator certification.

1.1.1 Safety Meetings

Safety meetings are used to keep employees informed of safe operating practices, new techniques and equipment, and workplace hazards. These meetings should be an everyday work activity. Regularly scheduled safety meetings are held to reinforce safety policy and procedures and to inform workers of any new industry standards or safety practices the company is implementing.

Daily meetings at the job site are short and topical and are intended to keep attention focused on task-related safety. Toolbox meetings are usually held first thing in the morning when everyone is alert and before people have a chance to spread out. Topics can include accident prevention, safe working practices, emergency procedures, tool and equipment operation, and personal protective equipment (PPE). The meeting discussions and demonstrations may be led by a safety officer or supervisor who is qualified to present safety material or information. There are many resources for prepared typical toolbox meetings. A good online resource can be found at **www.toolboxtopics.com**. The meeting subject and names of those in attendance are usually documented on a form (*Figure 1*).

In addition to toolbox meetings, companies often have formal training programs that operators are required to attend periodically. Topics of these meetings include first aid, emergency procedures, OSHA and other government safety requirements

JOBSITE SAFETY MEETING REPORT

DATE: _____ TIME: _____ SUPERVISOR: _____ PROJECT # _____

STANDARD TOPICS:

- ☐ EMERGENCY RESPONSE PROCEDURES   ☐ EQUIPMENT WALK AROUNDS
- ☐ HARD HATS / VESTS / CLOTHING   ○ FIRE EXTINGUISHER
- ☐ EYE / EAR / HAND PROTECTION   ○ BACK-UP ALARMS
- ☐ PROPER LIFTING TECHNIQUES

JOBSITE TOPICS DISCUSSED: _____

_____

_____

_____

MAIN POINTS EMPHASIZED: _____

_____

REVIEWED "MSDS" YES / NO (circle one)   SUBJECTS: _____

☞ _____

☞ _____

SUGGESTIONS MADE BY CREWS: _____

_____

_____

CORRECTIVE ACTION TAKEN: _____

_____

_____

OTHER COMMENTS: _____

_____

ATTENDANCE:

_____   _____   _____

_____   _____   _____

_____   _____   _____

_____   _____   _____

SUPERVISOR'S SIGNATURE _____

22210-12_F01.EPS

Figure 1 Safety meeting attendance form.

and regulations, hazardous materials handling, and licensing and certification requirements. It is your responsibility to attend these meetings and pay attention to what is being presented.

Many people do not like to go to meetings or training sessions because they think the meet-ings are a waste of time. Others are always eager to attend because they think they are getting out of work, and relaxing at the company's expense. Safety training sessions should be treated like any other part of the job. Keep focused and learn everything you can about the subject being discussed. In the case of safety information, it could save your life.

1.1.2 Safety Committees

Some employers have safety committees. If you work for such a company, you have an obligation to that committee to help maintain a safe working environment. This means following the safety committee's rules for proper working procedures and practices and reporting any unsafe equipment or conditions to the safety committee or the appropriate supervisor.

Creating a committee is a good way to get more people involved in safety activities. You may be asked to participate on such a committee—either to help out with particular project, or to represent a specific part of your organization. If you are asked to be on a safety committee, you should feel honored that your organization thinks of you as someone who is safety-conscious. Involvement with this effort gives you an opportunity to make an impact on your company's safety performance, and you will gain the satisfaction of having diversity in your job and seeing your ideas put into practice. Safety committee activities may include the following:

- Reviewing company safety policies and procedures.
- Conducting training in certain areas of expertise.
- Promoting safety awareness and holding safety campaigns.
- Assisting in the investigation of incidents and accidents.
- Recommending new safety training and information programs.
- Assisting in the recognition and/or correction of hazards.

1.2.0 Safety Investigations and Inspections

In spite of good education and careful work practices, accidents can happen. The term *accident* applies to those occurrences resulting in property damage, as well as those involving physical injury. Incidents, or near misses, are defined as any event that could have resulted in damage or physical injury, but did not. All accidents must be reported to your employer, who should ensure that your supervisor perform an accident investigation. Not all incidents are reported; those that are reported should be investigated just as if an accident had occurred.

Investigation of an incident or accident is similar to the investigation of a crime; the object is to solve the mystery, and then take a course of corrective action to prevent it from happening again. It is essential to find out what happened (investigate) and then change or isolate the conditions that allowed the accident to happen in the first place.

Productive accident investigation is not driven by a desire to hand out punishment. When human error is identified as the cause, it should be confronted in a problem-solving, non-threatening way. What is important is not fault finding, but that facts are discovered which result in corrective action.

Investigations of accidents and incidents may result in an OSHA inquiry, depending on the type of occurrence. Regardless of OSHA requirements, the company should carry out its own investigation to determine what went wrong, and why.

1.2.1 Accident and Incident Reports

Investigating an accident requires interviewing the people involved and reviewing the accident scene. This should be done promptly, with care and attention to detail, so that all the facts are uncovered. An accident report form can be extremely useful in guiding investigators through an investigation and providing information necessary for later analysis. *Figure 2* shows an example of an accident investigation report. The accident report form used in your internal investigations should contain enough information to help determine what happened and what should be done to prevent recurrence.

An accident report is similar to an incident report, but an accident report must include the name of the employee involved in the accident and the extent of his or her injuries, as well as the names of witnesses. There is no standard format for these forms, although much of the basic information, such as names, dates, locations, and accident type, is standard. Accident report forms vary from company to company. Some forms rely on a check-off system to categorize many of the items of information such as operation, accident type, location, and severity. By using this approach, data can be summarized quickly, and more direct comparisons can be made without having to interpret a narrative description of the accident.

The following types of questions should be asked to obtain information for the accident report:

- Who was involved in the accident? What do they do?

SUPERVISOR'S ACCIDENT INVESTIGATION REPORT

Type of Accident: ☐ Motor Vehicle

☐ Property Damage

☐ Equipment Damage

Project Name:_____

Project No._____

Equipment No._____

Personal Injury: ☐ Employee

☐ Non-Employee

Name of Employee_____   How long employed on crew_____

Job Assignment_____   Length of known prior experience_____

Date of Accident_____ Time_____   First reported to me on_____

Exact Place of Accident_____

Where was the supervisor at the time of the accident?_____

What happened?_____

_____

_____

_____

_____ Diagram on back ☐

Describe injury/damage_____

Was treatment provided: First Aid ☐   Doctor ☐   Admitted to Hospital ☐

What caused the accident to happen?_____

_____

Corrective action taken to prevent this happening again_____

_____

_____

Recommendations to avoid similar accidents_____

_____

What safety equipment was being used?_____

Witnesses (please print)_____

Number of accidents on my crew this year_____ For This employee_____

Supervisor's signature_____ Date_____

Supervisor's printed name_____

Project Manager's signature_____ Date_____

Manager's printed name_____

Use back of form for additional detail.

22210-12_F02.EPS

Figure 2 Accident report form.

- How much experience do they have at this particular job?
- Who witnessed the accident?
- When did the accident occur?
- Where was the accident?
- What was the extent of injury or damage?
- What actually happened? What exactly was the person (or persons) doing at the time of the accident. What task or step of the job cycle?
- What was the direct cause of the accident? Were there contributing factors?

- What corrective action (in the investigator's opinion) should be taken to prevent this accident from recurring?

Some suggestions for completing an accident report form are:

- *Identify the direct cause or causes* – The direct causes can include unsafe actions, unsafe conditions, or, most likely, a combination of both. Try to state the cause or causes in simple, direct terms. Avoid vague and general statements; the clearer your statements are, the easier it will be to determine the appropriate corrective action. Be sure that the information is accurate.
- *Identify all contributing factors* – The factors that contributed to the accident might include horseplay, fatigue, lack of attention, inexperience, shortcutting recommended procedures, improper use of tools, poor supervision, and so on. Again, try to state the factors in simple, clear terms and as accurately as possible.
- *Document the incident* – Documentation may be done in writing and with photos or video.
- *Recommend corrective action* – Corrective action must be taken to prevent a recurrence of this type of accident. Further training or retraining, changes in job procedures, environmental changes, equipment modifications, or improvements in housekeeping are examples of possible corrective actions.

Investigative efforts for an accident depend on several factors, including the magnitude, severity, and complexity of the occurrence. Some accidents are simple to solve, because a witness observed the procedure or act that caused it. Many of these cases can be associated with human error while performing a task or operating a piece of machinery. Other accidents are caused by the gradual failure of some component or material until finally the accident occurs; these accidents may be much more difficult to investigate. Any job-related injuries or illness must be logged and reported to OSHA by the safety director or manager.

Incidents and accidents should be investigated by the supervisor responsible for the person(s) involved and the area where the accident occurred. This is important because the immediate supervisor is the one who should implement a plan of corrective action. The supervisor would also be the person most familiar with the operation and the work environment where the accident occurred. The company safety officer should review the accident investigation report to ensure that a complete and proper investigation has taken place. Because safety officers are trained in accident investigation and analysis, their expertise can be a valuable tool in preventing a recurrence.

There is no such thing as a minor accident. Any accident is symptomatic of a larger problem that should be identified and corrected.

Incident reports are similar to accident reports. While an accident report documents events that caused property damage or bodily injury, incident reports document near misses or events that could have resulted in property damage or bodily injury. The incident report is not difficult to complete, and simply requires basic information about what happened. Since incidents are not accidents, it may not be necessary to report them to OSHA. Incident reports are very valuable, however, because they provide the safety officer with data that can be used to identify and analyze patterns of unsafe work activities before an accident occurs.

1.2.2 OSHA Inspections

To enforce the safety standards defined in the Occupational Safety and Health Act of 1970 (and later revisions), OSHA is authorized to conduct workplace inspections. Every employer covered by the Act is subject to inspection (states with approved safety programs are also authorized to conduct inspections). There are a number of situations in which OSHA will conduct a job-site inspection. For example, an inspection will be conducted any time there is an on-the-job fatality or when three or more workers are hospitalized.

OSHA is responsible for developing and implementing legally enforceable standards based on requirements in the law. States with OSHA-approved occupational safety and health programs must set standards that are at least as effective as the federal programs. It is the responsibility of employers to become familiar with standards applicable to their line of business, and to ensure that employees have (and use) appropriate personal protective equipment when required for safety. Employees must comply with all rules and regulations that affect their own actions and conduct.

Prior to any inspection, the OSHA Compliance Safety and Health Officer (CSHO) becomes familiar with as many relevant facts as possible about the workplace, taking into account such things as the history of the establishment, the nature of the business, and the particular safety standards that may apply. The inspection process is detailed in the OSHA standards and in numerous other OSHA publications.

A typical inspection takes place as follows:

- The CSHO shows their credentials. All CSHOs carry US Department of Labor credentials that can be verified by contacting the nearest OSHA office.
- Company and employee representatives are designated to accompany the CSHO on the inspection.
- The route and duration of the inspection is determined by the CSHO, who may stop and confer with employees in private if necessary. The CSHO observes general conditions, consults with employees, takes pictures and instrument readings, and examines certain records, such as the OSHA 300 log.
- At the closing conference, the CSHO discusses all unsafe and unhealthful conditions observed, and indicates the apparent violations for which the company may be cited.

After the inspection and the report of findings by the CSHO, OSHA's area director determines what citations (if any) will be issued, and what penalties will be imposed. Citations inform the employer and employees of the regulations and standards alleged to have been violated and the length of time proposed for their correction. The employer receives the citations and notices of proposed penalties by certified mail. The employer must post a copy of each citation at or near the place where the violation occurred and leave it there for three days or until the violation is corrected, whichever is longer.

During an OSHA inspection, employees should be polite and respectful to the CSHO. OSHA inspections are not the time to get back at your employer for some injustice, nor is it the time to cover up unsafe conditions. Always answer questions directly and honestly. Lying to the CSHO is just like lying to a police officer.

1.3.0 Hazardous Materials

OSHA Standard Part 1910.1200 is titled *Hazard Communication*, and it specifically requires that a material safety data sheet (MSDS) be maintained for hazardous chemicals in the workplace and that all MSDSs be available for everyone's use (*Figure 3*). Employers must establish a written, comprehensive hazardous communications (HAZCOM) program that includes provisions for container labeling, MSDSs, and employee safety training activities. OSHA even has documents to help your employer develop a HAZCOM program.

22210-12_F03.EPS

Figure 3 HAZCOM.

The program must include the following:

- A list of hazardous chemicals in each work area.
- The means the employer uses to inform employees of the hazards of nonroutine tasks.
- The procedure the employer uses to inform other companies of the hazards to which their employees may be exposed.

Employers must establish a training and information program for employees at the time of their initial assignment if they will be exposed to hazardous chemicals, and again whenever a new hazard is introduced into their work area. The discussion must include at least the following topics:

- The existence of the HAZCOM program and the requirements of the standard.
- The components of the HAZCOM program in the employees' workplaces.
- Operations in the work areas where hazardous chemicals are present.
- Location where the employer keeps the written hazard evaluation procedures, communications program, list of hazardous chemicals, and the required MSDS forms.

The employee training plan must consist of the following elements:

- A description of how the HAZCOM program is implemented in their workplace; how to read and interpret information on the labels and the MSDSs; and how employees can obtain and use the available hazard information.
- An explanation of the hazards of the chemicals in a specific work area. The hazards may be discussed by individual chemical, or by hazard categories such as flammability.

- A discussion of safety measures that employees must follow to protect themselves from these hazards.
- Specific procedures put into effect by the employer to provide protection such as engineering controls, work practices, and the use of personal protective equipment.
- Methods of observation (such as visual appearance or smell) that workers can use to detect the presence of a hazardous chemical.

Anyone who works on a construction site is likely to work around hazardous materials. Fluids such as gasoline, diesel fuel, and solvents used in the workshop are all considered hazardous materials under the OSHA HAZCOM Standard. This section has general information concerning how to act when using or working around hazardous material, and how to respond to accidents or spills.

Employers rely on the MSDS provided by the manufacturers of any hazardous materials used at the workplace, including materials used by subcontractors. The form and style of the MSDS may vary, but each sheet includes sections that detail information as required by the OSHA standard. These sheets contain information about the hazards of a product. They help determine safe handling practices and procedures, emergency response in the event of an accident, and waste disposal requirements. Material safety data sheets are normally kept with the HAZCOM program documentation and placed in work areas for easy reference. This should be the first stop before working with a new chemical.

An MSDS provides information the manufacturer considers necessary to determine what chemicals are in the product and the actions required to protect persons using the product. The sheet is divided into the following sections:

- Basic information, such as the manufacturer, the name of the hazardous material or substance, and an emergency telephone number.
- Hazardous ingredients in the substance, including information on exposure limits. Such limits may be expressed as Permissible Exposure Limit (PEL), Threshold Limit Value (TLV), or in other ways.
- Physical data describing how to identify the material or substance by observation or odor.
- Fire and exposure data describing, in technical terms, the degree of hazard and how to extinguish fires should they occur.
- Reactivity data indicating chemical stability of the substance and situations where a hazardous reaction can occur.

- Health hazard data describing, in technical terms, the degree of hazard, as well as emergency and first-aid procedures.
- Special protection information describing how to protect against hazardous exposure, what to do if a spill or leak occurs, and how to properly dispose of waste materials. Precautions describing any special requirements for handling or storing material and information not provided elsewhere.

Another source of hazardous chemical information is the log of hazardous materials and substances. It serves as a summary list of the individual MSDSs. The list includes the common or trade name, the manufacturer's name, and the chemical name of materials and substances onsite. The location and quantity of certain large volumes of highly toxic substances may also be recorded. The log is usually kept with the HAZCOM program documentation.

Employees should obtain an MSDS for any hazardous material they use. Anyone purchasing or using hazardous materials should verify that all containers are clearly labeled as to contents, and that the label contains the appropriate hazard warning and the name and address of the manufacturer.

> **WARNING!**
>
> No container of hazardous material should be accepted unless it is properly labeled by the dispensing person or agency.

Employers should provide training for all employees who work with or who will potentially be exposed to hazardous materials. Working safely with chemicals is a two-way street. The employer will provide access to needed information and proper protective gear, but it is up to the worker to handle the chemicals safely and use proper safety equipment and safe work procedures when working around chemicals. Always follow these rules when using hazardous materials:

- Know what chemicals you are working with and how strong (concentrated) they are. Use appropriate personal protective equipment as required.
- In case of skin or eye contact, flush with cool water for at least 15 minutes, but do not rub the skin or eyes. This can break the skin or delicate eye tissue and give the substance a path into the body.

- Keep different types of chemicals stored in different areas or in cabinets like the one shown in *Figure 4*.
- Clean up spills promptly. If you cannot identify a chemical or do not know how to handle it, check with your supervisor or safety officer before taking any action.
- Storage areas should be kept free of unnecessary items such as chemicals that present fire hazards. Paints, thinners, pesticides, solvents, and other chemicals should be kept in a well-ventilated area under lock and key.
- Any unknown chemical or unlabeled containers should be treated as extremely hazardous until it is identified.

1.4.0 Excavation Safety

Safety is crucial during any excavation job. Trenches are a common form of excavation. A trench is a narrow excavation made below the surface of the ground in which the depth is greater than the width and the width does not exceed 15 feet, as measured at the bottom of the trench. Working around trenches is one of the most dangerous situations for a construction worker. When earth is removed from the ground, the excavation walls are not supported, so any pressure on the ground around the trench can cause the walls to collapse. To prevent this, the walls need to be properly secured by shoring, sloping, or shielding.

One cubic yard of earth weighs about 3,000 pounds. That is the weight of a small car and is more than enough weight to seriously injure or kill a worker. In fact, each year in the United States, more than 100 people are killed and many more are seriously injured in cave-in accidents.

22210-12_F04.EPS

Figure 4 Safety cabinet.

When operating machinery around excavations, it is important to stay alert for workers on foot. Always keep co-workers in sight. If you lose track of a worker, stop the equipment until you are certain he or she is safe. The following guidelines must be enforced to ensure everyone's safety:

- Be alert. Watch and listen for possible dangers.
- Never enter an excavation without the approval of the competent person on site.
- Never operate your machinery above workers in an excavation.
- Ensure that the OSHA-approved competent person inspects the excavation daily for changes in the excavation environment, such as rain, frost, or severe vibration from nearby heavy equipment.
- Stay alert for other machinery and stay clear of any vehicle that is being loaded.
- Keep the excavated soil (spoil) at least two feet from the edge of the excavation.
- Stop work immediately if there is any potential for a cave-in. Make sure any problems are corrected before starting work again.
- Use shoring, trench boxes, benching, or sloping for excavations and trenches over five feet deep.

- Keep equipment back from the edge of the excavation to prevent cave-ins.

NOTE

For any excavation deeper than 20 feet, the shoring design must be developed and stamped by a professional engineer.

1.4.1 Indications of an Unstable Trench

A number of stresses and weaknesses can occur in an open trench or excavation. For example, increases or decreases in moisture content can affect the stability of a trench or excavation. The following sections discuss some of the more frequently identified causes of trench failure. These conditions are shown in *Figure 5*.

Tension cracks usually form a quarter to half way down from the top of a trench. Sliding or slipping may occur as a result of tension cracks. In addition to sliding, tension cracks can cause toppling. Toppling occurs when the trench's vertical face shears along the tension crack line and topples into the excavation. Subsidence is when pressure on the surface of the ground around the excavation stresses the excavation walls and can cause the wall to bulge. If uncorrected, this condition can cause wall failure and trap workers in the trench or topple equipment near the excavation. Bottom heaving is caused by downward pressure created by the weight of adjoining soil. This pressure causes a bulge in the bottom of the cut. These conditions can occur even when shoring and shielding are properly installed.

Another indication of an unstable trench is boiling. Boiling is when water flows upward into the bottom of the cut. A high water table is one of the causes of boiling. Boiling can happen quickly and can occur even when shoring or trench boxes are used. If boiling starts, stop what you are doing and leave the excavation area immediately.

1.4.2 Making the Excavation Safe

There are several ways to make an excavation site a safer place to work. Heavy equipment operators are called to install shoring or trenching systems, so it is important to be familiar with these procedures. Sometimes, plans call for the excavation walls to be sloped away from the excavation floor to relieve pressure and avoid cave-ins. Shoring, shielding, and sloping are different methods used to protect workers and equipment. It is important to recognize the differences between them.

Shoring in a trench is placed against the excavation walls to support them and prevent their movement and collapse. Shoring does more than provide a safe environment for workers in a trench. Because it restrains the movement of trench walls, shoring also stops the shifting of surrounding soil, which may contain buried utilities or on which sidewalks, streets, building foundations, or other structures are built.

Trench shields, also called trench boxes, are placed in unshored excavations to protect personnel from excavation wall collapse. They provide no support to trench walls or surrounding soil. But for specific depths and soil conditions, trench shields withstand the side weight of a collapsing trench wall to protect workers in the event of a cave-in.

1.4.3 Shoring and Shielding Systems

Shoring systems are metal, hydraulic, mechanical, or timber structures that provide a framework to support excavation walls. Shoring uses uprights, walers, and cross braces to support walls. Because of their great weight, these structures need to be lifted into the excavation with heavy equipment. *Figure 6* shows a shoring system in place.

When excavating near existing structures or performing short-term excavations, vertical sheeting may be used and supported with hydraulic walers. The walers support the vertical sheeting and are held in place against the trench walls by hydraulic spreaders (*Figure 7*). Other types of spreaders, including screw jacks and trench jacks, are also available.

Interlocking steel sheeting may be specified under certain conditions such as deep excavations and excavations near buildings or building foundations. Interlocking steel sheeting is commonly used on Department of Transportation (DOT) right-of-ways. It prevents damage to subbase pavement caused by vibration from vehicle traffic. Interlocking steel sheeting is required when working in waterways. Steel sheeting con-

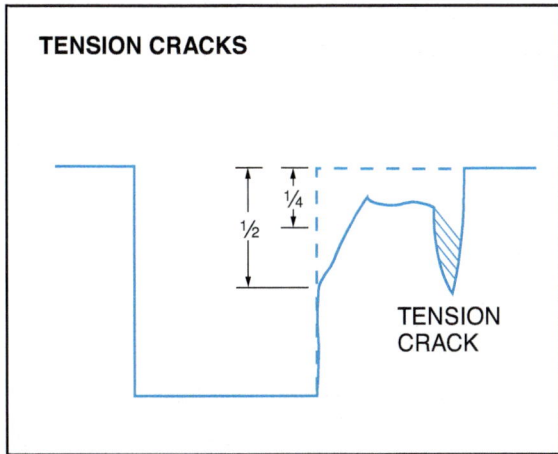

(1)

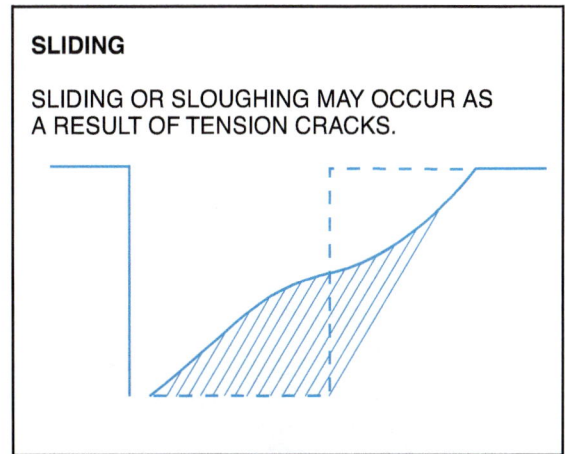

(2)

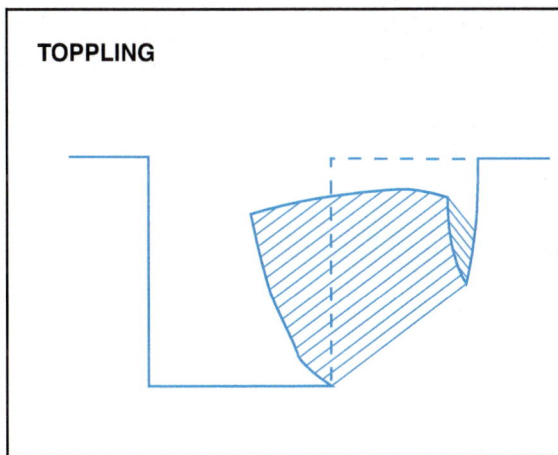

(3)

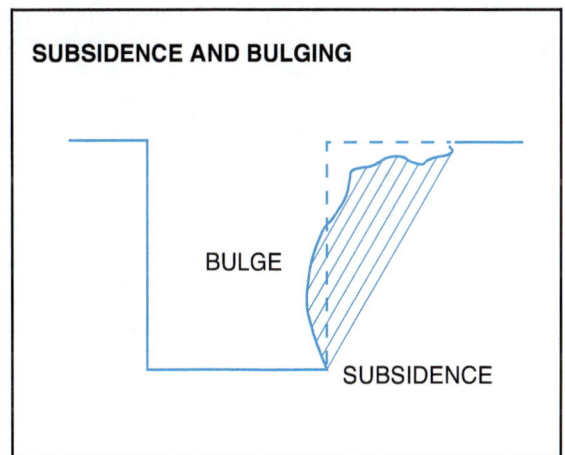

(4)

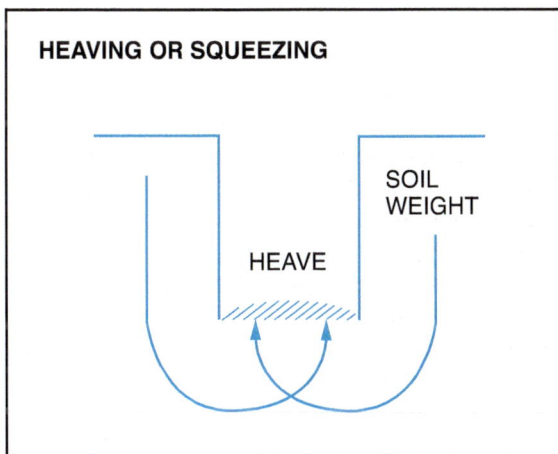

(5)

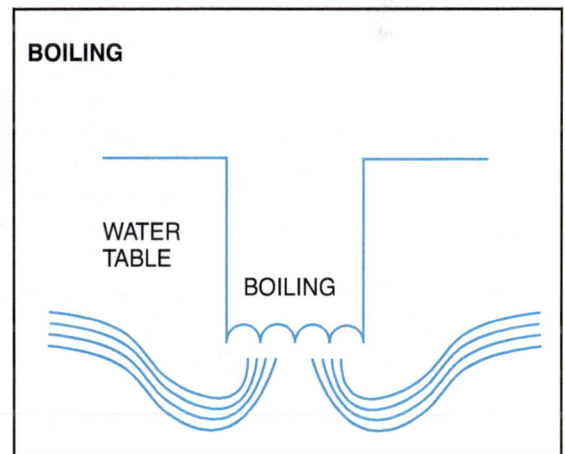

(6)

22210-12_F05.EPS

Figure 5 Indications of an unstable trench.

sists of interlocking panels of steel reinforced with cross members. Steel sheeting is engineered for a particular application. It must be installed precisely in accordance with the engineer's spec- ifications. Steel sheeting is commonly installed by driving it into the ground using a vibrating hydraulic hammer. It can also be driven with a drop hammer.

Figure 6 Shoring system in place.

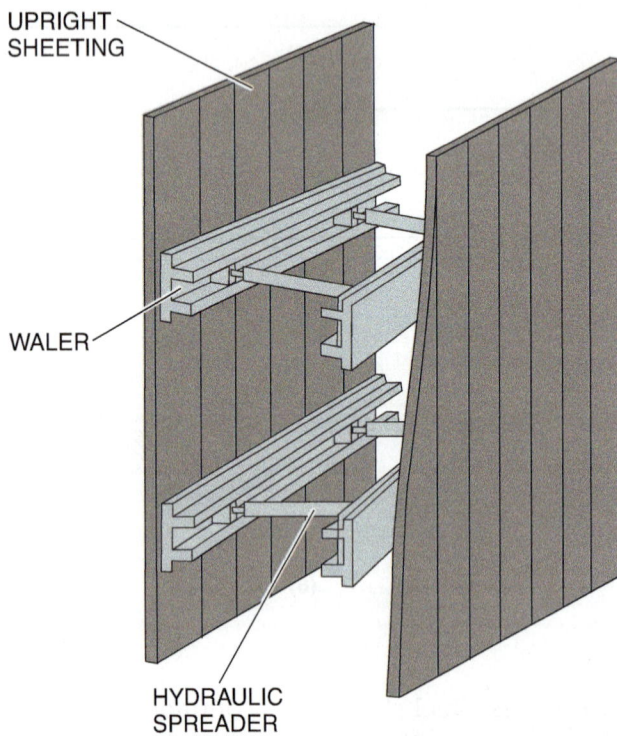

Figure 7 Vertical sheeting system.

To avoid accidents and injury when shoring an excavation, these special safety rules must be followed:

- Never install shoring while workers are in the trench.
- Install shoring by starting at the top of the excavation and working down.
- At large excavation sites ensure that all workers are a safe distance from the installation site when placing shoring. Never move any shoring components over workers.
- The cross braces must be level across the trench. The cross braces should exert the same amount of pressure on each side of the trench.
- The vertical uprights must be pushed flat against the excavation wall.
- All materials used for shoring must be thoroughly inspected before use, must be in good condition, and must have an engineered spec sheet showing shoring limitations.
- Shoring is removed by starting at the bottom of the excavation and going up.
- The vertical supports are pulled out of the trench from above.
- Every excavation must be backfilled immediately after the support system is removed.

A shielding system is a structure that is able to withstand the forces imposed on it by a cave-in, and to protect employees within the excavation. Shields can be permanent structures or can be designed to be portable and moved along as work progresses. The shielding system is also known as a trench box or trench shield. *Figure 8* shows workers protected by a trench box. If the trench will not stand long enough to excavate for the shield, the shield can be placed high and pushed down as material is excavated.

The excavated area between the outside of the trench box and the face of the trench should be as small as possible. The space between the trench box and the excavation side must be backfilled to prevent side-to-side movement of the trench box. Remember that job-site and soil conditions, as well as trench depths and widths, determine what type of trench protection system should be used. A single project can include several depth or width requirements and varying soil conditions. This may require that several different protective systems be used for the same site. A registered engineer must certify shields for the existing soil conditions. The certification must be available at the job site during assembly of the trench box. After assembly, certifications must be made available upon request.

If used correctly, trench boxes protect workers from the dangers of a cave-in. All safety guidelines for excavations also apply to trench boxes. Follow these safety guidelines when using a trench box:

- Be sure that the vertical walls of a trench box extend at least 18 inches above the lowest point where the excavation walls begin to slope. *Figure 9* shows proper trench box placement.
- Never enter the trench box during installation or removal.
- Never enter an excavation behind a trench box (between the excavation wall and the trench box).
- Never enter an excavation after the trench box has been moved.
- Backfill the excavation as soon as the trench box has been removed.
- If a trench box is to be used in a pit or a hole, all four sides of the trench box must be shielded.

22210-12_F08.EPS

Figure 8 Workers in a trench box.

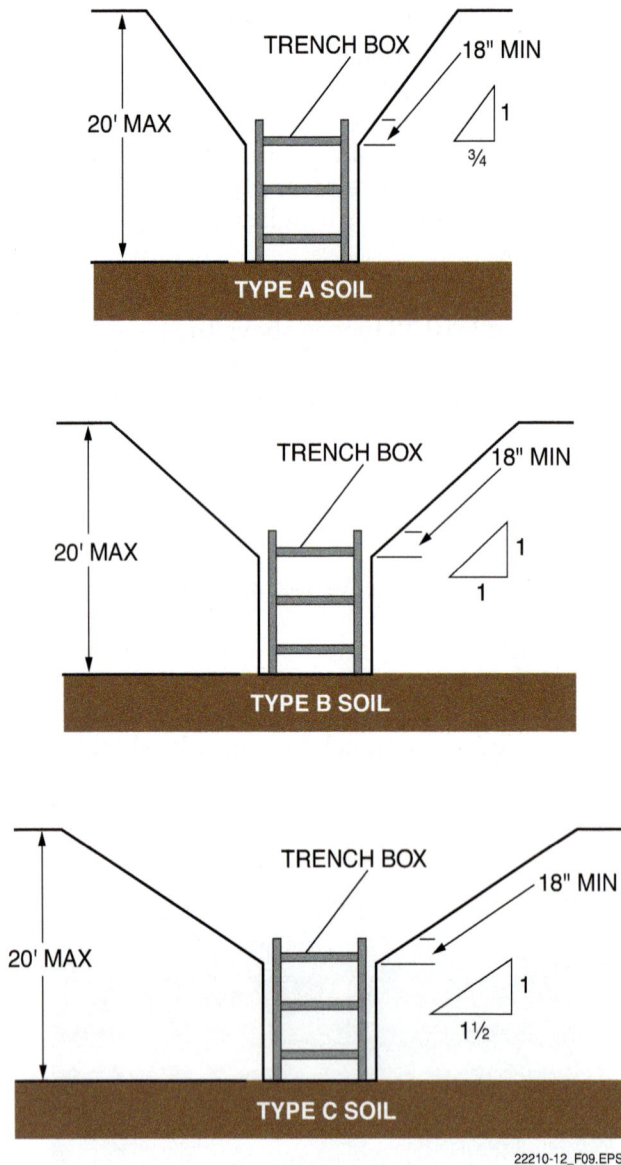

Figure 9 Proper trench box placement.

Remember that trench boxes are designed to protect workers in the excavation. They do not stabilize the walls of the trench, so be alert for changes in the ground surrounding the trench— they could indicate that a cave-in is imminent.

1.5.0 Transporting Equipment Safely

Heavy equipment is usually transported to job sites on flatbed trailers known as low-boys. The driver of the transport vehicle must have a CDL. Loading and securing heavy equipment is a critical operation that calls for knowledge, training, skill, and experience. Never attempt to load or unload equipment without the proper training. The loading procedure for all heavy equipment is not the same, so specific training is needed to safely load and unload each piece of equipment.

Always consult the manufacturer's instructions in the operator's manual for the equipment being transported before you transport any equipment. Follow these general guidelines when loading heavy equipment for transportation to another location:

- Ensure that the transport equipment is in good working order and that tires are inflated according to the manufacturer's specifications. Test lights and turn signals before starting to load the equipment.
- Use loading ramps and other equipment that are adequately rated for the weight of the equipment.
- Use ramps with a surface that gives the equipment being loaded good traction.
- Ensure that the transport vehicle is specified for the load.
- Chock the wheels of the transport vehicle before loading the equipment.

Soil Freezing

As technology becomes available, new methods and practices may be used to help ensure excavation safety. Soil freezing is one such technology. Although this exciting technology has been used with success on some jobs, its use is not widespread. It involves laying chiller pipes at the perimeter of the excavation site, and then a freezing solution is circulated through the pipes, freezing the surrounding soil. The system is left in place for several weeks until the desired level of freezing is reached. Then normal excavation work is performed inside the frozen area. The chiller system is left in place until excavation work is completed. When the pipes are removed, the soil thaws over the course of several weeks until the normal temperature is reached. This method has the added advantage of controlling groundwater flow.

- Inspect tie-down material before and after its use. Discard any worn or damaged material.
- Ensure that the equipment is securely fastened to the transport vehicle in all directions to prevent any movement (*Figure 10*). Use chains, wire rope, or other device as specified by the transporter manufacturer's instruction manual.
- Be certain to use protection on sharp edges to avoid damaging the tie-down material and the equipment.
- Use proper warning signs and flags.
- Secure all attachments and loose gear to prevent shifting of the load during transportation.
- If possible, have another competent person inspect the vehicle after it is loaded to be certain the payload is secure.

Often equipment is delivered at night or early in the morning so the customer can get maximum use during business hours. Keep in mind that loading and unloading equipment is a noisy job. Some areas have noise ordinances that forbid construction noise during the night. Be certain of local laws before attempting to load or unload after normal business hours.

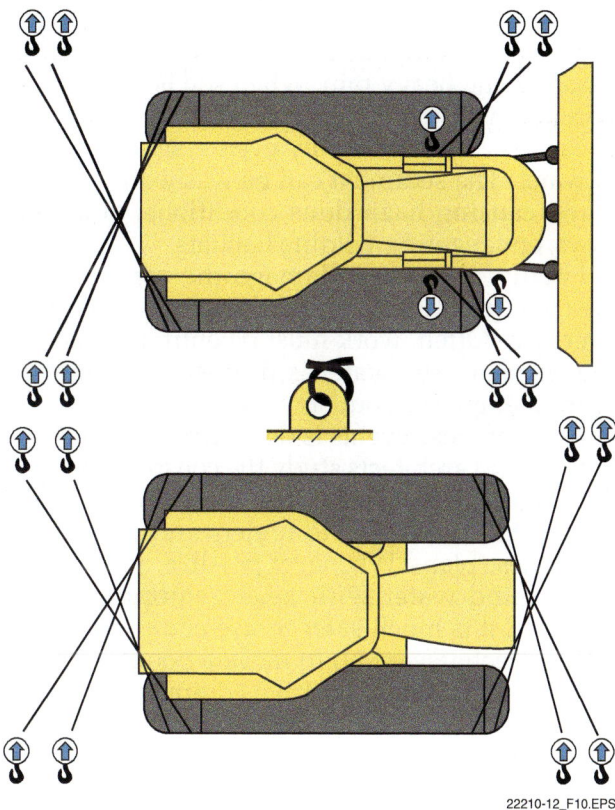

22210-12_F10.EPS

Figure 10 Machine transport.

Additional Resources

OSHA Trench Safety Quick Card: **www. osha.gov/Publications/quickcard/trench- ing_en.pdf**

1.0.0 Section Review

1. Which of the following is one of the responsibilities of a company safety officer?

   a. Approves the construction plans and drawings
   b. Analyzes work practices to improve work conditions
   c. Requests that the local OSHA office visit the site
   d. Checks out all heavy equipment before it is used

2. There must be an accident investigation whenever ____.

   a. a near miss occurs
   b. there is an employee complaint
   c. the local OSHA inspector requests one
   d. there is injury or property damage reported

3. An MSDS must be available for ____.

   a. all chemicals ever used by the company
   b. all chemicals in the OSHA Standards
   c. only those chemicals that are combustible
   d. all hazardous chemicals in the work area

4. One cubic yard of earth weighs approximately ____.

   a. 1,000 pounds
   b. 2,000 pounds
   c. 3,000 pounds
   d. 4,000 pounds

5. When loading equipment for transport, ensure that the equipment is secured ____.

   a. side to side
   b. front to back
   c. in all directions
   d. with the parking brake

2.0.0 Controlling Water at a Job Site

Objective 2

Describe the methods used to control water on job sites.

a. Explain the importance of maintaining proper drainage on a job site.
b. Describe the methods used to control ground water and surface water.
c. Describe the safety practices and construction methods used when working around bodies of water.

Trade Terms

Aquifer: An underground layer of water-bearing permeable rock or unconsolidated materials through which water can easily move.

Berm: A raised bank of earth.

Dewater: To remove water from a site.

Erosion: The removal of soil from an area by water or wind.

Groundwater: Water beneath the surface of the ground.

Sedimentation: Soil particles that are removed from their original location by water, wind, or mechanical means.

Stormwater: Water from rain or melting snow.

Sump: A small excavation dug below grade for the purpose of draining or retaining subsurface water. The water is then usually pumped out of the sump by mechanical means.

Swale: A shallow trench used to direct the flow of water.

Unchecked water flow can damage excavations, stop work, make work more difficult, and cause accidents. For these reasons, a great deal of time and effort goes into the project's dewatering plans. Federal and local agencies regulate all aspects of water control, so it is very important that you know and follow the plans developed by dewatering experts.

A construction site may contain surface water, groundwater, or both. Surface water comes from rain and snow, while groundwater exists below the surface of the ground. Surface water slowly seeps into the groundwater, thus recharging groundwater supplies. Some groundwater exists in underground aquifers. Groundwater sometimes breaks through the surface of the earth as a freshwater stream.

A great deal of planning is needed to complete a job safely in a wet environment. Construction workers face a number of hazards when working around water. First, water from either the ground or the surface can weaken or flood excavations, endangering workers. Trenches must be inspected for safety and integrity daily and after any event that could weaken them, such as rain.

Second, drinking water comes from groundwater aquifers, so it is important to protect groundwater from contamination. When surface water washes across construction debris, such as oil, gasoline, chemicals, and other contaminates, those contaminants can make their way into groundwater supplies. Workers can help to protect surface water and groundwater by performing maintenance in the designated area, avoiding chemical spills, and placing all trash in its proper receptacle.

Third, graded soil is susceptible to erosion and sedimentation. Normally, grass and other vegetation cover ungraded earth, holding the soil in place during heavy rain. When soil is graded, the vegetation is stripped away and the exposed soil is easily washed away by even a small amount of water. The sediment can be washed into roadways, causing hazardous conditions, or waterways, endangering wildlife habitats.

Finally, water can damage the construction area, contributing to scheduling delays and cost overruns. Often, work must be entirely halted at the site until the water is drained, temporarily putting employees out of work.

None of these events is desirable, so project engineers and architects study the construction area and make plans to minimize the flow of water. It is up to the workers to understand and implement these plans. Whenever possible, avoid working around water with heavy equipment. For example, it is much safer to use equipment with a long reach (*Figure 11*) to move a boulder from a streambed than it is to drive equipment into the water. Driving equipment into the water places the operator at risk of drowning. It places the equipment at risk of tipping over due to an underwater obstacle or depression. It also places the environment at risk because any soil or chemicals clinging to the equipment can be washed off into the water.

Figure 11 Long-reach excavator.

2.1.0 Establishing Proper Drainage

Proper drainage is a very important part of the construction process. It helps to maintain the stability of highway roadbeds and building foundations. Drainage is also necessary to reduce or eliminate erosion on embankments and slopes. If the job site has a stormwater control plan, it is your responsibility to ensure that you are familiar with its contents.

Drainage work required during construction falls into three broad categories: natural and constructed drainage; control ditches; and drains and collection systems. Natural drainage and constructed drains are placed to make sure water is drained away from the structure. Control ditches are constructed to keep water from entering the construction area and causing damage to work. Drains and collection systems collect unwanted water within the construction site and provide a method of removal, allowing the soil or other material to dry out and not be saturated.

2.1.1 Natural Drainage and Drains

Roads are always crowned and sloped so that water drains from the road and minimizes water ponding on the surface. Shoulders should be graded to slope as much or more than the road in order to keep water flowing to the ditches. For example, a paved roadway with an 11-foot lane and 4-foot shoulder should have a total crown (from center line to outside edge of shoulder) of not less than 3.5 inches. Roads with steeper grades may require higher crowns, because the water tends to flow down the road rather than across the crown.

Ditches and channels must be dug and maintained to avoid damage to the roadway. Their primary function is to carry water away from the

roadway for absorption, or to another area, such as a detention basin. Ditches must be properly shaped for safety and ease of maintenance, as well as proper water-flow and erosion control. A ditch should be at least one foot below the bottom of the roadway base to properly drain the pavement.

It is very important that water flows through the ditches and does not pond. Ponding water may saturate the subsurface material beneath a roadway, preventing it from draining during a storm. Ditches with at least a 1-percent grade are required to ensure proper flow.

Ponding water on a roadway surface is a result of poor grading and compaction and the lack of drainage ditches. Ponding water in ditches is the result of insufficient grading or poor definition of the grade line.

Springs or seepage areas under the road require special treatment. Rock-filled trenches known as French drains, or perforated pipes, are used to drain subsurface water into ditches or streams. Pipe culverts should be opened as soon as they are finished to help control water flow.

Some soils used in embankments are subject to high capillary action, which means that water easily creeps up towards the surface; this creeping action is also called wicking. To reduce this process, a granular blanket of sand or gravel is placed over these soils. This blanket, between the embankment soils and the subgrade, help protect the subgrade material from the water wicking upward. Requirements for a granular blanket are usually specified in the plans.

2.1.2 Control Ditches

Control ditches (*Figure 12*), also called intercepting ditches, need to be placed wherever there is a possibility of water intrusion from a source outside the construction area. This is a precautionary action based on knowledge of the terrain as well as possible rain and flooding conditions. These ditches drain water away from the construction site and deposit it in a nearby drainage ditch.

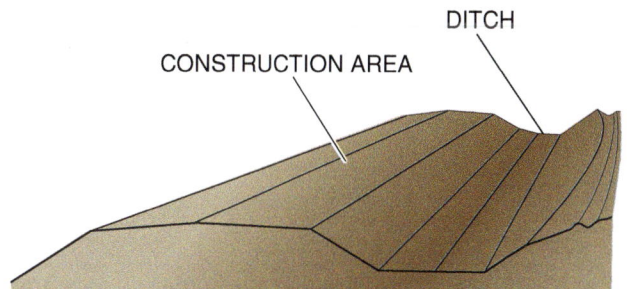

Figure 12 Control ditch.

Intercepting ditches for highway construction should be placed where the original ground outside of the finished cut sections slopes toward the center line. Seeping water in cut sections may indicate the need for a control ditch. The project engineer should be informed of these conditions, so intercepting ditches or another corrective action can be started.

2.1.3 Drains and Collection Systems

Excavation of pits and trenches can run into trouble because of unexpected pockets of water or a high groundwater table. When excavating below the water table, there is normally some intrusion into the excavated area. Because this water is below the natural ground line, it cannot be channeled out by digging a control ditch or another drain.

Channels must be dug in the bottom of the excavation to collect the water and direct it into a sump, where it can be pumped out. The sump should be placed as close as possible to the greatest source of the flow. The channel depth below the excavation floor should be about four feet, but depends on the pumping equipment being used. When a large amount of water flow exists, you should increase the number of sumps, not their size. Figure 13 provides two examples of different types of drainage in an excavation.

The first example (Figure 13A) shows water coming from only one face of an excavation. A drain along this wall channels the water into sumps at either end. Pump-out is done from the sumps. The second example (Figure 13B) is similar, except that the water comes off two faces of the excavation. In this case, the drains are designed to channel water into a sump in the corner between the two faces. Pump-out is done from this one point.

When water is coming in from the bottom of the excavation, the bottom must be graded to drain to each side. Then a layer of stone and damp-proof paper is placed to provide a stable work platform. Channels are dug on the sides of the excavation to collect the water and the water is pumped out by the sumps. The sumps always need to be placed as close to the major source of water flow, using multiple sumps if necessary.

Soil conditions dictate the method of constructing drains. Drainage layouts are governed not only by soil conditions, but also by the location, direction, and quantity of flow. Drains should be limited in width and depth to minimum requirements. They should intercept the ground water as close to its source as possible.

2.2.0 Controlling Ground Water and Surface Water

The US Environmental Protection Agency (EPA) requires that contractors disturbing one acre or more of land obtain a National Pollutant Discharge Elimination System (NPDES) stormwater permit before work begins on the site. The purpose of the permit is to ensure that the contractor has a workable plan to prevent pollution and to control stormwater runoff from the site, as well

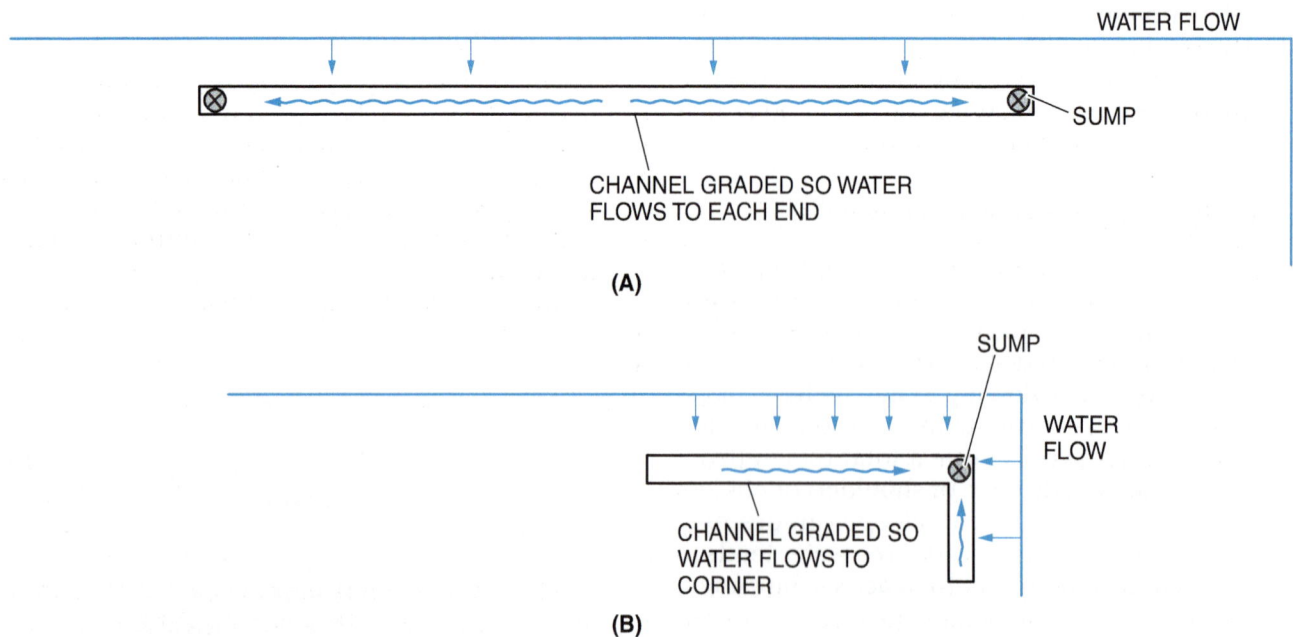

Figure 13 Sump drainage.

as to prevent erosion. Heavy equipment operators need to know and understand the site's erosion, sediment, and runoff-prevention plans.

Contractors need to use adequate measures to prevent erosion and sedimentation caused by stormwater runoff. Erosion is the eating away of soil by water or wind, while sedimentation refers to soil that has been moved from its original place by wind, water, or other means. Stormwater runoff is water from rain and snow that is shed from the ground rather than being absorbed. Runoff increases as an area is developed with roads and buildings because there is less available ground surface to absorb water.

Sediment is soil that has been moved from its original place, so the best way to avoid sediment is to prevent erosion. In earthmoving, some erosion is unavoidable, so most sites use various devices, such as installing silt fences (*Figure 14*) and blocking stormwater drains (*Figure 15*) to trap displaced soil. Other methods, such as berms and swales are used to guide and trap runoff. Workers can help prevent sedimentation by avoiding walking or driving across disturbed soil, washing soil off truck tires before leaving the site, and limiting vehicle operation to approved haul roads.

Water that flows through the ground is called groundwater. Any excavation, even one of only a few feet, has the potential of being affected by and affecting groundwater. Before a large project is approved, an environmental survey may be required. The survey includes a map of known groundwater hazards, such as buried chemical storage tanks, or landfills containing possible pollutants. In many jurisdictions, there are legal requirements for plans to remedy such hazards, and these plans must be supplied to and approved by the governing agencies before digging begins. Failing to follow the approved plan can make your employer liable for legal action and fines.

It is normal to encounter groundwater in deep trenches and excavations, especially where the groundwater is close to the surface. This is known as a high water table. The depth of the water table varies widely and depends on the amount of annual precipitation, the terrain, and the amount of water pumped from it for drinking, irrigation, and other uses. When groundwater begins to enter the trench or excavation, it causes several problems.

22210-12_F14.EPS

Figure 14 Silt fence.

22210-12_F15.EPS

Figure 15 Stormwater drain barrier.

First, the soil of the trench walls becomes more likely to collapse or to flow into the bottom of the trench. Second, the water makes it very difficult to keep digging. Further, when digging through water, some amount will fill the bucket, increasing the weight and forcing the equipment to use more fuel.

Soil helps protect groundwater by filtering chemicals and debris before it reaches the aquifer. Since excavations remove soil, chemicals that normally would not reach the underground level can contaminate the water. On many construction jobs, monitoring wells are dug to check the quality of groundwater, especially where blasting or toxic chemicals may be involved.

The easiest and best way to control water damage at an excavation is to prevent the water from entering the site. If this is not possible, control measures need to be as close to the source of water as possible. Various types and sizes of ditches are used to collect and channel the flow of water away from excavation areas. Some of these ditches are permanent and some temporary. All of them, however, are designed and specified according to the permeability of the soil, conditions at the site, and on the ability to handle the predicted quantity of water. Some common ditch configurations are shown in *Figure 16A*.

More active measures are sometimes needed to evacuate water from an area. In this case, the plans may call for sumps. Sumps are open pits or holes constructed to collect water (see *Figure 16B*). *Figure 17* shows examples of methods used to control ground water in road construction.

> **NOTE**
>
> A stilling basin is designed to dissipate the energy of fast-moving water in order to prevent erosion.

Pumps are placed in the sump to remove the water mechanically. The requirements for sumps and pumps vary with the depth and the flow rate of the water. A variety of pumps are described in the following section.

2.2.1 Equipment Used to Control Water

Portable pumps are widely used in construction for various applications, including draining water from excavations. The three most common types of portable pumps are trash pumps, mud pumps, and submersible pumps. These pumps can be pneumatic, electric, or gasoline engine-driven pumps.

Trash pumps – Trash pumps are centrifugal pumps specifically designed for use with water containing solids such as sticks, stones, sand, gravel, and other foreign materials that would clog a standard centrifugal pump. Trash pumps are adequate for pumping water from a maximum depth of 20 feet. To remove water from a depth greater than 20 feet, a submersible pump should be used. *Figure 18* shows a portable trash pump.

> **CAUTION**
>
> Any time that pumps are to be left outside during cold weather, remove all water from the pump to prevent the water from freezing inside and cracking the pump housing.

Trash pumps are continuous flow pumps that use centrifugal force generated by rotation. The liquid enters the impeller at the center, or the eye, and the rotation of the impeller blades causes a rotary motion of the liquid. Centrifugal force moves the liquid away from the center. As this happens, the liquid's velocity increases until it is finally released through the discharge outlet.

Mud pumps – Most mud pumps (*Figure 19*) are diaphragm pumps designed for use where low, continuous flow is required for highly viscous fluids, such as thick mud, air mixed with water, and water containing a considerable amount of solids or abrasive materials.

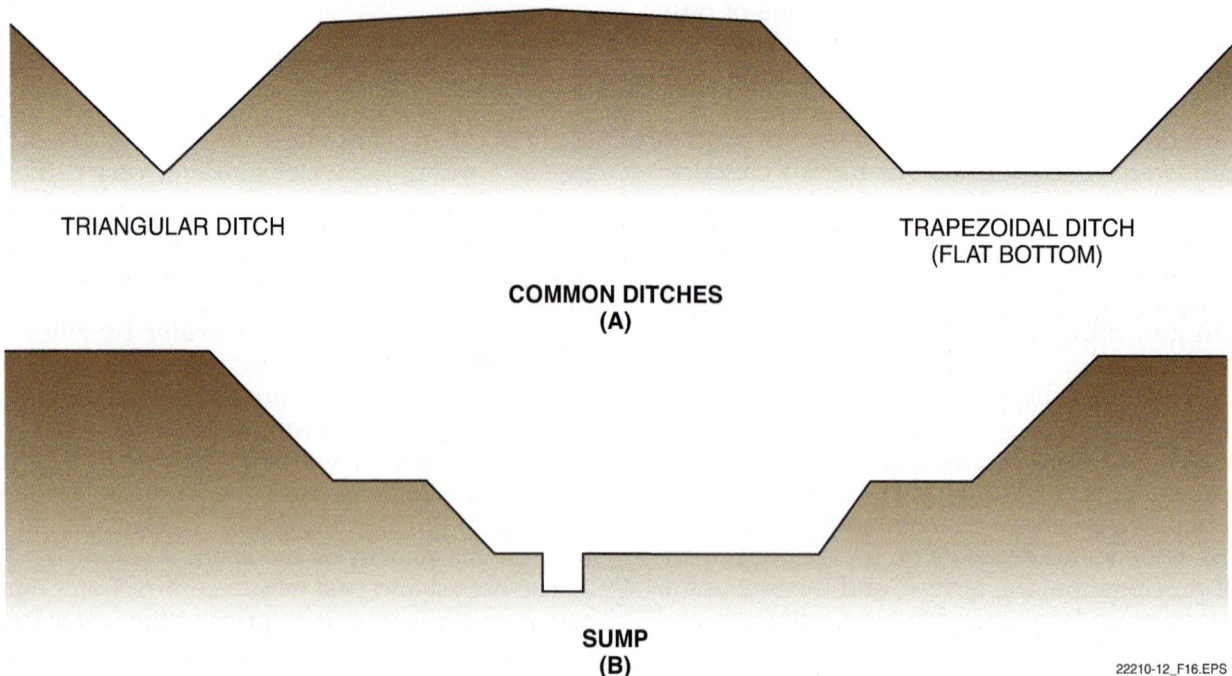

TRIANGULAR DITCH

TRAPEZOIDAL DITCH
(FLAT BOTTOM)

COMMON DITCHES
(A)

SUMP
(B)

22210-12_F16.EPS

Figure 16 Groundwater control.

(A) STILLING BASIN

(B) INLET SEDIMENT TRAP

(C) TEMPORARY SLOPE DRAIN PIPE

(D) SLOPE DRAIN PIPE AND SILT CHECK DAM

22210-12_F17.EPS

Figure 17 Examples of groundwater control methods.

Most diaphragm pumps use hydraulic pressure delivered by a piston to actuate the flexible diaphragm. The side of the diaphragm that produces the pumping action is called the power side, and the side doing the pumping is called the fluid side. A suction check valve and a discharge check valve open and close to move fluids through the pump. *Figure 20* shows the operation of a diaphragm pump.

Submersible pumps – A submersible pump (*Figure 21*) is encased with its motor in a protective housing that enables the entire unit to operate underwater. Submersible pumps are used in many residential, commercial, and industrial wells. Submersible pumps can be controlled by an optional float switch located inside the well. The float switch controls the pump by manual or automatic control switches outside the well. The float switch is a mercury-to-mercury switch that is controlled by the position of an attached float.

As long as the water level is at a height adequate enough to remove water from the well, the pump remains on. When the water level falls too low for pumping, the float drops and turns the pump off. As the water level increases, the float rises and turns the pump on again. *Figure 22* shows a submersible pump controlled by a float switch.

2.2.3 Digging in Wet Soil

Water always seeks the lowest level, so it is not unusual for the bottom of an excavation to be covered with water from either ground or surface water sources. A small amount of water can actually be helpful to ensure a level bottom grade. However, too much water saturates the soil and makes work hazardous.

If a large amount of water is found when digging out the soil, the job can take much longer than expected and the equipment is forced to

Figure 18 Portable trash pump.

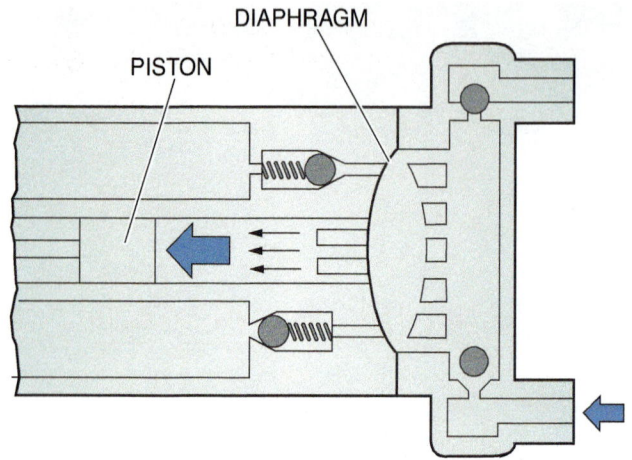

Figure 19 Mud pump.

SUCTION STROKE

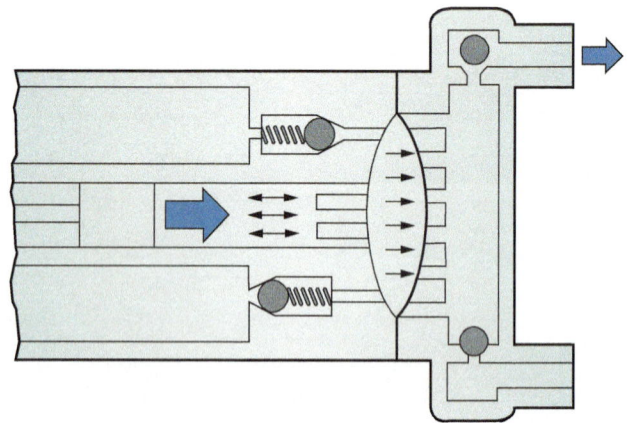

DISCHARGE STROKE

Figure 20 Operation of diaphragm pump.

Figure 21 Submersible pump.

work harder, thus increasing cost. The water can run back into the excavation, setting up a never-ending cycle of digging. Soil mixed with water is more liquid than dry soil so spoil piles may slide back into the excavation. It may be necessary to move them further from the open excavation than planned, thus increasing cycle time.

Under some conditions, the best solution would be to pump the water out of the excavation before digging. Pumping is an extra step that delays excavations and increases costs, so there are a number of other actions that may get the job done. These include compartment digging and using a spoil barrier.

Compartment digging – On a large excavation where a small amount of water is present, the water can be confined in a compartment, and thus kept out of the bucket. A hole must be dug in the exca-

PUMP CORD

PUMPING
RANGE

PUMP ON

PUMP OFF

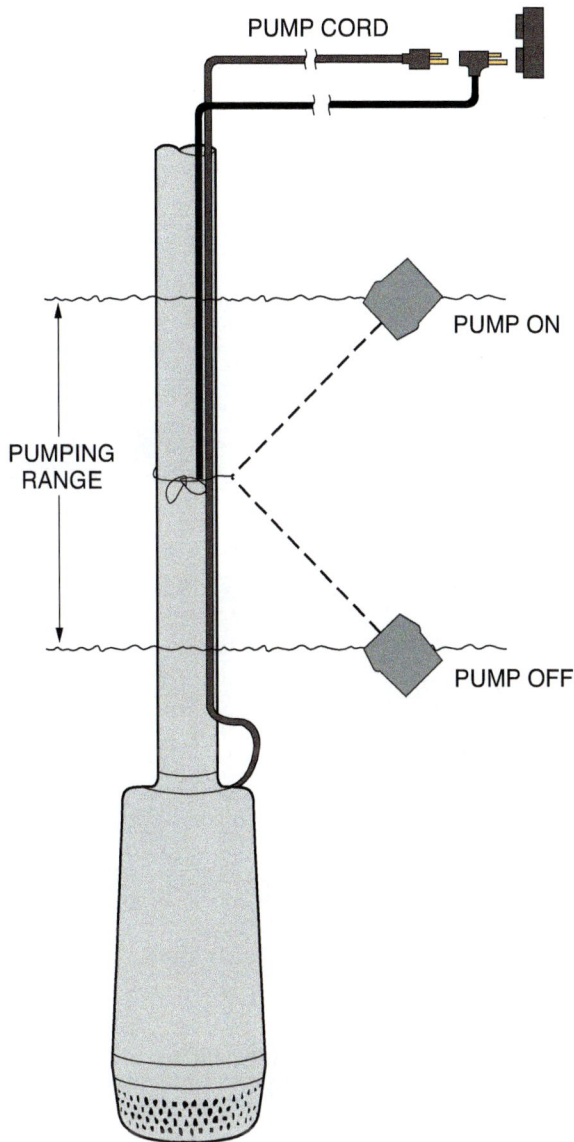

22210-12_F22.EPS

Figure 22 Submersible pump controlled by float switch.

vation large enough to hold the water. The floor of the excavation is sloped so the water runs into the hole. When the water has drained into the hole, the remainder of the excavation can be dug to grade, leaving a wall of soil between the water compartment and the area where digging is being done. Once at grade, break through the wall and allow the water to flow into the deep side of the excavation, then dig the compartment floor to grade. This process may need to be repeated several times, alternating sides to achieve the correct grade.

Spoil barrier – Muddy spoil is a problem. It can easily flow back into the excavation, increasing the amount of work that needs to be done. The best way to handle muddy spoil is to move the spoil pile far from the excavation. When space is limited, this is not a possibility. When dry soil is available, it is possible to build a dike with dry soil that

will act as a barricade between the excavation and the spoil. In this case, the dry soil can be piled a safe distance from the excavation, and the muddy spoil can be placed behind it. Keep in mind that this method will increase the cycle time (and thus costs), because the equipment needs to travel further from the excavation to dump the spoil.

> **NOTE**
>
> When muddy spoil is loaded onto a truck for immediate transport from the site, ensure that the floor of the bed is covered with dry soil so the load does not stick when it is dumped.

2.3.0 Working Around Water

Some construction projects are performed on or near water. Anyone working in such an environment needs to follow some basic safety guidelines. Sometimes work needs to be done on building or bridge foundations that are underwater. In this case, there must be some means of draining and controlling water so that workers can accomplish their work safely. Two structures that are used under these conditions are coffer dams and caissons.

2.3.1 Safety Guidelines

The first rule to working safely around water is to stay as far away from the water as possible. Heavy equipment operations are risky on level dry ground, but operations can become perilous near the water. When working around water, follow these guidelines:

- Understand that some municipalities have laws and regulations about heavy equipment entering water, such as streams, lakes, and rivers. Ask your supervisor for instructions before entering water.

> **NOTE**
>
> Special permits may be required when working in or around water. Check federal, state, and local regulations.

- Select the correct equipment for the job. If equipment is available that permits the work to be done from dry ground, use it, but never overextend the reach of any equipment.
- Before approaching any water body with heavy equipment, carefully examine the terrain for obstacles, depressions, and soft spots. Walk the area if possible. Observe the water current before driving the equipment into the water.

- Before approaching the water, inspect the equipment to ensure that it is in good working order and clean of mud or any chemical residues such as oil that can be washed off into the water. Make a plan about what to do in case of mechanical problems.
- Before entering the water, assess the task. Think about what could go wrong and be prepared.
- While in the water with the equipment, use the buddy system. Have a responsible person nearby who can help in case there is a problem.
- When moving equipment through water, move cautiously. Water masks obstacles and hazards.
- Once in the water, perform the task as quickly and as safely possible to minimize the time spent in the water.
- Special precautions are required when working around open water. These precautions include wearing a flotation device, having a throw ring available, and having a boat of some type to perform a rescue in case of a man-overboard situation.

2.3.2 Coffer Dams

A coffer dam is a temporary structure used to keep water out of a construction site so that work can be done in a dry area. Although most coffer dams are built to create a dry work space in a waterway, some are used to prevent ground water from entering work sites. This is common in the construction of roadways, as well as buildings that need deep foundations. *Figure 23A* shows a coffer dam under construction. *Figure 23B* shows the same coffer dam completed and in the process of being drained.

Coffer dams are generally built in place, but may be prefabricated and dropped into place using cranes. That type is known as a *gravity dam*. Once coffer dam construction is complete, water is pumped out of it. *Figure 24* shows workers inside the drained coffer dam previously seen in *Figure 23*. Coffer dams used in the construction of piers, docks, bridges and the like are usually made from sheet piles that fit together tightly enough to keep water infiltration at a minimum. If it is necessary to make the work area completely dry, there are various methods available to seal the interlocking joints in the sheet piles.

Sheet pile systems are popular because the piles are easy to install and remove and can be reused. A braced coffer dam like the one shown in *Figure 23* is made of a single row of sheet piles. Because the coffer dam must be able to withstand the pressure of the surrounding water, it is braced with walers

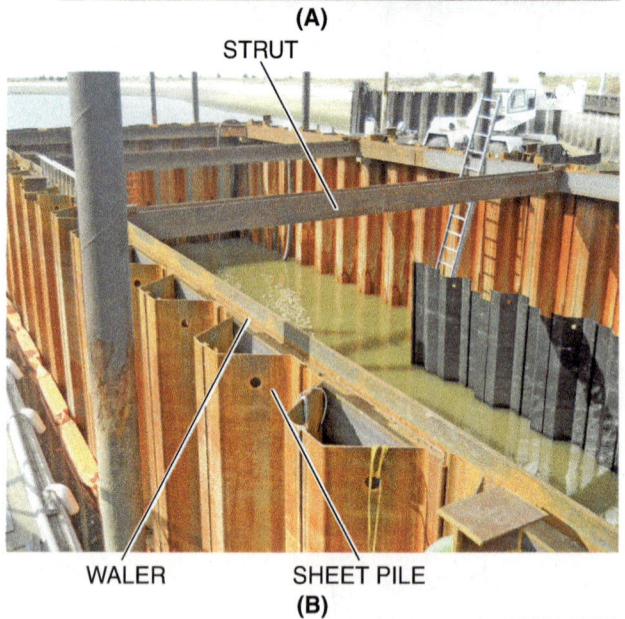

(A)

STRUT

WALER   SHEET PILE

(B)

22210-12_F23.EPS

Figure 23 A coffer dam.

22210-12_F24.EPS

Figure 24 Completed coffer dam.

and struts. Braced coffer dams are commonly used for bridge piers and other structures in shallow water. The coffer dam in *Figures 23* and *24* is being used for construction of a boat-launching ramp.

A double-walled coffer dam (*Figure 25*) has two parallel rows of sheet piles. The sheet piles are driven into the bottom and connected with heavy-duty tie rods. The space between rows is often filled with earth removed from the excavation. The walls of double-wall coffer dams are often spaced widely apart so that the space between them can be used as a roadway for construction vehicles. Double-walled coffer dams are used in locations where a large area is to be excavated.

When metal sheeting is used to build a coffer dam, the sheets are usually lifted into place with a crane and then driven into the floor of the waterway with a vibratory hammer pile driver (*Figure 26*). The sheets must be driven in to a point that is solid enough to prevent seepage, and then supported with bracing to counteract the force of the water. Once the area is enclosed, the water is pumped out so that work can begin. An excavator with the bucket removed is sometimes fitted with a special attachment and used to drive sheet piles. This approach is common in situations where there is not enough overhead space to accommodate a crane, such as the underside of an elevated roadway.

Wooden piles can be used instead of metal piles. Construction is done in a similar fashion. Earth and rock coffer dams are also used. An earthen dam is built by dumping earth fill into the water and shaping it to surround the construction site. This is slow work and requires a lot of fill, so it is not used often. Earthen coffer dams are usually restricted to shallow areas in waterways that have a very low current, because the earth tends to be washed away in strong currents.

2.3.3 Caissons

Caisson is French for box. Caissons are self-contained boxes or chambers that are used so construction work can take place underwater. Caissons are lifted into the water with a crane or other suitable equipment, floated on the water to the place they are needed, then sunk, and pumped dry. Caissons that are used in shallow water are usually open at the top and bottom, but caissons that are used in deep water can be quite sophisticated—these caissons are completely enclosed and pressurized with air to keep water out. *Figure 27* shows an example of a caisson being used for construction of a bridge abutment.

22210-12_F25.EPS

Figure 25 Double-wall coffer dam.

22210-12_F26.EPS

Figure 26 A crane driving sheet piles for a coffer dam.

WARNING!

Coffer dams and caissons should be treated as confined spaces, in that they must be monitored for proper oxygen levels and the presence of toxic gases or other airborne contaminants. It is possible that toxic gases such as methane can bubble up from the bottom of the excavation, for example. If any airborne contaminants exist in the excavation, personnel must be equipped with the proper respiratory equipment. Multigas monitors should be used to monitor for the presence of toxic gases.

22210-12_F27.EPS

Figure 27 Example of a caisson.

Additional Resources

Soil properties that affect groundwater:
www.co.portage.wi.us/groundwater/undrstnd/soil.htm

2.0.0 Section Review

1. In what situation would a road be built without a crown?

   a. In a subdivision
   b. In the desert
   c. On steep hills
   d. Never

2. Trash pumps are adequate for pumping water from a maximum depth of _____.

   a. 5 feet
   b. 10 feet
   c. 20 feet
   d. 50 feet

3. Which of the following is used to keep water away from a bridge foundation while it is being repaired?

   a. Trench box
   b. Shielding system
   c. Detention system
   d. Coffer dam

3.0.0 ESTABLISHING GRADE ON A JOB SITE

Objective 3

Explain how grades are established on a job site.
  a. Describe how to set grades from a benchmark.
  b. Describe how grades are set for highway construction.
  c. Describe how grades are set for building construction.
  d. Explain how grading operations are performed.
  e. Describe the use of stakeless and stringless grading systems.

Performance Tasks 1 and 2

Interpret layout and marking methods to determine grading requirements and operation.

Set up a level and determine the elevations at three different points, as directed by the instructor.

Trade Terms

Balance point: The location on the ground that marks the change from a cut to a fill. On large excavation projects there may be several balance points.

Boot: A special name for laths that are placed by a grade setter to help control the grading operation. The boot can also be the mark on the lath, usually 3, 4, or 5 feet above the finish grade elevation, which can be easily sighted. This allows the grade setter to check the grade alone instead of having to use another person to hold a level rod on the top of the grade stake.

Stations: Designated points along a line or a network of points used to survey and lay out construction work. The distance between two stations is normally 100 feet or 100 meters, depending on the measurement system used.

String line: A tough cord or small diameter wire stretched between posts or pins to designate the line and elevation of a grade. String lines take the place of hubs and stakes for some operations.

Topographic survey: The process of surveying a geographic area to collect data indicating the shape of the terrain and the location of natural and man-made objects.

Finish grading is normally done with graders. Dozers and scrapers are also used in grading operations. The site plan for a project shows the topographical features of a site, as well as project information such as the building outline, existing and proposed contours, and other information that is relevant to the project. Based on this information, survey teams set grade stakes. Effective grading operations depend on proper staking of the project and the ability of the equipment operators to follow the grading instructions on the stakes. The grading work begins after a survey crew places the initial grade stakes. Each phase of construction may require more grading work, so the survey team sets new stakes at each phase.

Before construction starts, a topographic survey and property survey of the site are performed. This information is used to develop the project plans. The plans include information about elevation requirements at the construction site. Based on the plan, stakes are set by the project survey team to inform workers of the earthwork needed to bring the building site to the specified elevations. Methods for staking grades vary greatly with the type, location, and size of project. Methods used by different engineering and construction organizations also vary; some companies use electronic equipment, while others use manual instruments. Regardless of how the stakes are set, unless the project is very small, they be set in reference to benchmarks.

3.1.0 Using Benchmarks to Establish Grade

Benchmarks are used to show a precise elevation in relation to sea level. On building projects, benchmarks are used as reference points from which grade stakes are set. Benchmarks may be permanent or temporary and can be set by federal or local governments or by the project survey team. Benchmarks must not be disturbed. It is very important to know the location of any benchmarks in the area so that you can avoid them with your equipment.

Control Points

Good practice dictates that a control point should always be placed so that three other control points are visible from it at all times. This is done in case the control point becomes covered up or destroyed. If this happens, two tapes can be stretched from the other points in order to relocate the control point.

3.1.1 State Plane Coordinates

Benchmarks on the state plane coordinate system are permanent reference points set by the federal government. They are permanent markers placed in specially designed locations and have specific reference information engraved on a bronze cap (*Figure 28*). These markers are part of the National Geodetic Survey, which is part of the National Oceanic and Atmospheric Administration (NOAA). These markers are frequently accompanied by a witness post to make them easier to find.

The markers provide an accurate reference point for other surveying activities within that general area. Many such markers are located throughout the United States. They are usually set in concrete on the ground to help preserve them. It is important that the markers not be moved or otherwise disturbed. If a permanent marker is damaged it may not be used as a benchmark. It is important to report it to a supervisor or the project engineer. The supervisor or engineer will notify the proper agency so that it may be reset.

States may set similar markers in accordance with their own specifications. State markers are also intended as permanent markers, so if there is one in the work area that is not recorded on the project plan, do not disturb it. Put some type of stake or marker next to it so that you are able to find it again, and notify the project engineer. The engineer will notify the proper authorities to see if the monument needs to be relocated by a survey crew before any further work is done in that area.

3.1.2 Permanent Project Benchmarks

Sometimes a project is developed over a long period. If additional construction is scheduled in the future, there needs to be a control point established close to the construction site so that new construction work can be easily referenced to the previous work. If no other benchmarks are close by, then a permanent project benchmark must be established. This benchmark may be a special cap of some type set in concrete on the ground or on the side of a structure, such as a building or bridge abutment.

3.1.3 Temporary Project Benchmarks

Temporary project benchmarks can be placed in the ground, on a structure, or on a large sturdy tree. They are used only for the specific project. The location and elevation of these benchmarks is usually derived from permanent benchmarks on the state plane coordinate system described earlier. *Figure 29* shows an example of how temporary project benchmarks are placed.

3.2.0 Highway and Other Horizontal Construction

For rough grading on highway and other horizontal construction, there are several methods to set grade control depending on the type of construction and the surrounding terrain. The types of stakes used in highway construction depends on the number of lanes and the existence of curbs and medians.

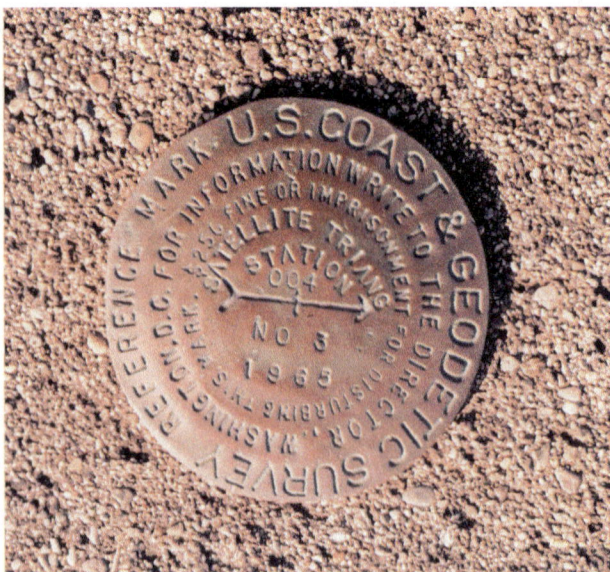

22210-12_F28.EPS

Figure 28 Benchmark of the National Geodetic Survey.

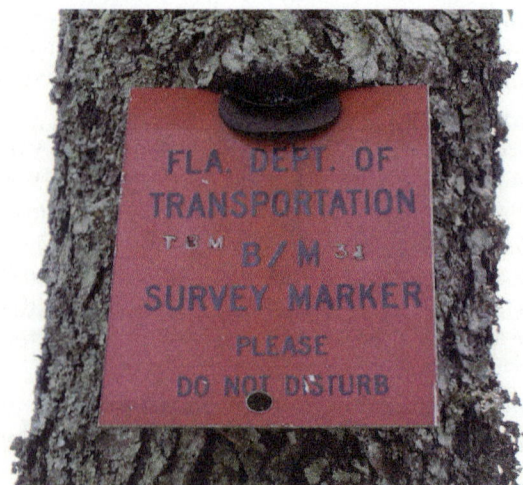

22210-12_F29.EPS

Figure 29 Temporary project benchmark.

3.2.1 Setting Highway Grade Control

For road construction, reference stakes are located outside the construction limits, and are the first stakes set by the project survey team. They are used by the contractor as a reference to perform rough grading. These stakes are located outside the work area, so it is unlikely that they will be damaged when the area is being prepared for construction.

The center line stakes are set by the survey crew using benchmarks and control points. Because the location and elevation of each control point is known, all center line stakes can be referenced directly or by referencing one stake to the stake behind it until the center line is tied into another control point further up the line. This tie-in to a second control point provides a double-check on the center line staking process, as well as any other stakes that must be precisely set. *Figure 30* shows an example of a highway plan and profile sheet that includes references to a benchmark used on this job. The benchmark is located at station 109+16.00.

All members of the construction team use the drawing set, so sometimes the plan and profile drawings look confusing (refer to *Figure 30*). The profile portion of the road project is shown

at the bottom of the drawing. On the right side of the sheet, the drawing shows the center line profile of the road. The drawing ends at station 115+00 with an elevation of about 2,842 feet (read from the elevation scale located on the right side of the drawing). The center drawing also shows the center line profile of the road. It starts at station 115+00 at an elevation of about 2,842 feet. The center line profile ends at station 110+20.89, which is written on the drawing, at an elevation of about 2,847 feet. The left side of the drawing does not show the roadway center line. The two profile curves shown on this portion of the drawing represent the sides of the road. The bottom drawing is the right side of the road and the top one is the left side of the road.

If reference stakes are used for rough grading, they can be placed relative to the center line stakes or referenced directly from the benchmarks. They are placed at the right-of-way limits at the distance from the center line shown in the plans. Because the stakes are placed at the right-of-way limit, there is little chance that they will be disturbed during earthmoving operations. These stakes are usually placed at each station (every 100 feet) but may be placed at closer intervals on corners and alignment changes.

The cut or fill amount is indicated on each stake. This is the vertical distance between the

profile grade stake and a line on the stake. *Figure 31* shows a typical cross-section of a cut. The stake shown in the figure identifies the station, amount of cut, and the point from which the cut or fill is measured. Most slope stakes would be marked in a similar manner.

Slope stakes, which can be used as cut and fill stakes, are set by the survey team at points where the cut slopes and fill slopes intersect a hinge point. For rough grading, it is unlikely that the survey team will set cut and fill stakes, but the team will probably use some other type of marker to guide equipment operators in the initial cut and fill operation. Such temporary markers have no information written on them, but should be familiar to the equipment operator. If you are confused about the stakes, talk to our supervisor before you begin grading work. Finish grade stakes show information about the elevation and offset distance.

To visualize the grade for a center line, look at the profile sheet in the plans. *Figure 32* shows an example of the profile sheet and the cut and fill areas of the roadway. From Station 10 to Station 40+31.05, the proposed grade is higher than the existing grade, so the area needs to be filled. At Station 40+31.05 the existing grade crosses the proposed grade and both are maintained until Station 50+10.00. This area is called the balance point—it needs neither cut nor fill. At Station 50+10.00, the existing elevation is greater than the proposed elevation, which will require a cut to achieve the desired elevation.

The slopes of the proposed grade are also shown, along with the reference axis. The vertical axis on the profile sheet shows the elevation. Usually, the elevation is shown at a much greater scale than the horizontal axis. The horizontal axis shows the stations along the center line. Therefore, you can identify the points on the natural ground where the elevation of the proposed grade is the same as the natural ground.

3.2.2 Finish Grade Reference Points

When the roadbed is close to the proposed grade, the survey crew places finish grade stakes to guide the final grading operation. If any type of stabilization of the subgrade is required, then the subgrade has to be mixed, regraded, and compacted to the proper density. After compaction, the subgrade must again be shaped to conform to the finish grades and cross-sections shown in the plans. This finish work is often performed by motor graders.

By setting a string line at a convenient height above each grade stake, measurements can easily

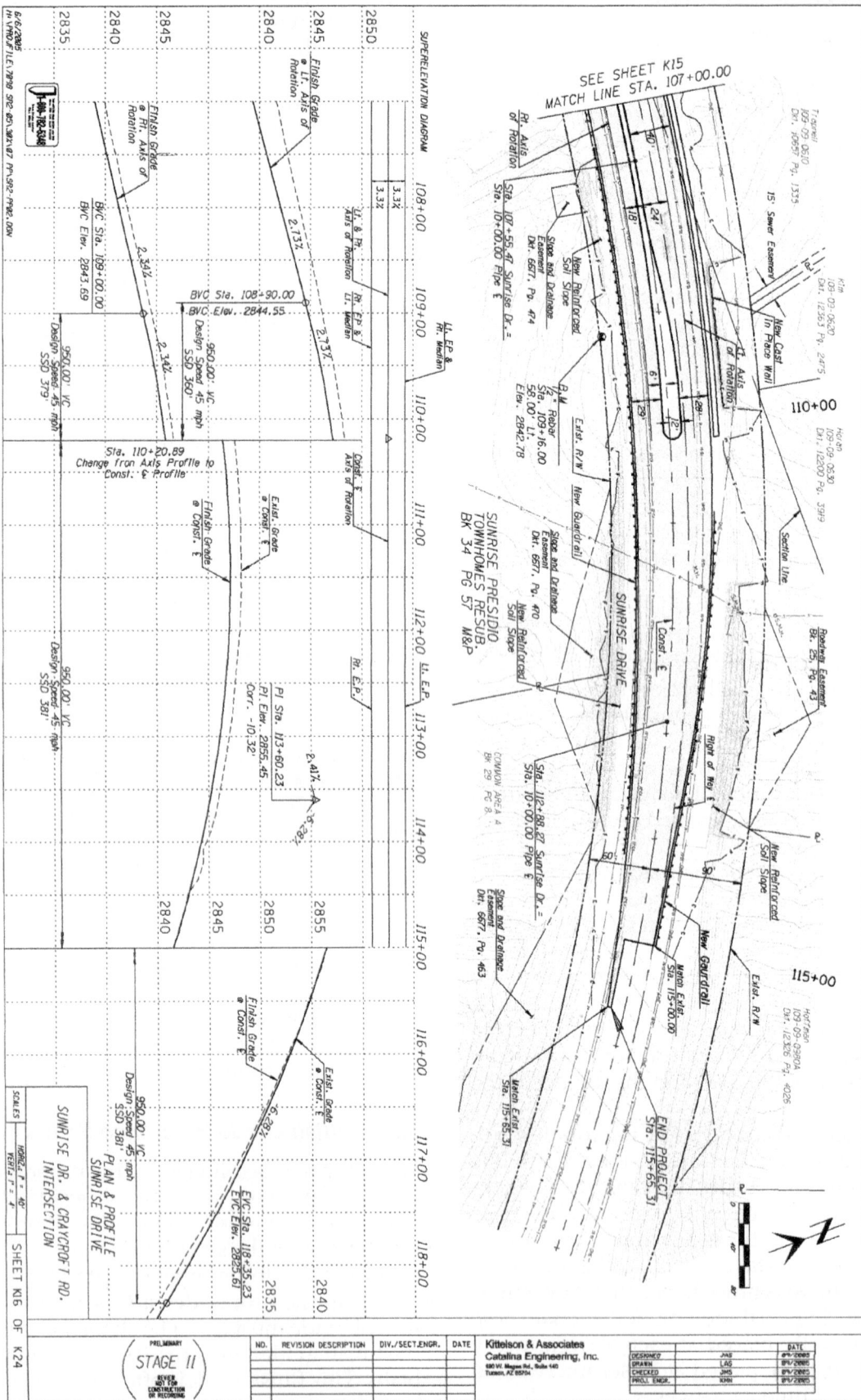

Figure 30 Highway plan sheet showing benchmark reference.

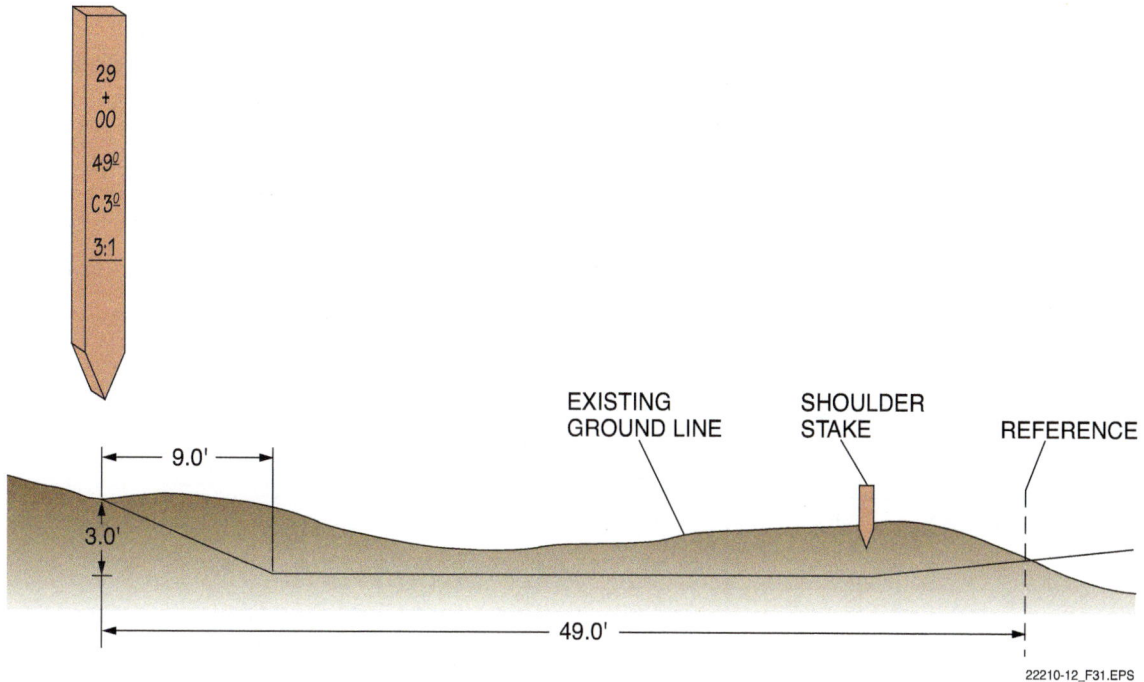

Figure 31 Cross-section of a cut.

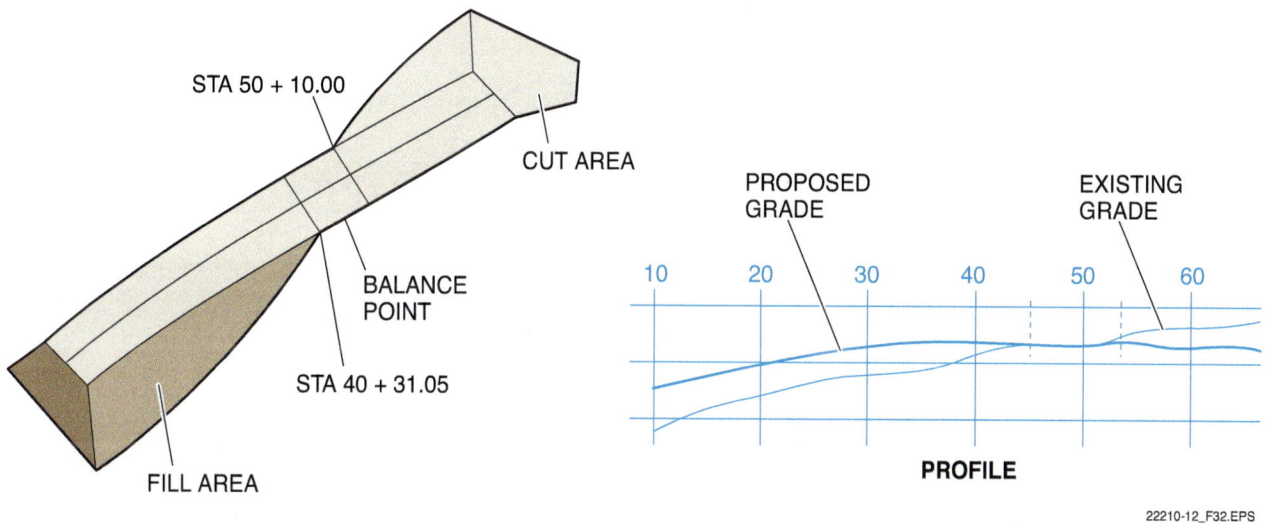

Figure 32 Example of cut and fill sections.

be made down to the finished subgrade surface. To check the lines of the completed subgrade, little flags or other markers can be attached to the string line at the edge of the pavement, shoulder, or other appropriate points.

3.3.0 Building Foundations and Pads

Site plans vary for buildings and other structures. A large building or development will usually have a detailed plan that shows existing elevations and the finished elevation of structure's foundation. All site plans should show the general outline of the structure and the property boundaries (*Figure 33*). On small jobs, a site plan may not have de-

tailed elevations, so you may need to ask the site supervisor for verbal instructions before you begin earthwork operations.

Study the site plan before starting work in order to verify the locations of the property boundaries and to avoid crossing them. Do not cross a property line without permission. The site plan may also show the existing elevation of the property, so the operator or the site supervisor can determine the amount of excavation needed to ensure that the finished elevation of the building foundation is correct.

For structures with foundations, temporary stakes or markers are usually set at the corners as a rough guide for starting the excavation. These

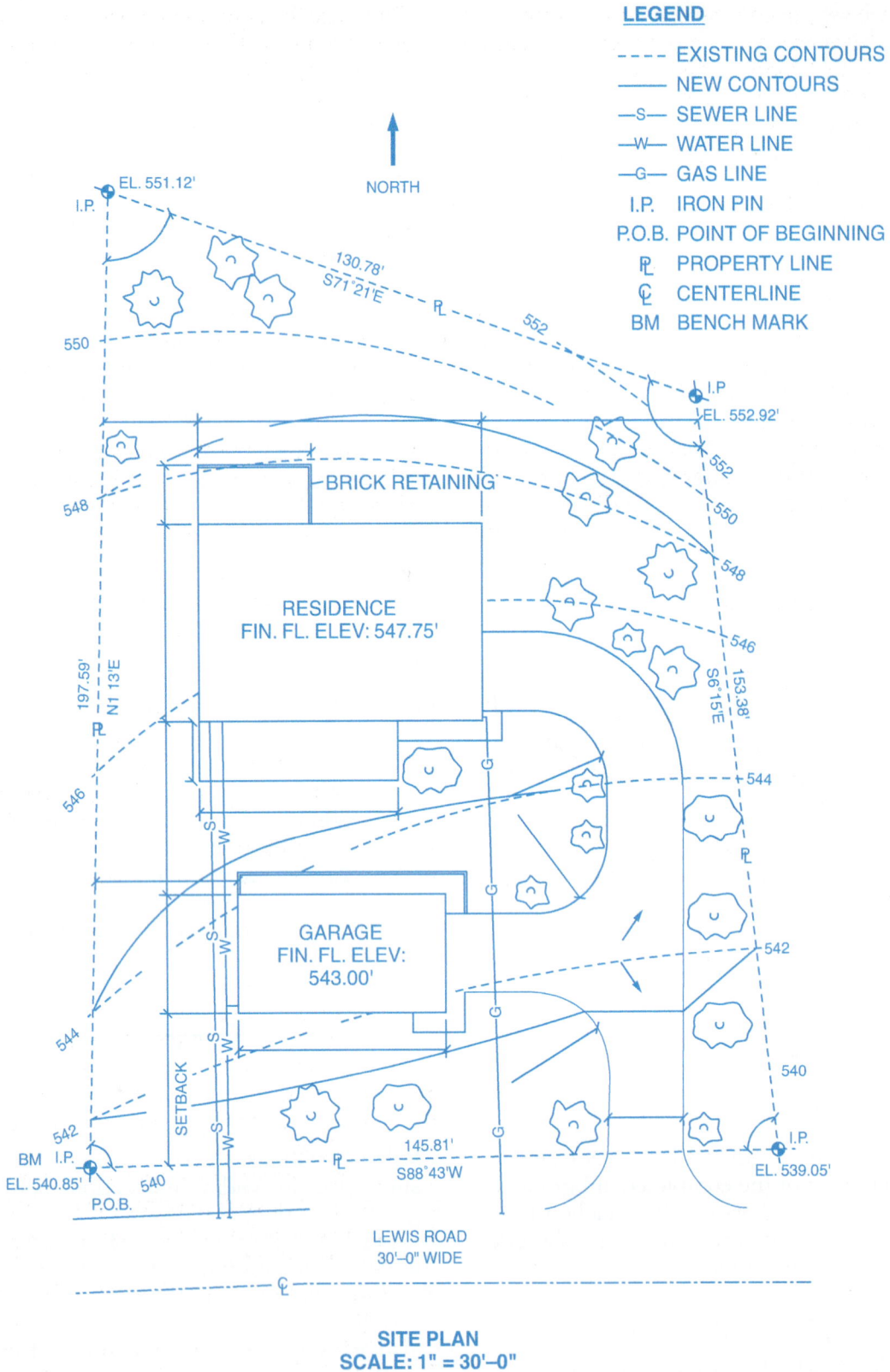

LEGEND

- - - - EXISTING CONTOURS
——— NEW CONTOURS
—S— SEWER LINE
—W— WATER LINE
—G— GAS LINE
I.P. IRON PIN
P.O.B. POINT OF BEGINNING
℞ PROPERTY LINE
℄ CENTERLINE
BM BENCH MARK

NORTH

EL. 551.12'
I.P.

130.78'
S71°21'E
℞
552

550

548

197.59'
N1°13E
℞

546

544

542

540

BM I.P.
EL. 540.85'

P.O.B.

I.P
EL. 552.92'

552

550

548

546

153.38'
S6°15'E

544

℞

542

540

I.P.
EL. 539.05'

BRICK RETAINING

RESIDENCE
FIN. FL. ELEV: 547.75'

GARAGE
FIN. FL. ELEV:
543.00'

SETBACK

S
W

S
W

S
W

G

G

G

G

145.81'
S88°43'W
℞

LEWIS ROAD
30'–0" WIDE

℄

**SITE PLAN
SCALE: 1" = 30'–0"**

22210-12_F33.EPS

Figure 33 Site plan showing topographical features.

markers are set outside the construction limits or area of disturbance, but close enough to be convenient. Permanent control points are set to allow for measurements during the construction process.

Stakes or other markers are set on important lines to mark clearly the limits of the work. For small building foundations and trenches, grade information is usually provided using batter boards.

3.3.1 Establishing the Proper Grade

To establish the proper grade for an excavation or fill, the grade setter must locate a benchmark or hub set by the surveyor. (Typically, a surveyor will set hubs for a project.) The grade setter will then begin setting boots. A boot is a mark on a stake or lath at a convenient height that can be used as a reference elevation relative to the finished subgrade, pavement, or structure foundation. *Figure 34* shows a typical boot placed by the grade setter. The boot will be 3, 4, or 5 feet above the finished grade. It provides a convenient height for the grade setter to view with a hand level.

For subdivision, office park, and commercial sites, the grade setter sets boots based on the finish grade, then adds the curb or road subgrade information when checking grade from these boots. After placing each grade mark, the cut needed to reach subgrade will be written on all the cut stakes.

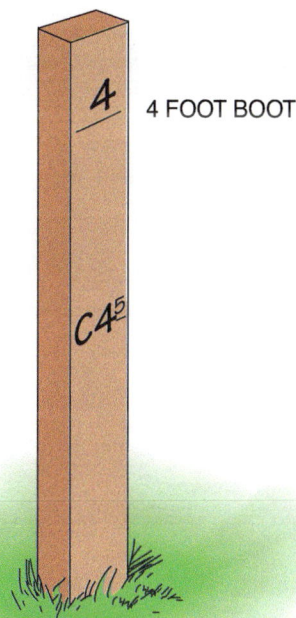

22210-12_F34.EPS

Figure 34 Grade setter boot.

This way the equipment operator can see what cut is needed directly without having to add the curb and road section height to the surveyor's cut marks.

3.4.0 Grading Operations

Some grading operations are automated. A computer-controlled grader is driven over the work site, and the computer adjusts the level and angle of the blade based on a topographical survey and Global Positioning System (GPS) coordinates. For grading operations that are not automated, the successful and efficient completion of the job relies on the skills of the equipment operators. The information presented in this section can provide you with knowledge that can help you develop grading skills, but the only way that you can actually develop grading skills is through experience. Keep the following points in mind while you are learning to cut and fill grades:

- One of the most common problems new operators have is that they are reluctant to make deep cuts. While you are learning this is wise, but skilled operators use their equipment to the fullest. Experienced operators are able to cut to grade with a minimum of passes by keeping the blade or bucket full at most times.
- After each pass, evaluate your work. Did you leave too many uncut areas that require a second or third pass? If so, adjust the blade or bucket attachment.
- When cutting close to grade, it may be wise to make smaller cuts and more passes, rather than risk cutting too much. It is better to make a second pass with a blade than it is to cut too much material, which will require that fill and a compactor be brought in to fix the cut.
- Always follow the site plans. Many sites have a stormwater control plan to help prevent erosion and sedimentation. Be sure to know and follow the plans. When you are not sure, ask your supervisor.
- Even if the site does not have a stormwater control plan, avoid driving the equipment across graded areas to avoid increased sedimentation.
- Never cut in an area that does not need grading. This can increase sedimentation and erosion.
- When performing rough grading or filling, be sure to leave enough material to trim during finish grading.
- When filling areas, thin layers of fill are easier to spread, have fewer air holes, and compress more easily than thick layers of fill.

Motor graders are specially designed to be used in many grading operations. All graders have blades that are used to cut and level grading material. The angle and height of the blade can be adjusted from the cab and the frame is designed to keep the blade stable even when driving over rough terrain. The grader's front wheels can be tilted to the right or left, which helps to increase the steering ability of the grader while cutting.

3.4.1 Grader Blade

The grader blade (*Figure 35*) is made up of a moldboard, endbits, and a cutting edge. The moldboard is shaped so that grading material roll and mixes as it is cut. The end bits and cutting edge protect the moldboard edges from the abrasive action of the grading material. Inspect the blade daily and replace the endbits and/or cutting edge as necessary.

The grader blade may be positioned in the center of the grader frame or off to one side (*Figure 36*). The blade may be set perpendicular to the frame or it may be angled. When the blade is angled, the front edge is called the toe and the rear edge is the heel. Grading material spills off the heel of the blade.

> **CAUTION**
>
> When the blade is set at a sharp angle, it may hit a tire when the grader is turned, causing tire damage. Use caution when operating a grader.

The pitch of the blade can be adjusted for the desired results. For most grading operations, the blade will be upright (*Figure 37A*), but for more cutting, the blade may be pitched back slightly (*Figure 37B*). To increase the mixing of the graded material, the blade is pitched slightly forward (*Figure 37C*). Pitching the blade forward sharply (*Figure 37D*) will help to compact the graded material and ensure that low spots are filled.

3.4.2 Grader Wheels

A grader's front wheels are positioned to help stabilize the grader while cutting, and to make it easier to steer while grading. The wheels are usually tilted in the direction of the heel of the blade. In

22210-12_F36.EPS

Figure 36 Grader with blade positioned slightly to the side.

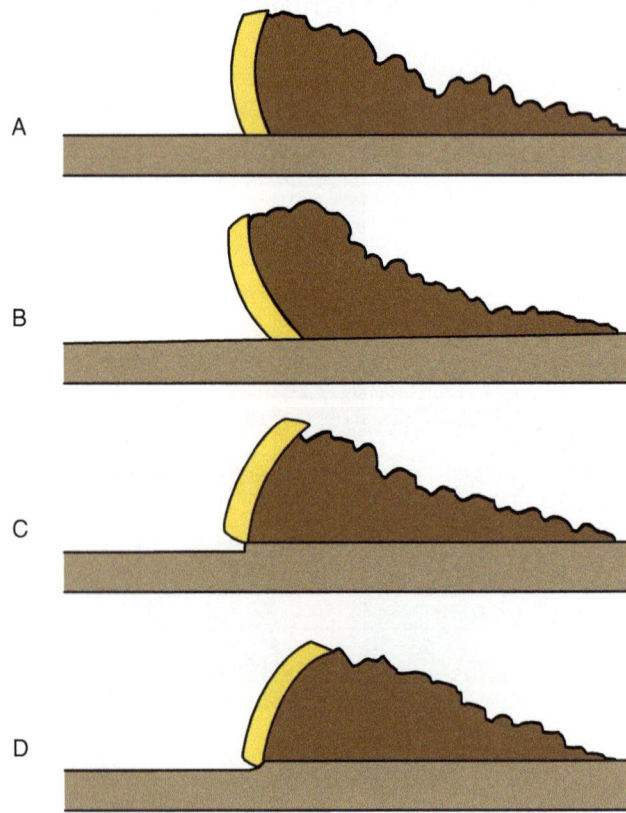

22210-12_F37.EPS

Figure 37 Blade pitch.

22210-12_F35.EPS

Figure 35 Grader blade.

Figure 38, the grader is forming the ditch slope. The blade is pushing dirt on the right side. The force of the dirt on the blade would normally tend to pivot the grader into the ditch. By tilting the wheels away from the ditch, the loader stays on a straight course.

3.4.3 Cutting a Ditch

Most grading operations require several passes to complete. As discussed previously, the most efficiency is achieved when the operator is skillful enough to keep the blade full and to complete the operation with a minimum number of passes. Experienced operators can often cut a ditch with two or three passes. These operators are familiar with the equipment and have developed a sense for the soil conditions. Initially, it is best to remove the smallest amount of material on your first pass and then gradually increase the depth of your cuts until you are comfortable with your skills.

When cutting a ditch, the first pass is called the marking cut (*Figure 39A*). It is a 3- to 4-inch deep cut made with the toe of the blade and it is used as a guide to help you to cut a straight ditch. In the second cut, the blade is angled more and positioned over the marking cut (*Figure 39B*). When the second cut is made, more material is deposited onto the road from the heel of the blade. At some point, the cut graded material needs to be spread toward the center of the road—this is called shoulder pickup (*Figure 39C*). These steps are repeated until the ditch is cut to the desired grade.

3.5.0 Stakeless and Stringless Grading Systems

At one time, site layout and grading were done by manually setting stakes and strings. This method

(A) MARKING CUT

(B) DEEPER CUT

(C) SHOULDER PICKUP

22210-12_F39.EPS

Figure 39 Cutting a ditch.

is being rapidly replaced with methods that rely on laser and satellite technology that make it possible for grading operations to be partially or fully automated. Using computer technology, entire projects can be completed without setting a single grade stake. Automatic grade control systems can use lasers, the GPS, or both to get a job done faster and more accurately than conventional staking, thus saving time and money.

3.5.1 Laser-Based Automatic Grade Control Systems

Laser-based automatic grade control systems use complicated technology, but their operation is very simple. *Figure 40* shows a typical automatic grade control system. Signals from an off-board laser transmitter are used as the reference. The grader or bulldozer is outfitted with electronic components that permit a computer processor to determine

22210-12_F38.EPS

Figure 38 Tires tilted to cut a ditch.

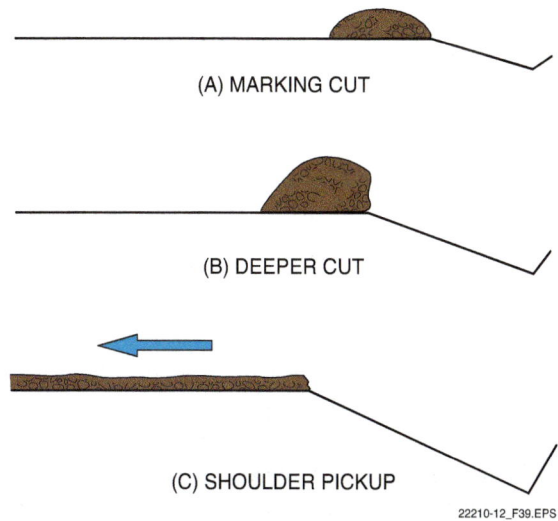

22210-12_F40.EPS

Figure 40 Caterpillar Accugrade laser grade control system.

its exact location on a site and the position of the blade. As the equipment moves throughout the site, the computer determines its exact position and the elevation of the blade. This information is then transmitted to an in-cab display (*Figure 41*) so the operator can make necessary blade adjustments.

The complexity and degree of automation vary among systems. Some systems use a three-dimensional digitized site plan so the computer processor can calculate the required position of the blade. In a fully automated system, a signal is sent to the grader or bulldozer and the blade position is adjusted automatically.

3.5.2 GPS-Based Automatic Grade Control Systems

GPS-based automatic grade control systems operate in a similar fashion to the laser-based systems, except the reference signal is from the GPS. The GPS was developed by the US military to provide precise position information for vehicle, troop, and equipment movement. It is made up of a series of satellites that circle Earth at various orbits. These satellites broadcast radio signals that contain information about satellite locations for 24 hours a day.

GPS-based systems are especially helpful on jobs that require a great deal of contouring. These systems compare the blade position to a three-dimensional digitized site plan to determine whether the blade position requires any adjustments. *Figure 42* shows a typical GPS-based automatic grade control system. The information can be displayed on an in-cab display (*Figure 43*). The operator can make the required blade adjustments or a signal can be sent to the grader or bulldozer control system to automatically adjust the blade.

22210-12_F42.EPS

Figure 42 Caterpillar Accugrade GPS grade control system.

Figure 41 Caterpillar Accugrade laser grade control system display and control.

22210-12_F43.EPS

Figure 43 Caterpillar Accugrade GPS grade control system display and control.

Additional Resources

Excavators Handbook Advanced Techniques for Operators. Reinar Christian. Addison, IL: The Aberdeen Group, A division of Hanley-Wood, Inc.
GPS Systems for the Earthmoving Contractor: **www.constructionequipment.com/gps-systems-earthmoving-contractor**

3.0.0 Section Review

1. A permanent benchmark shows a point's
   _____.
   a. exact elevation
   b. precise location
   c. exact azimuth
   d. precise value

2. A balance point is _____.
   a. the point of an excavation that changes from a cut to a fill
   b. the edge of a batter board where a level can be balanced
   c. when a job has used all of the cut material for fill
   d. the center of a benchmark that is marked by a tack

3. When you are working on a grading job for a new building, it is important that you know where the property boundaries are so that you can _____.
   a. park your equipment over the line
   b. be sure to grade right up to the line
   c. stay 10 feet from the boundary line
   d. avoid crossing over the property line

4. On a grader blade, which of the following is adjustable?
   a. Toe
   b. Heel
   c. Blade pitch
   d. Cutting edge

5. GPS-based systems are especially helpful on jobs that _____.
   a. are close to water
   b. require a great deal of contouring
   c. require three or more graders
   d. do not have a power supply

SECTION FOUR

4.0.0 PIPE-LAYING OPERATIONS

Objective 4

Describe grading and installation practices for pipe-laying operations.
   a. Explain how grades are established for pipe-laying operations.
   b. Describe the equipment and methods used to lay pipe.

Trade Term

Bedding material: Select material that is used on the floor of a trench to support the weight of pipe. Bedding material serves as a base for the pipe.

Setting the proper grade for pipe placement in trenches is very important because the drainage through the pipe is controlled by gravity. After the pipe is laid and the trench is backfilled, it is expensive to re-excavate if it will not drain properly. In addition, an uneven grade will put extra pressure on a pipe, causing it to break or leak at the joints.

> **WARNING!**
> Make sure the trench is barricaded if required and that proper protection is provided for anyone working in the trench.

4.1.0 Setting the Trench Grade

Setting the proper grade for pipe or other conduits can be accomplished in different ways. The most popular method involves using a laser level. Other methods include the use of a string line. Regardless of the grading method, a survey crew establishes hubs that can be used as control points for setting conduit or pipe. Before these hubs are set, a decision is made about how far the offset should be and on which side of the trench it should be placed. The width of the trench and the direction the spoil will be thrown usually determines these two factors.

Often, some type of bedding material is used to help set the pipe to the proper grade. This material can be any material that does not readily compress. However, it is important that the material uniformly support the pipe to avoid damaging it. Many workers find that it is helpful to use a shovel or long pole to position the pipe once it is in the trench. The pipe should be placed so that it contacts the bedding all at once.

Once the pipe is positioned, the trench can be backfilled. Filling the trench with thin layers of fill rather than large loads helps to ensure that the pipe is not pushed off grade. It also helps to reduce air pockets and decrease the need for compaction.

4.1.1 Setting Grade for Pipe Using a Laser

The center line for a pipe trench is located on the ground with stakes or other marks at 50-foot intervals where the grade is uniform. In rough terrain, stakes are set at a much closer interval. Hubs are set parallel to the center line but far enough away so they are not disturbed by the excavation process. These stakes are placed using a manual transit or electronic survey device.

When the trench has been excavated, a laser pipe level (*Figure 44*) can be used to provide a grade line at some elevation above the bottom of the trench. The pipe laser receiver is set up at one end of the segment, and the transmitter is set at the other end of the segment to check the grade. The laser level can be used first to check the rough grading of the trench, and then to recheck it after the pipe has been placed in the trench, and before the trench is backfilled.

> **CAUTION**
>
> A laser level may not be a good choice if the work site is dusty or foggy. The laser beam is bent by dust and water particles and may read the incorrect grade. In addition, some laser levels do not operate well when part or all of the trench or pipe is very hot. Check the manufacturer's instructions for more information.

4.1.2 Setting Grade Using a String Line

String lines can be used to set grades if no laser level or other surveying instrument is available. String lines are more effective in adverse environmental conditions such as heavy fog or dust that can hinder the operation of a laser level.

The initial setup for the excavation is similar to the procedure described earlier. Once the offset has been determined, the survey crew places the required number of stakes. This information can then be transferred to the string line setup, as shown in *Figure 45*. It is necessary to decide how high above the trench the string line should be placed over the top of the trench and how far apart the batter boards need to be spaced. A recommended maximum space for the batter boards is 25 feet. The string line sags if support is spaced any further apart. The height of the string line depends on the terrain and the amount of slope required at the bottom of the trench. The height of the string line is set by measuring a given elevation above the required grade of the trench plus the offset distance for the line.

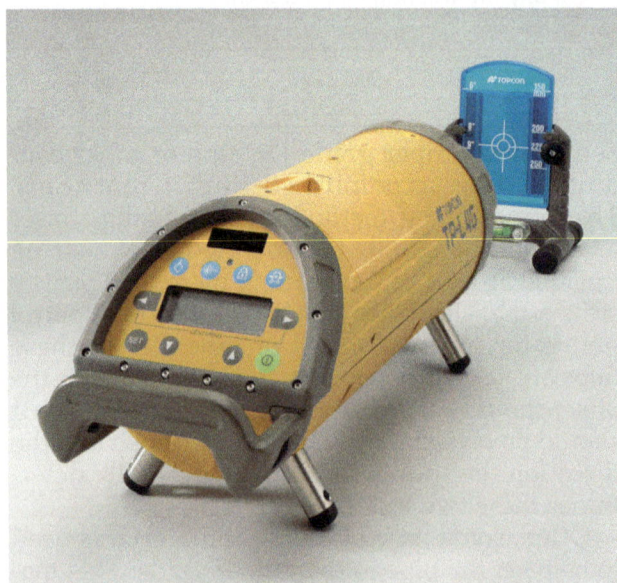

22210-12_F44.EPS

Figure 44 A laser pipe level.

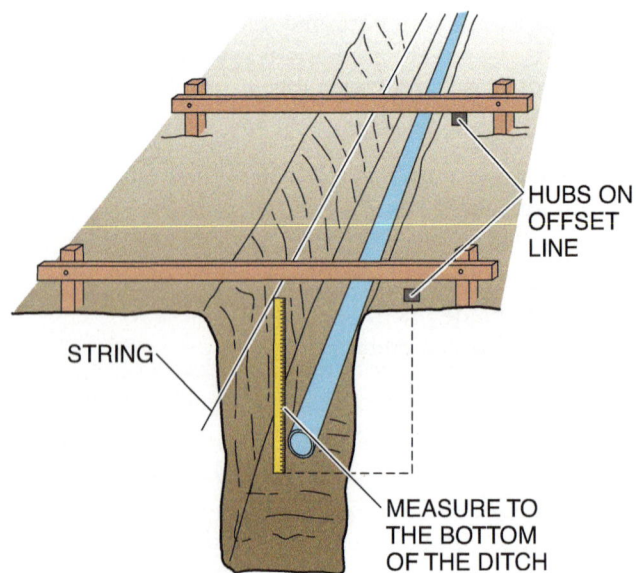

HUBS ON OFFSET LINE

STRING

MEASURE TO THE BOTTOM OF THE DITCH

22210-12_F45.EPS

Figure 45 Using a string line to check grades.

To determine the exact elevation of the pipe, measure the distance from the string to the bottom of the trench, and subtract the height of the batter boards from the hub. Then subtract the result from the hub elevation. Repeat this process for each station.

When setting any string line, keep three stations set up along the line so you can sight down the line to correct errors or movement that might occur. As you sight down the string line, each station should blend with the next, with no sudden rise or dip from station to station. If you spot a sudden grade change, check all measurements. If you cannot find any error in your work, check the rate of slope shown on the plans. If the plans do not show a sudden grade change, it is possible the surveyors made an error that has been carried through to the grade line.

4.2.0 Laying Pipe

On large building sites, cranes are usually available to move and place lengths of pipe. But on most jobs, the pipeline is away from other construction, and it may not be practical to keep a crane around to lay pipe. On these sites, a backhoe, excavator, or other equipment may be used to lift and place pipe. There are a number of attachments available to make pipe laying easy. But even if your equipment does not have an attachment, you can still lay pipe (*Figure 46*).

Figure 46 Excavator laying pipe.

Use approved lifting devices to lift the pipe. Be certain to select equipment that is strong enough to lift the pipe safely. You also need several shackles (*Figure 47*) and rigging hooks (*Figure 48*).

> **WARNING!**
> Always make sure the rigging is rated for the load, rated for overhead lifting, and properly tagged. The working load limit (WLL) should be stamped on all items of rigging equipment. WLL was formerly called SWL (safe working load), which may appear on some rigging equipment.

Before starting, look over the job site for potential obstructions. Remember that once the pipe is lifted, it oscillates, creating a dangerous pendulum effect. Minimize the number of workers in the area. You must always know the location of any workers in the area. Ideally, set the pipe pieces as close to the trench as possible to minimize the length of time the pipe is in the air.

Because the pipe may not have attachment points, a choker hitch may be used to attach the sling to the pipe. Chances are good that a rigger will be attaching the pipe to the lifting device, but it is necessary to know how it is done so you can recognize improper rigging. Never lift pipe if you believe it is improperly rigged.

> **WARNING!**
> Remember that the choker hitch decreases the capacity of the sling by a minimum of 25 percent. Keep this in mind when selecting equipment.

The bucket of the backhoe or excavator has an eye or a pin to which the cable or chain can be attached using a shackle. Short lengths of pipe can be transported with a single cable or choker hitch (*Figure 49*), but any pipes over 12 feet in length should use a double choker hitch (*Figure 50*). Be certain that all hitch hardware is oriented properly. Also make sure that the load is balanced to minimize oscillations and to make the pipe easier to place.

Pipe should be lifted slowly and smoothly. Do not raise the pipe any higher than necessary. To help ensure stability, make sure a tag line is used so workers do not need to be close to the pipe. As the pipe is being lowered into the trench, the ground worker can control the alignment of the pipe with the tag line. The pipe must be held close to the grade line so that the entire length of the pipe contacts the ground simultaneously. If all of the weight rests on one end of the pipe, it will

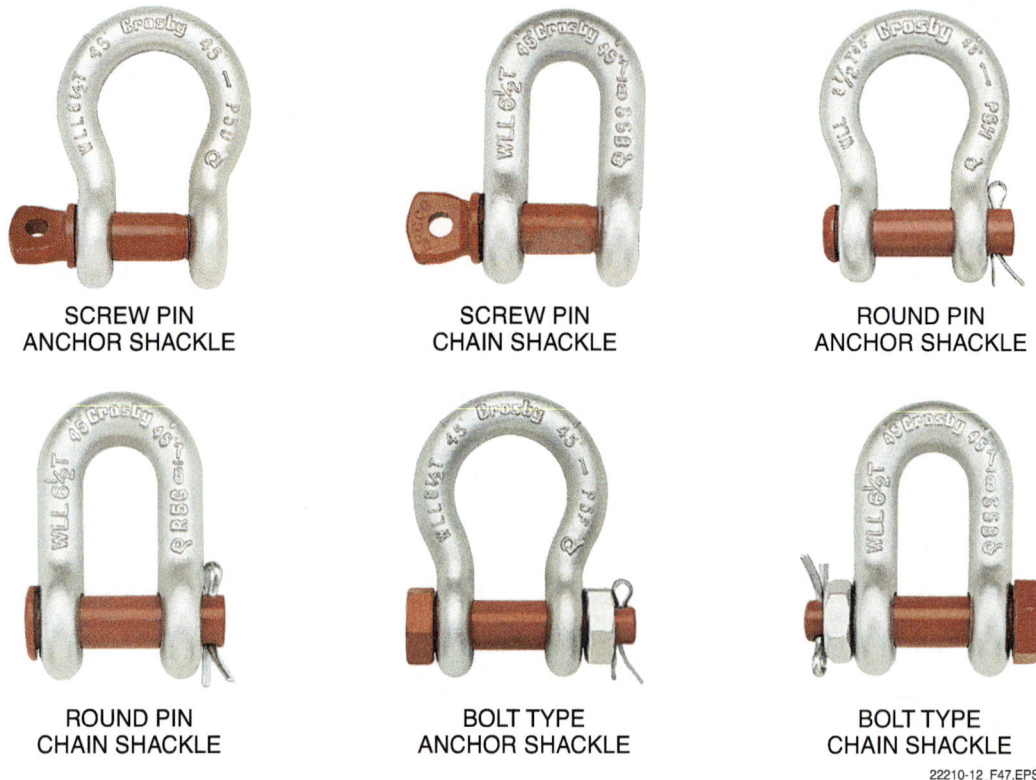

SCREW PIN
ANCHOR SHACKLE

SCREW PIN
CHAIN SHACKLE

ROUND PIN
ANCHOR SHACKLE

ROUND PIN
CHAIN SHACKLE

BOLT TYPE
ANCHOR SHACKLE

BOLT TYPE
CHAIN SHACKLE

22210-12_F47.EPS

Figure 47 Shackles.

compact the soil in the trench. This can change the grade of the trench floor.

After each section of pipe is in place, a pipe level or builder's level is used to check the grade. At the end of each workday, the last joint of pipe is placed into position and the end is closed with a cap or a blind flange to prevent water, animals, or contaminants from entering the pipe. When a run of pipe is finished, a shutoff valve or blowoff assembly is attached. The run is flushed to remove any foreign material.

4.2.1 Using a Dozer Side Boom

Pipe booms or side booms are used extensively for heavy hoisting and carrying, particularly in pipeline work. The boom is attached to one side of the bulldozer with the hoisting mechanism and counterweights attached on the opposite side. A power takeoff drives two drums that are controlled through separate clutches and brakes. One controls the boom height while the other controls the load line that is attached to a hoist block. The counterweight is hinged so that it can be brought in close to the dozer for traveling and handling light loads. It can be extended away from the dozer by hydraulic pistons to counterbalance heavy loads on the boom.

There are two approaches to determining the safe operation of a dozer with a side boom. Both have to do with the load rating of the boom.

Where stability governs lifting performance, load rating at the minimum load overhang may be based on structural competence rather than stability. The margin of stability for the determination of load ratings, with booms of specific lengths at a given load for the various types of tractor mountings, is established by taking a percentage of the loads that will produce a condition of tipping.

Where structural competence governs lifting performance, load ratings are governed by the stability of the side boom tractor. This is the load required to tip the side boom tractor at a given load overhang.

The following procedure summarizes the steps for handling a load with a side boom:

- Make sure the weight of the load does not exceed the maximum load capacity.
- Ensure that the hoist rope is not kinked or twisted on itself.
- Ensure that the load is secured and balanced in the sling or lifting device before it is lifted more than a few inches off the ground.

EYE HOOK

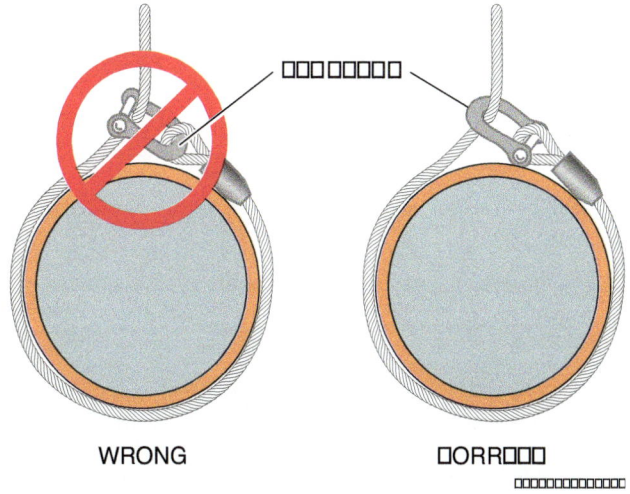

WRONG   □ORR□□□

Figure 49 Choker hitch.

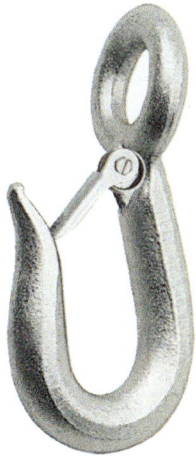

ROUND REVERSE
EYE HOOK

SLIDING CHOKER
HOOK

GRAB HOOK

SHORTENING
CLUTCH

22210-12_F48.EPS

Figure 48 Rigging hooks.

- When two or more side boom tractors are used to lift one load, one designated person should be responsible for the operation. This person should analyze the operation and instruct all personnel involved in the proper positioning, rigging of the load, and the movements to be made.

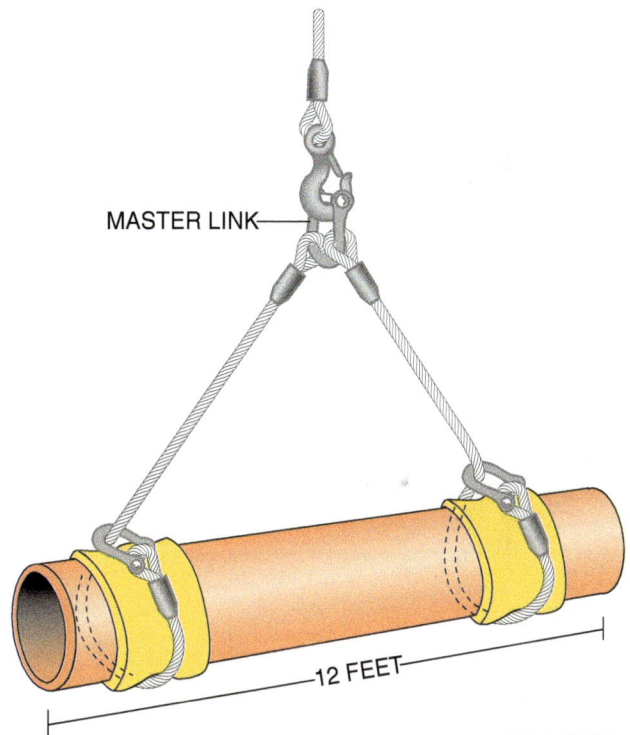

MASTER LINK

12 FEET

22210-12_F50.EPS

Figure 50 Double choker hitch.

- When operating at a fixed boom overhang, the bottom hoist pawl or other positive locking device should be engaged.

WARNING!

Be especially cautious when operating the equipment near electric power lines. The boom could come in contact with the lines and cause electrocution to the operator. Maintain the required minimum distance.

Additional Resources

Pipe & Excavation Contracting. Dave Roberts. Carlsbad, CA: Craftsman Book Company.

4.0.0 Section Review

1. Placement of bedding material in a trench _____.

   a. makes it easier to work in the bottom of the trench
   b. supports the pipe and helps achieve the proper grade
   c. is purely cosmetic—it makes the trench look better
   d. makes it easier for the inspector to move the pipe

2. A single choker hitch can be used to lift pipe no longer than _____.

   a. 6 feet
   b. 12 feet
   c. 18 feet
   d. 24 feet

SUMMARY

The most important part of any safety program is the individual worker. A company cannot have a good safety record without safety conscious employees. An employer is required to ensure that employees have safe equipment and a safe environment in which to work, but safety is ultimately up to the employee. Safe workers know, understand, and use safety procedures and devices that their employer has made available. Employees need to incorporate safe habits into their work on a daily basis, including attending required safety training sessions, participating in accident investigations, and learning how to work safely around hazardous materials.

During excavation work, it is extremely important to be aware of any potential hazards in the area. This means constantly monitoring the area for signs of instability and potential failures, as well as keeping co-workers in sight when they need to work near your equipment. It is essential to follow construction specifications whenever it is necessary to install excavation protective systems, such as shoring, trench boxes, or sloping, which are designed to ensure the safety of those working in an excavation.

Poor drainage on a construction site can cause major problems. If surface water is a problem, temporary drainage ditches or channels can be constructed to drain the water to retaining ponds. If groundwater problems are encountered in a cut area, control (intercept) ditches can be built to collect the outflow and reduce the possibility of damage to the construction site. In excavated pits or trenches, drains can be installed to channel water to sumps where it can be held until removed by pumping.

Equipment operators sometimes are required to read and use project plans in their work. At the beginning of a project, the grading supervisor or project engineer will usually discuss the grading and excavation operation with the equipment operators to familiarize them with the area. Operators must know the basic steps to go from natural ground to the finish grade, be able to follow the operation on the plans, and understand the work-flow at the site. You will need to work with the plan and profile sheet, the cross-section plans, and the grading plans to get information about grading requirements on the job. Each of these sheets gives specific information about the grading requirements and the elevations of the grade, so you must be able to read and understand each part in order to do your job.

Most projects require the project survey team to set reference points, center lines, slope stakes, and grading stakes for equipment operators and other workers to follow. After these controls are set, the contractor may use his or her own personnel to set the detailed stakes that direct the grading operations. Equipment operators must be aware of the survey requirements and various survey activities going on throughout the duration of the project. They must understand the information on the stakes and be able to visualize the completed grading project based on the many stakes that are placed. In some cases, an equipment operator may be asked to assist a grade setter or surveyor in staking activities. Although operators need to understand basic surveying functions, it is not their responsibility to set stakes independently.

Initial surveying requirements involve setting grades and control points from benchmarks or known reference points. The benchmarks are usually permanent markers installed by federal or state governments. If one of these markers is found on the construction site, it should not be disturbed until the proper authorities are notified and a decision is made about relocation.

Heavy equipment, such as excavators and backhoes, is commonly used to place pipe in trenches once the trenches have been dug. Establishing and maintaining the correct grading of the trench is necessary to ensure that the liquid in the pipe will flow naturally.

On today's construction sites, grading is often controlled by lasers or GPS. Operators must know how to interact with these tools in order to work effectively.

Review Questions

1. Who has the most effect on worker safety?
    a. Supervisor
    b. Employer
    c. OSHA
    d. Worker

2. Toolbox safety training sessions should be chaired by a safety officer or _____.
    a. any safety committee member at the site
    b. supervisor qualified to discuss the subject
    c. the local OSHA agent
    d. a professional presenter

3. One function of the safety committee is _____.
    a. reviewing all accident reports
    b. selecting the company safety officer
    c. performing pop inspections
    d. promoting safety awareness

4. Last week five workers were injured in an on-the-job accident. All are still hospitalized. This is cause for a(n) _____.
    a. OSHA inspection
    b. plant shutdown
    c. increase in health insurance premiums
    d. investigation by the local newspaper

5. If you cannot identify a chemical in an unlabeled container, you can assume it is safe.
    a. True
    b. False

6. Material safety data sheets must be available for use by _____.
    a. everyone on the work site
    b. the work site supervisor
    c. OSHA inspectors only
    d. the site safety officer

7. If a worker finds a partially full, unlabeled container on a job site, what should he or she do?
    a. Do nothing. The container is unlabeled, so it is not hazardous.
    b. Since you do not know what is in the container, pour it down the drain.
    c. Treat the container as if it contains an extremely hazardous chemical.
    d. Dial 9-1-1 and call in the local fire department's HAZMAT team.

8. If a person's eyes come in contact with a hazardous chemical, the correct course of action is _____.
    a. rub it out with a rag
    b. wait until medical aid is available
    c. apply compresses and seek medical aid
    d. flush the eyes with cool water for 15 minutes

9. A trench is defined as an excavation in which the depth is no greater than the width and the width is no greater than _____.
    a. 5 feet
    b. 10 feet
    c. 15 feet
    d. 20 feet

10. One cubic yard of earth weighs approximately the same amount as a _____.
    a. gallon of water
    b. backhoe
    c. small car
    d. bag of cement

11. The foreman sent you to work with your backhoe near an excavation site. You notice a worker in the excavation below you, so you _____.
    a. keep the worker in sight as you complete your work
    b. ignore him because the foreman must know the worker is there
    c. stop work until the worker leaves the excavation
    d. attract the worker's attention so he knows to be careful

12. To relieve pressure on the walls and prevent material from falling into the excavation, the excavation walls are _____.
    a. sloped
    b. moistened
    c. shielded
    d. vibrated

13. A device laid up against the excavation wall to support the wall is called _____.
    a. shoring
    b. a waler
    c. shielding
    d. sloping

14. To protect workers from a cave-in in an excavation, use a _____.

    a. slide
    b. waler
    c. trench box
    d. ladder

15. When a trench box is removed from an excavation, it is important to _____.

    a. inspect the trench box for damage
    b. immediately backfill the trench
    c. clean the dirt from the trench box
    d. check the stability of the trench

16. In a braced coffer dam, the walers are the _____.

    a. walls of the coffer dam
    b. braces that go across the dam
    c. connections between the sheet piles
    d. horizontal support members around the walls

17. Caissons are used _____ .

    a. so that work can take place underwater
    b. to shore up the sides of a narrow trench
    c. when coffer dams are not practical
    d. as a shield to prevent water pollution

18. Grade stakes are set by _____.

    a. the project survey team to inform equipment operators of grading needs
    b. government workers to inform the survey team of a site's exact elevation
    c. equipment operators to inform the project survey team of existing elevations
    d. government workers to inform equipment operators of grading needs

19. The posts the National Oceanic and Atmospheric Administration (NOAA) sets near its benchmarks to make them easier to find are called _____.

    a. witness posts
    b. goal post
    c. fence posts
    d. trial stakes

20. Whenever you find a permanent benchmark that is damaged, you _____ .

    a. need to stop work and repair it immediately
    b. must report it to the site supervisor or engineer
    c. can ignore it because there are plenty of other markers
    d. stop work since the grade will be incorrect

21. A temporary benchmark is often placed _____ .

    a. on a structure
    b. in a foundation of concrete
    c. on the site's office trailer
    d. in the general work area

22. To check the grade of an excavation, the grade setter can set grade boots that he or she can _____ .

    a. sight with a hand level
    b. use with a pipe level
    c. use to check subgrades only
    d. look at through a laser level

23. The motor grader blade is made up of a moldboard, end pieces, and a cutting edge. Every day the end pieces and cutting edge should be _____ .

    a. replaced
    b. used
    c. inspected
    d. polished

24. The first pass when cutting a ditch is used _____ .

    a. to make a marking cut
    b. for shoulder pick-up
    c. to level the ditch area
    d. to spread loose material

25. A double choker hitch is used to lift pipe longer than _____ .

    a. 6 feet
    b. 12 feet
    c. 18 feet
    d. 24 feet

Trade Terms Introduced in This Module

Aquifer: An underground layer of water-bearing permeable rock or unconsolidated materials through which water can easily move.

Balance point: The location on the ground that marks the change from a cut to a fill. On large excavation projects there may be several balance points.

Bedding material: Select material that is used on the floor of a trench to support the weight of pipe. Bedding material serves as a base for the pipe.

Berm: A raised bank of earth.

Boot: A special name for laths that are placed by a grade setter to help control the grading operation. The boot can also be the mark on the lath, usually 3, 4, or 5 feet above the finish grade elevation, which can be easily sighted. This allows the grade setter to check the grade alone instead of having to use another person to hold a level rod on the top of the grade stake.

Competent person: A person who is capable of identifying existing and predictable hazards in the area or working conditions that are unsanitary, hazardous, or dangerous to employees, and who has the authority to take prompt corrective measures to fix the problem.

Cross braces: The horizontal members of a shoring system installed perpendicular to the sides of the excavation, the ends of which bear against either uprights or walers.

Dewater: To remove water from a site.

Erosion: The removal of soil from an area by water or wind.

Groundwater: Water beneath the surface of the ground.

Sedimentation: Soil particles that are removed from their original location by water, wind, or mechanical means.

Shielding: A structure that is able to withstand the forces imposed on it by a cave-in and thereby protect employees within the structure.

Stations: Designated points along a line or a network of points used to survey and lay out construction work. The distance between two stations is normally 100 feet or 100 meters, depending on the measurement system used.

Stormwater: Water from rain or melting snow.

String line: A tough cord or small diameter wire stretched between posts or pins to designate the line and elevation of a grade. String lines take the place of hubs and stakes for some operations.

Subsidence: Pressure created by the weight of the soil pushing on the walls of the excavation. It stresses the excavation walls and can cause them to bulge.

Sump: A small excavation dug below grade for the purpose of draining or retaining subsurface water. The water is then usually pumped out of the sump by mechanical means.

Swale: A shallow trench used to direct the flow of water.

Topographic survey: The process of surveying a geographic area to collect data indicating the shape of the terrain and the location of natural and man-made objects.

Uprights: The vertical members of a trench shoring system placed in contact with the earth and usually positioned so that individual members do not contact each other. Uprights placed so that individual members are closely spaced, in contact with, or interconnected to each other, are often called sheeting.

Walers: Horizontal members of a shoring system or coffer dam placed parallel to the excavation face whose sides bear against the vertical members of the shoring system or the earth. Also, supports for piles in a coffer dam.

Additional Resources

This module presents thorough resources for task training. The following resource material is suggested for further study.

OSHA Trench Safety Quick Card: **www.osha.gov/Publications/quickcard/trenching_en.pdf**

Soil properties that affect groundwater: **www.co.portage.wi.us/groundwater/undrstnd/soil.htm**

Excavators Handbook Advanced Techniques for Operators. Reinar Christian. Addison, IL: The Aberdeen Group, A division of Hanley-Wood, Inc.

GPS Systems for the Earthmoving Contractor:

   www.constructionequipment.com/gps-systems-earthmoving-contractor

Pipe and Excavation Contracting. Dave Roberts. Carlsbad, CA: Craftsman Book Company.

Figure Credits

Section Review Answers

Answer	Section Reference	Objective
Section One		
1-1. b	1.1.0	1a
1-2. d	1.2.0	1b
1-3. d	1.3.0	1c
1-4. c	1.4.0	1d
1-5. c	1.5.0	1e
Section Two		
2-1 d	2.1.1	2a
2-2 c	2.2.1	2b
2-3 d	2.3.2	2c
Section Three		
3-1 a	3.1.0	3a
3-2 a	3.2.1	3b
3-3 d	3.3.0	3c
3-4 c	3.4.1	3d
3-5 b	3.5.2	3e
Section Four		
4-1 b	4.1.0	4a
4-2 b	4.2.0	4b

NCCER CURRICULA — USER UPDATE

NCCER makes every effort to keep its textbooks up-to-date and free of technical errors. We appreciate your help in this process. If you find an error, a typographical mistake, or an inaccuracy in NCCER's curricula, please fill out this form (or a photocopy), or complete the online form at **www.nccer.org/olf**. Be sure to include the exact module ID number, page number, a detailed description, and your recommended correction. Your input will be brought to the attention of the Authoring Team. Thank you for your assistance.

Instructors – If you have an idea for improving this textbook, or have found that additional materials were necessary to teach this module effectively, please let us know so that we may present your suggestions to the Authoring Team.

NCCER Product Development and Revision

13614 Progress Blvd., Alachua, FL 32615

Email: curriculum@nccer.org
Online: www.nccer.org/olf

❏ Trainee Guide   ❏ Lesson Plans   ❏ Exam   ❏ PowerPoints   Other _____

Craft / Level: _____   Copyright Date: _____

Module ID Number / Title: _____

Section Number(s): _____

Description: _____

_____

_____

_____

Recommended Correction: _____

_____

_____

_____

Your Name: _____

Address: _____

_____

Email: _____   Phone: _____

22207-13

Excavation Math

OVERVIEW

Math is an everyday reality in excavation work. Operators must be able to determine a volume of soil to remove, calculate load weights, and calculate the area of a cut or fill.

Module Seven

Trainees with successful module completions may be eligible for credentialing through NCCER's National Registry. To learn more, go to **www.nccer.org** or contact us at **1.888.622.3720**. Our website has information on the latest product releases and training, as well as online versions of our *Cornerstone* newsletter and Pearson's product catalog.

Your feedback is welcome. You may email your comments to **curriculum@nccer.org**, send general comments and inquiries to **info@nccer.org**, or fill in the User Update form at the back of this module.

This information is general in nature and intended for training purposes only. Actual performance of activities described in this manual requires compliance with all applicable operating, service, maintenance, and safety procedures under the direction of qualified personnel. References in this manual to patented or proprietary devices do not constitute a recommendation of their use.

Objectives

When you have completed this module, you will be able to do the following:

1. Explain how to use formulas.
   a. Explain the sequence of operations in solving a problem using a formula.
   b. Explain how squares and square roots are derived.
   c. Define angles and identify the types of angles.
2. Explain how math is used to solve right triangle problems.
   a. Explain how to determine the length of a slope.
   b. Explain how a building is laid out using right triangle math.
3. Define area and explain why determining the area of a space is required.
   a. Determine the area of squares and rectangles.
   b. Determine the area of a triangle.
   c. Determine the area of a trapezoid.
   d. Determine the area of a circle.
4. Define volume and explain the purpose of calculating volume.
   a. Calculate the volume of a cube.
   b. Calculate the volume of a prism.
   c. Calculate the volume of a cylinder.
   d. Describe the estimating process used to determine the volume and weight of simple and complex excavations.

Performance Task

Under the supervision of the instructor, you should be able to do the following:

1. Using information provided by the instructor, calculate the volume and weight of a given excavation project.

Trade Terms

Average
Constant
Hypotenuse
Parallel

Parallelogram
Quadrilateral
Squared
Variable

Required Trainee Materials

1. Pencil and paper
2. Ruler
3. Calculator

Industry Recognized Credentials

If you are training through an NCCER-accredited sponsor, you may be eligible for credentials from NCCER's Registry. The ID number for this module is 22207-13. Note that this module may have been used in other NCCER curricula and may apply to other level completions. Contact NCCER's Registry at 888.622.3720 or go to **www.nccer.org** for more information.

Contents ———————————————————————

Topics to be presented in this module include:

Figures and Tables ———————————————————————

SECTION ONE

1.0.0 WORKING WITH FORMULAS AND EQUATIONS

Objective 1

Explain how to use formulas.
  a. Explain the sequence of operations in solving a problem using a formula.
  b. Explain how squares and square roots are derived.
  c. Define angles and identify the types of angles.

Trade Terms

Constant: A value in an equation that is always the same; for example pi is always 3.14.

Variable: A value in an equation that depends on the factors being considered; for example, the lengths of the sides of a triangle may vary from one triangle to another.

Squared: Multiplied by itself.

During excavations, heavy equipment operators need to understand how to calculate the volume of material to be removed from or brought into a site. Once the volume of material is determined, it is necessary to calculate its weight. Knowing the volume and weight of excavation material is necessary for many reasons, but the most important is safety. Because heavy equipment is rated by weight and volume, it is important to make sure that a load does not exceed the capacity of the machine being used. Overloading a vehicle makes it harder to operate and places workers in danger of an accident. In addition, overloading a vehicle can damage it, requiring costly repairs.

In studying this module, resist the desire to skip practice problems. Math is just like anything else—skill comes with practice. Initially, calculations may seem difficult, but practice will bring steady improvement.

Excavations require determining the volume of a cut or fill area. To determine the volume of any excavation, you will first calculate the area of the shape.

Math is governed by formulas, which are equations made up of letters and symbols. The letters represent values and symbols (mathematical signs such as + and ×) define what to do. Study the following formula for calculating the area of a circle:

$$a = \pi r^2$$

In this case, the letter a means area, π means pi (pronounced *pie*), and the letter r means radius. To read this formula, you would say, "Area is equal to pi times the radius squared." When you see an expression such as πr^2 in a formula, it means that you must multiply the values. In this case, you multiply the radius squared by π.

Some values in formulas are **variables**. This means that the value is not a single number, so any number can replace a variable. In this sample formula, r is a variable. No matter what size a circle is, and thus no matter what its radius is, its area can be determined with this formula.

Some values, however, are **constant**. Pi is a constant. Its value is always 3.14159, often rounded off to 3.14. Whenever you see the word pi or its symbol (π) in a formula, you know to replace it with 3.14.

Equations are collections of numbers, symbols, and mathematical operators connected by equal signs (=). Everything on the left of the equal sign must match the right side. Consider the following equation for calculating the area of a rectangle:

$$\text{Area} = l \times w$$

In this case, the letter l means length and the letter w means width. The formula means multiply the length by the width. It can also be written as lw, without the multiplication sign. No multiplication sign is required when the intended relationship between symbols and letters is clear. For example, 2l means two times l.

Complicated equations must be solved by performing the indicated operations in a prescribed order: parentheses, exponents, multiply and divide, and add and subtract (PEMDAS). Always move from left to right when performing multiplication and division, and addition and subtraction. For example, the following equation can result in a number of answers if the PEMDAS order is not followed:

$$(3 + 3) \times 2 - 6 \div 3 + 1 = ?$$

Step 1 Parentheses:

$$\underline{(3 + 3)} \times 2 - 6 \div 3 + 1 = ?$$

Step 2 Multiply and divide:

$$\underline{6 \times 2} - 6 \div 3 + 1 = ?$$
$$12 - \underline{6 \div 3} + 1 = ?$$

Step 3 Add and subtract:

$$12 - 2 + 1 = ?$$

$$10 + 1 = ?$$

$$\text{Result} = 11$$

When none of the numbers are grouped within parentheses, the process is as follows:

$$3 + 3 \times 2 - 6 \div 3 + 1 = ?$$

Step 1 Multiply and divide:

$$3 + \underline{3 \times 2} - 6 \div 3 + 1 = ?$$

$$3 + 6 - \underline{6 \div 3} + 1 = ?$$

Step 2 Add and subtract:

$$\underline{3 + 6} - 2 + 1 = ?$$

$$\underline{9 - 2} + 1 = ?$$

$$\underline{7 + 1} = ?$$

$$\text{Result} = 8$$

1.1.0 Squares and Square Roots

The formula for the area of a circle is one of the formulas that involve squared numbers. You will also work with formulas in which you must find the square root of a given number.

Converting Fractions to Decimals

When dimensions are stated in inches and fractions of inches, it is easier to work with them if you convert the dimensions to decimal inches. Adding fractions can be difficult, and often results in errors. Decimals are much easier to add. With some practice, you can do most of the conversions in your head.

A square is the product of a number or quantity multiplied by itself. For example, the square of 6 means 6 × 6. To denote a number as squared, simply place the exponent 2 above and to the right of the base number. An exponent is a small figure or symbol placed above and to the right of another figure or symbol to show how many times the latter is to be multiplied by itself. For example:

$$6^2 = 6 \times 6 = 36$$

$$6^3 = 6 \times 6 \times 6 = 216$$

The square root of a number is the divisor which, when multiplied by itself (squared), gives the number as a product. Extracting the square root refers to a process of finding the equal factors which, when multiplied together, return the original number. The process is identified by the radical symbol [√]. This symbol is a shorthand way of stating that the equal factors of the number under the radical sign are to be determined. Finding the square root is necessary in many calculations, including those involving right triangles.

For example, $\sqrt{16}$ is read as the square root of 16. The number consists of the two equal factors 4 and 4. Thus, when 4 is squared, it is equal to 16. Again, squaring a number simply means multiplying the number by itself.

The number 16 is a perfect square. Numbers that are perfect squares have whole numbers as the square roots. For example, the square roots of perfect squares 4, 25, 36, 121, and 324 are the whole numbers 2, 5, 6, 11, and 18, respectively.

Squares and square roots can be calculated by hand, but the process is very time consuming and subject to error. Most people find squares and square roots of numbers using a scientific calculator like the one shown in *Figure 1*. To find the square of a number, the calculator's square key [x^2] is used. When pressed, it takes the number shown in the display and multiplies it by itself. For example, to square the number 4.235, you would enter 4.235, press the [x^2] key, then read 17.935225 on the display.

Similarly, to find the square root of a number, the calculator's square root key [√] or [$\sqrt{x}$] is used. When pressed, it calculates the square root of the number shown in the display. For example, to find the square root of the number 17.935225, enter 17.935225; press the [√] or [$\sqrt{x}$] key, then read 4.235 on the display.

Figure 1 Scientific calculator.

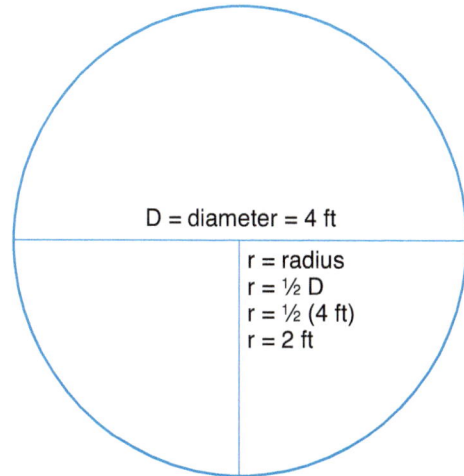

Figure 2 Access cover.

1.2.0 Using Formulas to Solve Problems

To solve problems using formulas, enter the measurements into the equation that makes up the formula. For example, let's say you need to know the area of a round access cover located in one area of a job site. See *Figure 2*. You would measure across the cover and determine that its diameter is four feet. Its radius (r) is half the size of the diameter, or two feet. To determine the area of the round access cover, this number can be plugged into the formula for the area of a circle (area = πr^2) in place of the letter r. Look at the formula with the numbers that represent the variable (r) and the constant (π) plugged into the equation:

$$a = \pi r^2$$

$$a = (3.14)(2^2)$$

Notice that 3.14 and 2^2 have been placed inside parentheses. When numbers are enclosed in parentheses, calculations inside each set of parentheses must be finished before completing any other calculations. In this case, you will first find 2^2 by multiplying 2×2:

$$a = (3.14)(4)$$

$$a = 12.56 \text{ square feet}$$

The parentheses in this equation are a grouping symbol. Grouping symbols are just like punctuation in writing. Imagine how hard it would be to read this module if there were no periods or commas to tell you where to stop or pause. Grouping symbols help make sense of an equation, just like punctuation helps make sense of a sentence. Grouping symbols identify numbers that belong together and which functions to perform first. It is important to pay attention to how terms are grouped in a formula and to do the calculations in the right order. In more complex problems, additional types of grouping symbols, such as square brackets [ ] and braces { }, are used.

1.3.0 Angles

Angles are measured in degrees and can be related to a circle. As shown in *Figure 3*, a circle contains 360 degrees. It can be evenly divided into four parts that each contain 90 degrees. An angle is made when two straight lines meet (*Figure 4*). The point where they meet is called a vertex (point B in *Figure 4*). The two lines are the sides of the angle. These lines are called the rays of the angle. The angle is the amount of opening that exists between the rays. It is measured in degrees. Two ways are commonly used to identify angles. One is to assign a letter to the angle, such as angle D shown in *Figure 4*. This is written: ∠D. The other way is to name the two end points of the rays and put the vertex letter between them; for example, ∠ABC. When showing the angle measure in degrees, it should be written inside the angle, if possible. If the angle is too small to show the measurement, it may be put outside of the angle and an arrow drawn to the inside.

CIRCLE = 360°

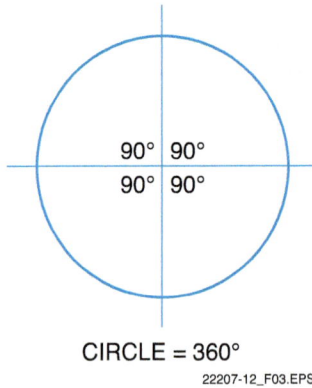

22207-12_F03.EPS

Figure 3 360 degrees in a circle.

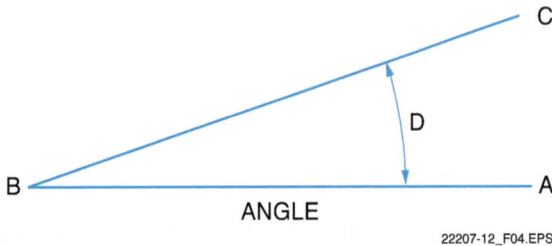

ANGLE

22207-12_F04.EPS

Figure 4 An angle.

There are several kinds of angles:

- *Right angle* – A right angle has rays that are perpendicular to one another (*Figure 5A*). The measure of this angle is always 90 degrees. The right angle is used often in construction, so remember that it is always 90 degrees.
- *Straight angle* – A straight angle (*Figure 5B*) does not look like an angle at all. The rays of a straight angle lie in a straight line, and the angle measures 180 degrees.
- *Acute angle* – An acute angle has less than 90 degrees (*Figure 5C*).
- *Obtuse angle* – An obtuse angle is greater than 90 degrees, but less than 180 degrees (*Figure 5D*).
- *Adjacent angles* – When three or more rays meet at the same vertex, the angles formed are adjacent (next to) one another. In *Figure 6A*, the angles ∠ABC and ∠CBD are adjacent angles. The ray BC is common to both angles.

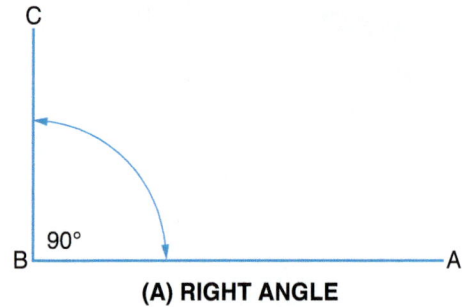

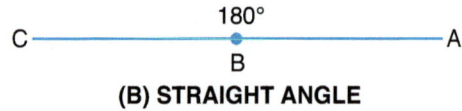

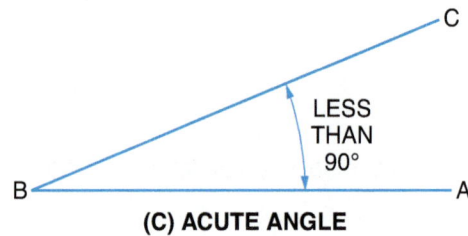

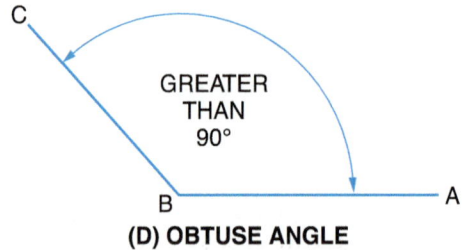

(A) RIGHT ANGLE

(B) STRAIGHT ANGLE

(C) ACUTE ANGLE

(D) OBTUSE ANGLE

22207-12_F05.EPS

Figure 5 Right, straight, acute, and obtuse angles.

- *Complementary angles* – Two adjacent angles that have a combined total measure of 90 degrees. In *Figure 6B*, ∠DEF is complementary to ∠FEG.
- *Supplementary angles* – Two adjacent angles that have a combined total measure of 180 degrees. In *Figure 6C*, ∠HIJ is supplementary to ∠JIK.

(A) ADJACENT ANGLES

(B) COMPLEMENTARY ANGLES

(C) SUPPLEMENTARY ANGLES

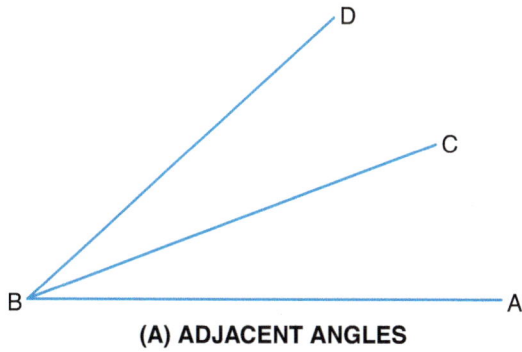

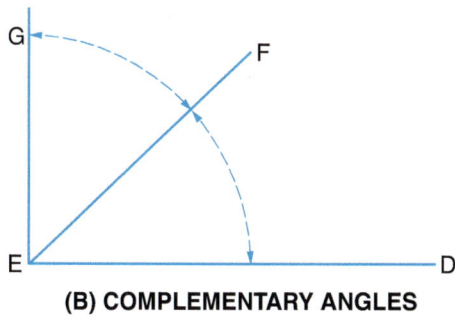

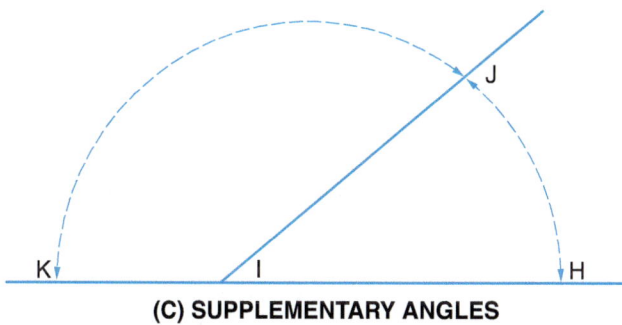

22207-12_F06.EPS

Figure 6 Adjacent, complementary, and supplementary angles.

Additional Resources

Applied Construction Math: A Novel Approach, Alachua, FL: NCCER.

1.0.0 Section Review

1. When solving an equation, the first step is to _____.

   a. multiply
   b. divide
   c. add
   d. subtract

2. The cube of 4 is _____.

   a. 12
   b. 16
   c. 48
   d. 64

3. An angle containing more than 90 degrees is known as a(n) _____.

   a. obtuse angle
   b. acute angle
   c. right angle
   d. adjacent angle

2.0.0 WORKING WITH RIGHT TRIANGLES

Objective 2

Explain how math is used to solve right triangle problems.

   a. Explain how to determine the length of a slope.
   b. Explain how a building is laid out using right triangle math.

Trade Terms

Hypotenuse: The long dimension of a right triangle and always the side opposite the right angle.

The right triangle is perhaps the most used shape in construction. Any vertical object or structure, such as a column or the face of a building, is part of a right triangle. If you draw an imaginary line from a point on the ground to the top of the structure, such as the top of a column or the roof of a building, that line forms the hypotenuse of a right triangle. The base of the triangle extends from the bottom of the pole or structure to the point on the ground that was the starting point for your line.

2.1.0 Right Triangle Calculations

Because the right triangle has one right angle, the other two angles are acute angles. They are also complementary angles, the sum of which equals 90 degrees. The right triangle has two sides perpendicular to each other, thus forming the right angle. To aid in writing equations, the sides and angles of a right triangle are labeled as shown in *Figure 7*. Normally, capital (uppercase) letters are used to label the angles and lowercase letters are used to label the sides. The third side, which is always opposite the right angle (C), is called the hypotenuse. It is always longer than either of the other two sides. The other sides can be remembered as a, for altitude, and b, for base. Note that the letters that label the sides and angles are opposite each other. For example, side a is opposite angle A, and so forth.

If the length of any two sides of a right triangle are known, the length of the third side can be determined using a rule called the Pythagorean theorem. It states that the square of the hypotenuse (c) is equal to the sum of the squares of the remaining two sides (a and b). Expressed mathematically:

$$c^2 = a^2 + b^2$$

This formula can be simplified to solve for the unknown side as follows:

$$a = \sqrt{c^2 - b^2}$$
$$b = \sqrt{c^2 - a^2}$$
$$c = \sqrt{a^2 + b^2}$$

For example, assume a right triangle with an altitude (side a) equal to 8' and a base (side b) equal to 12'. To find the length of the hypotenuse (side c), proceed as follows:

Right Triangles

Pythagoras was a Greek philosopher and mathematician who lived about 2,500 years ago. He is credited with developing the math for solving right triangle problems. Fifty years before Pythagoras, Thales, another Greek mathematician, figured out how to measure the height of the Egyptian pyramids using a technique that later became known as trigonometry. Trigonometry recognized that there is a relationship between the size of an angle and the lengths of the sides of a right triangle.

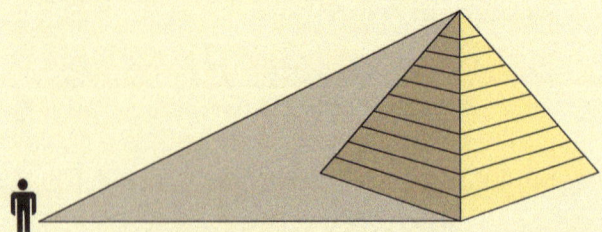

22207-12_SA01.EPS

Figure 7 Labeling of angles and sides of a right triangle.

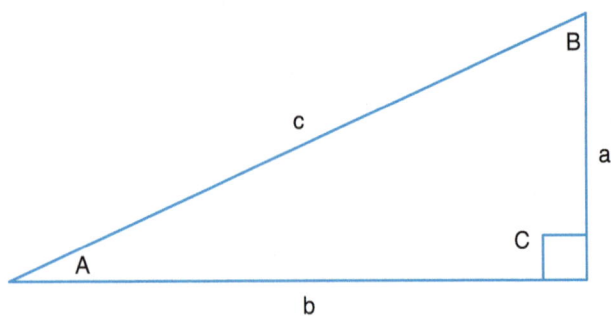

22207-12_F07.EPS

$$c = \sqrt{a^2 + b^2}$$

$$c = \sqrt{8^2 + 12^2}$$

$$c = \sqrt{64 + 144}$$

$$c = \sqrt{208}$$

$$c = 14.422'$$

To determine the actual length of the hypotenuse using this formula, it is necessary to calculate the square root of the sum of the sides squared. Fortunately, this is easy to do using a scientific calculator. On many calculators, you simply key in the number and press the square root [√] key. On some calculators, the square root does not have a separate key. Instead, the square root function is the inverse of the $[x^2]$ key, so you have to press [INV] or [2nd F], depending on the calculator, followed by $[x^2]$, to obtain the square root.

For an example of the Pythagorean theorem, let's say you know the length of a support cable attached to the top of a telephone pole. You wish to know the height of the pole, but measuring it from the ground would not be practical. You can, however, easily measure the distance between the end of the cable attached to the ground and the bottom of the pole—which forms the base of a right triangle.

If the length of the cable (the hypotenuse, or c from the equation) is 25' and the distance from the cable to the base of the pole is 10', the c and b parts of formula are known:

$$c^2 = a^2 + b^2$$

$$25^2 = a^2 + 10^2$$

$$625 = a^2 + 100$$

Subtracting 100 from both sides of the equation yields the following:

$$525 = a^2$$

The square root of 525, therefore, is the height of the pole (roughly 22.9').

2.2.0 Laying Out and Checking 90-Degree Angles Using the 3-4-5 Rule

The 3-4-5 rule describes a simple method for laying out or checking 90-degree angles (right angles) and requires only the use of a tape measure. The rule is based on the Pythagorean theorem and has been used in building construction for centuries. The numbers 3-4-5 represent dimensions in feet that describe the sides of a right triangle. Right triangles that are multiples of the 3-4-5 triangle, such as 9-12-15, 12-16-20, 15-20-25, and 30-40-50,

are commonly used. The specific multiple used is determined by the relative distances involved in the job being laid out or checked.

An example of the 3-4-5 rule using the multiples 15-20-25 is shown in *Figure 8*. In order to square or check a corner as shown in the example, first measure and mark 15' down the line in one direction, then measure and mark 20' down the line in the other direction. The distance measured between the 15' and 20' points must be exactly 25' to ensure that the angle is a perfect right angle.

The specific multiple used is determined mainly by the relative distances involved in the job being laid out or checked. It is best to use the highest multiple that is practical. When smaller multiples are used, any error made in measurement results in a much greater angular error.

Figure 9 shows an example of the 3-4-5 rule involving the multiple 48-64-80 (multiple of 16). In order to square or check a corner as shown in the

The 3-4-5 Rule

The 3-4-5 rule has been used in construction for centuries. It is a simple method for laying out and checking 90-degree angles (right angles). It requires only the use of a tape measure. Because they are easy to check, right triangles based on the 3-4-5 rule are commonly used in construction. Shown below is an example of the 3-4-5 rule using the multiples of 15-20-25. In order to check or square a corner, first measure and mark 15' down the line in one direction Then measure and mark 20' down the line in the other direction. The distance between the 15' and 20' points must be 25' for a perfect right triangle.

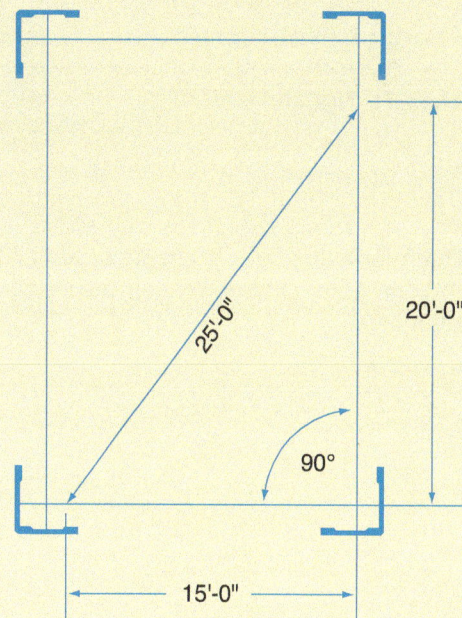

22207-12_SA02.EPS

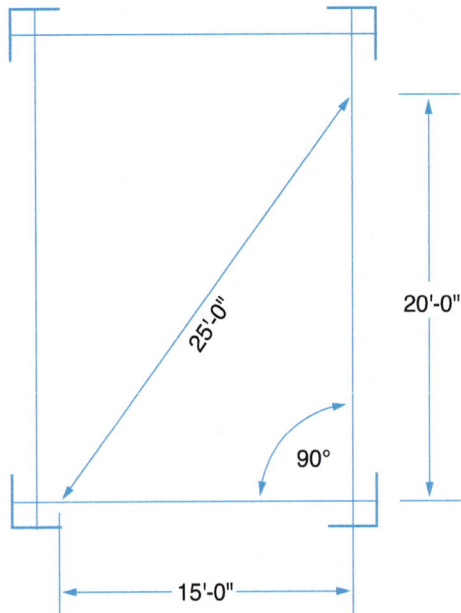

Figure 8 The 3-4-5 rule.

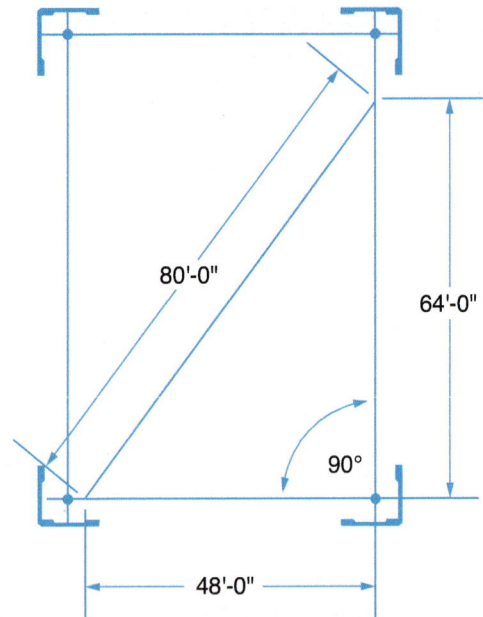

Figure 9 Checking a layout using the 3-4-5 rule.

example, first measure 48' down the line in one direction, then 64' down the line in the other direction. The distance measured between the 48' and 64' points must be exactly 80' if the angle is to be a perfect right angle. If the measurement is not exactly 80', the angle is not 90 degrees. This means that the direction of one of the lines or the corner point must be adjusted until a right angle exists.

It cannot be emphasized enough that exact measurements are necessary to get the desired results when using the 3-4-5 method of laying out or checking a 90-degree angle. Any error in the measurements of the distances will result in not establishing a right angle. If an existing 90-degree angle is being checked, inaccurate measurements may cause an unnecessary adjustment.

Additional Resources

Applied Construction Math: A Novel Approach. Alachua, FL: NCCER.

2.0.0 Section Review

1. The hypotenuse of a right triangle is 720 feet and one leg is 530 feet. What is the length of the other leg?

   a. 320 feet
   b. 487 feet
   c. 647 feet
   d. 720 feet

2. Using the 3-4-5 method, measurements that would be used to verify the squareness of the building lines for a 24' × 32' building are _____ .

   a. 21-28-35
   b. 24-32-40
   c. 24-32-48
   d. 20-40-60

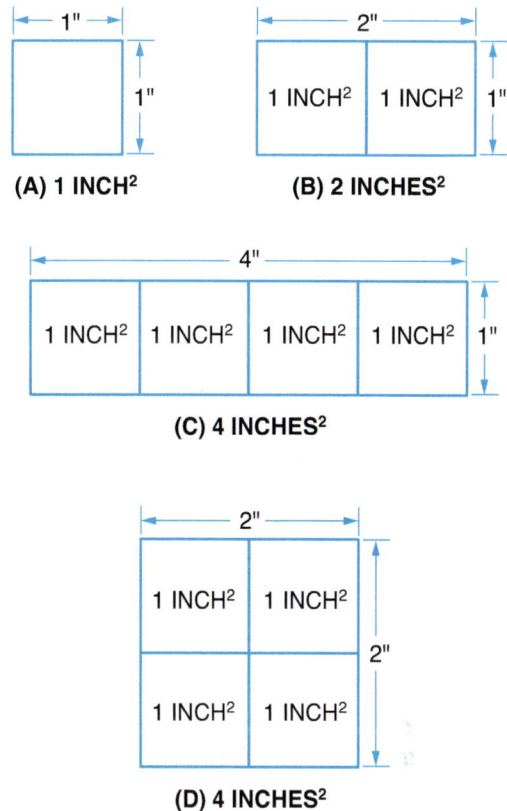

Figure 10 Square inches.

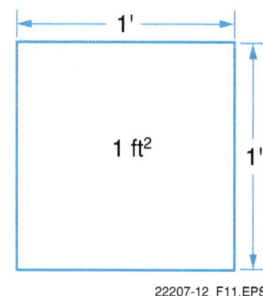

Figure 11 One square foot.

SECTION THREE

3.0.0 CALCULATING AREA

Objective 3

Define area and explain why determining the area of a space is required.
  a. Determine the area of squares and rectangles.
  b. Determine the area of a triangle.
  c. Determine the area of a trapezoid.
  d. Determine the area of a circle.

Trade Terms

Average: The middle point between two numbers or the mean of two or more numbers. It is calculated by adding all numbers together, and then dividing the sum by the quantity of numbers added. For example, the average (or mean) of 3, 7, 11 is 7 (3 + 7 + 11 = 21; 21 ÷ 3 = 7).

Parallel: Two lines that are always the same distance apart even if they go on into infinity (forever is called infinity in mathematics).

Parallelogram: A two-dimensional shape that has two sets of parallel lines.

Quadrilateral: A four-sided, closed shape with four angles whose sum is 360 degrees.

A rea is the measurement of the amount of space on a flat surface. Houses and apartments are advertised using area—area determines how much floor space is available. For example, a 12' × 12' room has 144 square feet of floor space, and that is its area.

Area is measured in square units, such as square inches, feet, yards, and miles (in the metric system, square centimeters, meters, and kilometers). This unit of measure is often written with the abbreviation inches^2, feet^2, meters^2, or in^2, ft^2, m^2.

A square inch is the area in a shape that is 1 inch long and 1 inch wide. Shapes that are measured in square units are called two-dimensional because they have only two measurements. *Figure 10A* shows a block that has an area of 1 square inch. The block is 1 inch long and 1 inch wide so it is a 1-inch square and has an area of 1 square inch. *Figure 10B* shows a larger block. That block is 2 inches long and 1 inch wide and contains two 1-inch squares, so it has an area of 2 square inches.

You have probably already guessed that *Figure 10C* shows a block that has an area of 4 square

inches, because it is 4 inches long and 1 inch wide and contains four 1-inch squares.

See *Figure 10D*. The block still contains four 1-inch squares so it has an area of 4 square inches just like *Figure 10C*, but this block is 2 inches long and 2 inches wide. (Calculating the area of three-dimensional objects is covered in later sections.) In *Figure 10D*, the shape of the surface does not matter—it is still the same area as the shape in *Figure 10C*.

Just as square inch is the area on the surface of a shape that is 1 inch long and 1 inch wide, a square foot is the area on the surface of a shape that is 1 foot long and 1 foot wide. See *Figure 11*.

When a shape is 1 yard long and 1 yard wide, it has the area of 1 square yard. Since 1 yard is equal to 3 feet, a square yard is 3 feet long and 3 feet wide. See *Figure 12*. When a square yard is divided into square foot blocks, there are 9 square foot blocks, so 1 square yard is equal to 9 square feet.

Because one foot equals 12 inches, a square foot is 12 inches long and 12 inches wide, and one square foot equals 144 square inches (*Figure 13*).

Area Exercises

Any flat shape can be measured in square units. To help you to understand this, perform the following exercise on a separate piece of paper.

1. Use a ruler to draw a block 4 inches long and 1 inch wide just like the one shown in *Figure 10C*. This block has an area of 4 square inches like the one shown in *Figure 10C*.

2. Use a ruler to draw a block 2 inches long and 2 inches wide just like the one shown in *Figure 10D*. This block has an area of 4 square inches like the one shown in *Figure 10D*.

3. Divide the figure drawn in exercise 1 in half so that there are two blocks 2 inches long and 1 inch wide. Since the whole block has an area of 4 square inches, half of the block must have an area of 2 square inches.

4. Divide the figure drawn in exercise 2 in half by drawing a line from its upper right corner to its lower left corner. Since the whole block has an area of 4 square inches, half of the block must have an area of 2 square inches.

5. Compare the lower half of the figure drawn in exercise 4 to the right half of the one drawn in exercise 3. Both have the same area of 2 square inches, but the shapes are quite different.

3.1.0 Calculating Area of Squares and Rectangles

Squares and rectangles are quadrilaterals. *Quad* means four and *lateral* means side, so a quadrilateral is a four-sided closed shape. *Figure 14* shows some common quadrilaterals. All quadrilaterals have four corners, with angles that add up to 360 degrees. Squares and rectangles are also parallelograms because they have two pairs of opposite parallel sides with angles that add up to 360 degrees. Of the shapes shown in *Figure 14*, all but the trapezoid are parallelograms.

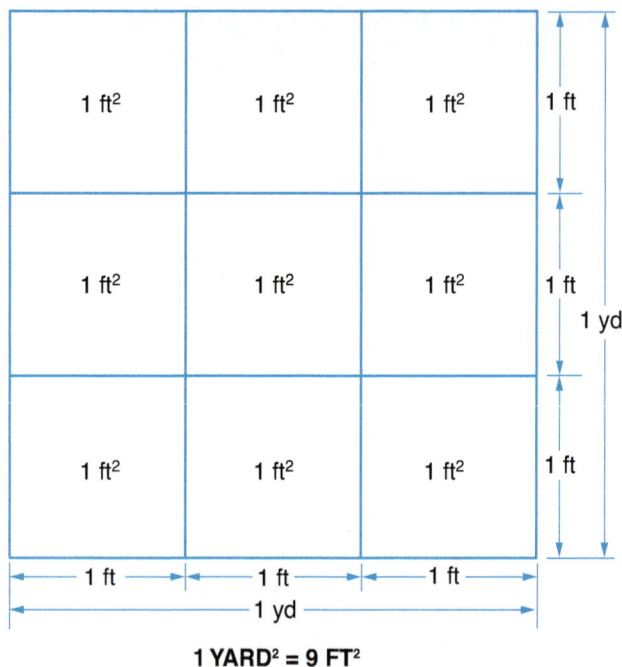

1 YARD² = 9 FT²

22207-12_F12.EPS

Figure 12 One square yard.

A rectangle (*Figure 15*) is a four-sided shape joined so that four 90-degree angles are formed. The sum of the four angles in any rectangle is 360 degrees. A rectangle has two pairs of equal sides. The longer side is called the length and is designated with the letter *l*, while the shorter side is called the width and is designated with the letter *w*. The area of a rectangle is calculated by multiplying the length times width (l × w or lw).

EACH BLOCK IS 1 INCH HIGH AND 1 INCH WIDE

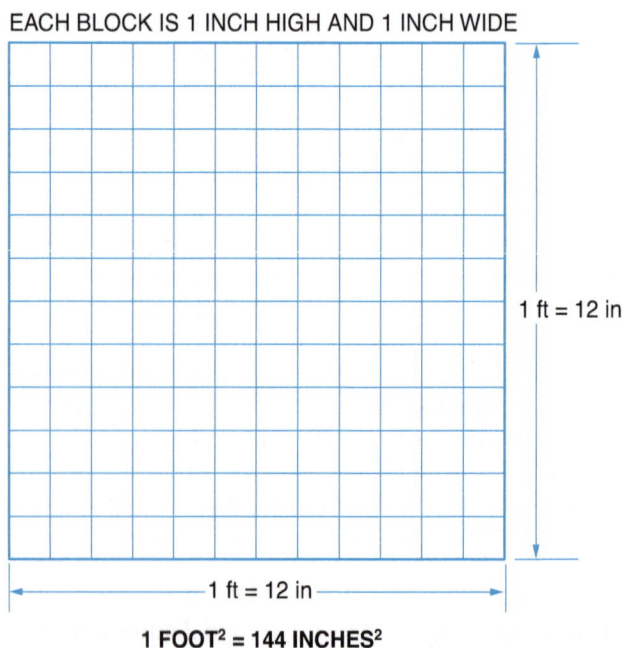

1 ft = 12 in

1 FOOT² = 144 INCHES²

22207-12_F13.EPS

Figure 13 144 square inches.

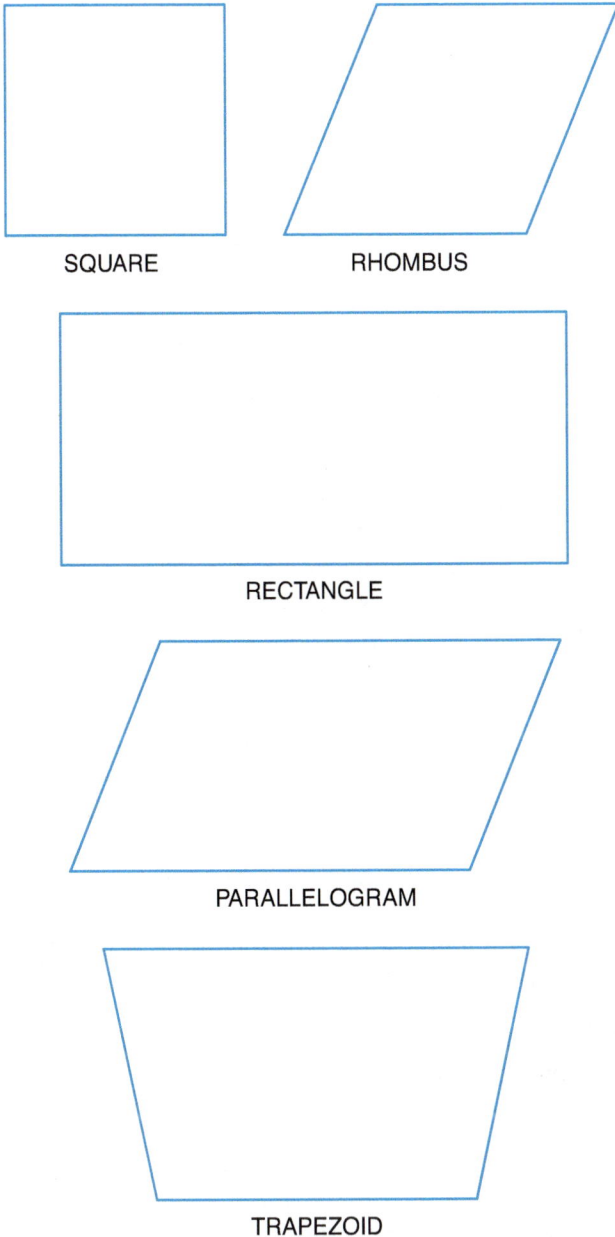

SQUARE · RHOMBUS

RECTANGLE

PARALLELOGRAM

TRAPEZOID

22207-12_F14.EPS

Figure 14 Quadrilaterals.

A rectangle with a length of 3 inches and width of 6 inches has an area of 18 square inches. It is calculated as follows:

$$\text{Area} = l \times w$$
$$\text{Area} = 3 \text{ inches} \times 6 \text{ inches}$$
$$\text{Area} = 18 \text{ inches}^2$$

When you are calculating the area of any object, the numbers multiplied together must be in the same units. For example, a rectangle with a length of 3 inches and a width of 1 foot has an area of 36 square inches, because 1 foot equals 12 inches. It is calculated as follows:

$$\text{Area} = 18 \text{ inches}^2$$
$$\text{Area} = 3 \text{ inches} \times 1 \text{ foot}$$
$$\text{Area} = 3 \text{ inches} \times 12 \text{ inches}$$
(convert 1 foot to 12 inches)
$$\text{Area} = 36 \text{ inches}^2$$

A square (*Figure 16*) is similar to a rectangle. It has four sides that are joined to form four 90-degree angles. The sum of these angles is 360 degrees—just like a rectangle—but all sides of the square are the same length. The sides of the square are labeled with a single letter. In this module, the sides of a square will be labeled with the letter *e*. The area of a square is calculated by multiplying two sides together ($e \times e$ or e^2).

A square that is 3 inches long has an area of 9 square inches, which is calculated as follows:

$$\text{Area} = e^2$$
$$\text{Area} = 3 \text{ inches} \times 3 \text{ inches}$$
$$\text{Area} = 9 \text{ inches}^2$$

A square that is 12 inches long has an area of 144 square inches, which is calculated as follows:

$$\text{Area} = e^2$$
$$\text{Area} = 12 \text{ inches} \times 12 \text{ inches}$$
$$\text{Area} = 144 \text{ inches}^2$$

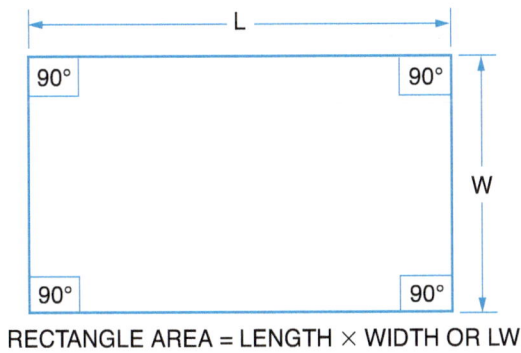

RECTANGLE AREA = LENGTH × WIDTH OR LW

22207-12_F15.EPS

Figure 15 A rectangle.

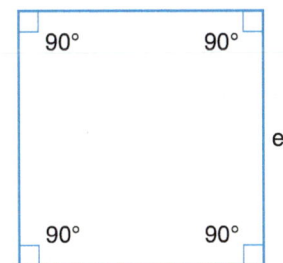

SQUARE AREA = e×e OR e²

22207-12_F16.EPS

Figure 16 A square.

The numbers quickly become large when working with inches, so it is better to work with smaller numbers by converting inches into feet or even yards. In the example above, 12 inches was the value used to arrive at the answer of 144 square feet. By converting the 12 inches to one foot, the area can be calculated as follows:

$$\text{Area} = e^2$$
$$\text{Area} = 1 \text{ foot} \times 1 \text{ foot}$$
$$\text{Area} = 1 \text{ foot}^2$$

These two calculations show that 1 square foot is the same as 144 square inches. If it is necessary to convert square feet to square inches, multiply the square feet by 144 as follows:

$$1 \text{ foot}^2 \times 144 = 144 \text{ inches}^2$$
$$2 \text{ feet}^2 \times 144 = 288 \text{ inches}^2$$
$$3 \text{ feet}^2 \times 144 = 432 \text{ inches}^2$$

To convert square inches to square feet, divide the square inches by 144 as follows:

$$144 \text{ inches}^2 \div 144 = 1 \text{ foot}^2$$
$$432 \text{ inches}^2 \div 144 = 3 \text{ feet}^2$$
$$729 \text{ inches}^2 \div 144 = 5\tfrac{1}{2} \text{ feet}^2$$

Complete the following three exercises by drawing the shape described and then calculating the area of each shape. Hint: always convert the measurements into the same units before calculating the area.

1. The building plans call for a building that is 40 feet long and 30 feet wide.

2. Outside the building, the parking pad is 15 feet by 15 feet.

3. The driveway leading up to the building is 12 yards long and 12 feet wide.

3.2.0 Calculating the Area of a Triangle

A triangle is a three-sided figure that has three angles. The angles in the triangle may vary, but the sum of the angles is always 180 degrees. Construction work involves the use of several types of triangles as shown in *Figure 17*:

- *Right triangle* – A right triangle has one 90-degree angle.
- *Equilateral triangle* – An equilateral triangle has three equal angles and three sides of equal length.

- *Isosceles triangle* – An isosceles triangle has two equal angles and two sides of equal length. An isosceles triangle can be divided into two equal right triangles.
- *Scalene triangle* – A scalene triangle has no equal angles or side lengths.

The right triangle is one of the most frequently used triangles in construction. A right triangle must have one 90-degree angle; the sum of the other two angles is 90 degrees. A right triangle is created when a square or rectangle is divided in half diagonally. This creates two identical triangles, each with half the area of the rectangle (*Figure 18*).

A triangle's length is called the base, which is abbreviated with the letter *b*. Its width is called the height, which is abbreviated with the letter *h*, so the formula to calculate the area of a triangle is ½*bh*. See *Figure 19*. This formula is used to calculate the area of all triangles.

In *Figure 20*, the four triangles all have the same base (7 inches) and the same height (6 inches), so the area of each triangle is the same (21 square inches), even though their appearances are very different. The area is calculated as follows:

$$\text{Area} = \tfrac{1}{2}bh$$
$$\text{Area} = \tfrac{1}{2}(7 \times 6)$$
$$\text{Area} = \tfrac{1}{2}(42)$$
$$\text{Area} = 21 \text{ inches}^2$$

Some trainees have trouble understanding how the same formula can be used to calculate the area of all triangles, but the exercises at the end of this section will help you to understand.

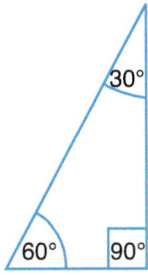

RIGHT TRIANGLE
$30° + 60° + 90° = 180°$

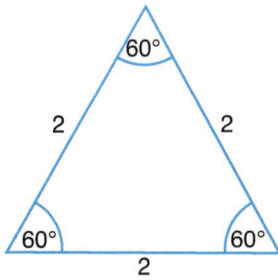

EQUILATERAL TRIANGLE
$60° + 60° + 60° = 180°$

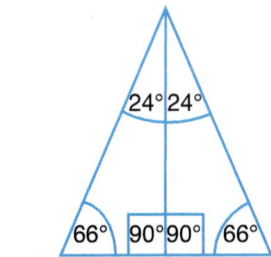

ISOSCELES TRIANGLE
$48° + 66° + 66° = 180°$

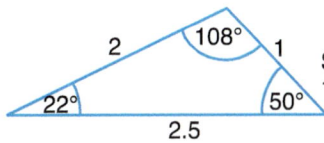

SCALENE TRIANGLE
$108° + 22° + 50° = 180°$

22207-12_F17.EPS

Figure 17 The sum of a triangle's three angles is always 180 degrees.

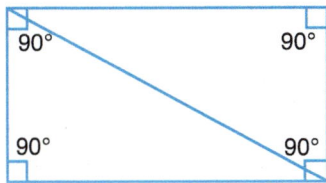

RECTANGLE
Surface Area = L × W, so the Surface
Area of these triangles must be ½ (L × W)

22207-12_F18.EPS

Figure 18 One rectangle, two triangles.

Triangles are related to excavation. See *Figure 21*. This highway job requires the removal of part of the existing ground to make an inslope and backslope. The slope stake directs a cut of 3 feet and a grade of 3:1. This means the ground must drop 1 foot every 3 feet from the stake until the cut depth is 3 feet. In this example, assume that

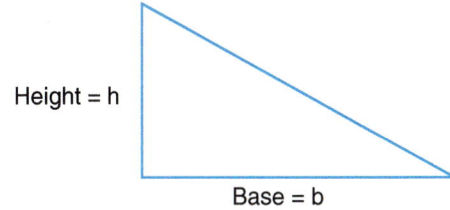

TRIANGLE
Surface Area = ½ (bh)

22207-12_F19.EPS

Figure 19 Triangle area.

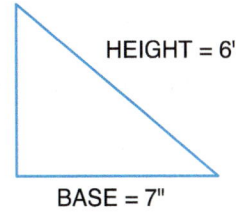
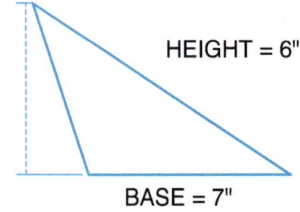

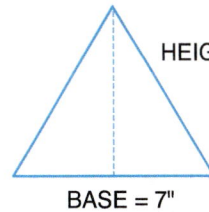
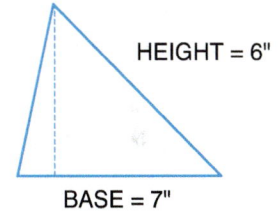

22207-12_F20.EPS

Figure 20 Different triangles, same area.

the backslope will be cut the same way. Look carefully at *Figure 21*. The proposed cut forms an inverted isosceles triangle.

Think about the stake directions. The maximum cut is 3 feet, so the height of the triangle must be 3 feet. A 3:1 slope means the ground must drop 1 foot every 3 feet from the stake. So to the point of maximum cut, the horizontal distance is 9 feet (3 feet + 3 feet + 3 feet). Since the backslope is being cut the same as the inslope, that side is 9 feet, too. So the base of the triangle is 18 feet. You now have all of the information needed to calculate the area of the proposed cut, which is as follows:

$$\text{Base} = 18 \text{ feet}$$
$$\text{Height} = 3 \text{ feet}$$
$$\text{Area} = \tfrac{1}{2}\,bh$$
$$= \tfrac{1}{2}(18 \times 3)$$
$$= \tfrac{1}{2}(54)$$
$$= 27 \text{ feet}^2$$

This process can be used to calculate the area of any cut that forms a triangle. See *Figure 22*. On this job, the inslope needs to be cut just like the one in

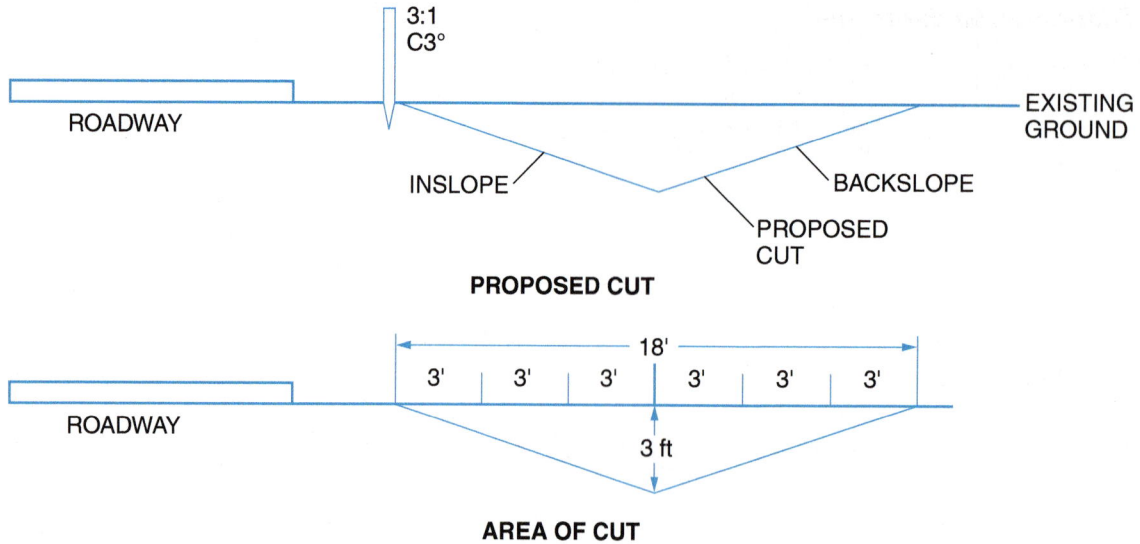

PROPOSED CUT

AREA OF CUT

22207-12_F21.EPS

Figure 21 Calculating the area of a proposed cut.

the previous example, but the backslope has a 6:1 slope from the ditch. The inslope is cut 9 feet from the stake to the maximum depth of 3 feet, just as in the previous example. The backslope needs a 6:1 ratio, so the level of the ground must decrease 1 foot for every 6 feet from the stake to the maximum cut of 3 feet. This means that it is 18 feet (6 + 6 + 6) from the backslope stake to the maximum cut. In this case, the cut forms an inverted scalene triangle.

The base of the triangle is 27 feet and the height is 3 feet. You now have all of the information you need to calculate the area of the proposed cut, which is as follows:

$$\text{Base} = 27 \text{ feet}$$

$$\text{Height} = 3 \text{ feet}$$

$$\text{Area} = \frac{1}{2}bh$$

$$= \frac{1}{2}(27 \times 3)$$

$$= \frac{1}{2}(81)$$

$$= 40\frac{1}{2} \text{ feet}^2$$

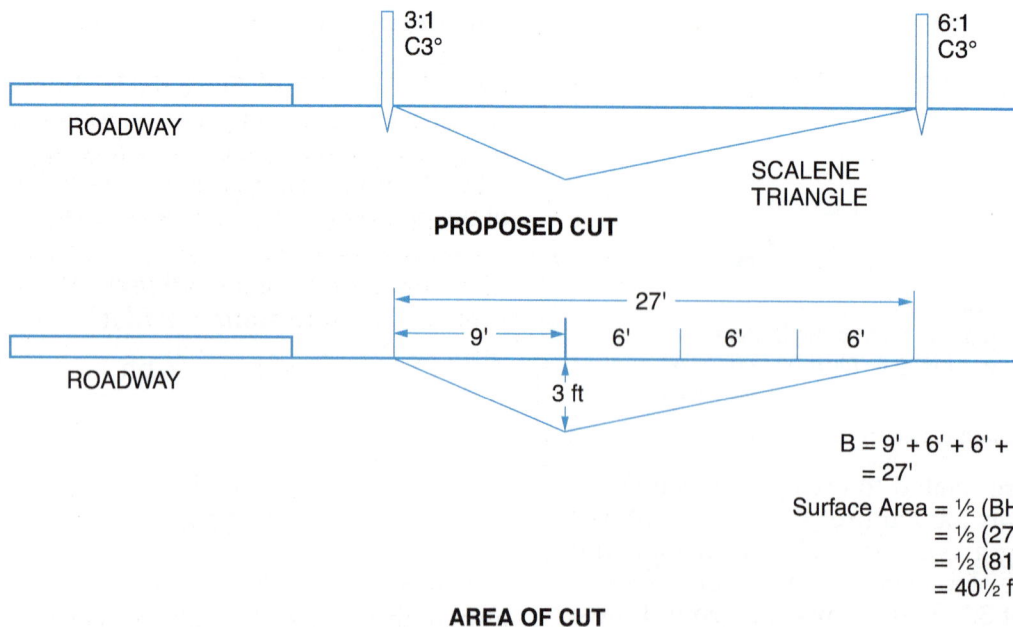

PROPOSED CUT

AREA OF CUT

$$B = 9' + 6' + 6' + 6'$$
$$= 27'$$
$$\text{Surface Area} = \frac{1}{2}(BH)$$
$$= \frac{1}{2}(27' \times 3')$$
$$= \frac{1}{2}(81 \text{ ft})$$
$$= 40\frac{1}{2} \text{ ft}^2$$

22207-12_F22.EPS

Figure 22 Calculating the area of another proposed cut.

NCCER – *Heavy Equipment Operations Level Two*

3.2.1 Triangle Area Exercises

The following exercises will provide a better understanding of how to calculate the area of a triangle. Read each question carefully and draw figures as necessary.

1. Find the following for *Figure 23*:
   a. Length =
   b. Width =
   c. Area =

2. Find the following for *Figure 24*:

   Triangle 1:
   a. Base =
   b. Height =
   c. Area =

   Triangle 2:
   d. Base =
   e. Height =
   f. Area =

   Triangles 1 and 2:
   g. Total area =

3. Find the following for *Figure 25*:

   Triangle 1:
   a. Base =
   b. Height =
   c. Area =

   Triangle 2:
   d. base =
   e. height =
   f. Area =

   Triangle 3:
   g. Base =
   h. Height =
   i. Area =

   Triangles 1, 2, and 3:
   j. Total area =

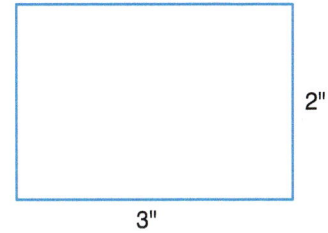

SA of a rectangle = LW

22207-12_F23.EPS

Figure 23 Triangle Exercise 1.

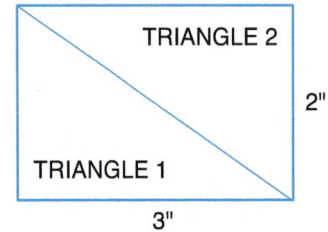

SA of a triangle = ½ BH

22207-12_F24.EPS

Figure 24 Triangle Exercise 2.

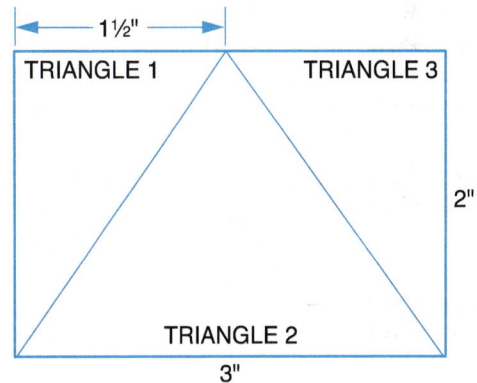

22207-12_F25.EPS

Figure 25 Triangle Exercise 3.

3.3.0 Calculating the Area of a Trapezoid

A trapezoid is also a quadrilateral. It has four sides with two parallel sides. The angles formed by the sides add up to 360 degrees. *Figure 26* shows some examples of trapezoids. Note that the two parallel lines are called base 1 and base 2. *Figure 27* shows the cross-section of a roadway with the shoulder slope, ditch, and backslope. The shape made by the two slopes and ditch is a trapezoid.

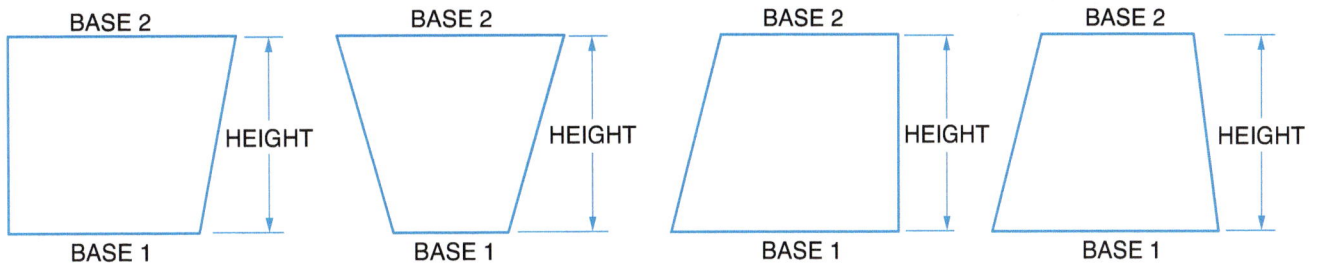

22207-12_F26.EPS

Figure 26 Trapezoids.

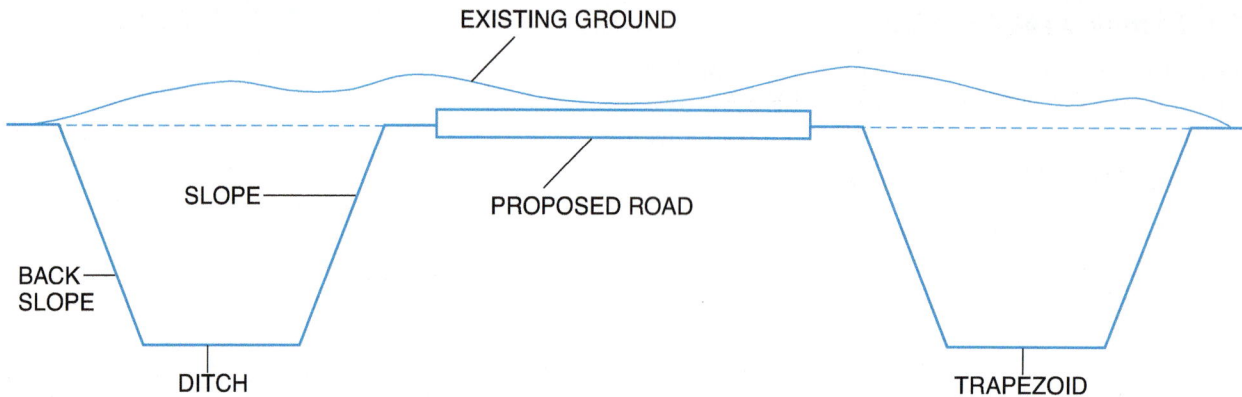

Figure 27 Trapezoids on a roadway.

Any trapezoid can be divided into a rectangle (or square) and at least one triangle. See *Figure 28*. The area of a trapezoid can be found by calculating the areas of the rectangle and triangle and adding the results together as shown in *Figure 29*, but there is an easier way. You can average the lengths of the two bases and then multiply the average by the height. The formula to calculate the area of a trapezoid is as follows:

$$Area = \frac{1}{2}(base\ 1 + base\ 2) \times height$$

Using the measurements of base 1 = 8 inches, base 2 = 3 inches, and height = 5 inches given in *Figure 30*, which are the same as the ones used in *Figure 29*, the area of the trapezoid is calculated as follows:

$$Area = \frac{1}{2}(base\ 1 + base\ 2) \times height$$

$$Area = \frac{1}{2}(8 + 3) \times 5$$

$$Area = \frac{1}{2}(11) \times 5$$

$$Area = 5\frac{1}{2} \times 5$$

$$Area = 27\frac{1}{2}\ inches^2$$

It may help to remember the trapezoid formula by relating it to the formula for a rectangle. The rectangle in *Figure 31* has a length of 3 inches and a width of 2 inches, so it has an area of 6 square

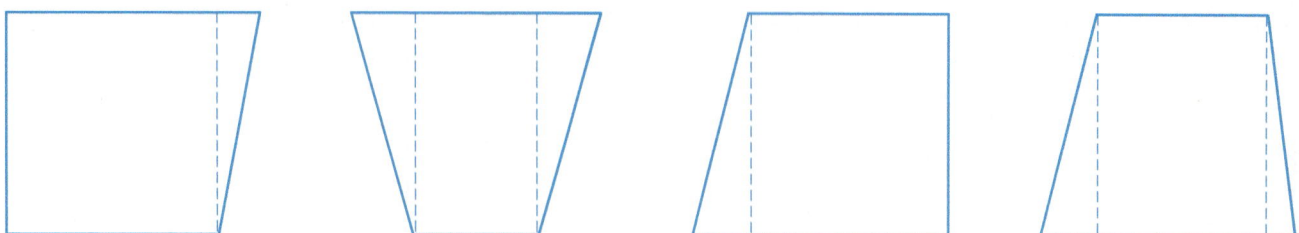

RECTANGLE
SA = LW
SA = 3" × 5"
  = 15 in²

TRIANGLE 1
SA = ½ (BH)
SA = ½ (3" × 5")
  = ½ (15")
  = 7½ in²

TRIANGLE 2
SA = ½ (BH)
SA = ½ (2" × 5")
  = ½ (10")
  = 5 in²

TRAPEZOID
SA = SA Rectangle + SA Triangle 1 + SA Triangle 2
SA = 15 in² + 7½ in² + 5 in²
  = 27½ in²

22207-12_F29.EPS

Figure 29 Trapezoid area using triangles and rectangles.

inches. Using the trapezoid formula for area, you need to use the length as base 1 and base 2, and calculate the average of the bases as follows:

$$\frac{1}{2}(3 + 3) = 3$$

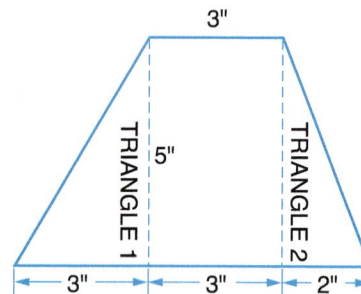

22207-12_F28.EPS

Figure 28 Trapezoids contain triangles and rectangles.

NCCER – *Heavy Equipment Operations Level Two*

BASE 2 = 3"

$$SA = \frac{1}{2}(Base_1 + Base_2) \times H$$
$$= \frac{1}{2}(8" + 3") \times 5"$$
$$= \frac{1}{2}(11") \times 5"$$
$$= 5\frac{1}{2} \times 5"$$
$$= 27\frac{1}{2}\ in^2$$

HEIGHT = 5"

BASE 1 = 8"

22207-12_F30.EPS

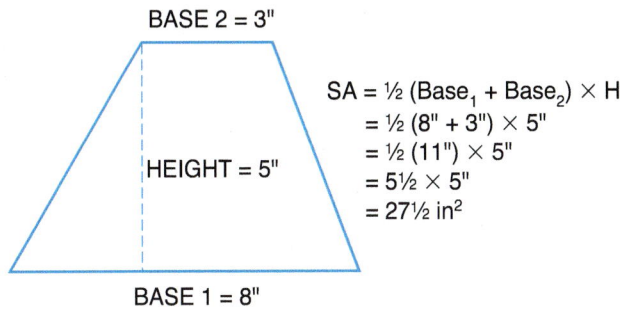

Figure 30 Trapezoid area using formula.

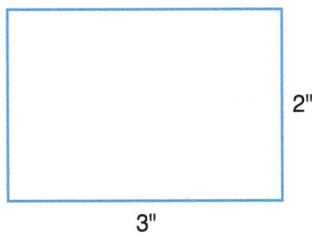

2"

3"

Rectangular Area = 3" × 2"
= 6 in²

Trapezoid Area = ½ (3" + 3") × 2"
= ½ (6") × 2"
= 3" × 2"
= 6 in²

22207-12_F31.EPS

Figure 31 Rectangle and trapezoid relationship.

Then use the width as the height and calculate the area as follows:

$$3 \times 2 = 6\ inches^2$$

3.3.1 Trapezoid Exercises

Refer to *Figure 32* and complete the following exercises. Remember to convert units of measure as required.

Scenario: A crew is working on a highway job and needs to cut a shoulder slope, ditch, and backslope into existing ground. The cross-section plan is shown in *Figure 32*. The cut is 3 feet, the ditch is 2 feet wide, and the slope for the shoulder slope and backslope is 1:1 to the edges of the ditch. (Hint: the plan specifies that a 3-foot cut be made from the shoulder into the existing ground; 1:1 means that for every one foot of travel there is a one-foot drop in elevation.)

1. Base 1 =

2. Base 2 =

3. Height =

4. Area =

SLOPE
1:1

22207-12_F32.EPS

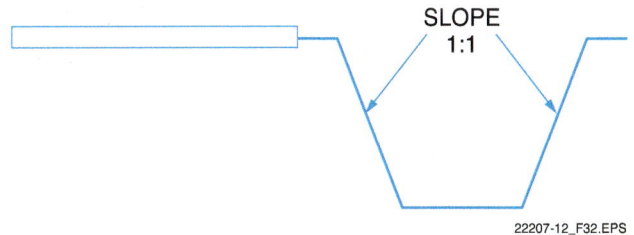

Figure 32 Trapezoid exercise cross-section plan.

3.4.0 Calculating the Area of a Circle

A circle is a single curved line that connects with itself (*Figure 33*). A circle also has these properties:

- All points on a circle are the same distance (equidistant) from the point at the center.
- The distance from the center to any point on the curved line, called the radius (r), is always the same.
- The shortest distance from any point on the curve through the center to a point directly opposite is called the diameter (d). The diameter is therefore equal to twice the radius (d = 2r).
- The distance around the outside of the circle is called the circumference. It can be determined by using the equation: circumference = πd, where π is a constant approximately equal to 3.14 and d is the diameter.
- A circle is divided into 360 parts, with each part called a degree. Therefore, one degree = $\frac{1}{360}$ of a circle. The degree is the unit of measurement commonly used in construction for measuring the size of angles.
- The total measure of all the angles formed by all consecutive radii equals 360 degrees.

The formula to calculate the area of a circle is πr^2. The π symbol represents pi, which is a constant of 3.14, and the r means radius. The area of a circle with a radius of 6 inches is calculated as follows:

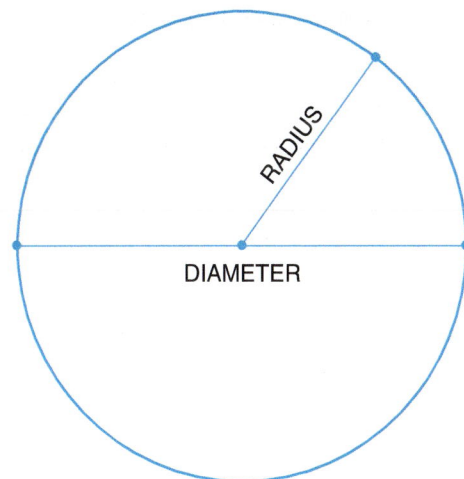

RADIUS

DIAMETER

22207-12_F33.EPS

Figure 33 A circle.

$$Area = \pi r^2$$
$$Area = (3.14)(6^2)$$
$$Area = (3.14)(36)$$
$$Area = 113.04 \text{ inches}^2$$

The area of a circle with a radius of 10 feet is calculated as follows:

$$Area = \pi r^2$$
$$Area = (3.14)(10^2)$$
$$Area = (3.14)(100)$$
$$Area = 314 \text{ feet}^2$$

The area of a circle with a diameter of 10 feet is calculated as follows:

$$Area = \pi r^2, \text{ and radius} = d \div 2,$$
$$\text{so } r = 10 \div 2 = 5 \text{ feet}$$
$$Area = (3.14)(5^2)$$
$$Area = (3.14)(25)$$
$$Area = 78\tfrac{1}{2} \text{ feet}^2$$

In earthwork, it will sometimes be easier to get the circumference of a circle than the diameter or radius—such as when a large depression needs to be filled. The formula for finding the diameter of a circle with the circumference is: $c \div \pi$, where c is circumference and π is 3.14. The area of a circle with a circumference of 31 feet is calculated as follows:

$$Diameter = \text{circumference}/\pi$$
$$Diameter = 31/3.14$$
$$Diameter = 9.87 \text{ feet}$$
$$Radius = 9.87/2$$
$$Radius = 4.94 \text{ feet}$$
$$Area = \pi r^2, \text{ and}$$
$$Area = (3.14)(4.94^2)$$
$$Area = (3.14)(24.4)$$
$$Area = 76.62 \text{ feet}^2$$

3.4.1 Circle Exercises

Complete the following exercises. Hint: when the circle diameter or circumference is given, be sure to calculate the radius.

1. What is the area of a circle with a diameter of 10 inches?

2. What is the area of a circle with a radius of 100 feet?

3. What is the area of a circle with a circumference of 628 feet?

4. What is the area of a circle with a radius of 7 feet?

5. What is the area of a circle with a diameter of 12 inches?

Additional Resources

Applied Construction Math: A Novel Approach.
Alachua, FL: NCCER.

3.0.0 Section Review

1. The area of a rectangle that is 15 feet wide by 12 feet deep is _____.

   a. 120 square feet
   b. 150 cubic feet
   c. 180 square feet
   d. 180 cubic feet

2. The area of triangle 3 in *Figure 1* is _____.

   a. 1 in²
   b. 2 in²
   c. 3 in²
   d. 4 in²

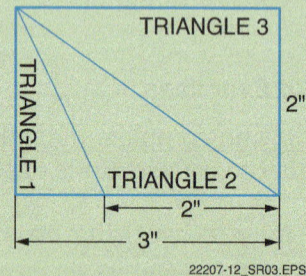

TRIANGLE 3 / TRIANGLE 1 / TRIANGLE 2 / 2" / 2" / 3"

22207-12_SR03.EPS

3. A trapezoid can be divided into a _____.

   a. square and a rectangle
   b. rectangle or square and at least one triangle
   c. triangle, a square, and a rectangle
   d. quadrilateral and a square

4. The formula for determining the circumference of a circle is _____.

   a. πr^2
   b. πd
   c. $2\pi r$
   d. πd^2

Section Four

4.0.0 Calculating Volume

Objective 4

Define *volume* and explain the purpose of calculating volume.

a. Calculate the volume of a cube.
b. Calculate the volume of a prism.
c. Calculate the volume of a cylinder.
d. Calculate the volume and weight of simple and complex excavations.

Performance Task 1

Using information provided by the instructor, calculate the volume and weight of a given excavation project.

Volume is the amount of space inside a three-dimensional object. Objects such as boxes, trash cans, coffee cups, and water pipes are three-dimensional objects. Any vessel that can hold a substance is a three-dimensional object, so it can be measured in terms of volume.

Three-dimensional objects have three measurements—length, width, and depth. The length is abbreviated with the letter *l*, width with the letter *w*, and depth with the letter *d*. Volume is measured in cubic units, so an object that is 1 inch in length, width, and depth has a volume of 1 cubic inch (see *Figure 34*), which can be abbreviated $inch^3$ and is calculated as follows:

$$\text{Volume} = l \times w \times d$$
$$\text{Volume} = 1 \text{ inch} \times 1 \text{ inch} \times 1 \text{ inch}$$
$$\text{Volume} = 1 \text{ inch}^3$$

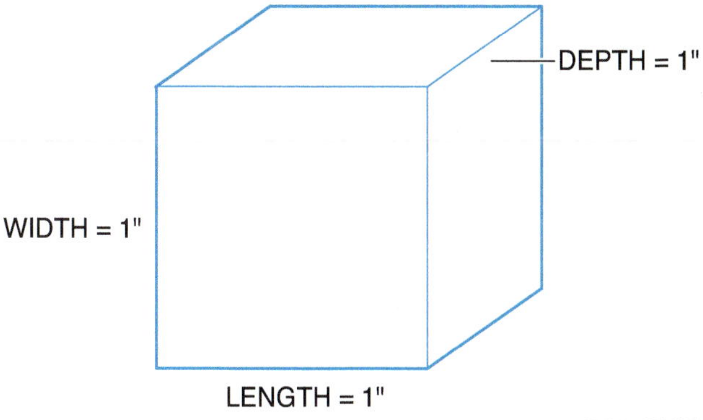

DEPTH = 1"
WIDTH = 1"
LENGTH = 1"
22207-12_F34.EPS

Figure 34 One cubic inch.

When two 1 cubic inch objects are placed together (see *Figure 35*), the volume is 2 cubic inches and is calculated as follows:

$$\text{Volume} = l \times w \times d$$
$$\text{Volume} = 2 \text{ inches} \times 1 \text{ inch} \times 1 \text{ inch}$$
$$\text{Volume} = 2 \text{ inches}^3$$

Just as with area, the same volume measurement can take many shapes. See *Figure 36A*. An object with an area of 2 cubic inches can be divided into two equal parts horizontally (*Figure 36B*), vertically (*Figure 36C*), and diagonally (*Figure 36D*) to form three differently shaped objects each with a volume of 1 cubic inch.

When calculating the volume of an object, all the numbers must be in the same unit of measure. That is, all numbers must be in inches, feet, or yards. It is a good idea to use the smallest measure you can. For example, 1 yard equals 3 feet, and 3 feet equal 36 inches, so the volume of a box that has a length, width, and depth of 1 yard can be calculated as follows:

Volume in yards:

$$\text{Volume} = 1 \text{ yd} \times 1 \text{ yd} \times 1 \text{ yd} = 1 \text{ yd}^3$$

Volume in feet:

$$\text{Volume} = 3 \times 3 \times 3 = 27 \text{ ft}^3$$

Volume in inches:

$$\text{Volume} = 36 \times 36 \times 36 = 46{,}656 \text{ in}^3$$

All of the above represent the same volume, but as you can see, it is much easier to work with the yard measure because it has the smallest numbers.

4.1.0 Cubes and Rectangular Objects

When a square is the base of a three-dimensional object, all the dimensions are equal and the object is called a cube (*Figure 37*). The volume of this object is calculated by multiplying its length by its width by its depth. Since all of the cube's dimensions are equal, each side can be abbreviated with the same letter (*e*), and the formula can be written e^3. The cube shown in *Figure 37* has a length of 2 feet, width of 2 feet, and depth of 2 feet, so its volume is calculated as follows:

$$\text{Volume} = l \times w \times d \text{ or } e^3$$
$$\text{Volume} = 2 \times 2 \times 2 \text{ or } 2^3$$
$$\text{Volume} = 8 \text{ feet}^3$$

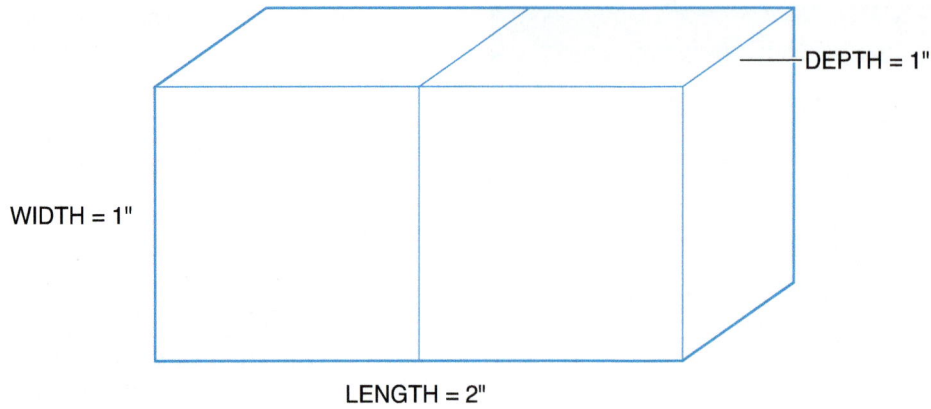

Figure 35 Two cubic inches.

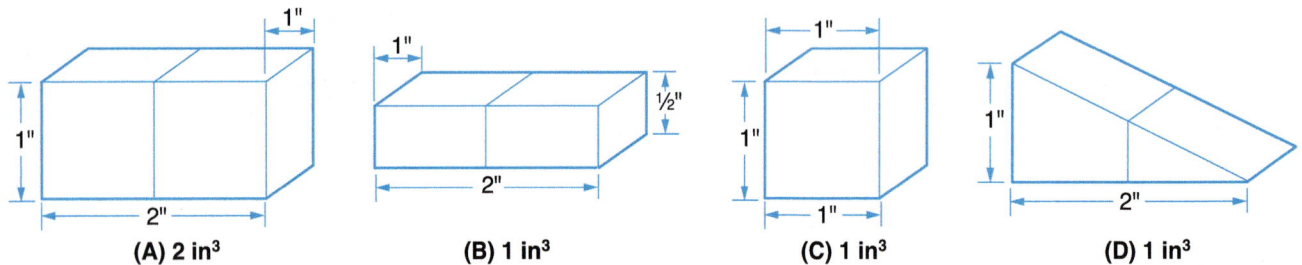

(A) 2 in³   **(B) 1 in³**   **(C) 1 in³**   **(D) 1 in³**

Figure 36 A cubic inch can be different shapes.

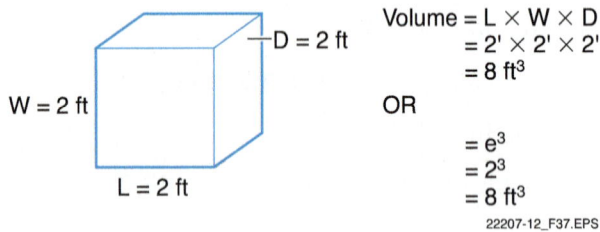

$$\text{Volume} = L \times W \times D$$
$$= 2' \times 2' \times 2'$$
$$= 8 \text{ ft}^3$$

OR

$$= e^3$$
$$= 2^3$$
$$= 8 \text{ ft}^3$$

Figure 37 A cube.

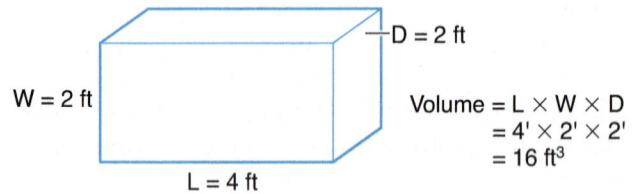

$$\text{Volume} = L \times W \times D$$
$$= 4' \times 2' \times 2'$$
$$= 16 \text{ ft}^3$$

Figure 38 Rectangular object.

When a rectangle is the base of a three-dimensional object, it is called a rectangular object (*Figure 38*). This object's volume is calculated by multiplying its length by its width by its depth, just like a cube. Since each dimension of a rectangle can be different, the length is abbreviated with the letter *l*, the width with the letter *w*, and the depth with the letter *d*. The object shown in *Figure 38* has a length of 4 feet, width of 2 feet, and depth of 2 feet, so its volume is calculated as follows:

$$\text{Volume} = l \times w \times d$$
$$\text{Volume} = 4 \times 2 \times 2$$
$$\text{Volume} = 16 \text{ feet}^3$$

Note that the formula for a cube and a rectangular object contains the formula for area (l × w). To find the volume of a cube, multiply its area by its depth. This will be true for most three-dimensional objects used to estimate excavations.

4.2.0 Prisms

A prism is a multi-sided three-dimensional object that must meet all of the following requirements:

- It has two bases.
- The bases are parallel.
- The bases are the same shape.
- The bases are the same size.
- The remaining sides must be parallelograms.

All of the objects shown in *Figure 39* are prisms. You will notice that rectangular objects and cubes are prisms. The other prisms you need to know about to estimate excavations have bases of triangles and trapezoids.

Recall that the volume of a rectangular object is found by multiplying the area of the rectangle by the depth of the object. The same is true with prisms. To find the volume of a prism, first define the shape of the base, find the area of the base, and then multiply the area by the depth. Use the following procedure to calculate the volume of a prism:

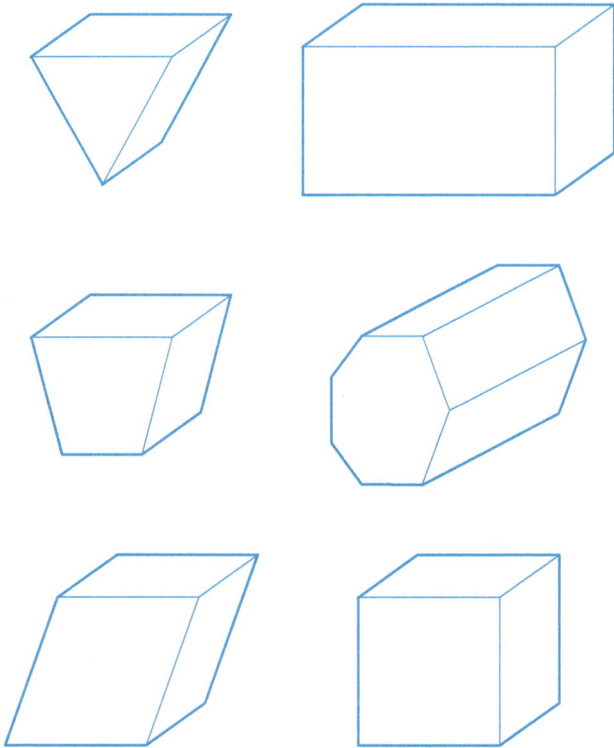

Figure 39 Prisms.

22207-12_F39.EPS

Step 1 Identify the shape of the base.

Step 2 Identify the dimensions of the shape.

Step 3 Calculate the area of the base.

Step 4 Calculate the volume of the prism.

Look carefully at the shape in *Figure 40A* and then perform Steps 1 through 4.

Step 1 Identify the shape of the base.

The base shape is a triangle.

Step 2 Identify the dimensions of the shape.

The dimensions are base = 3 inches, height = 2 inches, and depth = 4 inches.

Step 3 Calculate the area of the base.

$$\text{Area} = \tfrac{1}{2}\,bh$$
$$\text{Area} = \tfrac{1}{2}(3 \times 2)$$
$$\text{Area} = \tfrac{1}{2}(6)$$
$$\text{Area} = 3 \text{ inches}^2$$

Step 4 Calculate the volume of the prism.

$$\text{Volume} = \text{area} \times \text{depth}$$
$$\text{Volume} = 3 \text{ inches}^2 \times 4 \text{ inches}$$
$$\text{Volume} = 12 \text{ inches}^3$$

(A)

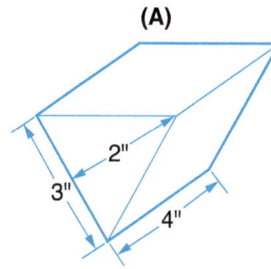

1. Base = Triangle
2. Triangle B = 3", and H = 2"
3. Surface Area = ½ BH
   $$= \tfrac{1}{2}\,(3" \times 2")$$
   $$= \tfrac{1}{2}\,(6")$$
   $$= 3 \text{ in}^2$$
4. Prism D = 4"
5. Volume = 3 in² × 4"
   $$= 12 \text{ in}^3$$

(B)

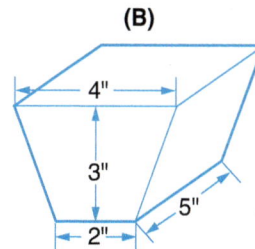

1. Base = Trapezoid
2. Triangle B_1 = 2", B_2 = 4", and H = 3"
3. Surface Area = ½ (2" + 4") × 3"
   $$= \tfrac{1}{2}\,(6") \times 3"$$
   $$= 3" \times 3"$$
   $$= 9 \text{ in}^2$$
4. Prism D = 5"
5. Volume = 9 in² × 5"
   $$= 45 \text{ in}^3$$

22207-12_F40.EPS

Figure 40 Finding the volume of a prism.

Repeat the procedure for the shape shown in *Figure 40B*.

Step 1 Identify the shape of the base.

The base shape is a trapezoid.

Step 2 Identify the dimensions of the shape.

The dimensions are base 1 = 2 inches, base 2 = 4 inches, height = 3 inches, and depth = 5 inches.

Step 3 Calculate the area of the base.

$$\text{Area} = \tfrac{1}{2}(\text{base 1} + \text{base 2}) \times \text{height}$$
$$\text{Area} = \tfrac{1}{2}(2 + 4) \times 3$$
$$\text{Area} = \tfrac{1}{2}(6) \times 3$$
$$\text{Area} = 9 \text{ inches}^2$$

Step 4 Calculate the volume of the prism.

$$\text{Volume} = \text{area} \times \text{depth}$$
$$\text{Volume} = 9 \text{ inches}^2 \times 5 \text{ inches}$$
$$\text{Volume} = 45 \text{ inches}^3$$

4.3.0 Cylinders

A cylinder is a three-dimensional object with a circle as its base (*Figure 41*). The formula to calculate the area of a circle is πr^2, and the formula to calculate the volume of a cylinder is $\pi r^2 h$.

Cylinder A:

$$\text{Diameter} = 4 \text{ feet, so radius} = 2 \text{ feet}$$
$$\text{Height} = 4 \text{ feet}$$
$$\text{Volume} = \pi r^2 h$$
$$= (3.14)(2^2)(4)$$
$$= (3.14)(4)(4)$$
$$= 50.24 \text{ feet}^3$$

Cylinder B:

$$\text{Diameter} = 2 \text{ feet, so radius} = 1 \text{ foot}$$
$$\text{Height} = 2 \text{ feet}$$
$$\text{Volume} = \pi r^2 h$$
$$= (3.14)(1^2)(2)$$
$$= (3.14)(1)(2)$$
$$= 6.28 \text{ feet}^3$$

Cylinder C:

$$\text{Radius} = 3 \text{ feet}$$
$$\text{Height} = 5 \text{ feet}$$
$$\text{Volume} = \pi r^2 h$$
$$= (3.14)(3^2)(5)$$
$$= (3.14)(9)(5)$$
$$= 141.3 \text{ feet}^3$$

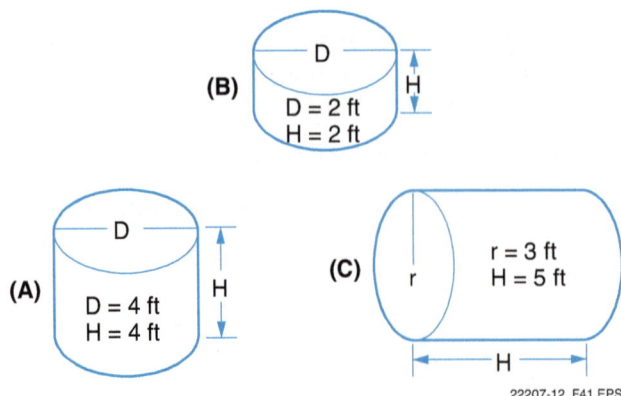

Figure 41 Cylinders.

4.3.1 Volume Exercises

The following exercises will help you to better understand how to calculate the volume of an object.

1. Calculate the volume for an object that is 2 feet in length, width, and depth.

2. An object has a rectangular base with a area of 6 square feet and a depth of 11 feet. Calculate the volume of the object.

3. Calculate the volume of an object with a triangular base of 2 feet and a height of 6 feet. The object has a depth of 6 inches.

4. An object has a base shape of a triangle with a base of 6 inches and a height of 1½ inches. The object is 3 inches deep. Calculate the volume of the object.

5. An object has a trapezoidal base with the dimensions of base 1 = 5 feet, base 2 = 11 feet, and height = 2 feet. The object's depth is 25 feet. Calculate the volume.

6. Calculate the volume of a cylinder with a diameter of 2 feet and a height of 6 feet.

7. Calculate the volume of the shape shown in *Figure 42*.

8. Calculate the volume of a cylinder that has a radius of 2 yards and a height of 3 yards.

4.4.0 Calculating the Volume and Weight of an Excavation

Excavations are measured in cubic yards of soil. You have already learned that 1 cubic yard contains 27 cubic feet and that those 27 cubic feet can take any three-dimensional shape and still have a volume of 1 cubic yard (*Figure 43*). Each of the shapes shown in the figure has a volume of 1 cubic yard or 27 cubic feet. Since volume is the measurement of length, width, and depth, all three dimensions must be considered to calculate volume.

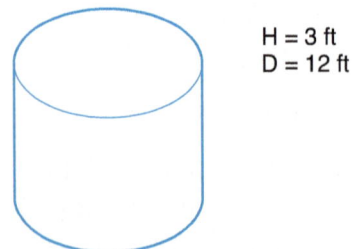

Figure 42 Cylinder Exercise 7.

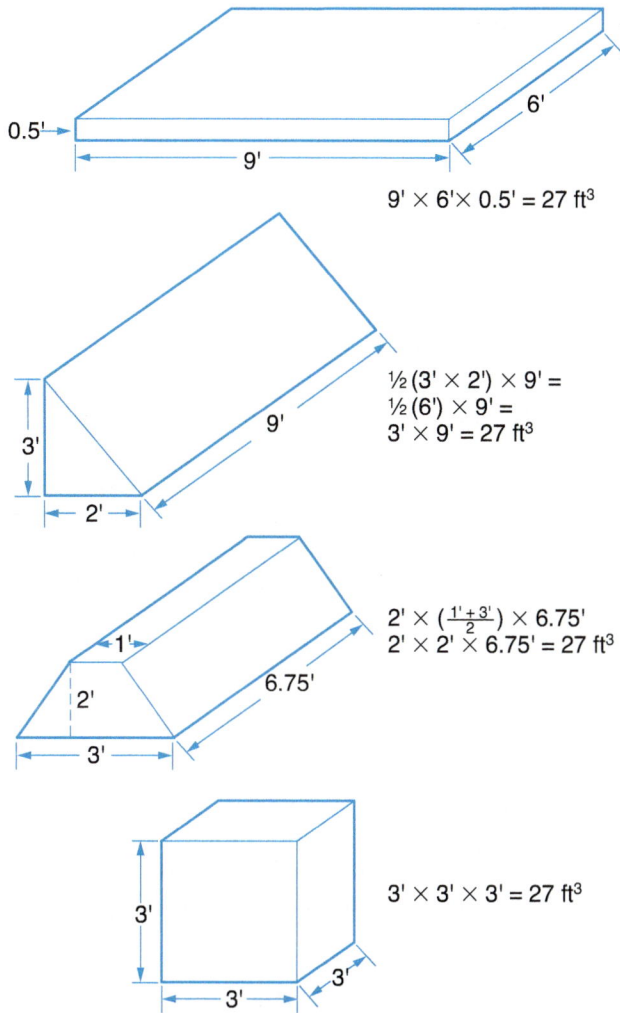

$9' \times 6' \times 0.5' = 27 \text{ ft}^3$

$\frac{1}{2}(3' \times 2') \times 9' =$
$\frac{1}{2}(6') \times 9' =$
$3' \times 9' = 27 \text{ ft}^3$

$2' \times \left(\frac{1' + 3'}{2}\right) \times 6.75'$
$2' \times 2' \times 6.75' = 27 \text{ ft}^3$

$3' \times 3' \times 3' = 27 \text{ ft}^3$

22207-12_F43.EPS

Figure 43 One cubic yard equals 27 cubic feet.

The capacity of heavy equipment is rated by both weight and volume. Operating an over-loaded vehicle creates the risk of an accident, so it is important to consider the weight of excavated material as well as the volume when loading the equipment. *Table 1* shows the weights of various materials. To calculate the weight of an excavation, multiply the volume in cubic yards by the material's weight per cubic yard.

Calculating volumes for cuts and fills can be time consuming, but by following a few easy steps, you can soon become skilled at it. First, to estimate excavation volume, assume that the ground is flat. (A way to calculate volume for uneven surfaces is described later in this module.) Second, study the area so it can be divided it into manageable shapes. Third, determine what information you need to make the volume calculations. Finally, gather the needed information. Once you know what shapes to use and have collected the dimensions, you can calculate the volume.

Table 1 Material Weights

Material	Weight in Pounds per Cu Yd (27 Cu Ft)
Clay, dry in lumps	1,701
Clay, compact	2,943
Earth, loamy, dry, loose	2,025
Earth, dry, packed	2,565
Earth, wet	2,970
Gravel, dry, loose	2,970
Gravel, dry, packed	3,051
Gravel, wet, packed	3,240
Limestone, fine	2,700
Limestone, 1½ to 2 inches	2,295
Limestone, above 2 inches	2,160
Sand, dry, loose	2,565
Sand, wet, packed	3,240

The construction plans shown in this section are simplified to make it easier to find the information needed to calculate volumes. Once you gain some experience, you will be able to quickly find the information on actual building plans. Look at the high-way plan in *Figure 44*. The cross-section is clearly a trapezoid. (It can also be divided into a rectangle and two triangles, but then three calculations must be performed. There is only one calculation needed if a trapezoid is used.) The three-dimensional object of a trapezoid is a prism. See *Figure 45*.

Since it is assumed that the existing grade is even, the job will involve only fill. To calculate the volume of a prism, measurements are needed for the following:

- Base 1
- Base 2
- Height
- depth

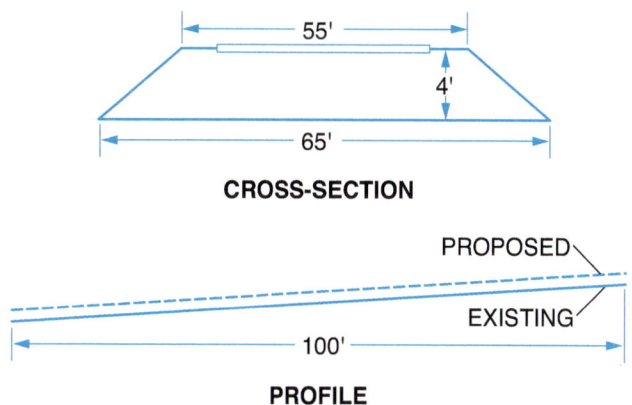

CROSS-SECTION

PROFILE

22207-12_F44.EPS

Figure 44 Highway plan.

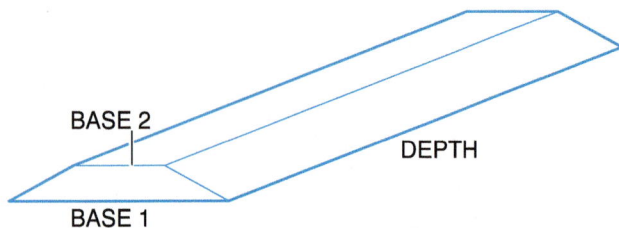

BASE 2

DEPTH

BASE 1

22207-12_F45.EPS

Figure 45 Highway plan as a prism.

Carefully examine the diagrams shown in *Figures 44* and *45*. The cross-section is a trapezoid with the following measurements:

Base 1 = 65 feet

Base 2 = 55 feet

Height = 4 feet

The profile shows that this section of roadway is 100 feet in length, so the trapezoid has a depth of 100 feet.

Example

Use the following dimensions to calculate the amount of fill needed to construct this roadway:

Base 1 = 65 feet

Base 2 = 55 feet

Height = 4 feet

Depth = 100 feet

The volume is calculated as follows:

Volume = ½ × (base 1 + base 2) × height × depth

Volume = ½ × (65 + 55) × 4 × 100

Volume = ½ × 120 × 4 × 100

Volume = 24,000 feet³

Volume = 24,000 ÷ 27 cubic feet/cubic yard

Volume = 888.9 yards³

When the weight of the fill is 3,200 pounds per cubic yard, the fill weight is calculated as follows:

888.9 × 3,200 = 2,844,480 pounds

As stated earlier, the only way to become proficient at performing estimating calculations is to do them. The following examples will help explain how to perform calculations for other excavation jobs. The examples become increasingly complex, so be sure to read each example and study the associated figures carefully to be sure you understand the result.

4.4.1 Excavating a Simple Foundation

Look at the foundation plan in *Figure 46*. The pad is a simple rectangular object. See *Figure 47*. The footer has the shape of a trapezoid, so its three-dimensional figure is a prism. (The footer can be divided into a rectangular solid and a prism, too).

Calculate the volume of the earth that needs to be excavated so the building can be constructed. To calculate the total volume, first calculate the volumes of all the objects in which the foundation has been divided and then add the volumes of the objects to arrive at the total excavation volume.

All the information needed to calculate soil excavations is usually not readily found on the building plans. Plans are drawn so that all construction tasks can be performed from a single set of plans. Do not become overwhelmed with all the information on the plans. Look for only the information that you need.

The measurements shown on these plans are in feet and inches, rather than the typical decimal to make the example easier to understand. Since the pad is a rectangular object, the length, width, and depth of the excavation are needed in order to calculate its volume. The length and width are the same as the pad dimensions, which are 30 feet by 30 feet. The pad is 8 inches thick. To calculate the beginning elevation, you need to know the pad's finished elevation, which is 626 feet and 6 inches, and the existing elevation of the building site, which is 626 feet and 4 inches (*Figure 48*). This means that the pad needs to begin at an elevation of 625 feet and 10 inches. This is 6 inches below the existing elevation, so the depth of the excavation is 6 inches.

Unsuitable Soil

Excavation will sometimes uncover soil that is unsuitable for use on the site because it contains too much moisture. This type of soil is easily recognized because the ground will be mushy and will ripple when equipment rolls over it. The initial testing by soil engineers should locate such soil, but patches of it may turn up during excavation. In such cases, the equipment operator should consult a supervisor or engineer before proceeding because it may be necessary to remove the soil and fill the area. In many cases, this work is beyond the scope of the contract, and would require a renegotiation to fund the additional work.

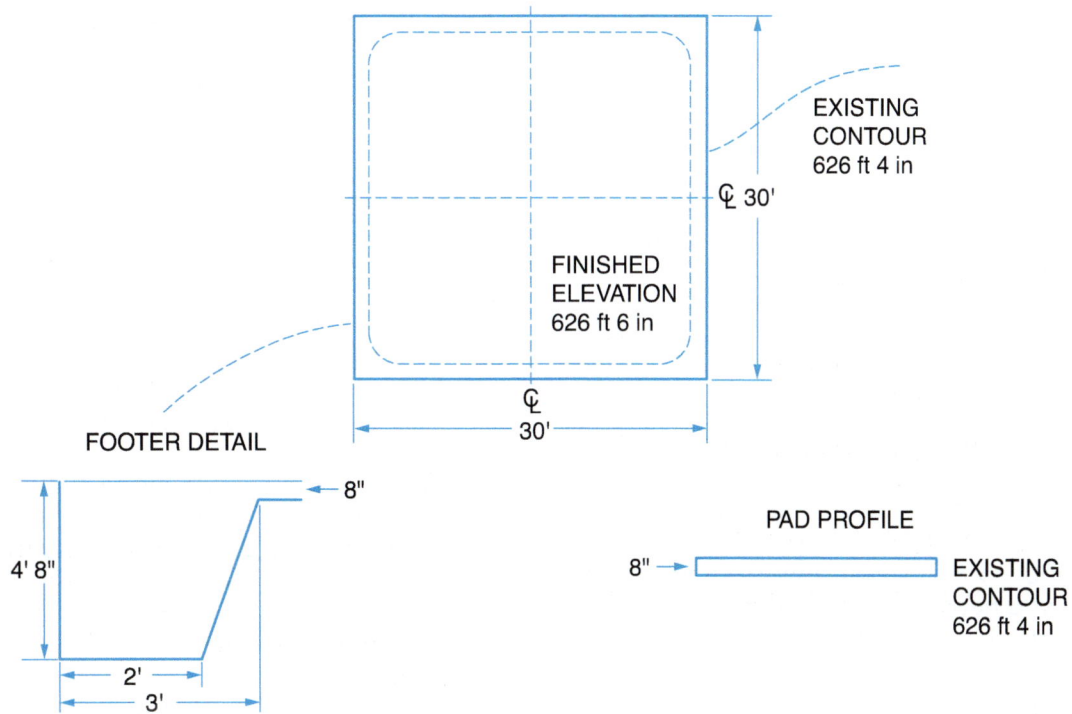

EXISTING
CONTOUR
626 ft 4 in

C 30'

FINISHED
ELEVATION
626 ft 6 in

C
30'

FOOTER DETAIL

8"

4' 8"

2'

3'

PAD PROFILE

8"

EXISTING
CONTOUR
626 ft 4 in

22207-12_F46.EPS

Figure 46 Foundation plan.

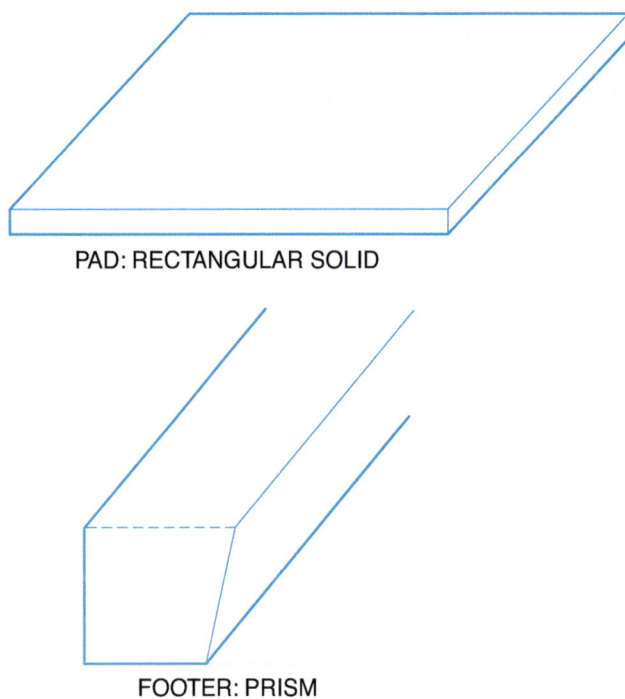

PAD: RECTANGULAR SOLID

FOOTER: PRISM

22207-12_F47.EPS

Figure 47 Foundation plan as a rectangular object and a prism.

You now have all of the information needed to calculate the excavation volume for the slab, which is as follows:

$$\text{Volume} = \text{length} \times \text{width} \times \text{depth}$$

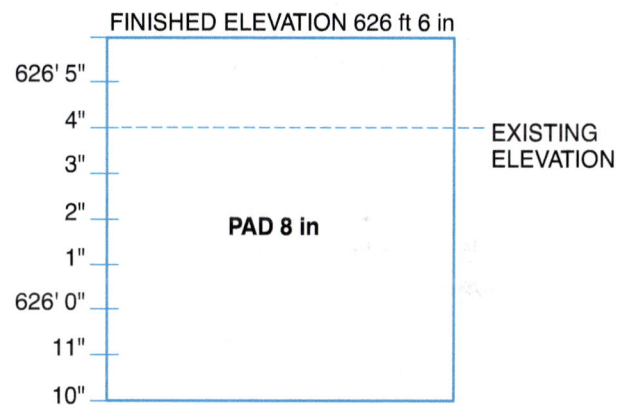

FINISHED ELEVATION 626 ft 6 in

626' 5"

4"

3"

2"

1"

626' 0"

11"

10"

EXISTING
ELEVATION

PAD 8 in

22207-12_F48.EPS

Figure 48 Existing elevation, finished elevation.

Length = 30 feet

Width = 30 feet

Depth = 6 inches = 0.5 feet
(convert inches to feet)

Volume = 30 × 30 × 0.5

Volume = 450 feet^3 ÷ 27 feet^3
(convert cubic feet to cubic yards)

Volume = 16.67 yards^3

Looking at the footer plans (footer detail in *Figure 46*) and the prism that represents the footer (*Figure 47*), it can be seen that the bottom base is 2 feet and the top base is 3 feet. The height is a little harder to calculate. The total height of the footer from its beginning to the finished surface of the pad is 4 feet 8 inches. The pad is 8 inches, so the height of the footer must be 4 feet. Therefore, the following dimensions are known:

Base 1 = 2 feet
Base 2 = 3 feet
Height = 4 feet

You still need to know the lengths of the footers to calculate volume. See *Figure 49*. There is a footer along each edge of the pad, so there are four footers. Each footer is 3 feet wide, so each footer is 27 feet long (30 − 3 = 27). You now have all the information you need to calculate the volume of the footer prisms. It is as follows:

Base 1 = 2 feet
Base 2 = 3 feet
Height = 4 feet
Length = 7 feet

$$\text{Volume} = \tfrac{1}{2} \times (\text{base 1} + \text{base 2}) \times \text{height} \times \text{depth}$$

$$= \tfrac{1}{2} \times (2 + 3) \times 4 \times 27$$

$$= \tfrac{1}{2} \times 5 \times 4 \times 27$$

$$= 270 \text{ feet}^3$$

$$= 270 \div 27 \text{ feet}^3$$

(convert square feet to cubic yards)

$$= 10 \text{ yards}^3 \text{ per footer}$$

Total excavation volume is calculated by adding the four footer volumes and the pad volume as follows:

$$\text{Total volume} = (4 \times 10 \text{ yards}^3) + 16.67 \text{ yards}^3$$

$$\text{Total volume} = 40 \text{ yards}^3 + 16.67 \text{ yards}^3$$

$$\text{Total volume} = 56.67 \text{ yards}^3$$

Referring to *Table 1*, the weight of the foundation excavation can be determined by multiplying the total excavation in cubic yards by the weight of the material per cubic yard. A comparison of the weight of loose earth, packed earth, and wet earth is as follows:

Earth (dry, loose) 2,025 lb/cu yd

$$56.67 \times 2,025 = 114,756.75 \text{ lbs}$$

Earth (dry, packed) 2,565 lb/cu yd

$$56.67 \times 2,565 = 145,358.55 \text{ lbs}$$

Earth (wet) 2,970 lb/cu yd

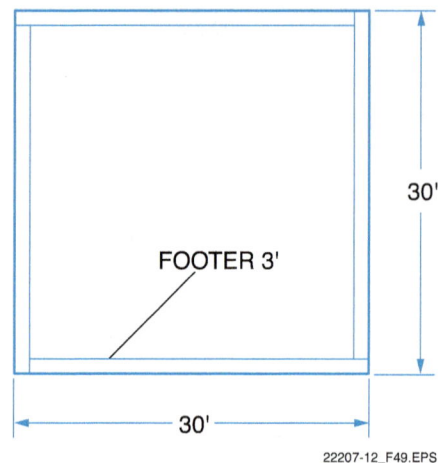

Figure 49 Footer length.

$$56.67 \times 2,970 = 168,309.9 \text{ lbs}$$

4.4.2 Excavating Slopes

The diagram in *Figure 50* shows a slope stake. It requires that a 2-foot cut be made at a 3:1 slope on level ground. This means for every 3 feet of horizontal travel, the elevation of the ground must drop 1 foot up to the desired 2-foot cut. This cut forms a triangular-shaped cut. The cut section is 100 feet long.

The prism's volume is calculated as follows (see *Figure 51*):

$$\text{Height} = 2 \text{ feet}$$
$$\text{Base} = 6 \text{ feet}$$
$$\text{Depth} = 100 \text{ feet}$$

$$\text{Volume} = \tfrac{1}{2} \times \text{base} \times \text{height} \times \text{depth}$$

$$\text{Volume} = \tfrac{1}{2} \times 6 \times 2 \times 100$$

$$= 600 \text{ feet}^3$$

$$= 600 \text{ feet}^3 \div 27 \text{ feet}^3$$

(convert square feet to cubic yards)

$$= 22.22 \text{ yards}^3$$

When the excavated material weights 1,200 pounds per cubic yard, the weight of the excavation material is calculated as follows:

$$22.22 \times 1,200 = 26,664 \text{ pounds}$$

4.4.3 Excavating a Complex Foundation

Study the foundation plan in *Figure 52*. This shape can be divided into three to five shapes. With a complex plan such as this one, time can be saved by calculating the area of the entire foundation before calculating the volume.

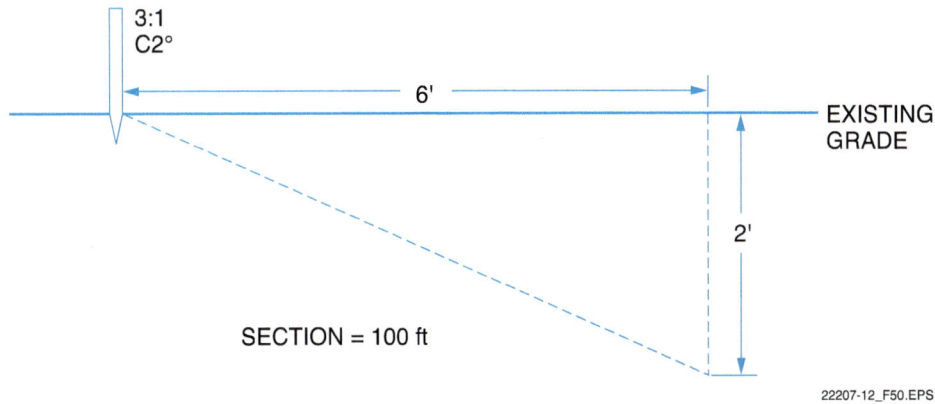

Figure 50 Excavating a slope.

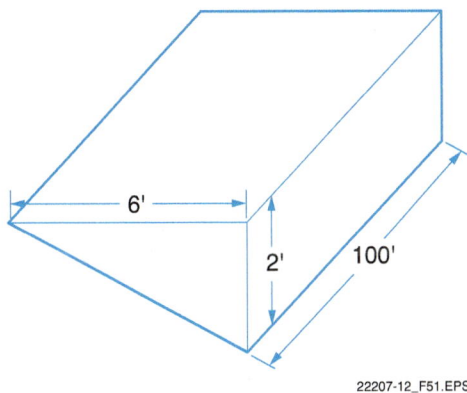

Figure 51 Excavating a slope as a prism.

Figure 53A shows three shapes: two trapezoids and one rectangle. *Figure 53B* shows five shapes: two triangles and three rectangles. It is better to see as many shapes as possible because building plans may not provide enough information to easily gather the measurements needed to calculate volume. In this case, the dimensions shown on *Figure 52* are sufficient to calculate the area of all of the shapes shown in *Figures 53A* and *53B*, so *Figure 53A* can be used because it has the fewest calculations.

When trying to gather information from the building plans, it is easy to become overwhelmed with all the information on the diagram. Concentrate on one shape at a time. For example, Shape 1 in *Figure 54* is a trapezoid; it is the left wing of the building shown in *Figure 52*. To calculate the area of Shape 1, the base 1, base 2, and height measurements are needed. Looking at *Figure 52*, it can be seen that base 1 is 80 feet and height is 40 feet. The base 2 measurement is not obvious so it needs to be calculated. The part of base 2 represented by the solid line is 20 feet and the part of base 2 represented by the dashed line is 30 feet, so the length is 50 feet.

Dealing with Contaminated Soil

When sites are being excavated, there is a possibility that the soil being removed is contaminated. The contamination can come from a variety of sources, including leachate from nearby landfills, contaminated surface water, or the prior use of the land as an industrial site. In the past, there were many manufacturing facilities that used toxic chemicals in their processes. These chemicals may have leached into the soil, causing it to become contaminated. Environmental laws regulate disposal of such soil, so it must be decontaminated before it can be moved to another location. Environmental testing is required to ensure that soil on a job site is not contaminated before it is excavated and removed to another location. Treatment approaches can include flushing contaminants out of the soil using water, chemical solvents, or air; destroying the contaminants by incineration; encouraging natural organisms in the soil to break them down; or adding material to the soil to encapsulate the contaminants and prevent them from spreading.

The dimensions for Shape 1 of *Figure 54* are as follows:

$$\text{Base 1} = 80 \text{ feet}$$
$$\text{Base 2} = 50 \text{ feet}$$
$$\text{Height} = 40 \text{ feet}$$

The area for Shape 1 is calculated as follows:

$$\text{Area} = \frac{1}{2}(\text{base 1} + \text{base 2}) \times \text{height}$$
$$\text{Area} = \frac{1}{2}(80 + 50) \times 40$$
$$\text{Area} = \frac{1}{2} \times 130 \times 40$$
$$\text{Area} = 2{,}600 \text{ feet}^2$$

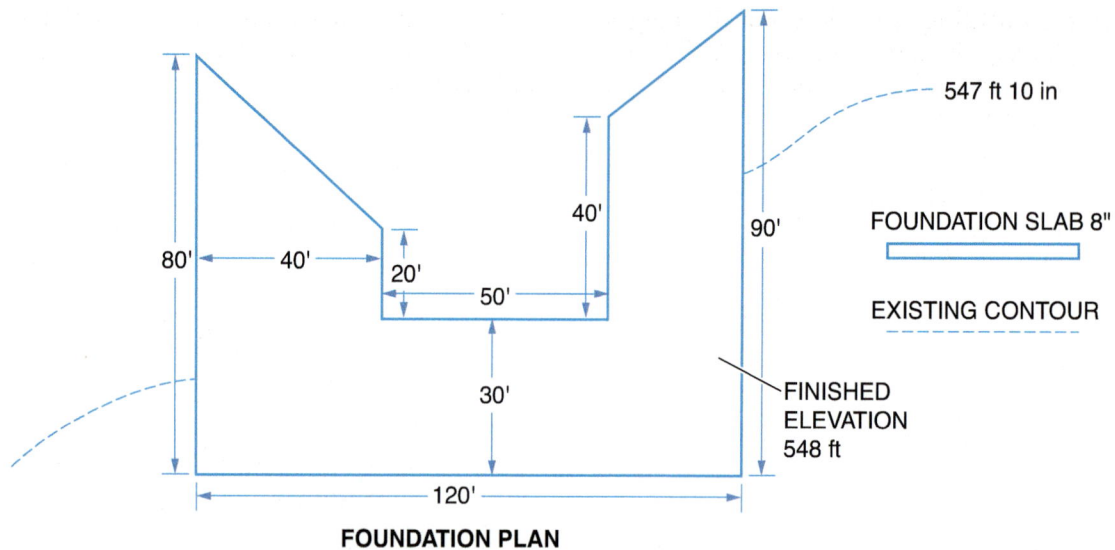

Figure 52 Complex foundation plan.

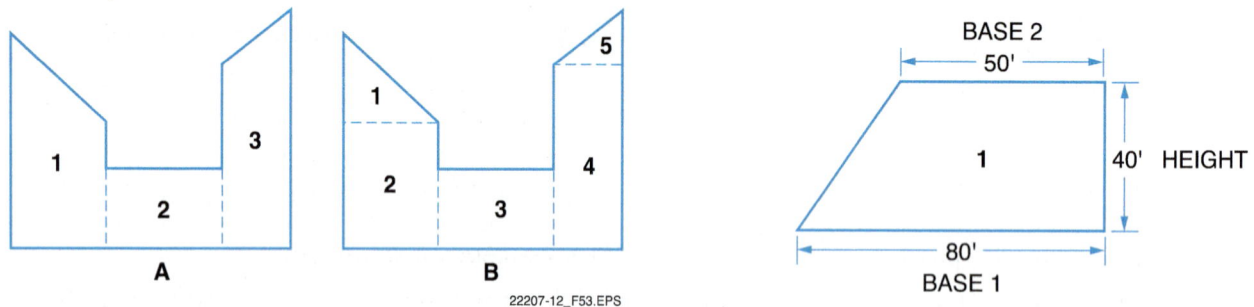

Figure 53 Complex foundation plan shapes.

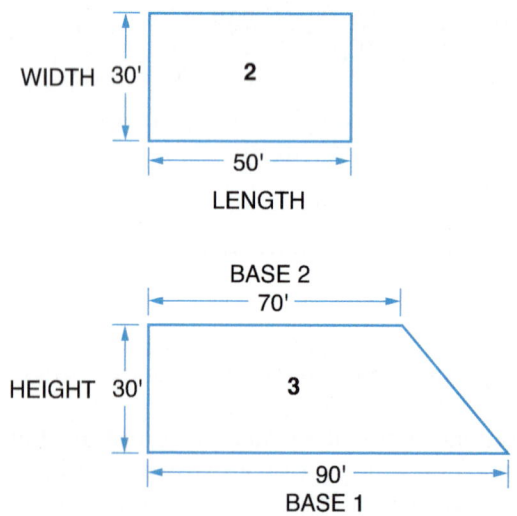

Figure 54 Three shapes.

Shape 2 in *Figure 54* is a little easier. Shape 2 is a rectangle and represents the main part of the building shown in *Figure 53*. You already used the width of the rectangle as part of base 2 in Shape 1, so you know the width is 30 feet. The length of 50 feet is shown at the top of the main building in *Figure 52*, so all the information needed to calculate the area of Shape 2, which is as follows, is available:

$$\text{Length} = 50 \text{ feet}$$
$$\text{Width} = 30 \text{ feet}$$
$$\text{Area} = \text{length} \times \text{width}$$
$$\text{Area} = 50 \times 30$$
$$\text{Area} = 1,500 \text{ feet}^2$$

Shape 3 in *Figure 54* is another trapezoid and represents the right wing of the building shown in *Figure 53*. Base 1 is on the far right side of the foundation and is 90 feet. Base 2 is made of the width of Shape 2 (30 feet) and the 40 feet measure on the left side of the wing (*Figure 53*), so base 2 is 70 feet.

Now the height of Shape 3 needs to be calculated. You know that the entire length of the face of the building is 120 feet (it is on *Figure 52*) and the face of the building is made up of the height of shape 1 (40 feet), the length of Shape 2 (50 feet) and the height of Shape 3. Since the dimensions of Shapes 1 and 2 are known the height of Shape 3 can be calculated as follows:

Face length =
height Shape 1 + length Shape 2
+ height Shape 3

120 feet = 40 feet + 50 feet + height Shape 3

120 feet = 90 feet + height Shape 3

120 feet − 90 feet =
90 feet + height Shape 3 − 90 feet

30 feet = height Shape 3

So the height of Shape 3 is 30 feet. This is all the information you need to calculate the area of Shape 3, as follows:

Base 1 = 90 feet
Base 2 = 70 feet
Height = 30 feet

Area = ½(base 1 + base 2) × height

So area = 2,400 square feet

You still need the depth of each object to calculate its volume. Look on the plans and find the thickness of the slab, the existing contour of the building site, and the finished elevation of the foundation. The slab is 8 inches thick and the finished elevation is 548 feet, so the slab must start at 547 feet 4 inches in elevation. Since the existing elevation is 547 feet 10 inches, 6 inches (or ½ foot) of earth needs to be excavated to achieve the finished elevation after the foundation is poured.

Now that depth has been determined, the volume of each object can be calculated. The answers are added together to get the total excavation volume.

Shape 1:

Volume = 2,600 feet^2 × ½ foot

Volume = 1,300 feet^3

Shape 2:

Volume = 1,500 feet^2 × ½ foot

Volume = 750 feet^3

Shape 3:

Volume = 2,400 feet^2 × ½ foot

Volume = 1,200 feet^3

Total volume:

1,300 feet^3 + 750 feet^3 + 1,200 feet^3 =
3,250 feet^3

3,250 feet^3 ÷ 27 feet^3 = 120.37 yards^3

Another way to figure out total volume is to add together the area of all three shapes and then multiply the total area by the depth. Calculate this yourself. You should get a total area of 6,500 square feet.

4.4.4 Complex Calculations

Up until now, the material has focused on how to calculate excavation volumes for uniformly shaped objects, but most building sites are irregularly shaped. There are a number of computer programs that can be used to calculate these volumes quickly and accurately, saving time and money. *Figure 55* shows a screen from one such program.

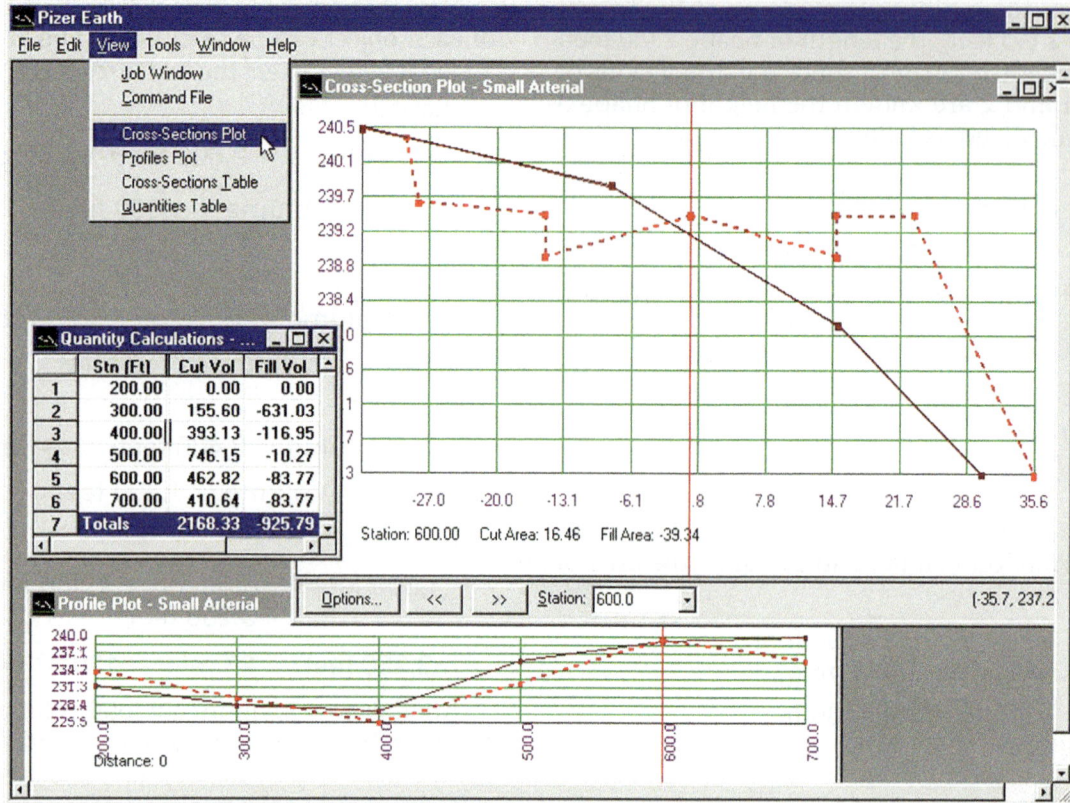

Figure 55 Cut and fill computer program.

You have already learned that you need to calculate the area of an object's base to calculate its volume. Another way to find the area of an irregular shape is to draw a grid over it. *Figure 56* is a benched trench. Each block of the grid in *Figure 57* represents 1 square foot. To estimate the area of the benched trench, you need to count the number of blocks (estimating the size of the partial blocks).

Usually, the ground at the building site is irregular (*Figure 58*), making it difficult to determine the excavation depth precisely. In this case, a more accurate estimation can be made by averaging the elevations of the site. This is done by measuring the elevations at several points and then adding these numbers and dividing by the number of points measured. In *Figure 59*, 14 points are measured; their sum is 93, so the average depth of the excavation is 6.64 inches, which is computed as follows:

$$93 \div 14 = 6.64 \text{ inches}$$

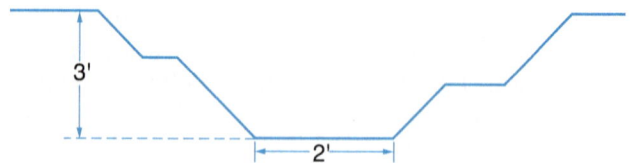

Figure 56 Benched trench.

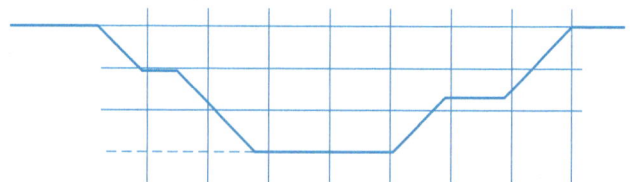

Figure 57 Benched trench with grid.

Figure 58 Irregular elevation.

NCCER – *Heavy Equipment Operations Level Two*

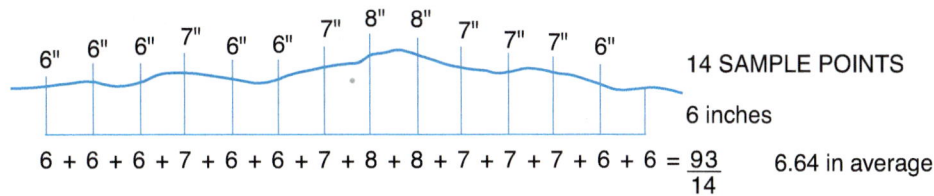

$$6 + 6 + 6 + 7 + 6 + 6 + 7 + 8 + 8 + 7 + 7 + 7 + 6 + 6 = \frac{93}{14}$$

14 SAMPLE POINTS

6 inches

6.64 in average

22207-12_F59.EPS

Figure 59 Irregular elevation averaged.

4.0.0 Section Review

1. In order to determine the volume of a cube, the product of the length and width are _____.

   a. divided by the depth
   b. multiplied by pi
   c. added to the depth
   d. multiplied by the depth

2. Which of the following is *not* required for a prism?

   a. Two bases
   b. Parallel bases
   c. It must be a cube
   d. Bases are the same size

3. A cylinder is a three-dimensional object with a circle as its base.

   a. True
   b. False

4. Which of the following values is *not* needed in order to calculate the volume of excavation material for a simple foundation?

   a. Depth
   b. Length
   c. Width
   d. Weight

SUMMARY

Calculating the volume and weight of excavations can be a time-consuming task, but it is necessary in order to determine the type of equipment needed on the job. To determine the volume of any excavation, you first calculate the area of a figure and then you use the area to calculate the volume of an object.

Excavations are three-dimensional, but they are usually based on one of several common two-dimensional shapes, such as the square, rectangle, triangle, or circle. Once you have determined the area of the base shape, the volume is calculated by multiplying the base area by the object's depth. Whenever you need to calculate the volume of a complex object, you need to break the object into familiar shapes and then calculate the volume of each shape. Once you have the volume of each shape, it is only a matter of adding them together to calculate the total volume of the object. Cut and fill volumes are measured in cubic yards, so after calculating the excavation volume, it may have to be converted to cubic yards.

The Pythagorean theorem is used to solve for unknown sides of a right triangle. In its basic form, this simple formula is designed to calculate the hypotenuse of the right triangle. It can also be manipulated to solve for any unknown side of the triangle. The 3-4-5 rule, which is derived from the Pythagorean theorem, is used to verify that the corner of a right triangle is perfectly square.

Overloading a vehicle can place the operator at risk for an accident and can also damage the vehicle. Therefore, it is important to use equipment that is rated for the weight of the material. Each type of material has its own weight, so once you have calculated the volume of an excavation, you will need to calculate its weight. This is done by multiplying the total volume in cubic yards by the material weight per cubic yard.

1. It is important to be able to calculate the volume and weight of cut and fill material because _____.
   a. the material cost is established by weight and volume
   b. all heavy equipment is rated by weight and volume
   c. it determines the number of workers needed on a job
   d. trucks use the least amount of fuel when overloaded

2. It is necessary to know the area of a shape in order to calculate its volume.
   a. True
   b. False

3. When an equation has numbers grouped in parentheses, that calculation is performed _____.
   a. first
   b. second
   c. third
   d. last

4. The symbol π in an equation is considered a _____.
   a. variable
   b. root
   c. constant
   d. radius

5. The square root of the number 25 is _____.
   a. 5
   b. 25
   c. 50
   d. 625

6. A right angle contains _____.
   a. 45 degrees
   b. 90 degrees
   c. 180 degrees
   d. 360 degrees

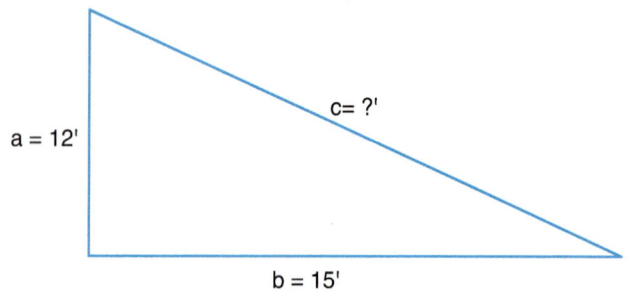

c= ?'

a = 12'

b = 15'

22207-12_RQ01.EPS

Figure 1

7. The length of the hypotenuse in *Figure 1* is _____ .
   a. 27 feet
   b. 19.2 feet
   c. 365 feet
   d. 14 feet

8. Two-dimensional objects can be measured in all of the following values, *except* square _____.
   a. inches
   b. feet
   c. yards
   d. radius

9. One square yard is equal to _____.
   a. 3 square feet
   b. 9 square feet
   c. 27 square feet
   d. 144 square feet

10. One side of a square is 6 inches, so its area is _____.
    a. 3 square inches
    b. 12 square inches
    c. 24 square inches
    d. 36 square inches

11. A rectangle has a length of 6 inches and a width of 2 inches, so its area is _____.
    a. 6 square inches
    b. 12 square inches
    c. 16 square inches
    d. 36 square inches

12. Calculate the area for a triangle with a base of 7 inches and a height of 4 inches.

    a. 12 square inches
    b. 14 square inches
    c. 22 square inches
    d. 28 square inches

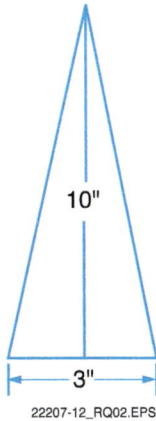

22207-12_RQ02.EPS

Figure 2

13. The area for the shape shown in *Figure 2* is _____.

    a. 3 square inches
    b. 15 square inches
    c. 30 square inches
    d. 45 square inches

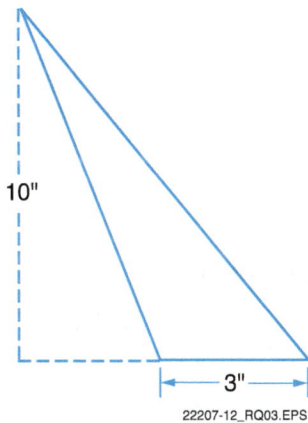

22207-12_RQ03.EPS

Figure 3

14. The area for the shape shown in *Figure 3* is _____.

    a. 3 square inches
    b. 15 square inches
    c. 30 square inches
    d. 45 square inches

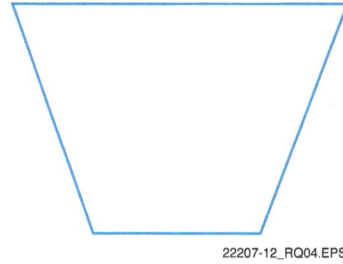

22207-12_RQ04.EPS

Figure 4

15. The shape shown in *Figure 4* is a _____.

    a. rhombus
    b. parallelogram
    c. trapezoid
    d. rectangle

16. A trapezoid's measurements are base 1 = 2 inches, base 2 = 4 inches, and height = 4 inches. What is its area?

    a. 3 square inches
    b. 12 square inches
    c. 30 square inches
    d. 45 square inches

17. The distance from the center of a circle to the edge of its curved line is called the _____.

    a. angle
    b. circumference
    c. diameter
    d. radius

18. A circle has a diameter of 2 feet. What is its area?

    a. 3.14 square feet
    b. 6.28 square feet
    c. 9.42 square feet
    d. 45.5 square feet

19. A backhoe loader's bucket can hold 7.0 cubic feet or up to 134 pounds, so its volume is _____.

    a. 7.0 cubic feet
    b. 19 pounds per foot
    c. 134 pounds
    d. 938 cubic feet per pound

20. A cube that has a dimension of 1 yard has a volume of _____.

    a. 1 cubic foot
    b. 9 cubic feet
    c. 27 cubic feet
    d. 2 cubic yards

21. A rectangular object has the dimensions of length = 2 feet, width = 4 feet, depth = 3 feet. It has a volume of _____.

    a. 6 cubic feet
    b. 8 cubic feet
    c. 12 cubic feet
    d. 24 cubic feet

22. A three-dimensional object with two identical triangular bases has the following dimensions: base = 9 inches, height = 10 inches, and depth = 2 feet. It has a volume of _____.

    a. 90 cubic inches
    b. 180 cubic inches
    c. 1,080 cubic inches
    d. 2,160 cubic inches

23. A three-dimensional object has a depth of 8 inches and a triangular base that has an area of 12 square inches, so it has a volume of _____.

    a. 8 cubic inches
    b. 20 cubic inches
    c. 48 cubic inches
    d. 96 cubic inches

24. A trapezoidal object has the following dimensions: base 1 = 2 feet, base 2 = 3 feet, height = 2 feet, and depth = 6 feet, so it has a volume of _____.

    a. 5 cubic feet
    b. 25 cubic feet
    c. 30 cubic feet
    d. 72 cubic feet

25. A cylinder has a diameter of 6 feet and a height of 2 feet, so its volume is _____.

    a. 12 cubic feet
    b. 28.3 cubic feet
    c. 56.5 cubic feet
    d. 226 cubic feet

26. A material has a weight of 1,700 pounds per cubic yard, and there are 54 cubic feet to move. What is the total weight of the material?

    a. 3,400 pounds
    b. 10,200 pounds
    c. 30,600 pounds
    d. 91,800 pounds

27. Wet excavation material is lighter than dry material.

    a. True
    b. False

28. An excavation has a volume of 203 cubic feet. The material weighs 1,200 pounds per cubic yard. What is the weight of the excavation?

    a. 2,700 pounds
    b. 9,000 pounds
    c. 81,120 pounds
    d. 243,600 pounds

29. A triangle-shaped slope excavation has a base of 4 feet, height of 6 feet, and length of 100 feet. Its volume is _____.

    a. 800 cubic feet
    b. 1,200 cubic feet
    c. 2,000 cubic feet
    d. 2,400 cubic feet

30. A complex excavation has a depth of 6 inches and can be broken into three shapes with areas of 100, 226, and 300 square feet. The total volume of the excavation is _____.

    a. 104 cubic feet
    b. 313 cubic feet
    c. 626 cubic feet
    d. 3,756 cubic feet

Average: The middle point between two numbers or the mean of two or more numbers. It is calculated by adding all numbers together, and then dividing the sum by the quantity of numbers added. For example, the average (or mean) of 3, 7, 11 is 7 (3 + 7 + 11 = 21; 21 ÷ 3 = 7).

Constant: A value in an equation that is always the same; for example pi is always 3.14.

Hypotenuse: The long dimension of a right triangle and always the side opposite the right angle.

Parallel: Two lines that are always the same distance apart even if they go on into infinity (forever is called infinity in mathematics).

Parallelogram: A two-dimensional shape that has two sets of parallel lines.

Quadrilateral: A four-sided, closed shape with four angles whose sum is 360 degrees.

Squared: Multiplied by itself.

Variable: A value in an equation that depends on the factors being considered; for example, the lengths of the sides of a triangle may vary from one triangle to another.

Additional Resources

This module presents thorough resources for task training. The following resource material is suggested for further study.

Applied Construction Math: A Novel Approach. Alachua, FL: NCCER.

Figure Credits

Reprinted courtesy of Caterpillar Inc., Module opener

Topaz Publications, Inc., Figure 1

Courtesy of Pizer Inc., Figure 55

Section Review Answers

Answer	Section Reference	Objective
Section One		
1 a	1.0.0	1a
2 d*	1.1.0	1b
3 a	1.3.0	1c
Section Two		
1 b*	2.1.0	2a
2 b*	2.2.0	2b
Section Three		
1 c*	3.1.0	3a
2 c*	3.2.0	3b
3 b	3.3.0	3c
4 b	3.4.0	3d
Section Four		
1 d	4.1.0	4a
2 c	4.2.0	4b
3 a	4.3.0	4c
4 d	4.4.1	4d

Section Review Calculations

1-1. $4 \times 4 \times 4 = 64$

2-1.
$$b = \sqrt{c^2 + a^2}$$
$$b = \sqrt{720^2 - 530^2}$$
$$b = \sqrt{518,400 - 280,900}$$
$$b = \sqrt{237,500}$$
$$b = 487'$$

SR01.EPS

2-2.
$24 \div 3 = 8$

$8 \times 4 = 32$

$8 \times 5 = 40$

$24 - 32 = 40$

3-1. $15 \times 12 = 180$

3-2. Triangle 3:

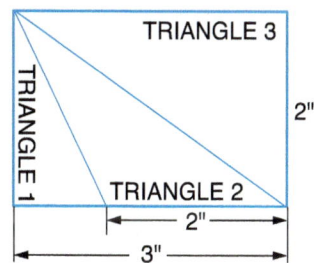

22207-12_SR03.EPS

base = 3 inches

height = 2 inches

Area = ½ × (3 × 2)

= ½ × 6

= 3 square inches

NCCER CURRICULA — USER UPDATE

NCCER makes every effort to keep its textbooks up-to-date and free of technical errors. We appreciate your help in this process. If you find an error, a typographical mistake, or an inaccuracy in NCCER's curricula, please fill out this form (or a photocopy), or complete the online form at **www.nccer.org/olf**. Be sure to include the exact module ID number, page number, a detailed description, and your recommended correction. Your input will be brought to the attention of the Authoring Team. Thank you for your assistance.

Instructors – If you have an idea for improving this textbook, or have found that additional materials were necessary to teach this module effectively, please let us know so that we may present your suggestions to the Authoring Team.

NCCER Product Development and Revision

13614 Progress Blvd., Alachua, FL 32615

Email: curriculum@nccer.org
Online: www.nccer.org/olf

❏ Trainee Guide   ❏ Lesson Plans   ❏ Exam   ❏ PowerPoints   Other _____

Craft / Level: _____   Copyright Date: _____

Module ID Number / Title: _____

Section Number(s): _____

Description: _____

_____

_____

_____

_____

Recommended Correction: _____

_____

_____

_____

_____

Your Name: _____

Address: _____

_____

Email: _____   Phone: _____

22209-13

Interpreting Civil Drawings

OVERVIEW

Civil drawings define the excavation and grading requirements for roads and building construction sites. The ability to interpret those drawings is important for all personnel involved in site work.

Module Eight

Objectives

When you have completed this module, you will be able to do the following:

1. Describe the types of drawings usually included in a set of plans and list the information found on each type.
   a. Explain the use of title sheets, title blocks, and revision blocks.
   b. Describe the types of drawings used in highway construction.
   c. Describe the types of drawings used in building site construction.
   d. Describe how as-built drawings are prepared.
2. Read and interpret drawings.
   a. Identify different types of lines and symbols used on drawings.
   b. Define common abbreviations used on drawings.
   c. Interpret building site and highway drawings to determine excavation requirements.
3. Explain specifications and the purpose of specifications.
   a. Identify the types of information contained in specifications.
   b. Explain the common format used in specifications.

Performance Tasks

Under the supervision of your instructor, you should be able to do the following:

1. Determine the scale of different drawings.
2. Interpret a set of drawings to determine the proper type and sequence of excavation and grading operations needed to prepare the site.

Trade Terms

Change order
Contour lines
Easement
Elevation view
Invert
Loadbearing

Monuments
Plan view
Property lines
Request for information
Setback
Uniform Construction Index

Industry Recognized Credentials

If you are training through an NCCER-accredited sponsor, you may be eligible for credentials from NCCER's Registry. The ID number for this module is 22209-13. Note that this module may have been used in other NCCER curricula and may apply to other level completions. Contact NCCER's Registry at 888.622.3720 or go to **www.nccer.org** for more information.

Contents ————————————————————

Topics to be presented in this module include:

Figures

1.0.0 TYPES OF CIVIL DRAWINGS

Objective 1

Describe the types of drawings usually included in a set of plans and list the information found on each type.

a. Explain the use of title sheets, title blocks, and revision blocks.
b. Describe the types of drawings used in highway construction.
c. Describe the types of drawings used in building site construction.
d. Describe how as-built drawings are prepared.

Trade Terms

Change order: A formal instruction describing and authorizing a project change.

Contour lines: Imaginary lines on a site/plot plan that connect points of the same elevation. Contour lines never cross each other.

Easement: A legal right-of-way provision on another person's property (for example, the right of a neighbor to build a driveway or a public utility to install water and gas lines on the property). A property owner cannot build on an area where an easement has been identified.

Elevation view: A drawing giving a view from the front or side of a structure.

Loadbearing: A base designed to support the weight of an object of structure.

Monuments: Physical structures that mark the locations of survey points.

Plan view: A drawing that represents a view looking down on an object.

Property lines: The recorded legal boundaries of a piece of property.

Request for information (RFI): A form used to question discrepancies on the drawings or to ask for clarification.

Setback: The distance from a property line in which no structures are permitted.

Every construction project is defined in detail by a set of drawings. In addition, many projects have written specifications that contain further details about the quality of work and the materials to be used. The drawings and specifications are prepared by architects and/or engineers and become part of the construction contract.

During the planning phase of a new construction project, architects and engineers develop detailed plans for the project using data gathered by the survey team. The survey team verifies the site boundaries, which are established by a licensed surveyor, and then set up the construction project's precise position in relation to the property boundaries. In addition, they establish the exact location of road and utility rights-of-way (ROW), easements, and other important features of the project. This information is reflected on the project drawings.

Equipment operators must be able to interpret construction drawings and specifications correctly. Failure to do so may result in costly rework and unhappy customers. Depending on the severity of the mistake, it can also expose you and your employer to legal liability.

An equipment operator's duties vary with the type and size of the construction project. Highway construction jobs may require the constant use of heavy equipment for hauling fill, cutting grades, leveling the roadbed, and spreading paving materials. Residential or commercial building projects may require the intense use of heavy equipment at the beginning of the project to prepare the site for construction and at end of the project to complete final grades and landscaping tasks.

Depending on the type of project, preparation of the project drawings may be the responsibility of an architect or an engineering firm. A complete set of project drawings is likely to include three classifications of drawings:

- *Architectural* – The architectural drawings deal primarily with the appearance and finish of the structure. Architectural drawings might include building elevations, floors plans, window and door details, and finish schedules.
- *Structural* – Structural drawings contain information specific to the structure, including foundations, footings, and other loadbearing structures.

- *Civil* – The civil drawings are of special interest to the heavy equipment operator because they deal with excavation and grading. These drawings cover site preparation, including clearing and grubbing; cut and fill plans; grading plans; underground utility plans; road profiles and sections; and paving plans.

Most drawings are drawn to scale, which is prominently displayed on the drawing. The scale used depends on the size of the project. A large project often has a small scale, such as 1" = 100', while a small project might have a large scale, such as 1" = 10'. The dimensions shown on drawings can be in engineering scale, which is in feet and tenths of a foot, or architectural scale, which is in feet, inches, and fractions of an inch. Plans used in excavation and grading work are likely to use engineering scale. It is common for a set of drawings to include on the first sheet a map that locates the project (*Figure 1*). In this case, the map also shows the detour that will be required while the work is being done.

The plans for a roadway project differ significantly from those for a building site. Because roadways can stretch across many miles, these plans need to consider the topography and soil conditions over long distances, while building plans need to consider the terrain over a comparatively small area. Equipment operators must understand how to read and interpret both types of plans in order to perform their jobs.

1.1.0 Title Sheets, Title Blocks, and Revision Blocks

A title sheet is normally placed at the beginning of a set of drawings or at the beginning of a major section of drawings. It may provide an index to the other drawings; a list of abbreviations used on the drawings and their meanings; a list of symbols used on the drawings and their meanings. The title sheet often contains other project data, such as the project location, the size of the land parcel, and the building size. It is important to use the title sheet that comes with the drawing in order to understand the specific symbols and abbreviations used on the drawings. These symbols and abbreviations may vary from job to job.

A title block or box (*Figure 2*) is normally placed on each sheet in a set of drawings. It is usually located in the bottom right-hand corner of the sheet, but this location can vary. The drawing set should be folded so that the title block faces up.

The title block serves several purposes in terms of communicating information. It contains the name of the firm that prepared the drawings, the owner's name, and the address and name of the project. It also gives locator information, such as the title of the sheet, the drawing or sheet number, the date the sheet was prepared, the scale, and the initials or names of the people who prepared and checked the drawing.

A revision block is normally shown on each sheet in a set of drawings. Typically, it is located in the upper right-hand corner or bottom right-hand corner of the drawing, near or within the title block. It is used to record any changes (revisions) to the drawing. An entry in the revision block usually contains the revision number or letter, a brief description of the change, the date, and the initials of the person making the revision(s). When using drawings, it is essential to note the revision designation on each drawing and use only the latest issue; otherwise, costly mistakes can result. If there is a conflict between drawings, or if you are in doubt about the revision status of a drawing, check with your supervisor to make sure that you are using the most recent version of the drawing. Also, check to see if any requests for information (RFI) or sketches have been included in the latest plan revision.

1.2.0 Highway Plans

Highway construction projects are called horizontal projects because they cover long distances and have almost no height. On highway jobs, the emphasis is on grading the roadbed and nearby areas according to engineering plans, which are

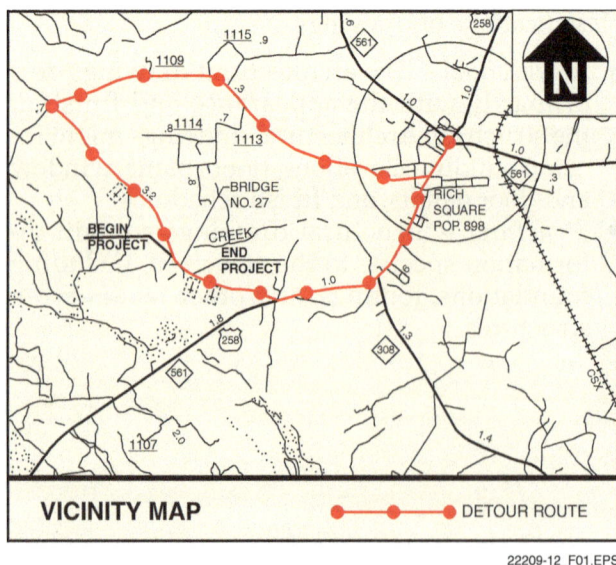

22209-12_F01.EPS

Figure 1 Example of a project location map.

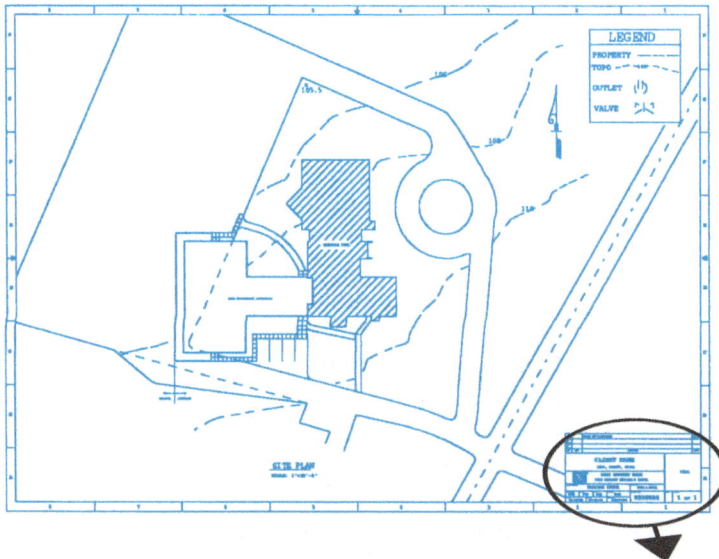

Figure 2 Title and revision blocks.

Technology at Work

Drawings have been used for centuries to define the construction of buildings, roads, bridges, and other structures. Until late in the twentieth century, the drawings were painstakingly done by drafters who worked with mechanical instruments at tall drawing boards. Today, most drawings are done using software in what is known as a computer-aided design, or CAD, system. This software greatly simplifies the process and makes the drawings easy to change. Many CAD systems are capable of producing three-dimensional renderings.

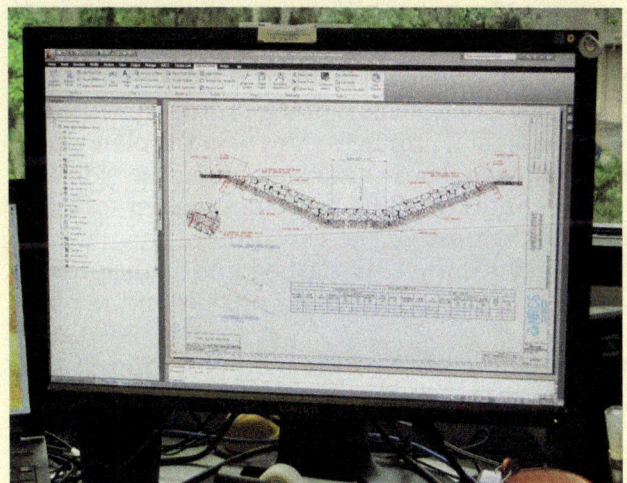

called plan and profile sheets and cross-section sheets. Equipment operators and other construction workers use these plans to obtain grading information at various times during the construction project.

Grade information will generally be transferred from the drawings to the grade stakes. The operator will read the grade stakes to accomplish the job. In some instances, however, it will be necessary for the operator to interpret the drawings.

1.2.1 Plan and Profile Sheets

Refer to this module's *Appendix. Figure A1* (Sheet C8) shows a typical roadway construction plan view and profile. For some projects, a whole sheet may be used for the plan view and a separate sheet used for the profile view. *Figure A1* covers 800 feet of the A-line road as measured from the center line of Highway 270, and 225 feet of the B-line road. *Figure A2* (Sheet C7) shows what are known as typical sections. The six sections reflect the design of the road for specific station-to-station segments of the roads. Note that the designs vary to some extent between the left and right sides of the roadways. The plan view shown in *Figure A1* is the view that would be seen looking down on the project from the top. The profile and section view shown in *Figure A2* is similar to a side elevation view. It shows the key elevations and slopes along the center line of the route. The main information shown on these sheets includes the following:

- *Direction* – The directional arrow always points north. Stations on the drawings are numbered from west to east and south to north.
- *Station numbers* – The station numbers listed along the bottom axis of the profile correspond to the station numbers shown on the plan view.
- *Elevations* – These are listed along the sides of the profile according to the designated scale.
- *Natural ground* – The elevation of the natural ground is drawn as a dashed line on the profile sheet, while contour lines appear on the plan view.
- *Planned grade* – The planned elevation of the grade is drawn as a continuous line on the profile.
- *Center line* – The center line of the roadway is plotted on the plan view.

- *Right-of-way* – The right-of-way limits are shown on the plan view.
- *Benchmarks* – Benchmarks, if any, are noted on the plan view.

The profile view provides information about the existing natural ground (or grade) and the planned final grade. It graphically shows each grade in relation to the other and gives a good picture of what types of excavations need to be done. In this case, all the work would be fill because the natural ground is below the required finish grade. When plotting the profile view, it is common to use a vertical scale much larger than the horizontal to make the elevation differences very clear.

1.2.2 Cross-Sections

Cross-section sheets are views of the construction as if the area was cut crosswise. For a highway, this would be like taking a knife and slicing across the road from one right-of-way line to the other and looking at the slice taken. The cross-section shows the layers of the road construction and the shapes of the side slopes and ditches.

Figure A2 shows typical sections or templates for a two-lane highway. It covers both the A and B roads shown in *Figure A1*. A typical section shows features, such as the slopes, ditches, and ramps, along with materials used to build up the roadway. However, most roads require many typical cross-sections because the terrain varies from point to point. In addition to the typical cross-sections, there may be an additional set of cross-section sheets showing the natural ground and the shape of the finish grade every 50 feet or other established distance. These sheets must be checked frequently to get grade information about the section of road that is being worked on because grade details can change as the terrain varies.

1.3.0 Building Construction Site Drawings

Every project requires a site plan to show the locations of buildings and other structures on the site. The site plan is often divided into additional plans that cover excavation, grading, utilities, and drainage. The site plan and its subordinate plans are the civil drawings of primary interest to heavy equipment operators. In addition to the site plans, there are some sheets of the architectural and structural drawings that may be of use during excavation and grading. These include the building elevations and the foundation plan.

Plan view drawings are drawings that show the site looking down from above. The object is projected from a horizontal plane. Typically, plan view drawings are made to show the overall construction (site plan), the structure's foundation (foundation plan), and the structure's floor plans.

1.3.1 Site Plans

Man-made and topographical (natural) features and other relevant project information, including the information needed to correctly locate structures on the site, are shown on a site plan. The site plan is sometimes called a plot plan. Man-made features include roads, sidewalks, utilities, and buildings. Topographical features include trees, streams, springs, and existing contours. Project information includes the building outline, general utility information, proposed sidewalks, parking areas, roads, landscape information, proposed contours, and any other information that conveys what is to be constructed or changed on the site. A prominently displayed north direction arrow is included for orientation purposes on site plans. Sometimes a site plan contains a large-scale map of the overall area that indicates where the project is located on the site. Examples of two different site plans are shown in *Figures 3* and *A3*. The plan in *Figure 3* is for a small site and shows topographical features in addition to the locations of structures. The plan in *Figure A3* is for a much larger site.

Typically, site plans show the following types of detailed information:

- Coordinates of control points or property corners
- Direction and length of property lines or control lines
- Description, or reference to a description, for all control and property monuments
- Location, dimensions, and elevation of the structure on site
- Finish and existing grade contour lines
- Finish elevations of building floors
- Location of utilities
- Location of existing elements such as trees and other structures

- Locations and dimensions of roadways, driveways, and sidewalks
- Names of all roads shown on the plan
- Locations and dimensions of any easements

Like other drawings, site plans are usually drawn to scale. The scale is prominently displayed on the drawing. The scale used depends on the size of the project. A project covering a large area typically has a small scale, such as 1" = 100', while a project on a small site might have a large scale, such as 1" = 10'.

> **NOTE**
>
> Not all drawings are made to scale. Those that are not scaled should be marked "NOT TO SCALE" or "NTS."

Normally, the dimensions shown on site plans are stated in feet and tenths of a foot (engineer's scale). However, some site plans state the dimensions in feet, inches, and fractions of an inch (architect's scale). Dimensions measured between the property lines and the structures are shown to verify that the locations of structures meet building code requirements. Building codes typically establish minimum setbacks that are measured from a property line. Front setbacks may be measured from the center line of the road rather than the property line. A front setback specifies the minimum distance that must be maintained between the property line (or the center line of the road) and the front of a structure (building line). Side and rear setbacks are also established by code. Building lines reflecting these setbacks are often included on the site plan (see *Figure 3*). Normally, side yard setbacks are specified to allow for access to rear yards and to reduce the possibility of fire spreading to adjacent buildings.

Site plans and survey maps often show areas of easement on the property. Easements are legal rights of persons other than the owner to use the property. The most common reason for an easement is to provide access to utility lines such as sewer, water, and electricity. Easements are also granted to municipalities for the purpose of maintaining drainage swales. The property owner

Blueprints

Many people still refer to construction drawings as blueprints, even though today's drawings are usually black ink on white paper. The term *blueprint* derives from the ammonia-based process once used to copy drawings, a process that turned the paper blue. Although it has been many years since this process was in common use, the term *blueprint* is still used by many people in the construction industry.

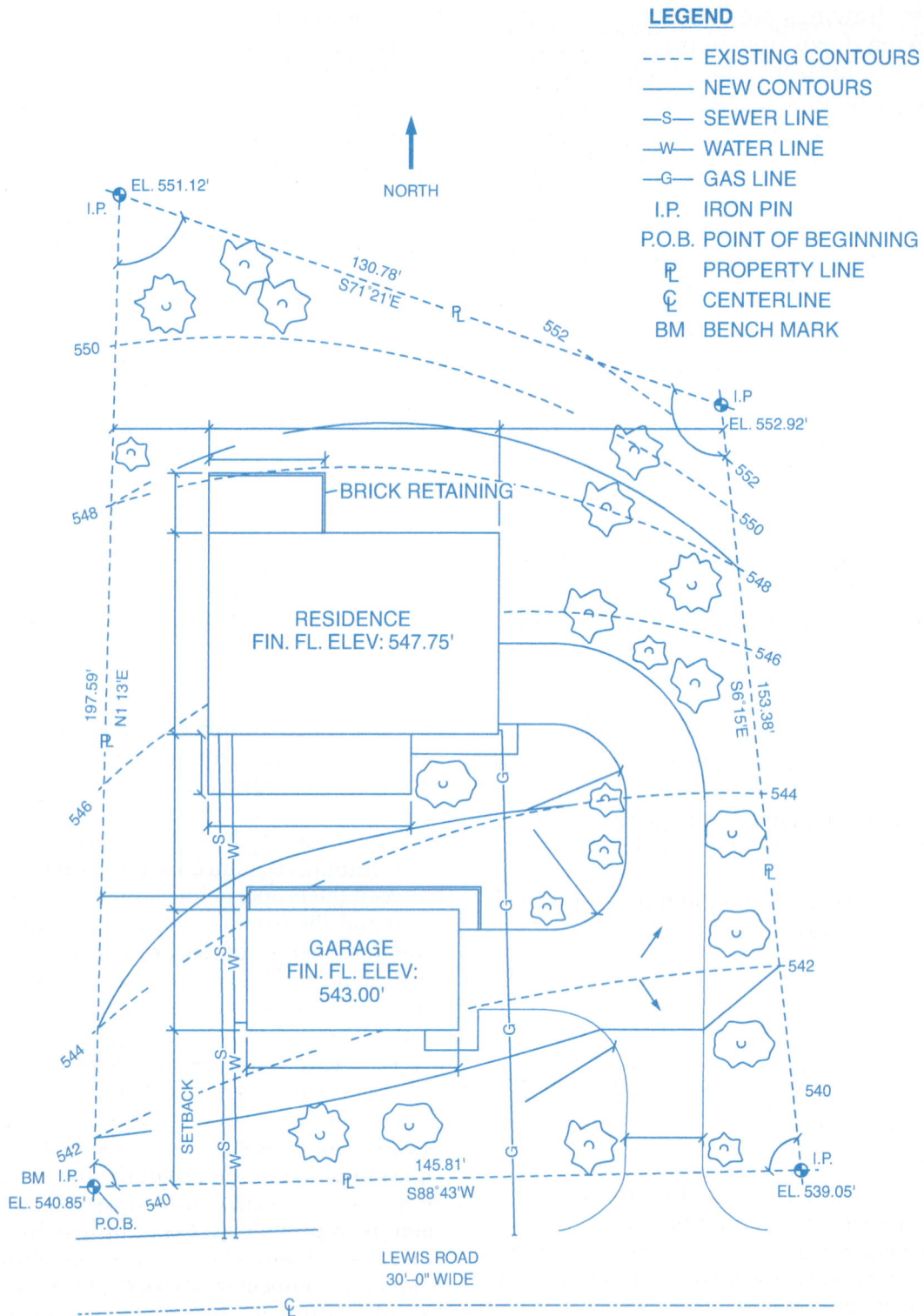

LEGEND

- – – – EXISTING CONTOURS
- ——— NEW CONTOURS
- —S— SEWER LINE
- —W— WATER LINE
- —G— GAS LINE
- I.P. IRON PIN
- P.O.B. POINT OF BEGINNING
- P.L. PROPERTY LINE
- C.L. CENTERLINE
- BM BENCH MARK

NORTH

EL. 551.12'
I.P.

130.78'
S71°21'E
P.L.

552

I.P
EL. 552.92'

552

550

BRICK RETAINING

548

546

548

550

RESIDENCE
FIN. FL. ELEV: 547.75'

197.59'
N1°13'E
P.L.

153.38'
S6°15'E

546

544

544

GARAGE
FIN. FL. ELEV:
543.00'

542

P.L.

544

540

542

SETBACK

542

BM I.P.
EL. 540.85'

540

P.O.B.

145.81'
S88°43'W
P.L.

I.P.
EL. 539.05'

LEWIS ROAD
30'–0" WIDE

C.L.

SITE PLAN
SCALE: 1" = 30'–0"

22209-12_F03.EPS

Figure 3 Example of a simple site plan showing topographical features.

is prohibited from building any structure on an easement or otherwise obstructing access to it.

Site plans show finish grades (also called elevations) for the site, based on data provided by a surveyor or engineer. It is necessary to know these elevations for grading the lot and for construction of the structure. Finish grades are typically shown for all four corners of the lot, as well as other points within the lot. Finish grades or elevations are also shown for the corners of the structure and relevant points within the building.

Heavy equipment operators need to pay particular attention to existing and proposed contour lines on these plans. These lines define how deep the cuts will be or how much fill is required to bring the site to the correct elevation.

It is important to study the site plans because they show details such as locations of property lines, survey markers, and utilities that help equipment operators avoid problems. Follow these rules while you are working:

- Be aware of the locations of property boundaries. Do not cross property boundaries unless you know that the owner has given the managers of the project an easement. Heavy equipment can damage terrain and underground structures such as drainage pipes, culverts, and septic tanks.
- Know the locations of surveyors marks. Do not operate heavy equipment in the location of a benchmark, monument, or control points until you are sure of its location. Damaging these references can cause costly delays.
- Know the locations of buried utilities. Do not operate heavy equipment near any buried utilities unless you are certain of their locations. Not only can hitting underground gas lines and power cables cause delay, it can be fatal. If the property you are working on was previously developed, the chances are it has buried utilities. These utilities must be located and marked before any excavation can begin.

Take a few moments to study the site plan shown in *Figure 3*. Find the North marker and review the legend to become familiar with the symbols used on the drawing. Look for property

Scaling Drawings

Measuring the length of a line on a drawing, then converting that measurement to an actual length is known as scaling. Scaling can be done using an engineer's scale, but the task can be simplified by using an electronic plan wheel scaler like the one shown. The device is first set to match the drawing scale. Then, as the scaler is rolled along a line on the drawing, its digital readout gives a direct reading of the length of the line.

22209-12_SA02.EPS

boundaries, utilities, and existing and proposed grades (contours).

All the finish grade references shown are keyed to a reference point, called a benchmark or job datum. This is a reference point established by the surveyor on or close to the property, usually at one corner of the lot. At the site, this point may be marked by a plugged pipe driven into the ground, a brass marker, or a wood stake. The location of the benchmark is shown on the plot plan with a grade figure next to it. This grade figure may indicate the actual elevation relative to sea level, or it may be an arbitrary elevation, such as 100.00' or 500.00'. All other grade points shown on the site plan, therefore, are relative to the benchmark. In *Figure 3*, this point is labeled P.O.B. for point of beginning and is located at the southwest corner of the property.

Enforcing Setbacks

Municipal inspectors can be very strict in enforcing setback requirements. If an addition to a building penetrates a setback, for example, the inspector may refuse to issue a certificate of occupancy until the problem is corrected. As a result, the property owner would be forced to change or remove the structure or obtain a variance by appealing to a special review board. Both methods can be costly and time consuming.

A site plan usually shows the finish floor elevation of the building. This is the level of the first floor of the building relative to the job-site benchmark. For example, if the benchmark is labeled 100.00' and the finish floor elevation indicated on the plan is marked 105.00', the finish floor elevation is 5' above the benchmark. During construction, many important measurements are taken from the finish floor elevation point.

On *Figure 3*, the benchmark is located at the southwest property corner and is 540.85' (P.O.B.). Since the finish floor elevation of the residence is 547.75', the finish floor elevation is 6.9' above the benchmark.

Depending on the size and complexity of the site, and sometimes on the requirements of the local municipality, the site plan may be subdivided to include one or more additional plans showing specific details of site preparation. These additional plans might include the following:

- Excavation plan
- Utility plan
- Grading plan
- Drainage plan

Excavation plan – Heavy equipment operators play a key role in the cut and fill work that needs to be done in preparing a job site for construction. The plans that guide cut and fill operations are part of the drawing set that defines a project. *Figure 4* shows cross-sections of a construction site. The red lines on the charts represent locations of rock; the blue lines are finish grade; and the green lines are existing grade. *Figure 5* is a cut and fill plan for the same site. It shows in detail how much cut and fill work is needed to complete the project. "F" represents fill, and each F bubble indicates how much fill is needed. For example, F1+72 means fill 1 foot, 7 tenths, and 2 hundredths. The "C" (cut) bubbles are interpreted the same way.

Utility plan – Most, if not all, sites have buried utilities such as sanitary sewer lines; stormwater drain lines; fresh water piping; natural gas lines; and any electrical and communications cabling. A utility plan is prepared to show the locations of these utilities, as well as the depth at which they are buried. If the property does not have access to some municipal services, the locations of on-site services such as wells, septic systems, and gas tanks are included on the plan. *Figure A4* is an example of a utility plan showing a water line. Additional sheets are used to show other utilities, such as sewer, electrical, and gas lines.

Grading and drainage plans – A grading plan is used to show how the surface of the construction site will be altered in order to accommodate the construction. One of the main functions of the grading plan is to ensure proper drainage of the site to the established stormwater removal system. This plan also helps to verify that drainage from the site under construction will not adversely affect adjoining properties. Municipal engineers and planning boards, who review and approve the site plans, will want to see that required swales, culverts, and drains have been accounted for. *Figure A5* shows an example of a grading plan. The contour lines on the plan represent the surface features of the site. The solid contour lines show the current topographical features, while the dashed lines show the planned configuration. Site layout crews will used this plan to place cut and fill stakes.

In some instances, a separate drainage plan is required to show how rainwater runoff will be contained on the property under construction. *Figure 6* is an example of a drainage plan. Project engineers will analyze the soil on the property under construction to determine its ability to absorb water. They will also take into account the amount of impervious area that will be created by the construction. The construction of buildings, roads, and parking lots, as well as the removal of trees, reduces the capacity of a site to absorb water. In such cases, it may be necessary to create holding areas for runoff water. Such areas are known as detention or retention ponds. A detention pond is an excavation intended to hold overflow water until it can be absorbed naturally or evaporate. A detention pond is dry much of the time. A retention pond is an excavation that holds water continuously. Retention ponds are often installed where there is a natural source of water, such as a spring. Retention ponds such as the one shown in *Figure 7* are often used as landscape features. Another method of controlling runoff water is to install drains that are connected through buried pipes to the stormwater system (*Figure 8*).

> **NOTE**
> On smaller sites, the grading, drainage, and utility plans may be included on the same drawing.

1.3.2 Foundation Plans

Foundation plans, such as the one shown in *Figure 9*, give information about the location and dimensions of footings, grade beams, foundation walls, stem walls, piers, equipment footings, and windows and doors. The specific information shown on the plan is determined by the type of construction involved, such as full-basement foundation,

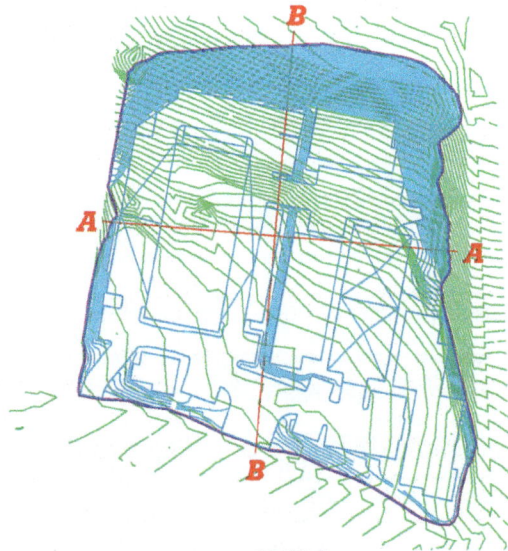

N.T.S.

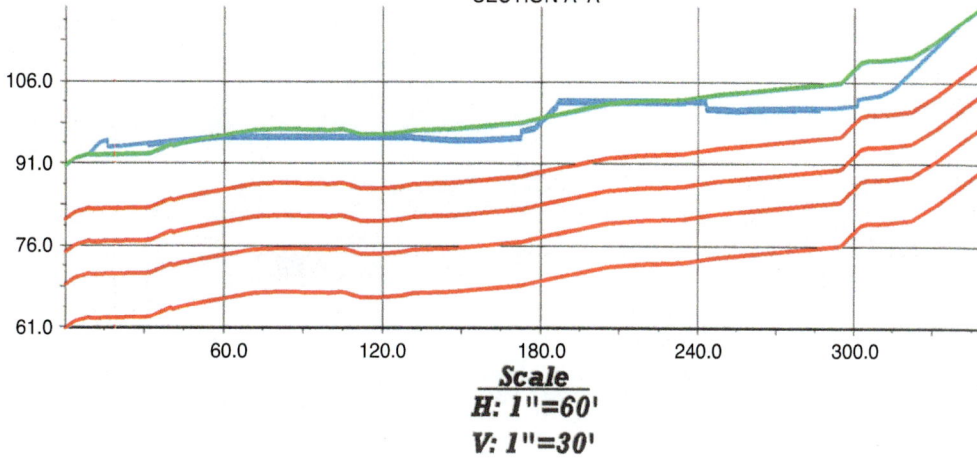

SECTION A–A

Scale
H: 1"=60'
V: 1"=30'

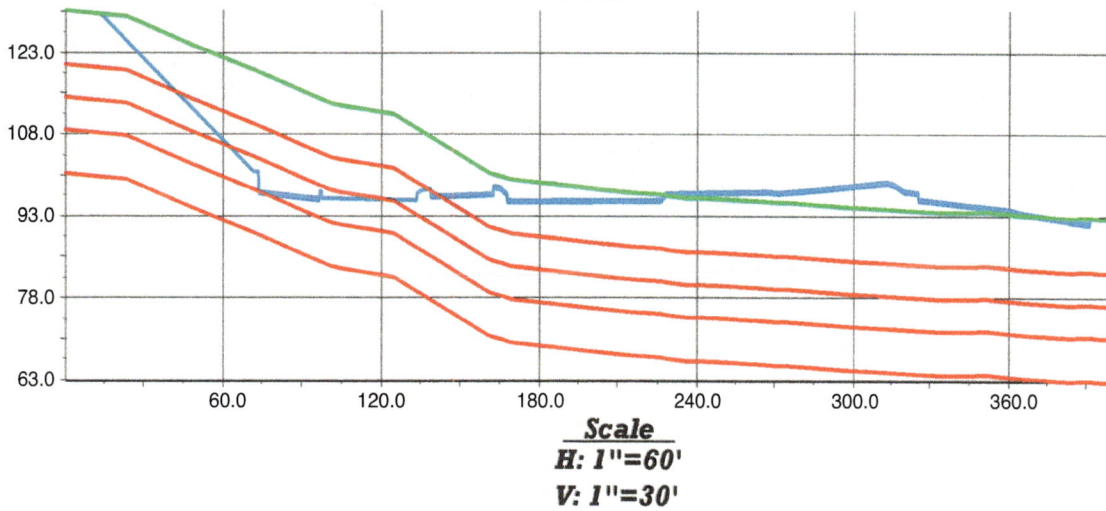

SECTION B–B

Scale
H: 1"=60'
V: 1"=30'

22209-12_F04.EPS

Figure 4 Site cross-sections.

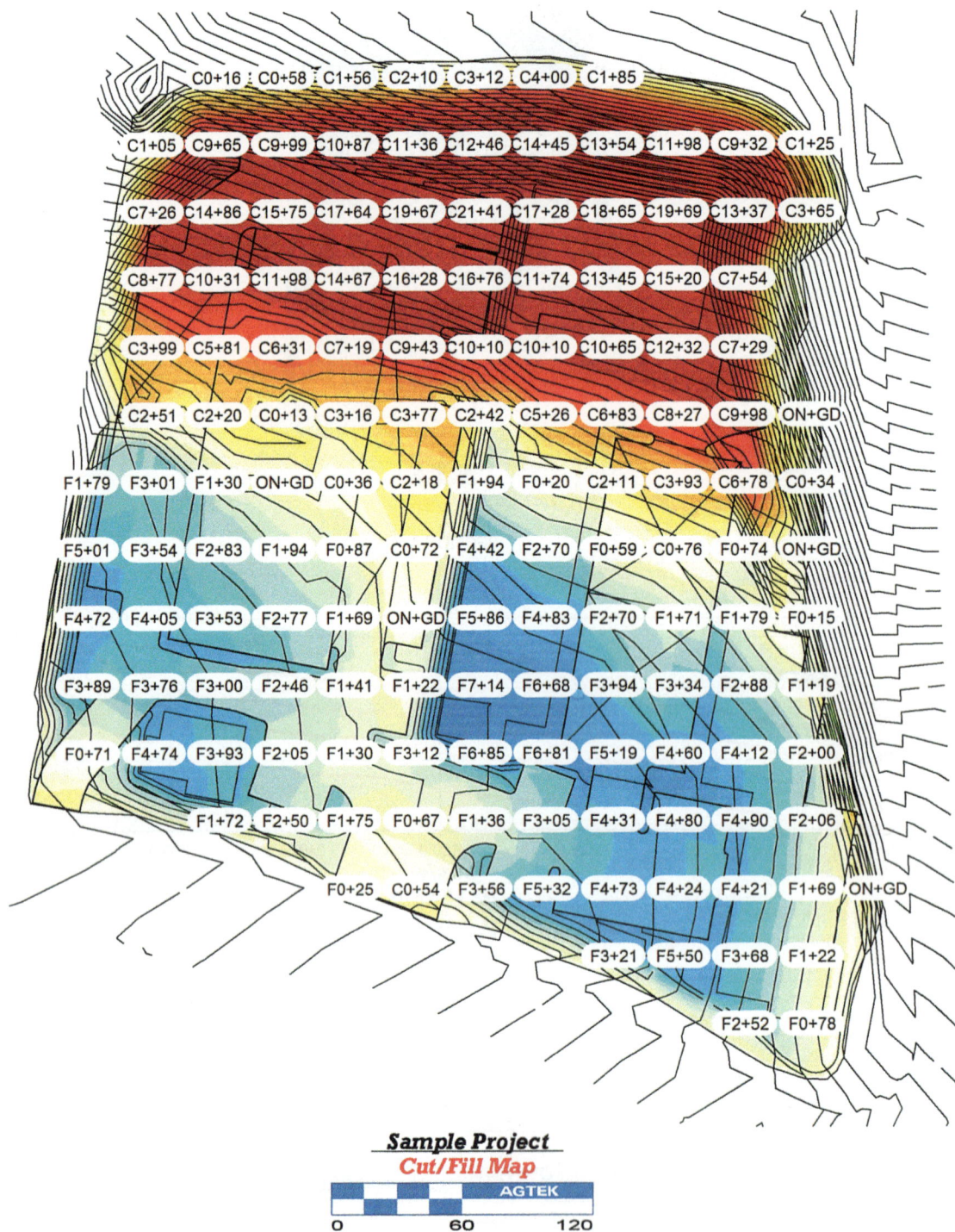

C0+16 · C0+58 · C1+56 · C2+10 · C3+12 · C4+00 · C1+85

C1+05 · C9+65 · C9+99 · C10+87 · C11+36 · C12+46 · C14+45 · C13+54 · C11+98 · C9+32 · C1+25

C7+26 · C14+86 · C15+75 · C17+64 · C19+67 · C21+41 · C17+28 · C18+65 · C19+69 · C13+37 · C3+65

C8+77 · C10+31 · C11+98 · C14+67 · C16+28 · C16+76 · C11+74 · C13+45 · C15+20 · C7+54

C3+99 · C5+81 · C6+31 · C7+19 · C9+43 · C10+10 · C10+10 · C10+65 · C12+32 · C7+29

C2+51 · C2+20 · C0+13 · C3+16 · C3+77 · C2+42 · C5+26 · C6+83 · C8+27 · C9+98 · ON+GD

F1+79 · F3+01 · F1+30 · ON+GD · C0+36 · C2+18 · F1+94 · F0+20 · C2+11 · C3+93 · C6+78 · C0+34

F5+01 · F3+54 · F2+83 · F1+94 · F0+87 · C0+72 · F4+42 · F2+70 · F0+59 · C0+76 · F0+74 · ON+GD

F4+72 · F4+05 · F3+53 · F2+77 · F1+69 · ON+GD · F5+86 · F4+83 · F2+70 · F1+71 · F1+79 · F0+15

F3+89 · F3+76 · F3+00 · F2+46 · F1+41 · F1+22 · F7+14 · F6+68 · F3+94 · F3+34 · F2+88 · F1+19

F0+71 · F4+74 · F3+93 · F2+05 · F1+30 · F3+12 · F6+85 · F6+81 · F5+19 · F4+60 · F4+12 · F2+00

F1+72 · F2+50 · F1+75 · F0+67 · F1+36 · F3+05 · F4+31 · F4+80 · F4+90 · F2+06

F0+25 · C0+54 · F3+56 · F5+32 · F4+73 · F4+24 · F4+21 · F1+69 · ON+GD

F3+21 · F5+50 · F3+68 · F1+22

F2+52 · F0+78

Sample Project
Cut/Fill Map

AGTEK

0   60   120

22209-12_F05.EPS

Figure 5 Cut and fill plan.

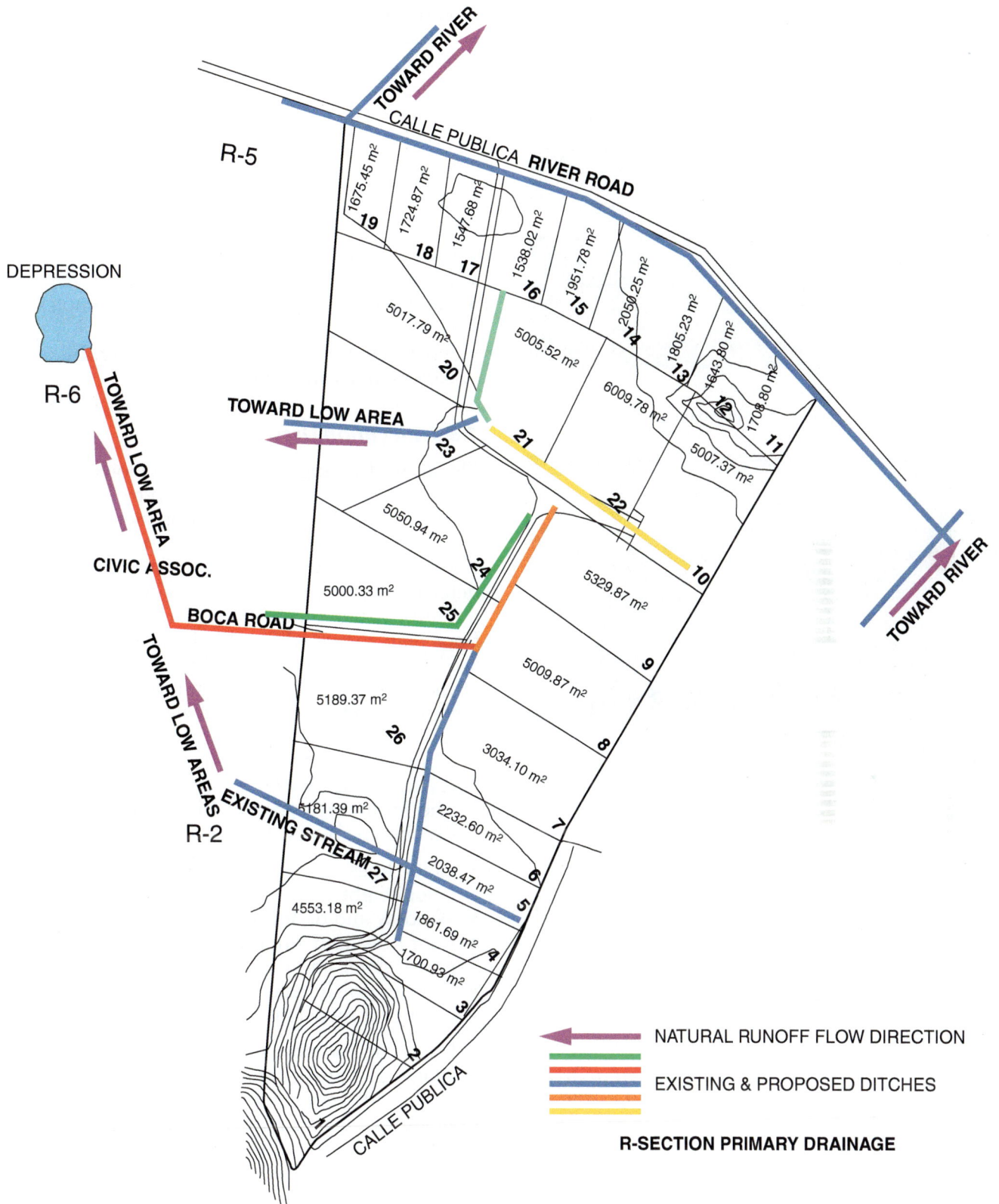

Figure 6 Example of a drainage plan.

Legend within figure:
- NATURAL RUNOFF FLOW DIRECTION
- EXISTING & PROPOSED DITCHES
- R-SECTION PRIMARY DRAINAGE

22209-12_F06.EPS

Figure 7 Retention pond in a residential subdivision.

Figure 8 On-site stormwater drain.

crawl space, or a concrete slab-on-grade level (*Figure 10*).

The following are types of information normally shown on foundation plans for full-basement and crawl space foundations:

- Location of the inside and outside of the foundation walls
- Location of the footings for foundation walls, columns, posts, chimneys, and fireplaces

Stormwater Detention

Sandy soil does not absorb water very well, so heavy rain tends to run off. Construction sites often require acres of detention ponds to compensate for the impervious area created by the construction of buildings, roads, and parking lots.

- Walls for entrance platforms (stoops)
- Notations for the strength of concrete used for various parts of the foundation and floor
- Notations for the composition, thickness, and underlaying material of the basement floor or crawl space surface

The types of information normally shown on foundation plans for slab-on-grade foundations include the following:

- Size and shape of the slab
- Exterior and interior footing locations
- Loadbearing surface (fireplace, for example)
- Notations for slab thickness
- Notations for wire mesh reinforcing, fill, and vapor barrier materials

1.3.3 Elevation Drawings

Elevation drawings are views that look straight ahead at a structure. The object is projected from a vertical plane. Typically, elevation views are used to show the exterior features of a structure so that the general size and shape of the structure can be determined. Elevation drawings clarify much of the information on the floor plan. For example, a floor plan shows where the doors and windows are located in the outside walls; an elevation view of the same wall shows actual representations of these doors and windows. *Figure 11* shows an example of a basic elevation drawing. Look for the existing and proposed grade elevations, identified with the arrows on *Figure 11*.

The following types of information are normally shown on elevation drawings:

- Grade lines
- Floor height
- Window and door types
- Roof lines and slope, roofing material, vents, gravel stops, and projection of eaves
- Exterior finish materials and trim
- Exterior dimensions

Unless one or more views are identical, four elevation views are generally used to show each exposure. With very complex buildings, more than four views may be required. Because elevation drawings often contain grade information, equipment operators may need to refer to them.

1.3.4 Soil Reports

Soil conditions are among the factors that determine the type of foundation best suited for a structure. This information can be vital in determining how you do your job. A structure that is built on soil that lacks consistent quality and compaction

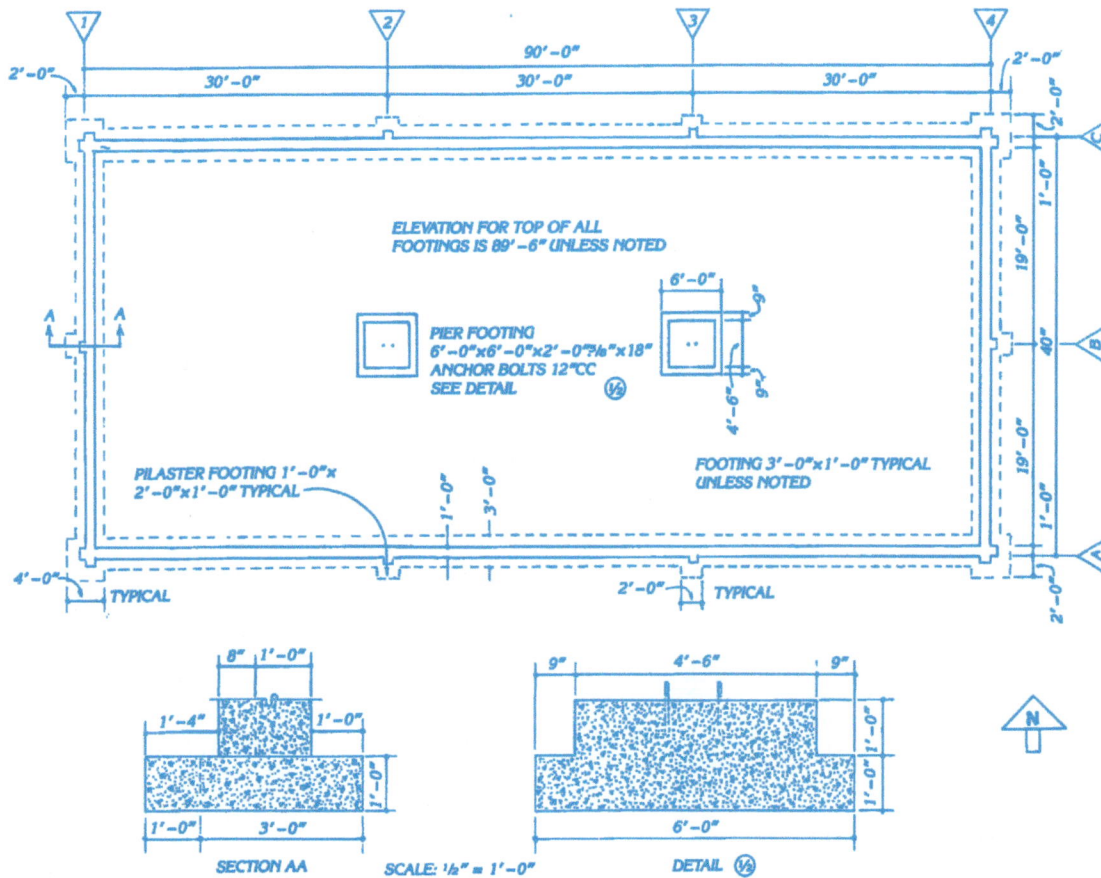

ELEVATION FOR TOP OF ALL FOOTINGS IS 89'–6" UNLESS NOTED

PIER FOOTING
6'–0"x6'–0"x2'–0"¾"x18"
ANCHOR BOLTS 12"CC
SEE DETAIL ½

PILASTER FOOTING 1'–0"x
2'–0"x1'–0" TYPICAL

FOOTING 3'–0"x1'–0" TYPICAL
UNLESS NOTED

TYPICAL

TYPICAL

SECTION AA   SCALE: ½" = 1'–0"   DETAIL ½

22209-12_F09.EPS

Figure 9 Foundation plan.

will settle unevenly. This can result in cracks in the foundation and structural damage to the rest of the building. Therefore, in designing the foundation for a structure, the architect must consider the soil conditions on the building site. Typically, the architect consults a soil engineer, who makes test bores of the soil on the building site and analyzes the samples. The results of the soil analysis are summarized in a soil report issued by the engineer. This report is often included as part of the drawing set. When using this information, consider all aspects of the soil report, including elevation of the water table. The type of soil on the job site will determine the types of equipment needed to do the job. For example, a backhoe can easily excavate sandy soil, but hard-packed clay may require some other equipment to break it up before it can be excavated with a backhoe.

1.4.0 As-Built Drawings

As-built drawings are formally incorporated into the drawing set to record changes made during construction. These drawings are marked up on the job by the various trades to show any differences between what was originally shown on a plan by the architect or engineer and what was actually built. Such changes result from the need to relocate equipment to avoid obstructions; relocate utilities; or because the architect has changed a certain detail in the site design in response to customer preferences. On many jobs, any such changes to the design can only be made after a change order has been generated and approved by the project engineer or other designated person. Depending on the complexity of the change, changes to the drawings are typically outlined with a unique design such as a cloud symbol. Changes should be made in red ink to make sure they stand out. Changes must be dated and initialed by the responsible party.

A supervisor or the project engineer will determine if it is necessary to deviate from the design plans for some reason, but it is part of the operator's job to make sure that the changes get marked on the as-built drawings. One of the most important entries on these plans is any deviation in the placement of underground utilities.

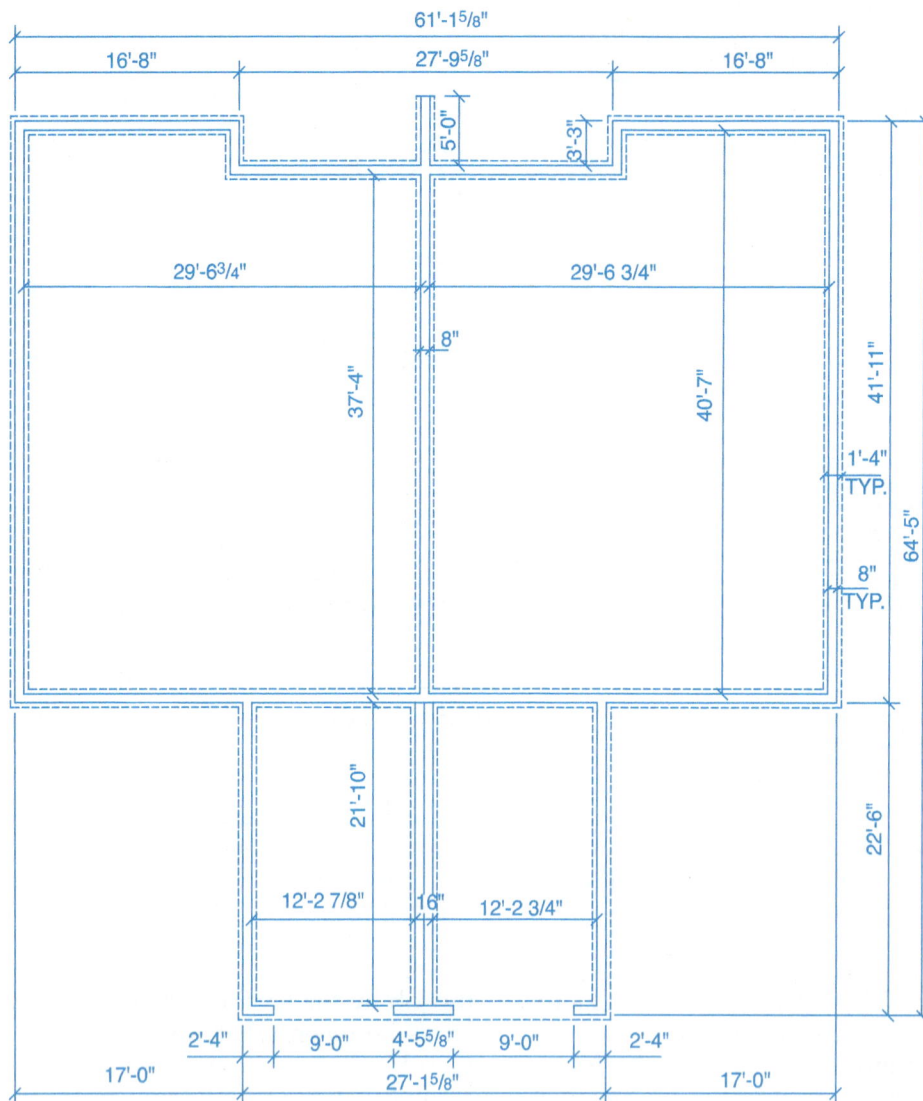

Figure 10 Slab-on-grade plan.

As-Built Drawings

The job specifications generally contain a section defining how as-built drawings are to be prepared. Changes made by various contractors are usually marked on a master set of drawings that have been set aside for that purpose. On some jobs, it may be necessary to obtain an approved change order before making any change to the drawings.

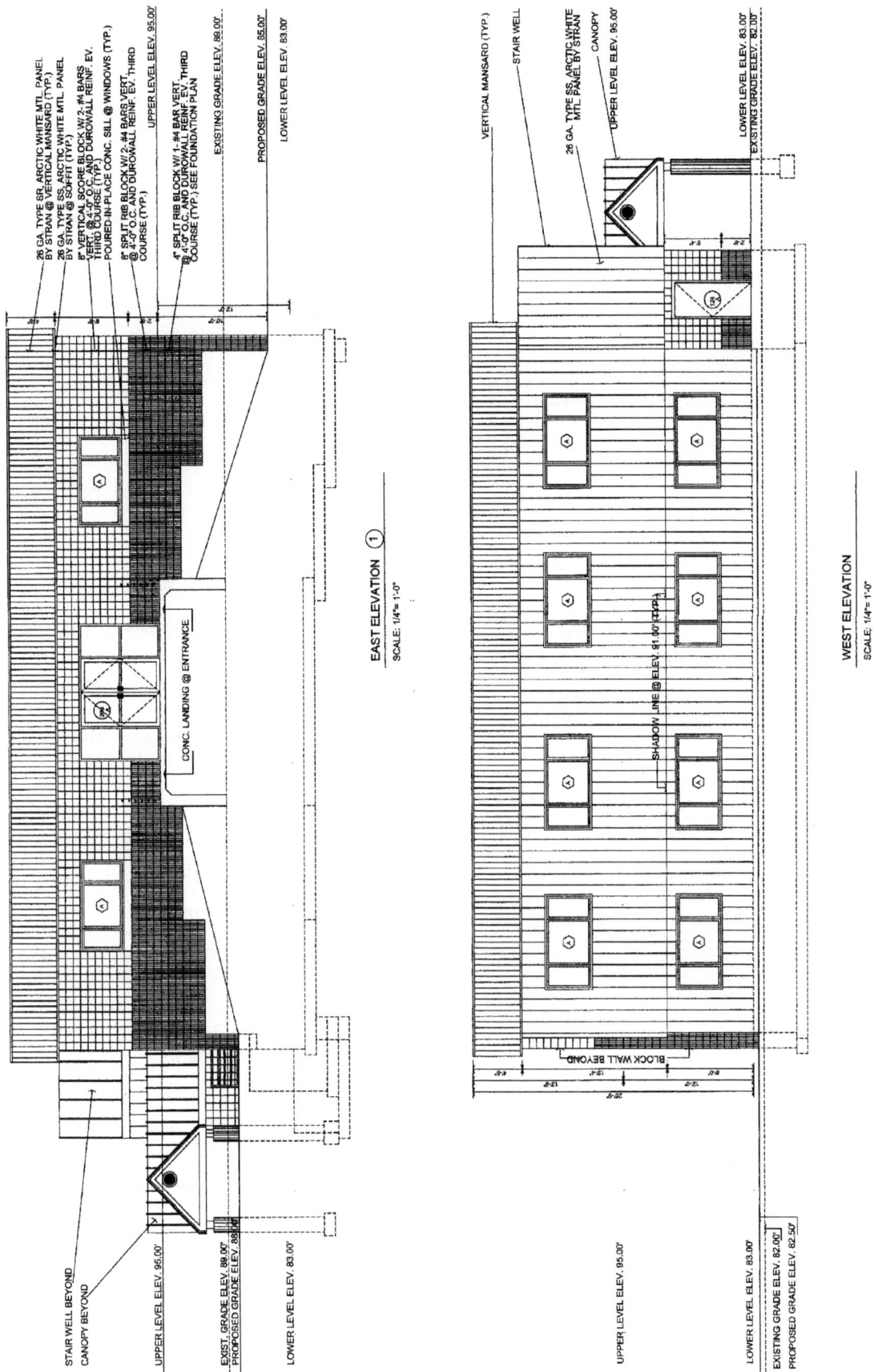

Figure 11 Elevation drawing.

EAST ELEVATION 1
SCALE: 1/4" = 1'-0"

WEST ELEVATION
SCALE: 1/4" = 1'-0"

22209-12_F11.EPS

Additional Resources

Surveying with Construction Applications, Barry
F. Kavanaugh; Pearson, Upper Saddle River, NJ.

1.0.0 Section Review

1. A list of the symbols and abbreviations used on a set of drawings can usually be found on the _____.

   a. title block
   b. revision block
   c. title sheet
   d. last sheet

2. The highway drawing that is like an elevation is the _____.

   a. plan view
   b. profile
   c. cross-section
   d. grading plan

3. A release that allows access to property owned by another is a(n) _____.

   a. easement
   b. setback
   c. right-of-passage
   d. right-of-way

4. The drawing set that incorporates the changes made during construction is known as the _____.

   a. master drawings
   b. final drawings
   c. revised drawings
   d. as-built drawings

SECTION TWO

2.0.0 READING AND INTERPRETING DRAWINGS

Objective 2

Read and interpret drawings.
   a. Identify different types of lines used on drawings.
   b. Define common abbreviations and symbols used on drawings.
   c. Interpret building site and highway drawings to determine excavation requirements.

Performance Tasks 1 and 2

Interpret a set of drawings to determine the proper type and sequence of excavation and grading operations needed to prepare the site. Determine the scale of different drawings.

Trade Terms

Invert: The lowest portion of the interior of a pipe, also called the flow line.

In order to read and interpret the information on drawings, it is necessary to learn the special language used in construction drawings. This section of the module describes the different types of lines, dimensioning, symbols, and abbreviations used on drawings. It also describes how to interpret the drawings used in highway and building site construction. When working with drawings, it is best to use a logical, structured approach. The following general procedure is suggested as a method of reading a set of drawings for maximum understanding:

Step 1 Acquire a complete set of drawings and specifications, including the title sheet(s), so that you can better understand the abbreviations and symbols used throughout the drawings.

Step 2 Read the title block. The title block defines what the drawing is about. Take note of the critical information such as the scale, date of last revision, drawing number, and architect or engineer. After using a sheet from a set of drawings, be sure to refold the sheet with the title block facing up.

Step 3 Find the north arrow. Always orient yourself to the structure. Knowing where north is enables you to more accurately describe the location of the building and other structures.

Step 4 Always be aware that the drawings work together as a group. The reason the architect or engineer draws plans, elevations, and sections is that drawings require more than one type of view to communicate the whole project. Learn how to use more than one drawing when necessary to find the information you need.

Step 5 Check the list of drawings in the set. Note the sequence of the various plans. Some drawings have an index on the front cover. Notice that the prints in the set are of several categories:
- Architectural
- Structural
- Civil
- Mechanical
- Electrical
- Plumbing
- Landscape

Step 6 Study the site plan to determine property boundaries and carefully note the location of any benchmarks. Further, determine the location of the building to be constructed, as well as the various utilities, roadways, and any easements. Note the various elevations and the existing and proposed contours.

Step 7 Check the floor plan for the orientation of the building. Observe the locations and features of entries, corridors, offsets, and any special features to get an idea of the finished construction.

Step 8 Check the foundation plan for the sizes and types of footings, reinforcing steel, and loadbearing substructures.

Step 9 Check the floor construction and other details relating to excavations.

Step 10 Study the utility drawings and structural plans for features that affect earthwork.

Step 11 Check the notes on various pages, and compare the specifications against the construction details.

Step 12 Browse through the sheets of drawings to become familiar with all the plans and details.

Step 13 Recognize applicable symbols and their relative locations in the plans. Note any special excavation details.

Step 14 After you are acquainted with the plans, walk the site so that you can relate the plans to the site.

2.1.0 Lines and Symbols

Many different types of symbols and lines are used in the development of a set of plans. Lines are drawn wide, narrow, dark, light, broken, and unbroken, with each type of line conveying a specific meaning. *Figure A6* shows the most common lines and symbols used on site drawings. The following describes the types of lines commonly found on drawings:

- *Object lines* – Heavier-weight lines used to show the main outline of the structure, including exterior walls, interior partitions, porches, patios, sidewalks, parking lots, and driveways.
- *Dimension and extension lines* – Provide the dimensions of an object. An extension line is drawn out from an object at both ends of the part to be measured to indicate the part being measured. Extension lines are not supposed to touch the object lines. This is so they cannot be confused with the object lines. A dimension line is drawn at right angles between the extension lines and a number placed above, below, or to the side of it to indicate the length of the dimension line. Sometimes a gap is made in the dimension line and the number is written in the gap.
- *Leader line* – Connects a note or dimension to a related part of the drawing. Leader lines are usually curved or at an angle from the feature being distinguished to avoid confusion with dimension and other lines.
- *Center line* – Designates the center of an area or object and provides a reference point for dimensioning. Center lines are typically used to indicate the centers of roadways and the center of objects such as columns, posts, footings, and door openings. On roadways, the center line is a common reference point.

Walking the Site

The plans provide a two-dimensional perspective, but walking the site is the best way to visualize the project. Most racecar drivers walk the track before a race. They can look at a map of the track and identify locations of the entry, apex, and exit points of each turn, but walking the course helps them visualize and memorize the track.

- *Cutting plane (section line)* – Indicates an area that has been cut away and shown in a section view so that the interior features can be seen. The arrows at the ends of the cutting plane indicate the direction in which the section is viewed. Letters identify the cross-sectional view of that specific part of the structure. More elaborate methods of labeling section reference lines are used in larger, more complicated sets of plans (*Figure 12*). The sectional drawing may be on the same page as the reference line or on another page.
- *Break line* – Shows that an object or area has not been drawn in its entirety.
- *Hidden line* – Indicates an outline that is invisible to an observer because it is covered by another surface or object that is closer to the observer.
- *Phantom line* – Indicates alternative positions of moving parts, such as a damper's swing, or adjacent positions of related parts. It may also be used to represent repeated details.
- *Contour line* – Contour lines are used to show changes in the elevation and contour of the land. The lines may be dashed or solid. Generally, dashed lines are used to show the natural or existing grade, and solid lines show the finish grade to be achieved during construction.

Symbols are used on drawings to show different kinds of materials, objects, fixtures, and structural members. The meanings of symbols and the types used are not standardized and can vary from location to location. A set of drawings generally includes a sheet that identifies the specific symbols used and their meanings (see *Figure A6*). When using any drawing set, always refer to this sheet of symbols to avoid making mistakes when reading the drawings .

Some symbols give a good idea of what the object it represents looks like, but others are used to show the position of the object. Examples of this type of symbol include door and window designators that refer to door and window schedules where the different types are described. Still other symbols are used to show the orientation of the object, showing the direction or side (north, south, front, back, and so on).

2.2.0 Abbreviations Used on Drawings

Many written instructions are needed to complete a set of construction drawings. It is impossible to print out all such references, so a system of abbreviations is used. By using standard abbreviations, such as BRK for brick or CONC for concrete, the architect ensures that the drawings will be accurately interpreted. *Figure A7* contains a list of abbreviations commonly used on site drawings. Note that some architects and engineers may use different abbreviations for the same terms. Normally, the title sheet in a drawing set contains a list of abbreviations used in the drawings. For this reason, it is important to get a complete set of drawings and specifications, including the title sheet(s), in order to better understand the exact abbreviations used. Some practices for using abbreviations on drawings are as follows:

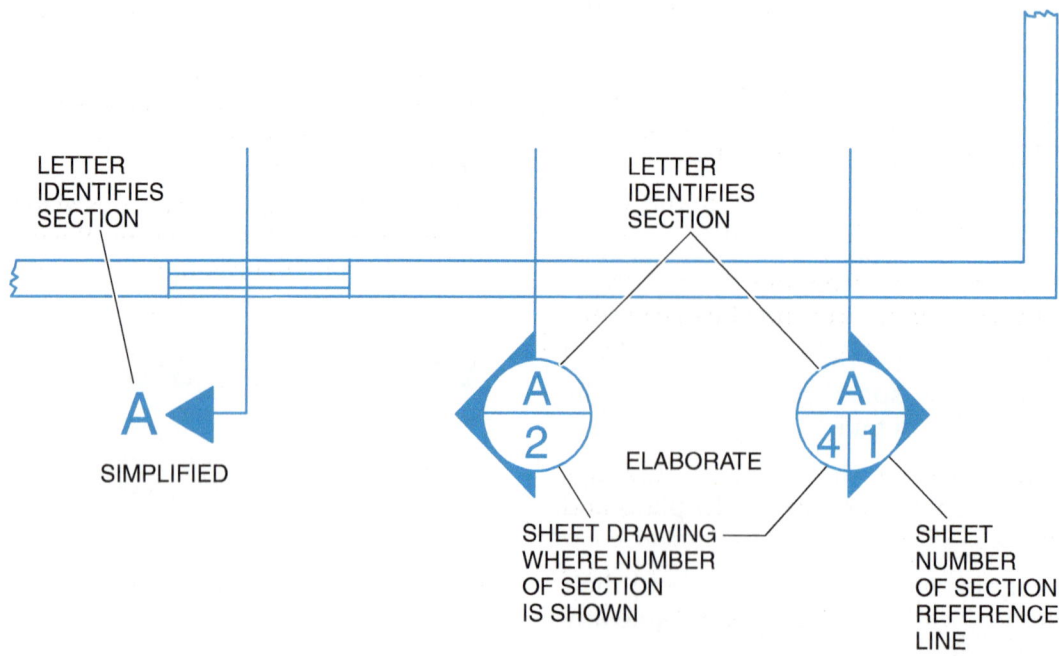

Figure 12 Methods of labeling section reference lines.

- Most abbreviations are capitalized.
- Periods are used when abbreviations might look like a whole word.
- Abbreviations are the same whether they are singular or plural.
- Several terms have the same abbreviations and can only be identified from the context in which they are found.

2.3.0 Interpreting Civil Drawings

The ability to read drawings is an important skill. A person in the construction industry who lacks this ability is limiting their growth potential and relies on others when it is necessary to use the project drawings. Learning to read drawings takes patience and practice, but it is a skill that can help improve productivity and reduce errors.

2.3.1 Site Plans

Site plans show the positions and sizes of all relevant structures on the site, as well as the features of the terrain. It would be difficult to show the amount of information on these drawings without using symbols and abbreviations. *Figure 13* shows symbols commonly used on site plans. It is important to take time to become familiar with these symbols so that you can quickly identify them on plans.

Each contour line across the plot of land represents a line of constant elevation relative to some point such as sea level or a local feature. *Figure 14* shows an example of a contour map for a hill. As shown, contour lines are drawn in uniform elevation intervals called contour intervals. Commonly used intervals are 1', 2', 5', and 10'.

On some plans and surveys, every fifth contour line is drawn using a heavier-weight line and is labeled with its elevation to help the user more easily determine the contour. This method of drawing contour lines is called indexing contours. The elevation is marked above the contour line, or the line is interrupted for it.

As shown in *Figure 14*, contour lines can form a closed loop within the map. These lines represent an elevation (hill) or depression. If you start at any point on these lines and follow their path, you eventually return to the starting point. A contour may close on a site plan or map, or it may be discontinued at any two points at the borders of the plan or map. Examples of this can be seen on the site plan shown in *Figure 3*. Such points mark the ends of the contour on the map, but the contour does not end at these points. The contour is continued on a plan or map of the adjacent land.

Some rules for interpreting contours include the following:

- Contour lines do not cross.
- Contour lines crossing a stream point upstream.
- The horizontal distance between contour lines represents the degree of slope. Closely spaced contour lines represent steep ground and widely spaced contour lines represent nearly level ground with a gradual slope. Uniform spacing indicates a uniform slope.
- Contour lines are at right angles to the slope. Therefore, water flow is perpendicular to contour lines.
- Straight contour lines parallel to each other represent man-made features such as terracing.

The existing contour is shown as a dashed line; the new or proposed contour is shown as a solid line (*Figure 15*). Both are labeled with the elevation of the contour in the form of a whole number. The spacing between the contour lines is at a constant vertical increment, or interval. The typical interval is five feet, but intervals of one foot are not uncommon for site plans requiring greater detail, or where the change in elevation is gradual.

A known elevation on the site that is used as a reference point during construction is called a benchmark. The benchmark is established in reference to the datum and is commonly noted on the site print with a physical description and its elevation relative to the datum. For example, "Northeast corner of catch basin rim—Elev. 102.34'" might be a typical benchmark found on a site plan. When individual elevations, or grades, are required for other site features, they are noted with a + and the grade. Grades vary from contours in that a grade has accuracy to two decimal places, whereas a contour is expressed as a whole number.

Always review materials symbols, dimensioning and scaling, and fundamental construction techniques before attempting to understand a full set

Drawing Revisions

Always be sure to use the latest set of drawings. When a set of construction drawings has been revised and reissued, the superseded set, or the sheets that were replaced, should be discarded. If they are kept for record purposes, they should be marked with a notation such as "Obsolete Drawing – Do Not Use." This will prevent them from being used in error.

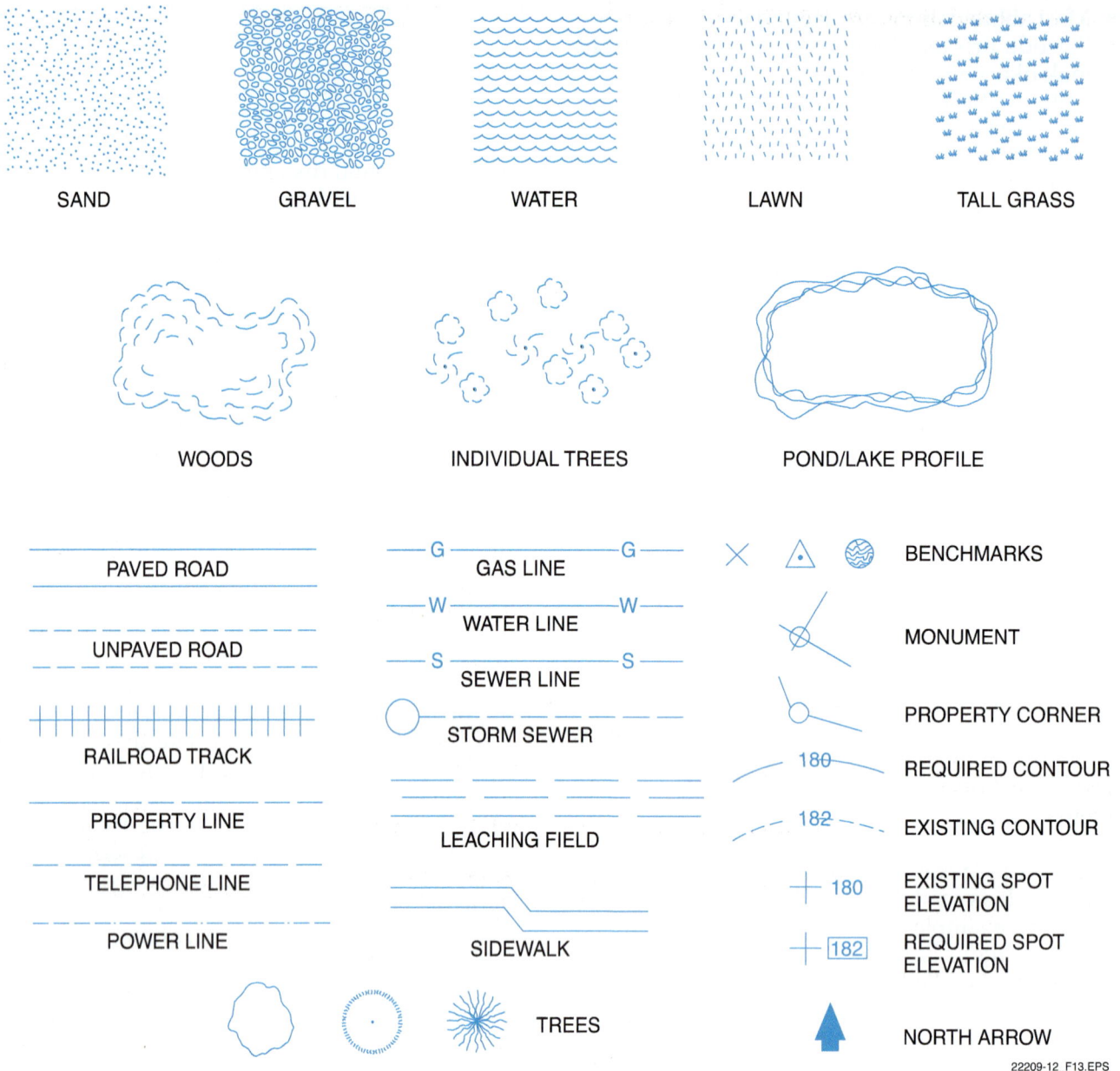

Figure 13 Common site/plot plan symbols.

SAND　　GRAVEL　　WATER　　LAWN　　TALL GRASS

WOODS　　INDIVIDUAL TREES　　POND/LAKE PROFILE

PAVED ROAD

UNPAVED ROAD

RAILROAD TRACK

PROPERTY LINE

TELEPHONE LINE

POWER LINE

G——GAS LINE——G

W——WATER LINE——W

S——SEWER LINE——S

STORM SEWER

LEACHING FIELD

SIDEWALK

TREES

BENCHMARKS

MONUMENT

PROPERTY CORNER

180 REQUIRED CONTOUR

182 EXISTING CONTOUR

180 EXISTING SPOT ELEVATION

182 REQUIRED SPOT ELEVATION

NORTH ARROW

22209-12_F13.EPS

of plans. Carefully review the site plan and get an overall concept of the work required. Look at the contours to determine where excavations are required. For example, if a point on *Figure 15* is at an existing elevation of about 68 feet and a finish elevation of 70 feet is required, 2 feet of fill is required at this location. It is often helpful to divide the site plan into sections or by grid lines to fully understand the amount of work required on the site.

Site plans are drawn using any convenient scale. This may be ⅛ inch to 1 foot, or it may be an engineering scale, such as 1 inch to 20 feet. Larger projects have several site plans showing different scopes of related or similar work. One such plan is the drainage and utility plan. Utility drawings show locations of the water, gas, sanitary sewer,

and electric utilities that will service the building. Drainage plans detail how surface water will be collected, channeled, and dispersed on-site or off-site. Drainage and utility plans illustrate in plan view the size and type of pipes, their length, and the special connections or terminations of the various piping. The elevation of a particular pipe below the surface is given with respect to its **invert**. The invert is the lowest point on the inside of the pipe. The invert is also referred to as the flow line. It is typically noted with the abbreviation for invert and an elevation, for example: INV. 543.15. An example of this can be seen in *Figure A4*. The inverts are shown at the intersections of pipes or other changes in the continuous run of piping, such as a manhole, catch basin, sewer manholes,

CONTOUR MAP OF HILL WITH 5' CONTOUR INTERVAL

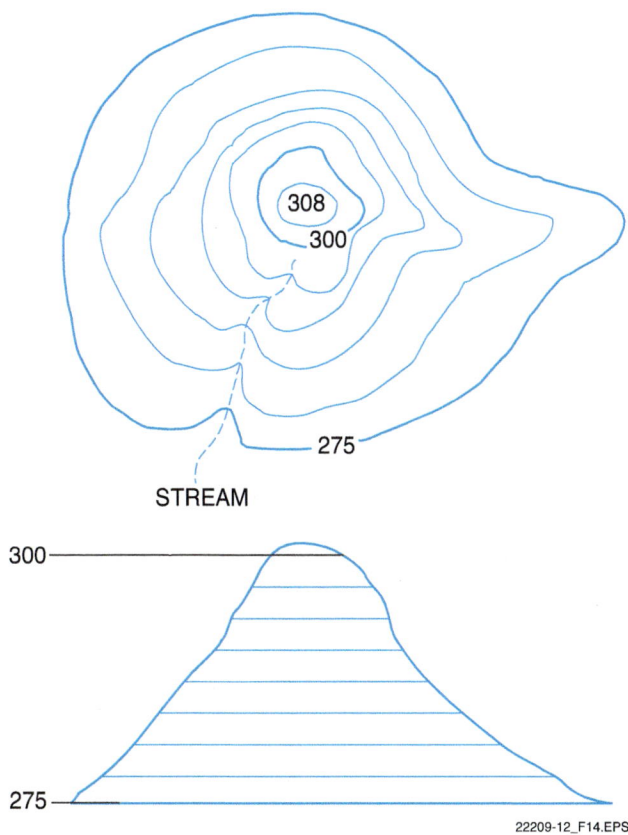

Figure 14 Contour map of a hill.

and so on. Inverts are usually given for piping that has a gravity flow or pitch. Using benchmarks, contours, or spot elevations, the distance of the piping below the surface and the direction of the flow can be quickly calculated.

For projects of a more complex nature, separate drawings showing various site improvements may be needed for clarification. Site improvements may include such items as curbing, walks, retaining walls, paving, fences, steps, benches, and flagpoles. *Figure A8* is one example of such a drawing showing sign placement and installation requirements.

2.3.2 Reading Highway Drawings

Highway construction and improvement projects are a source of employment for many heavy equipment operators. The drawings used for these projects are much different than those used for building sites. This section contains instructions for interpreting highway drawings in *Figures A1* and *A2*. These drawings are located in the *Appendix*.

Figure A2 is an example of a highway cross-section drawing. The sections show the finished road design for several segments of the road, including

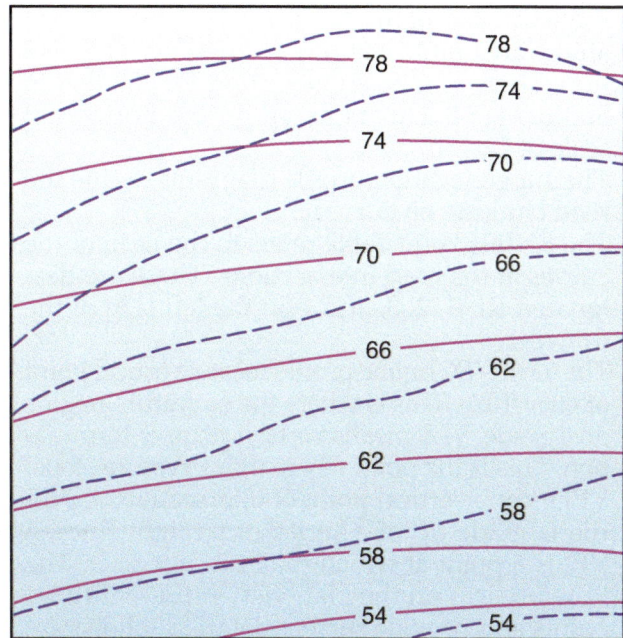

EXISTING GRADE – – – – – –
FINISHED GRADE ————————

Figure 15 Contour lines used for grading.

the pavement buildup and finish elevation. If the grade is the same on both sides of the road, it is only necessary to show one side on the drawing. In this case, both sides are shown because the grades vary and require different levels of cut and fill. The section table at the left of *Figure A2* shows the station segments to which the sections apply. For example, the design shown in Section C applies to four different segments on the left side of the A-line road.

> **NOTE**
> *Figure A1* only goes to Station 8+00, so some of the entries in the Section Table on *Figure A2* refer to another sheet of the drawing package (C6), which is not included here.

The distances shown on the *Figure A2* are measured from the center line of the road to the various outer points. Note that the A-line road is wider than the B-line road by 2.5 feet, as measured from the center line to the hinge point (also called the shoulder break). The Slope Rounding Chart is used to establish the rounding at the top and bottom of a slope in order to provide a finished appearance.

Referring to the profile and plan view on *Figure A1*, there are a few key points to be brought out:

• Station numbers appear along the upper edge of the plan view and the bottom of the profile.
• The starting point for construction is 11.11' from the center line of Highway 270.

- From the start of the A-line road (Station 0.11.11), all the area up to station 3+84 (384') is a fill section. There is then a cut section for 100 feet to Station 4+84. This is based on the horizontal scale of 1" = 80'.
- The superelevation profile diagrams define the road banking on curves.
- The Radius Point Table refers to the radii of the curves at the road intersections. These are designated RP#1, 2, 3, and 4 on the left side of the drawing.
- The term *VPC* on the profile means vertical point of curvature. This is where the curvature begins on a grade. VPT means vertical point of termination. This is the point where the curvature stops.
- VPI means vertical point of intersection. A profile is made up of a series of straight lines; a VPI is a point at which the lines intersect. The PI references on the plan view represent those points and provide further detail as follows:
  - *N northing, E easting* – These are numerical values for coordinate points that could be based on any number of coordinate systems.

They are used for layout of control points from known reference points, such as HP-12 and HP-13. Having two known points, the bearing and distances from either point to a layout control point can be calculated and surveyed in the field.
- *DC (degree of curvature)* – This is a value assigned by the design engineer to designate the sharpness of the curve.
- *R (radius)* – Since all curves are circular they have a radius, which varies with the degree of curvature.
- *T (Tangent)* – This is the length of the line from the PC to the PI and from the PI to the PC.
- *L (length of curve)* – This is the distance from the PC to the PT along the arc of the curve.
- The plan view shows two culverts. The one at Station 3+27 is a 24" diameter corrugated steel pipe (CSP) 71' in length. Its inlet elevation is 140' and its outlet elevation is 134'.
- Two benchmarks—HP-12 and HP-13—are available to serve as elevation reference points.

Additional Resources

Surveying with Construction Applications, Barry F. Kavanaugh; Pearson, Upper Saddle River, NJ.

Guideline for Preparation and Design of Construction Drawings, Reference Manual 10A;

US Department of the Interior National Park Service; *http://cadd.den.nps.gov/downloads/Support/RefMan10A.pdf*

2.0.0 Section Review

22209-12_SR01.EPS

1. The symbol shown in *Figure 1* represents a _____.

   a. soil boring
   b. property corner
   c. stone curb
   d. monument

2. The abbreviation CB stands for _____.

   a. curbstop valve
   b. cubic yards
   c. catch basin
   d. combined sewer

3. The horizontal distance between contour lines on a site plan represents the _____.

   a. degree of slope
   b. elevation
   c. amount of fill needed
   d. amount of cut needed

4. In order to determine the finish elevation of a road at a given station, it would be necessary to refer to the _____.

   a. cross-section
   b. profile
   c. plan view
   d. superelevation

3.0.0 SPECIFICATIONS

Objective 3

Explain specifications and the purpose of specifications.
  a. Identify the types of information found on specifications.
  b. Explain the common format used in specifications.

Trade Term

Uniform Construction Index: The construction specification format adopted by the Construction Specification Institute (CSI). Known as the CSI format.

Specifications, commonly called specs, are written instructions developed by architectural and engineering firms for use by the contractors and subcontractors involved in the construction. Specifications are just as important as the drawings in a set of plans. They furnish what the drawings cannot in that they define the quality of work to be done and the materials to be used. Specifications serve several important purposes:

- Clarify information that cannot be shown on the drawings
- Identify work standards, types of materials to be used, and the responsibility of various parties to the contract
- Provide information on details of construction
- Serve as a guide for contractors bidding on the construction job
- Serve as a standard of quality for materials and workmanship
- Serve as a guide for compliance with building codes and zoning ordinances
- Provide the basis of agreement between the owner, architect, and contractors in settling any disputes

Equipment operators use the specification to determine the quality of fill, the depth of top soil, special finish grading requirements, and landscaping requirements. The plans are often more specific to the job than the specifications. Therefore, notes on the plans may be considered by the architect/owner to be the true intent. Equipment operators must be very careful to watch for discrepancies between the plans and specifications and report them to a supervisor immediately.

3.1.0 Organization of Specifications

Specifications consist of various elements that may differ somewhat for particular construction jobs. Basically, two types of information are contained in a set of specifications: special and general conditions, and technical aspects of construction.

3.1.1 Special and General Conditions

Special and general conditions cover the nontechnical aspects of the contractual agreements. Special conditions cover topics such as safety and temporary construction. General conditions cover the following points of information:

- Contract terms
- Responsibilities for examining the construction site
- Types and limits of insurance
- Permits and payments of fees
- Use and installation of utilities
- Supervision of construction
- Other pertinent items

The general conditions section is the area of the construction contract where misunderstandings often occur. Therefore, these conditions are usually much more explicit on large, complicated construction projects. Part of a typical residential material specification is shown in *Figure 16*.

The earthwork sections of this specification are shown as Item 1 (Excavation) and under the headings *Other Onsite Improvements and Landscaping, Planting, and Finish Grading*.

> **NOTE**
> Residential specifications often do not spell out general conditions and are basically material specifications only.

3.1.2 Technical Aspects

The technical aspects section includes information on materials that are specified by standard numbers and by standard testing organizations such as the American Society for Testing and Materials (ASTM) International. The technical data section of specifications can be any of three types:

- *Outline specifications* – These specifications list the materials to be used in order of the basic parts of the job, such as foundation, floors, and walls.
- *Fill-in specifications* – This is a standard form filled in with pertinent information. It is typically used on smaller jobs.

UNITED STATES DEPARTMENT OF AGRICULTURE
U.S. DEPARTMENT OF HOUSING AND URBAN DEVELOPMENT-FEDERAL
HOUSING ADMINISTRATION
U.S. DEPARTMENT OF VETERANS AFFAIRS

FORM APPROVED
OMB NO. 0575-0042

☐ **Proposed Construction**

☐ **Under Construction**

DESCRIPTION OF MATERIALS

No. _____
(To be inserted by Agency)

Property address _____ City _____ State Oklahoma

Mortgagor or Sponsor _____   _____
(Name)   (Address)

Contractor or Builder _____   _____
(Name)   (Address)

INSTRUCTIONS

1. For additional information on how this form is to be submitted, number of copies, etc., see the instructions applicable to the FHA Application for Mortgage Insurance, VA Request for Determination of Reasonable Value or other, as the case may be.

2. Describe all materials and equipment to be used, whether or not shown on the drawings, by marking an X in each appropriate check-box and entering the information called for in each space. If space is inadequate, enter See misc.) and describe under item 27 or on an attached sheet: THE USE OF PAINT CONTAINING MORE THAN THE PERCENT OF LEAD BY WEIGHT PERMITTED BY LAW IS PROHIBITED.

3. Work not specifically described or shown will not be considered unless

required, then the minimum acceptable will be assumed. Work exceeding minimum requirements cannot be considered unless specifically described.

4. Include no alternates, or equal phrases, or contradictory items. (Consideration of a request for acceptance of substitute materials or equipment is not thereby precluded.)

5. Include signatures required at the end of this form.

6. The construction shall be completed in compliance with the related drawings and specifications, as amended during processing. The specifications include this Description of Materials and the applicable building code.

1. **EXCAVATION:**
Bearing soil, type Firm clay; Note: Where fill is in excess of 18", concrete piers to be installed

2. **FOUNDATIONS:** at 8' O.C. and the cost will be added to the contract.
Footings: concrete mix transite 14"x18" ftg ; strength psi 2500 PSI Reinforcing (4) 5/8" steel rebar
Foundation wall: material 2500 PSI concrete   concrete   Reinforcing _____
Interior foundation wall: material 2500 PSI concrete   Party foundation wall 2500 concrete
Columns: material and sizes _____   Piers: material and reinforcing _____
Girders: material and sizes _____   Sills: material W.Coast Utility Douglas Fir w/sill sealer
Basement entrance areaway _____   Window areaways _____
Waterproofing waterproof mix in concrete   Footing drains open mortar joints
Termite protection Pretreat soil _____ Chlordane and issue 5 year warranty.
Basementless space _____ ; foundation vents _____
Special _____ turbed soil.

─────────────────────────────

ion construction: 6"x8" poured monolithic with the s ___
floor: 4" concrete slab with 6x6-W1.4xW1.4 WWF smooth trowel finish.

TERRACES:
stoops: 4" concrete slab- smooth trowel finish.
patio: 4" concrete slab- smooth trowel finish. - see plans for size.

GARAGES: automatic garage door opener
foundation: 14"x 18" concrete footing with (4) 5/8" rebars; 6" concrete stem wall
floor: 4" concrete with 6x6-W1.4xW1.4 WWF; smooth trowel finish floors.
interior: 3/8" prefinished Sheetrock on walls; texture and paint 1/2" Sheetrock

WALKS AND DRIVEWAYS: on ceiling
                    see
Driveway: width plans ı; base material tamped earth; thickness 4 ı; surfacing material concrete ; thickness 4
Front walk: width 36" ; material concrete ; thickness 4 ı. Service walk: width _____ ; material _____ ; thickness _____
Steps: material _____ ; treads _____ ı; risers _____ ı. Check walls _____

OTHER ONSITE IMPROVEMENTS:
(Specify all exterior onsite improvements not described elsewhere, including items such as unusual grading, drainage structures, retaining walls, fence, railings, and accessor structures.)

NOTE: All dimensions to be rechecked on site prior to beginning construction by
builder and builder shall be responsible for the same.

LANDSCAPING, PLANTING, AND FINISH GRADING:
Topsoil _____ ı thick: ☐ front yard: ☐ side yards; ☐ rear yard to _____ feet behind main building.
Lawns *(seeded, sodded, or sprigged)*:. ☐ front yard _____ ; ☐ side yards _____ ; ☐ rear yard _____
Planting: ☐ as specified and shown on drawings; ☐ as follows:
_____ Shade trees, decidous. _____ ı caliper.   _____ Evergreen trees _____ ſ to _____ ſ, B & B.
_____ Low flowering trees, decidous. _____ ſ to _____ ſ   _____ Evergreen shrubs _____ ſ to _____ ſ, B & B.
_____ High-growing shrubs, decidous. _____ ſ to _____ ſ   _____ Vines, 2-years _____
_____ Medium-growing shrubs, decidous, _____ ſ to _____ ſ   _____
_____ Low-growing shrubs, decidous. _____ ſ to _____ ſ   _____

IDENTIFICATION. This exhibit shall be identified by the signature of the builder, or sponsor, and/or the propsed mortgagor if the latter is known at the time of application.

Date _____   Signature _____

Signature _____

4

22209-12_F16.EPS

Figure 16 Parts of a typical materials specification.

- *Complete specifications* – For ease of use, most specifications written for large construction jobs are organized in the Construction Specification Institute format called the Uniform Construction Index. This is known as the CSI format and is explained in the next section.

3.2.0 Format of Specifications

The most commonly used specification-writing format in North America is the *MasterFormat*™. This standard was developed jointly by the Construction Specifications Institute (CSI) and Construction Specifications Canada (CSC). In this format, the specifications are divided into a series of sections dealing with the construction requirements, products, and activities. Using this format makes it easy to write and use the specification, and it is easily understandable by the different trades.

For many years prior to 2004, the organization of construction specifications and suppliers catalogs was based on a standard with 16 sections, otherwise known as divisions, where the divisions and their subsections were individually identified by a five-digit numbering system. The first two digits represented the division number and the next three individual numbers represented successively lower levels of breakdown. For example, the number 13213 represents division 13, subsection 2, sub-subsection 1 and sub-sub-subsection 3. In this older version of the standard, electrical systems, including any electronic or special electrical systems, were lumped together under Division 16 – *Electrical*. Today, specifications conforming to the 16-division format may still be in use, so it is a good idea to be familiar with both formats.

In 2004, the *MasterFormat*™ standard underwent a major change. What had been 16 divisions was expanded to four major groupings and 49 divisions with some divisions reserved for future expansion. The first 14 divisions are essentially the same as the old format. Subjects under the old Division 15 – *Mechanical* have been relocated to new divisions 22 and 23. The basic subjects under old Division 16 – *Electrical* have been relocated to new divisions 26 and 27. In addition, the numbering system was changed to 6 digits to allow for more subsections in each division, which allows for finer definition. In the new numbering system, the first two digits represent the division number. The next two digits represent subsections of the division and the two remaining digits represent the third level sub-subsection numbers. The fourth level, if required, is a decimal and number added to the end of the last two digits. This allows tasks to be divided into finer definitions. For example, Division 31 is entitled *Earthwork* (see *Figure 17*), so much of your work will fall under this division. Use *Figure 17* to look up code 312219.13 and you will find it relates specifically to spreading and grading topsoil.

MasterFormat® 2012 – Numbers and Titles April 2012

31 00 00	**Earthwork**
31 01 00	**Maintenance of Earthwork**
31 01 10	Maintenance of Clearing
31 01 20	Maintenance of Earth Moving
31 01 40	Maintenance of Shoring and Underpinning
31 01 50	Maintenance of Excavation Support and Protection
31 01 60	Maintenance of Special Foundations and Load Bearing Elements
31 01 62	Maintenance of Driven Piles
31 01 62.61	Driven Pile Repairs
31 01 63	Maintenance of Bored and Augered Piles
31 01 63.61	Bored and Augered Pile Repairs
31 01 70	Maintenance of Tunneling and Mining
31 01 70.61	Tunnel Leak Repairs
31 05 00	**Common Work Results for Earthwork**
31 05 13	Soils for Earthwork
31 05 16	Aggregates for Earthwork
31 05 19	Geosynthetics for Earthwork
31 05 19.13	Geotextiles for Earthwork
31 05 19.16	Geomembranes for Earthwork
31 05 19.19	Geogrids for Earthwork
31 05 19.23	Geosynthetic Clay Liners
31 05 19.26	Geocomposites
31 05 19.29	Geonets
31 05 23	Cement and Concrete for Earthwork
31 06 00	**Schedules for Earthwork**
31 06 10	Schedules for Clearing
31 06 20	Schedules for Earth Moving
31 06 20.13	Trench Dimension Schedule
31 06 20.16	Backfill Material Schedule
31 06 40	Schedules for Shoring and Underpinning
31 06 50	Schedules for Excavation Support and Protection
31 06 60	Schedules for Special Foundations and Load Bearing Elements
31 06 60.13	Driven Pile Schedule
31 06 60.16	Caisson Schedule
31 06 70	Schedules for Tunneling and Mining
31 08 00	**Commissioning of Earthwork**
31 08 13	Pile Load Testing
31 08 13.13	Dynamic Pile Load Testing
31 08 13.16	Static Pile Load Testing
31 09 00	**Geotechnical Instrumentation and Monitoring of Earthwork**
31 09 13	Geotechnical Instrumentation and Monitoring
31 09 13.13	Groundwater Monitoring During Construction
31 09 16	Foundation Performance Instrumentation
31 09 16.26	Bored and Augered Pile Load Tests
31 10 00	**Site Clearing**
31 11 00	**Clearing and Grubbing**
31 12 00	**Selective Clearing**

111

22209-12_F17A.EPS

Figure 17 CSI Division 31 – *Earthwork* (1 of 6).

31 13 00	**Selective Tree and Shrub Removal and Trimming**
31 13 13	Selective Tree and Shrub Removal
31 13 16	Selective Tree and Shrub Trimming
31 14 00	**Earth Stripping and Stockpiling**
31 14 13	Soil Stripping and Stockpiling
31 14 13.13	Soil Stripping
31 14 13.16	Soil Stockpiling
31 14 13.23	Topsoil Stripping and Stockpiling
31 14 16	Sod Stripping and Stockpiling
31 14 16.13	Sod Stripping
31 14 16.16	Sod Stockpiling

31 20 00   Earth Moving

31 21 00	**Off-Gassing Mitigation**
31 21 13	Radon Mitigation
31 21 13.13	Radon Venting
31 21 16	Methane Mitigation
31 21 16.13	Methane Venting
31 22 00	**Grading**
31 22 13	Rough Grading
31 22 16	Fine Grading
31 22 16.13	Roadway Subgrade Reshaping
31 22 19	Finish Grading
31 22 19.13	Spreading and Grading Topsoil
31 23 00	**Excavation and Fill**
31 23 13	Subgrade Preparation
31 23 16	Excavation
31 23 16.13	Trenching
31 23 16.16	Structural Excavation for Minor Structures
31 23 16.26	Rock Removal
31 23 19	Dewatering
31 23 23	Fill
31 23 23.13	Backfill
31 23 23.23	Compaction
31 23 23.33	Flowable Fill
31 23 23.43	Geofoam
31 23 33	Trenching and Backfilling
31 24 00	**Embankments**
31 24 13	Roadway Embankments
31 24 16	Railway Embankments
31 25 00	**Erosion and Sedimentation Controls**
31 25 14	Stabilization Measures for Erosion and Sedimentation Control
31 25 14.13	Hydraulically-Applied Erosion Control
31 25 14.16	Rolled Erosion Control Mats and Blankets
31 25 24	Structural Measures for Erosion and Sedimentation Control
31 25 24.13	Rock Barriers
31 25 34	Retention Measures for Erosion and Sedimentation Controls
31 25 34.13	Rock Basins

31 30 00   Earthwork Methods

31 31 00	**Soil Treatment**

112

22209-12_F17B.EPS

Figure 17 CSI Division 31 – *Earthwork* (2 of 6).

31 31 13	Rodent Control
31 31 13.16	Rodent Control Bait Systems
31 31 13.19	Rodent Control Traps
31 31 13.23	Rodent Control Electronic Systems
31 31 13.26	Rodent Control Repellants
31 31 16	Termite Control
31 31 16.13	Chemical Termite Control
31 31 16.16	Termite Control Bait Systems
31 31 16.19	Termite Control Barriers
31 31 19	Vegetation Control
31 31 19.13	Chemical Vegetation Control

31 32 00   Soil Stabilization

31 32 13	Soil Mixing Stabilization
31 32 13.13	Asphalt Soil Stabilization
31 32 13.16	Cement Soil Stabilization
31 32 13.19	Lime Soil Stabilization
31 32 13.23	Fly-Ash Soil Stabilization
31 32 13.26	Lime-Fly-Ash Soil Stabilization
31 32 16	Chemical Treatment Soil Stabilization
31 32 16.13	Polymer Emulsion Soil Stabilization
31 32 17	Water Injection Soil Stabilization
31 32 19	Geosynthetic Soil Stabilization and Layer Separation
31 32 19.13	Geogrid Soil Stabilization
31 32 19.16	Geotextile Soil Stabilization
31 32 19.19	Geogrid Layer Separation
31 32 19.23	Geotextile Layer Separation
31 32 23	Pressure Grouting Soil Stabilization
31 32 23.13	Cementitious Pressure Grouting Soil Stabilization
31 32 23.16	Chemical Pressure Grouting Soil Stabilization
31 32 33	Shotcrete Soil Slope Stabilization
31 32 36	Soil Nailing
31 32 36.13	Driven Soil Nailing
31 32 36.16	Grouted Soil Nailing
31 32 36.19	Corrosion-Protected Soil Nailing
31 32 36.23	Jet-Grouted Soil Nailing
31 32 36.26	Launched Soil Nailing

31 33 00   Rock Stabilization

31 33 13	Rock Bolting and Grouting
31 33 23	Rock Slope Netting
31 33 26	Rock Slope Wire Mesh
31 33 33	Shotcrete Rock Slope Stabilization
31 33 43	Vegetated Rock Slope Stabilization

31 34 00   Soil Reinforcement

31 34 19	Geosynthetic Soil Reinforcement
31 34 19.13	Geogrid Soil Reinforcement
31 34 19.16	Geotextile Soil Reinforcement
31 34 23	Fiber Soil Reinforcement
31 34 23.13	Geosynthetic Fiber Soil Reinforcement

31 35 00   Slope Protection

31 35 19	Geosynthetic Slope Protection
31 35 19.13	Geogrid Slope Protection
31 35 19.16	Geotextile Slope Protection

113

22209-12_F17C.EPS

Figure 17 CSI Division 31 – *Earthwork* (3 of 6).

MasterFormat® 2012 – Numbers and Titles April 2012

31 35 19.19 Slope Protection with Mulch Control Netting
31 35 23 Slope Protection with Slope Paving
31 35 23.13 Cast-In-Place Concrete Slope Paving
31 35 23.16 Precast Concrete Slope Paving
31 35 23.19 Concrete Unit Masonry Slope Paving
31 35 26 Containment Barriers
31 35 26.13 Clay Containment Barriers
31 35 26.16 Geomembrane Containment Barriers
31 35 26.23 Bentonite Slurry Trench

31 36 00 Gabions

31 36 13 Gabion Boxes
31 36 19 Gabion Mattresses
31 36 19.13 Vegetated Gabion Mattresses

31 37 00 Riprap

31 37 13 Machined Riprap
31 37 16 Non-Machined Riprap
31 37 16.13 Rubble-Stone Riprap
31 37 16.16 Concrete Unit Masonry Riprap
31 37 16.19 Sacked Sand-Cement Riprap

31 40 00 Shoring and Underpinning

31 41 00 Shoring

31 41 13 Timber Shoring
31 41 16 Sheet Piling
31 41 16.13 Steel Sheet Piling
31 41 16.16 Plastic Sheet Piling
31 41 19 Metal Hydraulic Shoring
31 41 19.13 Aluminum Hydraulic Shoring
31 41 23 Pneumatic Shoring
31 41 33 Trench Shielding

31 43 00 Concrete Raising

31 43 13 Pressure Grouting
31 43 13.13 Concrete Pressure Grouting
31 43 13.16 Polyurethane Pressure Grouting
31 43 16 Compaction Grouting
31 43 19 Mechanical Jacking

31 45 00 Vibroflotation and Densification

31 45 13 Vibroflotation
31 45 16 Densification

31 46 00 Needle Beams

31 46 13 Cantilever Needle Beams

31 48 00 Underpinning

31 48 13 Underpinning Piers
31 48 19 Bracket Piers
31 48 23 Jacked Piers
31 48 33 Micropile Underpinning

31 50 00 Excavation Support and Protection

31 51 00 Anchor Tiebacks

31 51 13 Excavation Soil Anchors
31 51 16 Excavation Rock Anchors

114

22209-12_F17D.EPS

Figure 17 CSI Division 31 – *Earthwork* (4 of 6).

31 52 00	**Cofferdams**
31 52 13	Sheet Piling Cofferdams
31 52 16	Timber Cofferdams
31 52 19	Precast Concrete Cofferdams

31 53 00	**Cribbing and Walers**
31 53 13	Timber Cribwork

31 54 00	**Ground Freezing**

31 56 00	**Slurry Walls**
31 56 13	Bentonite Slurry Walls
31 56 13.13	Soil-Bentonite Slurry Walls
31 56 13.16	Cement-Bentonite Slurry Walls
31 56 13.19	Slag-Cement-Bentonite Slurry Walls
31 56 13.23	Slag-Cement-Bentonite Slurry Walls
31 56 13.26	Pozzolan-Bentonite Slurry Walls
31 56 13.29	Organically-Modified Bentonite Slurry Walls
31 56 16	Attipulgite Slurry Walls
31 56 16.13	Soil-Attipulgite Slurry Walls
31 56 19	Slurry-Geomembrane Composite Slurry Walls
31 56 23	Lean Concrete Slurry Walls
31 56 26	Bio-Polymer Trench Drain

31 60 00	**Special Foundations and Load-Bearing Elements**

31 62 00	**Driven Piles**
31 62 13	Concrete Piles
31 62 13.13	Cast-in-Place Concrete Piles
31 62 13.16	Concrete Displacement Piles
31 62 13.19	Precast Concrete Piles
31 62 13.23	Prestressed Concrete Piles
31 62 13.26	Pressure-Injected Footings
31 62 16	Steel Piles
31 62 16.13	Sheet Steel Piles
31 62 16.16	Steel H Piles
31 62 16.19	Unfilled Tubular Steel Piles
31 62 19	Timber Piles
31 62 23	Composite Piles
31 62 23.13	Concrete-Filled Steel Piles
31 62 23.16	Wood and Cast-In-Place Concrete Piles

31 63 00	**Bored Piles**
31 63 13	Bored and Augered Test Piles
31 63 16	Auger Cast Grout Piles
31 63 19	Bored and Socketed Piles
31 63 19.13	Rock Sockets for Piles
31 63 23	Bored Concrete Piles
31 63 23.13	Bored and Belled Concrete Piles
31 63 23.16	Bored Friction Concrete Piles
31 63 26	Drilled Caissons
31 63 26.13	Fixed End Caisson Piles
31 63 26.16	Concrete Caissons for Marine Construction
31 63 29	Drilled Concrete Piers and Shafts
31 63 29.13	Uncased Drilled Concrete Piers

115

22209-12_F17E.EPS

Figure 17 CSI Division 31 – *Earthwork* (5 of 6).

MasterFormat® 2012 – Numbers and Titles April 2012

31 63 29.16	Cased Drilled Concrete Piers
31 63 33	Drilled Micropiles
31 64 00	**Caissons**
31 64 13	Box Caissons
31 64 16	Excavated Caissons
31 64 19	Floating Caissons
31 64 23	Open Caissons
31 64 26	Pneumatic Caissons
31 64 29	Sheeted Caissons
31 66 00	**Special Foundations**
31 66 13	Special Piles
31 66 13.13	Rammed Aggregate Piles
31 66 15	Helical Foundation Piles
31 66 16	Special Foundation Walls
31 66 16.13	Anchored Foundation Walls
31 66 16.23	Concrete Cribbing Foundation Walls
31 66 16.26	Metal Cribbing Foundation Walls
31 66 16.33	Manufactured Modular Foundation Walls
31 66 16.43	Mechanically Stabilized Earth Foundation Walls
31 66 16.46	Slurry Diaphragm Foundation Walls
31 66 16.53	Soldier-Beam Foundation Walls
31 66 16.56	Permanently-Anchored Soldier-Beam Foundation Walls
31 66 19	Refrigerated Foundations
31 68 00	**Foundation Anchors**
31 68 13	Rock Foundation Anchors
31 68 16	Helical Foundation Anchors
31 70 00	**Tunneling and Mining**
31 71 00	**Tunnel Excavation**
31 71 13	Shield Driving Tunnel Excavation
31 71 16	Tunnel Excavation by Drilling and Blasting
31 71 19	Tunnel Excavation by Tunnel Boring Machine
31 71 23	Tunneling by Cut and Cover
31 72 00	**Tunnel Support Systems**
31 72 13	Rock Reinforcement and Initial Support
31 72 16	Steel Ribs and Lagging
31 73 00	**Tunnel Grouting**
31 73 13	Cement Tunnel Grouting
31 73 16	Chemical Tunnel Grouting
31 74 00	**Tunnel Construction**
31 74 13	Cast-in-Place Concrete Tunnel Lining
31 74 16	Precast Concrete Tunnel Lining
31 74 19	Shotcrete Tunnel Lining
31 75 00	**Shaft Construction**
31 75 13	Cast-in-Place Concrete Shaft Lining
31 75 16	Precast Concrete Shaft Lining
31 77 00	**Submersible Tube Tunnels**
31 77 13	Trench Excavation for Submerged Tunnels
31 77 16	Tube Construction (Outfitting Tunnel Tubes)
31 77 19	Floating and Laying Submerged Tunnels

116

22209-12_F17F.EPS

Figure 17 CSI Division 31 – *Earthwork* (6 of 6).

3.0.0 Section Review

1. In a specification, safety is covered under _____.

   a. special conditions
   b. general conditions
   c. technical details
   d. its own section

2. In a specification, information on earthmoving requirements are likely to be found in _____.

   a. the general conditions
   b. Division 16
   c. Division 26
   d. Division 31

SUMMARY

Construction plans show where a project will be located and how it will be built. All members of the construction team use these drawings, so they are often crowded with information and can be confusing to read. Regardless of that, part of the operator's job is to be able to find the information needed to complete the excavation and grading work. The main concern will be to interpret existing and proposed elevation readings in order to ensure that the site is prepared for construction according to the specification. Proper leveling and grading are vital to the successful comple-

tion and durability of any construction project. Highways and buildings are only as stable as the ground they are built on, so the grading, cut and fill, and compacting tasks are completed as called for on the plans.

In addition to the drawings, most jobs have specifications that provide detailed information not included on the drawings. Specifications define and clarify the scope of a job and identify specific materials and components to be used in the construction.

1. The revision status block of a drawing appears only on the first sheet of the drawing set.

    a. True
    b. False

2. The elevation of the existing ground is shown as a _____.

    a. solid line on the cross-section drawing
    b. dashed line on the plan view
    c. solid line on the profile drawing
    d. dashed line on the profile drawing

3. An operator has been assigned to a new job site and needs to become acquainted with the project. What is the best plan to study to identify the location of existing roads, easements, and utility information, as well as proposed construction?

    a. Structural drawings
    b. Floor plan
    c. Foundation drawings
    d. Site plan

4. Elevation drawings show _____.

    a. a straight ahead view of a building
    b. the natural grade of a project
    c. the proposed grade of a project
    d. the height of the roof line

5. A line on a drawing representing an object or area that is *not* shown in its entirety is known as a _____.

    a. phantom line
    b. cutting plane
    c. leader line
    d. break line

6. Contour lines show elevation changes on diagrams. Existing contours are usually shown as _____ lines and proposed finish grade contours are shown as _____ lines.

    a. heavy; light
    b. dashed; solid
    c. solid; dashed
    d. light; heavy

7. Contour lines that make closed loops on a plan represent _____.

    a. property boundaries
    b. depressions or hills
    c. building foundations
    d. roadway direction

8. Widely spaced contour lines represent _____.

    a. steep terrain
    b. hilly terrain
    c. rocky terrain
    d. level terrain

9. What drawing symbol represents a property line?

    a. + + + + + PPPPPPP++++++++
    b. __ __
    c. P
    d. _ _ _ _ _ _ b _ _ _ _ _ _ _ _ _ _

10. Refer to *Figure 1*. Which of these symbols is used to represent a benchmark?

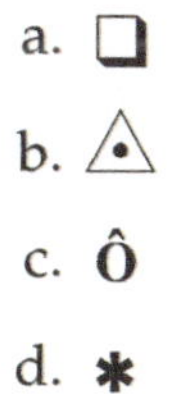

    a. ❑
    b. △ (with dot)
    c. Ô
    d. ✻

    22209-12_RQ01.EPS

    Figure 1

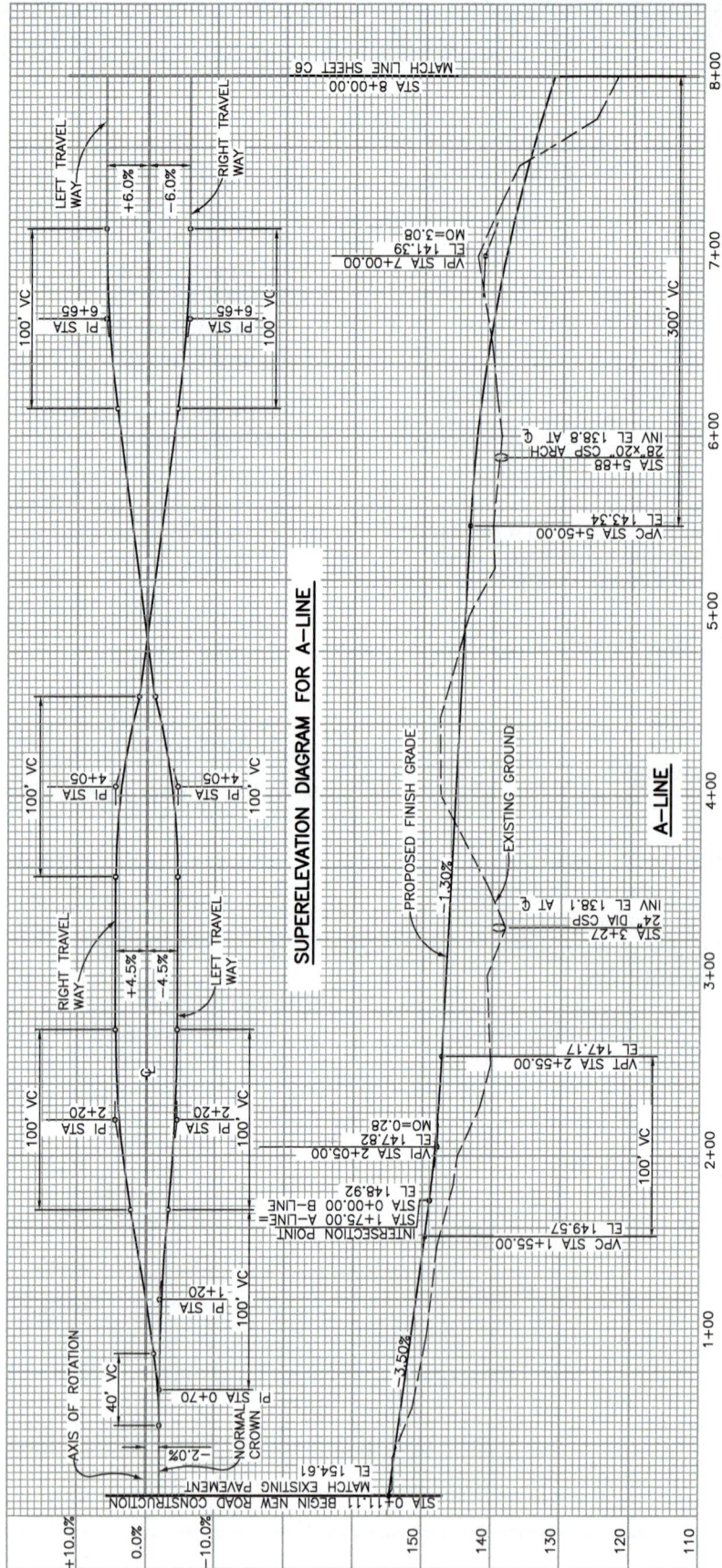

SUPERELEVATION DIAGRAM FOR A-LINE

A-LINE

Figure 2

22209-12_RQ02.EPS

11. On *Figure 2*, the elevation of the finished roadway at Station 7+00 is approximately _____ .

   a. 130'
   b. 135'
   c. 138'
   d. 141'

12. On *Figure 2*, the total change in elevation from the beginning of the A-line road to its end is approximately _____ .

   a. 45'
   b. 33'
   c. 28'
   d. 23'

13. Specifications are used to expand on the drawing set. When there is a difference between a drawing and a specification, you should _____ .

   a. notify your supervisor
   b. follow the drawing set
   c. follow the specification
   d. notify the owner

14. When you are new to a job and need to review the plans, you need to look at only the site plan, since all of the grading information can be found on it.

   a. True
   b. False

15. The CSI *MasterFormat*™ of 2004 identifies how many divisions?

   a. 16
   b. 25
   c. 32
   d. 49

Trade Terms Introduced in This Module

Change order: A formal instruction describing and authorizing a project change.

Contour lines: Imaginary lines on a site/plot plan that connect points of the same elevation. Contour lines never cross each other.

Easement: A legal right-of-way provision on another person's property (for example, the right of a neighbor to build a driveway or a public utility to install water and gas lines on the property). A property owner cannot build on an area where an easement has been identified.

Elevation view: A drawing giving a view from the front or side of a structure.

Invert: The lowest portion of the interior of a pipe, also called the flow line.

Loadbearing: A base designed to support the weight of an object of structure.

Monuments: Physical structures that mark the locations of survey points.

Plan view: A drawing that represents a view looking down on an object.

Property lines: The recorded legal boundaries of a piece of property.

Request for information (RFI): A form used to question discrepancies on the drawings or to ask for clarification.

Setback: The distance from a property line in which no structures are permitted.

Uniform Construction Index: The construction specification format adopted by the Construction Specification Institute (CSI). Known as the CSI format.

Appendix

EXAMPLES OF CIVIL DRAWINGS

The *Appendix* contains drawings that are referenced in the text. The drawings are included with this Trainee Guide in oversize form to make them more readable. The appendix contains the following drawings:

- Figure A1 – Sample Roadway Plan and Profile (C8) – 499/41001
- Figure A2 – Sample Roadway Cross-Sections (C7) – 499/41001
- Figure A3 – Sample Project Overview Site Plan (G1) – 499/41001A
- Figure A4 – Sample Water Line Plan and Profile (C9) – 499/41001
- Figure A5 – Sample Parking Area Grading Plan (C5) – 499/41001
- Figure A6 (1 of 2) – Sample Symbol Sheet (C2) – 499/41001
- Figure A6 (2 of 2) – Sample Mapping Symbols (C3) – 499/41001
- Figure A7 – Sample Abbreviation Sheet (C1) – 499/41001
- Figure A8 – Sample Plan and Details for Roadway Signs and Pavement Markings (C12) – 499/41001

> **NOTE**
>
> These drawings are taken from a book of sample drawings produced by the National Park Service. They are not necessarily related to the same project.

Additional Resources

This module presents thorough resources for task training. The following resource material is suggested for further study.

Surveying with Construction Applications, Barry F. Kavanaugh; Pearson, Upper Saddle River, NJ.
Guideline for Preparation and Design of Construction Drawings, Reference Manual 10A;
US Department of the Interior National Park Service; **http://cadd.den.nps.gov/downloads/Support/ RefMan10A.pdf**

Figure Credits

Reprinted courtesy of Caterpillar Inc., Module opener

Courtesy of Carolina Bridge Company, Inc., Figure 1

Topaz Publications, Inc., SA01, SA02, Figures 7 and 8

John Hoerlein, Figure 3

Courtesy of Mark Jones, Figures 4 and 5

Courtesy of Michael Reif, **nosarasprings.com**, Figure 6

The Construction Specifications Institute (CSI), Figure 17

MasterFormat® Numbers and Titles used in this book are from *MasterFormat*®, published by The Construction Specifications Institute (CSI) and Construction Specifications Canada (CSC), and are used with permission from CSI. For those interested in a more in-depth explanation of *MasterFormat*® and its use in the construction industry visit **www.masterformat.com** or contact:

The Construction Specifications Institute

110 South Union Street, Suite 100

Alexandria, VA 22314

800-689-200; 703-684-0300; **www.csinet.org**

Courtesy of National Park Service Denver Service Center, Appendix

Section Review Answers

Answer	Section Reference	Objective
Section One		
1 c	1.1.0	1a
2 b	1.2.1	1b
3 a	1.3.1	1c
4 d	1.4.0	1d
Section Two		
1 b	2.1.0, Figure A6	2a
2 c	2.2.1, Figure A7	2b
3 a	2.3.1	2c
4 b	2.3.2	2c
Section Three		
1 a	3.1.1	3a
2 d	3.2.0	3b

NCCER CURRICULA — USER UPDATE

NCCER makes every effort to keep its textbooks up-to-date and free of technical errors. We appreciate your help in this process. If you find an error, a typographical mistake, or an inaccuracy in NCCER's curricula, please fill out this form (or a photocopy), or complete the online form at **www.nccer.org/olf**. Be sure to include the exact module ID number, page number, a detailed description, and your recommended correction. Your input will be brought to the attention of the Authoring Team. Thank you for your assistance.

Instructors – If you have an idea for improving this textbook, or have found that additional materials were necessary to teach this module effectively, please let us know so that we may present your suggestions to the Authoring Team.

NCCER Product Development and Revision

13614 Progress Blvd., Alachua, FL 32615

Email: curriculum@nccer.org
Online: www.nccer.org/olf

❑ Trainee Guide   ❑ Lesson Plans   ❑ Exam   ❑ PowerPoints   Other _____

Craft / Level: _____   Copyright Date: _____

Module ID Number / Title: _____

Section Number(s): _____

Description: _____

_____

_____

_____

_____

Recommended Correction: _____

_____

_____

_____

_____

Your Name: _____

Address: _____

_____

Email: _____   Phone: _____

38102-18
Rigging Practices

OVERVIEW

Rigging is the preparation of a load for movement, as well as preparation of the hardware and other components used to connect the load to a crane. Rigging is associated with all types of cranes, and rigging skills are also required to move and position equipment inside buildings and other areas where cranes are not involved. This module will provide insight into rigging hardware, lifting slings and their proper use, and various types of rigging equipment.

Module Nine

Trainees with successful module completions may be eligible for credentialing through the NCCER Registry. To learn more, go to **www.nccer.org** or contact us at 1.888.622.3720. Our website has information on the latest product releases and training, as well as online versions of our *Cornerstone* magazine and Pearson's product catalog.

Your feedback is welcome. You may email your comments to **curriculum@nccer.org**, send general comments and inquiries to **info@nccer.org**, or fill in the User Update form at the back of this module.

This information is general in nature and intended for training purposes only. Actual performance of activities described in this manual requires compliance with all applicable operating, service, maintenance, and safety procedures under the direction of qualified personnel. References in this manual to patented or proprietary devices do not constitute a recommendation of their use.

Objectives

When you have completed this module, you will be able to do the following:

1. Identify and describe various types of rigging hardware.
   a. Identify and describe various hooks, shackles, eyebolts, and clamps.
   b. Identify and describe various lugs, turnbuckles, plates, and spreader beams.
2. Identify and describe various types of slings and sling hitches.
   a. Identify and describe wire-rope slings and their proper care.
   b. Identify and describe synthetic slings and their proper care.
   c. Identify and describe chain slings and their proper care.
   d. Explain the significance of sling angles and describe common hitches.
   e. Describe how to properly rig and handle piping materials and rebar.
   f. Identify and describe how to use taglines and knots for load control.
   g. Identify common rigging-related safety precautions.
3. Identify and describe how to use various types of hoisting and jacking equipment.
   a. Identify and describe how to use manual and powered hoisting equipment.
   b. Identify and describe how to use jacks.

Performance Tasks

Under the supervision of your instructor, you should be able to do the following:

1. Inspect various types of rigging components and report on the condition and suitability for a task.
2. Configure a sling to produce a single-wrap basket hitch.
3. Configure a sling to produce a double-wrap basket hitch.
4. Configure a sling to produce a single-wrap choker hitch.
5. Configure a sling to produce a double-wrap choker hitch.
6. Select the correct tagline for a specified application.
7. Tie specific instructor-selected knots.
8. Select, inspect, and demonstrate the safe use of the following rigging equipment:
   - Block and tackle
   - Chain hoist
   - Ratchet-lever hoist
   - One or more types of jack

Trade Terms

Basket hitch
Bird caging
Blind hole
Bridle hitch
Center of gravity (CG)
Choker hitch
Equalizer beams
Equalizer plates
Gantry
Hauling line
Independent wire rope core (IWRC)

Minimum breaking strength (MBS)
Parts of line
Rated load
Rigging links
Saddle
Sling angle
Spreader beams
Spur track
Tagline
Vertical hitch

Industry Recognized Credentials

If you are training through an NCCER-accredited sponsor, you may be eligible for credentials from NCCER's Registry. The ID number for this module is 38102-18. Note that this module may have been used in other NCCER curricula and may apply to other level completions. Contact NCCER's Registry at 888.622.3720 or go to **www.nccer.org** for more information.

Note

This module provides instruction and information about common rigging equipment and hitch configurations, but it does not provide any level of rigging certification. Any questions about rigging procedures and/or certification should be directed to an instructor or supervisor.

Contents

Figures and Tables

1.0.0 RIGGING HARDWARE

Objective

Identify and describe various types of rigging hardware.

a. Identify and describe various hooks, shackles, eyebolts, and clamps.
b. Identify and describe various lugs, turnbuckles, plates, and spreader beams.

Performance Task

1. Inspect various types of rigging components and report on the condition and suitability for a task.

Trade Terms

Blind hole: A hole that does not penetrate the material completely, leaving a hole with a bottom.

Bridle hitch: A type of hitch that consists of two or more slings that support the load attached to a common lifting point.

Center of gravity (CG): The point at which the entire weight of an object is considered to be concentrated, such that supporting the object at this specific point would result in its remaining balanced in position.

Equalizer beams: Beams used to distribute the load weight on multi-crane lifts. The beam attaches to the load below, with two or more cranes attached to lifting eyes on the top.

Equalizer plates: A type of rigging plate that has three or more holes, used to level loads when sling lengths are unequal.

Minimum breaking strength (MBS): The amount of stress required to bring a rigging component to its breaking point. The MBS is a factor in determining a component's rated load capacity.

Rated load: The maximum working load permitted by a component manufacturer under a specific set of conditions. Alternate names for rated load include *working load limit (WLL)*, *rated capacity*, and *safe working load (SWL)*.

Rigging links: Links or plates with two holes used as termination hardware to appropriate lifting points.

Saddle: The portion of a hook directly below the center of the lifting eye.

Sling angle: The angle formed by the legs of a sling with respect to the horizontal plane when tension is placed on the rigging.

Spreader beams: Beams or bars used to distribute the load of a lift across more than one point to increase stability. Spreader beams are often used when the object being lifted is too long or large to be lifted from a single point, or when the use of slings around the load may crush the sides.

Hardware used in rigging includes items such as slings, hooks, shackles, eyebolts, spreader beams, equalizer beams, and blocks. There are also many unique pieces of hardware designed for specific applications. These hardware items must be carefully matched to the load to be lifted to ensure the safety of the load and all workers in the area. Careful inspection and maintenance of all lifting hardware is essential to safe and effective material movement. Rigging hardware should always be inspected before each use.

To understand rigging hardware and applications, it is important to be familiar with terms related to the capacity of an object to withstand weight or force. Many of these terms are used interchangeably. The term rated load for rigging components can be defined simply as the weight that a component can safely lift without fear of breaking, based on manufacturer testing. Other terms for rated load that are used by manufacturers in their product specifications include *rated capacity* and *working load limit (WLL)*.

> **NOTE**
>
> *ASME Standard B30.10*, which is devoted exclusively to hooks, uses the term *rated load* to describe the maximum amount of weight or force that can be safely applied to a component. For this reason, rated load will be used primarily throughout this module.

To determine the weight a component can lift without fear of breaking and ensure that rigging hardware provides safe service, testing is done to determine the breaking point. This weight is referred to as the minimum breaking strength (MBS). The rated load for a component is determined by applying a factor to the MBS. It is not uncommon for the rated load to be only 20 percent of the MBS, resulting in a safety factor of 5.

Safe Working Load (SWL)

The term *safe working load (SWL)*, sometimes seen as *normal working load (NWL)*, was the primary term used for years to describe the safe load capacity of a component or piece of equipment such as a sling. However, the definition was not very specific, especially on a global scale, and legal entanglements developed.

Although the term may still be heard and encountered in writing, the United States stopped using the term in the 1990s. Some years later, the International Organization for Standardization (ISO) and European standards-setting organizations also left the term behind. However, there are numerous organizations and websites that continue to use the term as if nothing has changed. As a general rule, the term *working load limit (WLL)* has replaced safe working load in most applications related to crane operations.

For example, if the MBS of a hook is 5,400 lbs (2,449 kg) and the factor applied is 20 percent, the rated load would be 1,080 lbs (490 kg). As a result, the proper application of rigging hardware means that the load imposed should always be well below the breaking point. Safety factors may vary, so it is important not to assume that the rated load should be $\frac{1}{5}$ of the MBS. Always work within the specified rated load of the component.

> **CAUTION**
>
> Always refer to the manufacturer's instructions for all types of rigging equipment and its proper application.

1.1.0 Hooks, Shackles, Eyebolts, and Clamps

Hooks and shackles are essential items for rigging almost every load. Eyebolts and clamps are also commonly used components for lifts. These devices come in different types, and there are a variety of guidelines that must be observed for their safe operation.

1.1.1 Hooks

A rigging hook is typically used to attach a sling to a load. The eye hook (*Figure 1*) is the most common type. Hooks used for rigging must be equipped with safety latches to prevent a connection from slipping off of the hook if any slack develops in the sling. The capacity of a rigging hook is determined by its material of construction, size, and physical dimensions. Information about a specific hook's capacity is always available from the manufacturer or authorized distributor.

Always inspect hooks before each use. Look for wear in the saddle of the hook. Wear in a hook should not exceed 10 percent of the original hook

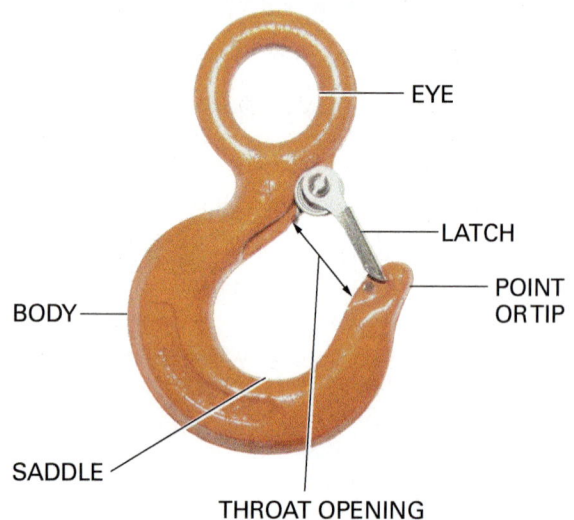

Figure 1 Typical eye hook.

dimensions. *ASME Standard B30.10* also requires that the hook be removed from service if the throat opening has enlarged 5 percent from its original size, to a maximum of $\frac{1}{4}$" (6.4 mm). *Figure 2* shows some of the inspection points for a hook.

Some other problems that result in the hook being removed from service include the following:

- Missing or illegible manufacturer's identification or rated load information
- Any visually apparent amount of bending or twisting of the hook
- Cracks, nicks, or gouges that could compromise its strength
- Damaged or inoperative latch that does not properly close the throat of the hook
- Any sign of modifications such as grinding, drilling, or machining

Never use a sling on a hook if the sling eye is marginal in size and must be forced over the hook. The body diameter of the hook should fit easily into the sling eye. (Note that information regarding the proper fit of the sling eye to other

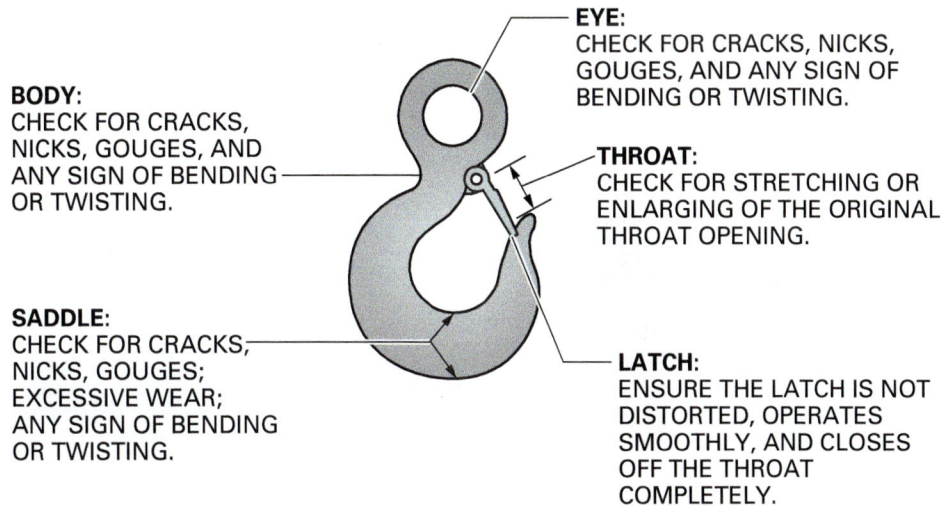

BODY:
CHECK FOR CRACKS, NICKS, GOUGES, AND ANY SIGN OF BENDING OR TWISTING.

EYE:
CHECK FOR CRACKS, NICKS, GOUGES, AND ANY SIGN OF BENDING OR TWISTING.

THROAT:
CHECK FOR STRETCHING OR ENLARGING OF THE ORIGINAL THROAT OPENING.

SADDLE:
CHECK FOR CRACKS, NICKS, GOUGES; EXCESSIVE WEAR; ANY SIGN OF BENDING OR TWISTING.

LATCH:
ENSURE THE LATCH IS NOT DISTORTED, OPERATES SMOOTHLY, AND CLOSES OFF THE THROAT COMPLETELY.

Figure 2 Rigging hook inspection points.

lifting hardware varies depending on the type of sling and hitch used. Slings and hitches will be discussed in more detail in later sections.)

The rated load of the hook is accurate only when the load is suspended from the saddle of the hook. The saddle is the portion of the hook that is directly beneath the center of the lifting eye. If the load is applied anywhere between the saddle and the hook tip, the rated load is reduced considerably, as shown in *Figure 3*. Point loading, also shown in the figure, is not acceptable.

1.1.2 Shackles

A shackle is used to attach an item to a load or to attach slings together. It can be used to attach the end of a wire rope to an eye fitting, hook, or other type of connector. Shackles used for lifting are made of forged steel and are sized by the diameter of the steel in the body, or bow section, of the shackle. However, the rated load capacity, pin size, and distance between the two lugs are often provided in the catalog data. Shackles are made with either screw pins or round pins as a means of safe closure, as shown in *Figure 4*. Screw-pin shackles, the most popular type, are threaded and have threaded pins that screw directly into the body of the shackle. No nut or cotter pin for security is required. A screw-pin shackle designed for synthetic web slings that allows the material to lie flat in the shackle throat is also shown in *Figure 4*.

A round-pin shackle body is not threaded. The round pin itself is threaded, but it is designed to pass completely through the shackle body. The threaded pin then receives a nut and a cotter pin to keep the nut from loosening. This type may also be referred to as a *safety shackle*, since the pin is more secure.

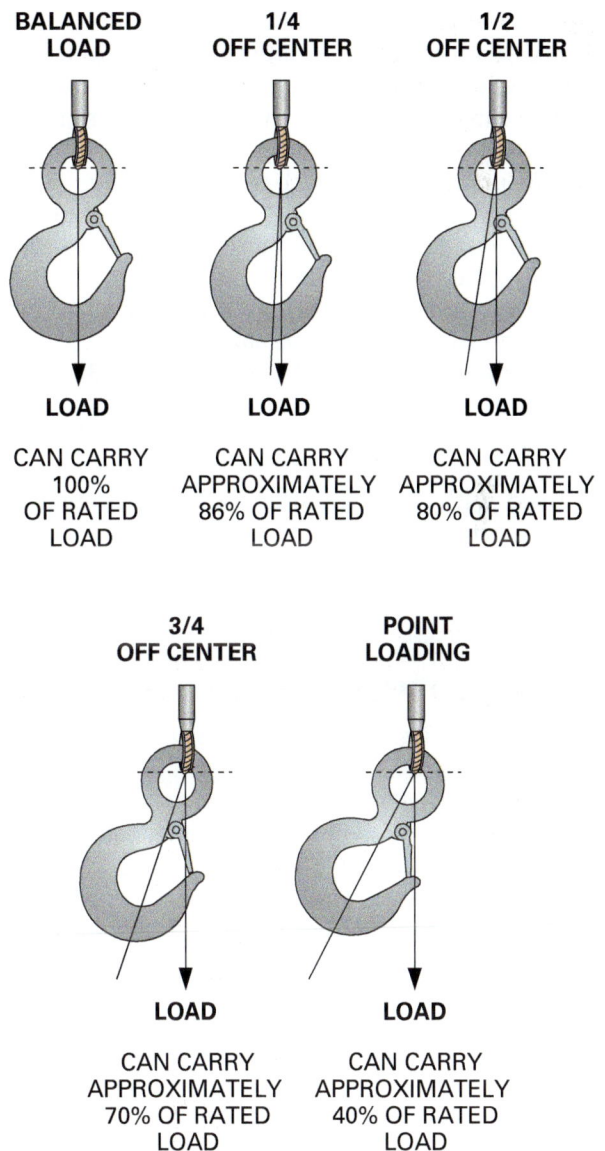

BALANCED LOAD
LOAD
CAN CARRY 100% OF RATED LOAD

1/4 OFF CENTER
LOAD
CAN CARRY APPROXIMATELY 86% OF RATED LOAD

1/2 OFF CENTER
LOAD
CAN CARRY APPROXIMATELY 80% OF RATED LOAD

3/4 OFF CENTER
LOAD
CAN CARRY APPROXIMATELY 70% OF RATED LOAD

POINT LOADING
LOAD
CAN CARRY APPROXIMATELY 40% OF RATED LOAD

Figure 3 Balanced and off-center loads.

(A) SCREW-PIN SHACKLE

(B) SCREW-PIN SHACKLE FOR SYNTHETIC WEB SLINGS

(C) ROUND-PIN SHACKLE (SAFETY SHACKLE)

Figure 4 Screw-pin and round-pin shackles.

The threaded pin of a screw-pin shackle must be fully seated in order for the shackle to function at its full rated capacity. In the field, the threaded pin is often left loose to make it easier to remove later, since lifting the load tends to tighten the pin. When it is tight, a tool is needed to loosen it. However, capacity can be significantly affected by leaving the pin loose, and vibration may cause the pin to back out and fall away. The pin must be fully seated in all cases to be safe.

When using shackles, be certain that all pins are straight, all screw pins are completely seated, and cotter pins are used with all round-pin shackles. It is a good practice to replace cotter pins used with shackles after each use to ensure their integrity.

Shackle pins should never be replaced with a common bolt. Common bolts are not hard enough and cannot take the stress normally applied to a shackle pin. Shackles that are stretched, or that have crowns or pins worn more than 10 percent of their original size should be removed from service.

Only shackles with suitable load ratings can be used for lifting. Like hooks, shackles must have the rated load information on the body. When using a shackle on a hook, the pin of the shackle should be hung on the hook, while the load is placed on the bow of the shackle (*Figure 5*). Spacers, such as large washers, can be used on the pin, on each side of the hook, to keep the shackle centered on it. Never use a screw-pin shackle in a situation where the pin can roll as the load sways, as shown in *Figure 6*.

Hook Integrity

The crane hook is a crucial part of many lifts that is used over and over again. Hook failure can happen at any time and may be caused by a number of factors, including cumulative fatigue, overloading, and mechanical abuse such as a free fall to a hard surface.

Hooks can be visually inspected in the field, but they should also be periodically removed, disassembled, and tested to ensure their integrity. In addition to the thorough examination of areas that are not always visible to the operator, various nondestructive testing techniques such as dye-penetrant, magnetic-particle and magnetic-rubber tests can be conducted. These techniques can help detect fatigue and damage that lead to failure. This photo shows a magnetic-particle test being conducted on a hook. Magnetic-particle, or magnaflux, testing can reveal surface flaws, as well as flaws slightly below the surface of ferrous metals.

Figure Credit Konecranes Americas, Inc.

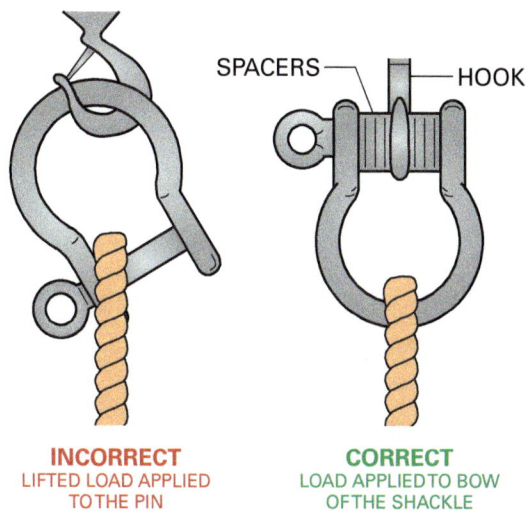

INCORRECT
LIFTED LOAD APPLIED
TO THE PIN

CORRECT
LOAD APPLIED TO BOW
OF THE SHACKLE

Figure 5 Shackle positioning.

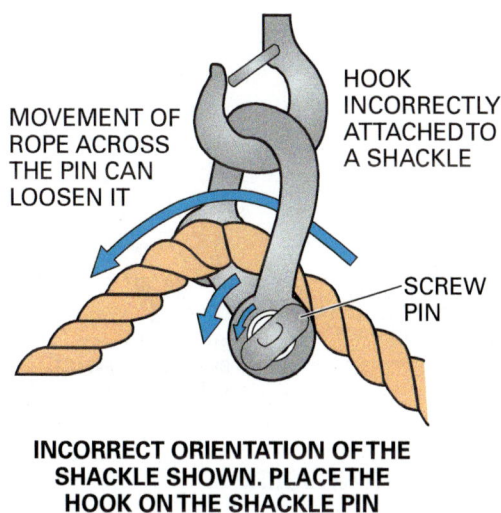

MOVEMENT OF
ROPE ACROSS
THE PIN CAN
LOOSEN IT

HOOK
INCORRECTLY
ATTACHED TO
A SHACKLE

SCREW
PIN

**INCORRECT ORIENTATION OF THE
SHACKLE SHOWN. PLACE THE
HOOK ON THE SHACKLE PIN**

Figure 6 Incorrect attachment of shackle causing pin to roll.

1.1.3 Eyebolts

Eyebolts (*Figure 7*) are often attached to heavy loads to aid in their handling and hoisting. Eyebolts can either have shoulders or be shoulderless. The shouldered type is recommended for use in hoisting applications because it is stronger and resists bending when being pulled from an angle. The shoulder provides a broad surface area of contact between the bolt eye and the load, stabilizing the bolt. The shoulderless type is designed only for lifting a vertical load.

Regardless of the type used, the rated load of all eyebolts however, is reduced during angular loading. Loads should always be applied in the same plane as the eye to reduce the chance of bending. Aligning the eye of the bolt with the shackle or other connector is particularly important when a bridle hitch is used. Proper and improper orientation of the eyebolt is shown in the bottom left corner of *Figure 7*.

When installed, the shoulder surface of the eyebolt must make full contact with the load surface. Washers or other suitable spacers may be used to ensure that the shoulders are in firm contact with the working surface. A threaded blind hole used for an eyebolt should have a minimum depth of $1\frac{1}{2}$ times the bolt diameter to ensure sufficient thread engagement. However, this does not ensure that the threaded portion of the hole is deep enough to allow the shoulder to make firm contact with the surface. For shouldered eyebolts, a blind hole should be deep enough so that the threaded portion is deeper than the length of the threaded eyebolt shank. This prevents the eyebolt from bottoming out or reaching the end of the threads before it is properly seated.

Swivel hoist rings, also shown in *Figure 7*, may be used instead of eyebolts. These devices swivel to the desired lift position and therefore do not require any load-rating reduction due to an angular pull.

1.1.4 Beam and Plate Clamps

Beam clamps (*Figure 8*) are used to connect hoisting devices to beams so the beams can be lifted and positioned. Observe the following guidelines when using beam clamps:

- Do not use clamps unless they are designed, load tested, and stamped by an engineer. Homemade clamps should never be used.
- Ensure that the clamp fits the beam and has the necessary rated load capacity.
- Ensure that the clamp is securely fastened to the beam.
- Be careful when using beam clamps where angular lifts are required. Most are designed for straight vertical lifts only.
- Be certain the rated load appears on the beam clamp and is legible. Never load a beam clamp beyond its rated load capacity.
- Attach rigging to the beam clamp using a shackle; do not place a hoist hook directly in the beam clamp lifting eye.
- Examine the clamp for the following defects, and remove it from service if any are present:
  - Jaws of the beam clamp have been opened more than 15 percent of their normal opening
  - Lifting eye worn, bent, or elongated
  - Excessive rust or corrosion
  - Capacity and beam size information unreadable

Plate clamps (*Figure 9*) attach quickly to structural steel plates to allow for easier rigging attachment and handling of the plate. There are two basic types of plate clamps: serrated-jaw clamps and screw clamps.

CORRECT
SHOULDERLESS EYE AND RING BOLTS ARE DESIGNED FOR VERTICAL LOADS ONLY.

INCORRECT
IF SHOULDERLESS EYEBOLTS AND RINGBOLTS ARE PULLED AT AN ANGLE AS SHOWN, THEY WILL EITHER BEND OR BREAK.

SHOULDERLESS EYEBOLT

SHOULDERLESS RINGBOLT

SHOULDERLESS EYEBOLT

SHOULDERLESS RINGBOLT

RESULTS IN

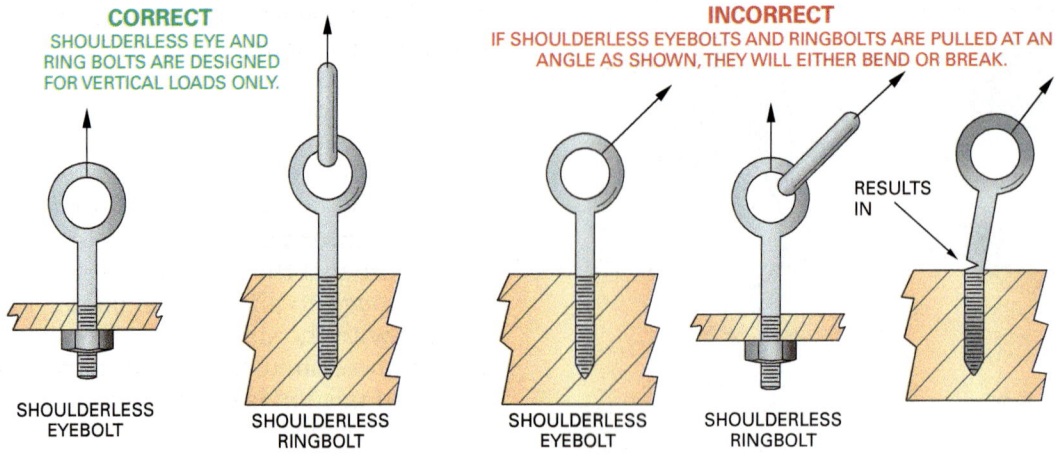

(A) USE OF EYEBOLTS

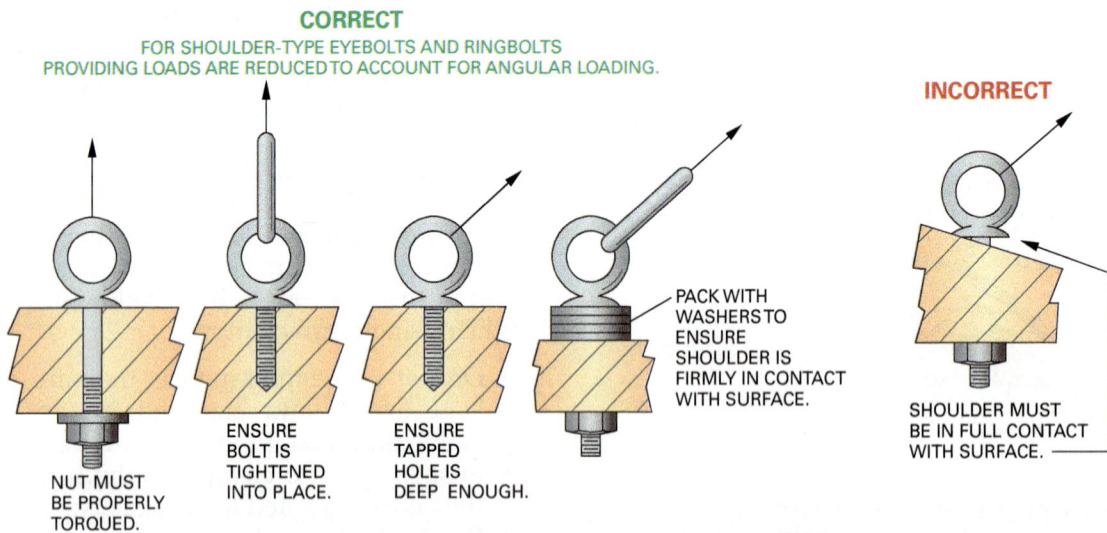

CORRECT
FOR SHOULDER-TYPE EYEBOLTS AND RINGBOLTS PROVIDING LOADS ARE REDUCED TO ACCOUNT FOR ANGULAR LOADING.

INCORRECT

NUT MUST BE PROPERLY TORQUED.

ENSURE BOLT IS TIGHTENED INTO PLACE.

ENSURE TAPPED HOLE IS DEEP ENOUGH.

PACK WITH WASHERS TO ENSURE SHOULDER IS FIRMLY IN CONTACT WITH SURFACE.

SHOULDER MUST BE IN FULL CONTACT WITH SURFACE.

(B) USE OF SHOULDER-TYPE EYEBOLTS AND RINGBOLTS

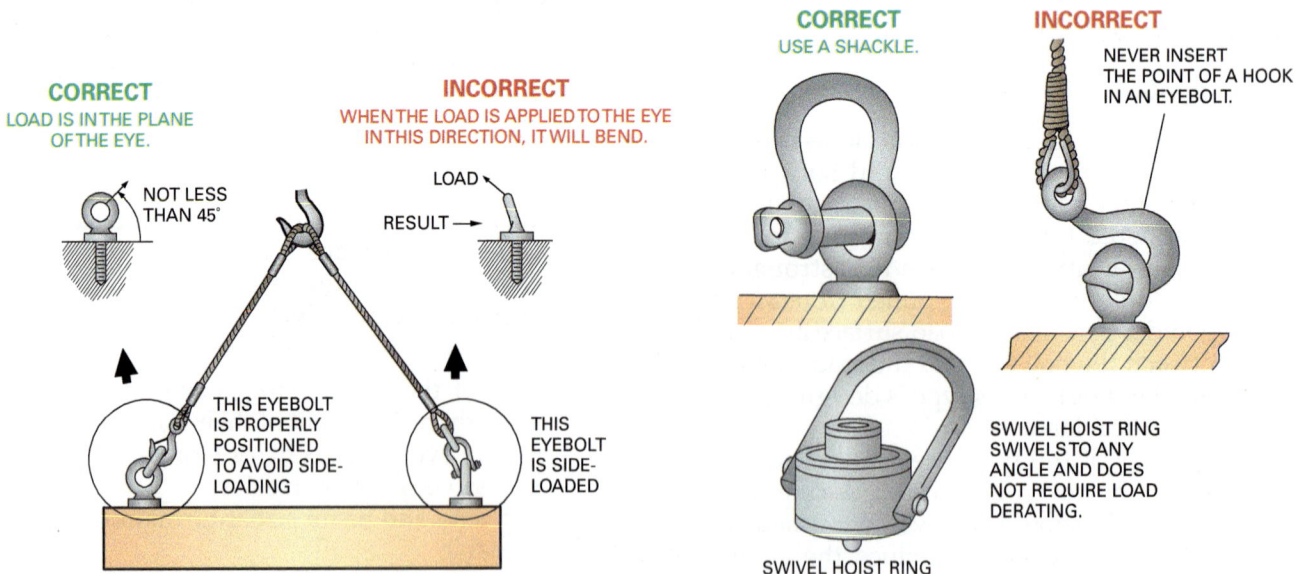

CORRECT
LOAD IS IN THE PLANE OF THE EYE.

NOT LESS THAN 45°

INCORRECT
WHEN THE LOAD IS APPLIED TO THE EYE IN THIS DIRECTION, IT WILL BEND.

LOAD

RESULT

CORRECT
USE A SHACKLE.

INCORRECT
NEVER INSERT THE POINT OF A HOOK IN AN EYEBOLT.

THIS EYEBOLT IS PROPERLY POSITIONED TO AVOID SIDE-LOADING

THIS EYEBOLT IS SIDE-LOADED

SWIVEL HOIST RING

SWIVEL HOIST RING SWIVELS TO ANY ANGLE AND DOES NOT REQUIRE LOAD DERATING.

(C) ORIENTATION OF EYEBOLTS

Figure 7 Eyebolt installation and lifting criteria.

Figure 8 Typical beam clamp application.

Serrated-jaw clamps are designed to grip a single plate for hoisting and are available with a locking device. The jaws of a serrated-jaw clamp are cam-operated; as the plate tries to escape the jaws, they rotate down slightly and press more tightly against the plate. Note that all plate clamps are designed to lift only one plate at a time. Always follow the manufacturer's recommendations for their use and remain within the rated load of the clamp.

Plate clamps must be examined before use and removed from service if any of the following defects are present:

- Identifying information and/or the rated load is absent or unreadable
- Distortion of the opening or wear of the jaw teeth
- Cracks in body
- Loose or damaged rivets

- Lifting eye worn, bent, or elongated
- Excessive rust or corrosion

1.2.0 Lugs, Turnbuckles, Plates, and Beams

There are many different types of hardware used for both common and unique rigging applications. Some of these include lifting lugs, turnbuckles, plates, and spreader beams. Riggers and crane operators must be familiar with different varieties of these components and know how to use them safely and effectively.

1.2.1 Lifting Lugs

Lifting lugs are often welded or bolted to an object by the manufacturer so that it can be lifted or moved more easily. A welded lifting lug is shown in *Figure 10*. They are typically designed and located to suspend a load near its center of gravity (CG) and support it safely. Lifting lugs should be used for straight, vertical lifting only. They are more likely to bend or fail if they are side-loaded. When lifting lugs are field-installed, consideration must be given to how they will be used in lifting and from what direction the stress will be applied.

1.2.2 Turnbuckles

Turnbuckles are available in a variety of sizes. They are used to adjust the length of rigging connections without twisting the cables or ropes. Two common types of turnbuckles use eyes and jaws as terminations (*Figure 11*). They can be used in any combination. Turnbuckles with hooks are also available but should not be used for rigging purposes as they can easily be disconnected and

| (A) HORIZONTAL SERRATED-JAW CLAMP | (B) VERTICAL SERRATED-JAW CLAMP | (C) SCREW CLAMP |

Figure 9 Plate clamps.

Figure 10 Welded lifting lug on a spreader beam.

have lower rated load capacities than the other types. The rated load for turnbuckles is based on the diameter of the threaded rods.

Observe the following guidelines and considerations when selecting turnbuckles:

- Turnbuckles should be made of forged steel and should not be welded.
- Do not use turnbuckles with open hooks for rigging loads.
- When using turnbuckles with multi-leg slings, do not use more than one turnbuckle per leg.

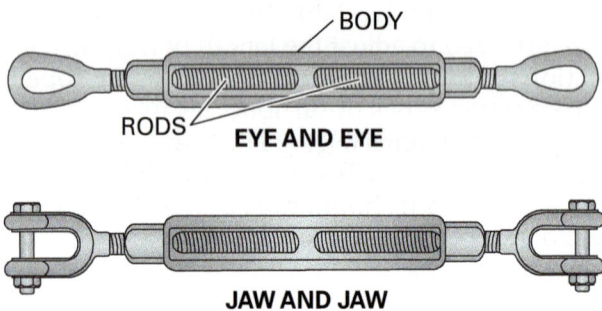

Figure 11 Turnbuckles.

- Do not use jamb nuts on turnbuckles that do not come equipped with them.
- Turnbuckles should not be overtightened. Perform tightening with a wrench of the proper size, using only as much force as a person can achieve by hand, without applying added leverage.

When inspecting turnbuckles, check for bent threaded rods and thread damage, and look for cracks in the body, threaded rods, and terminations.

1.2.3 Rigging Plates and Links

Rigging plates and links (*Figure 12*) are made for specific uses. The holes in the plates or links may be different sizes and may be placed in different locations in the plates. This creates a piece of connecting hardware that can be used for rigging the same type of objects repetitively. Plates with two holes are called rigging links. Plates with three or more holes are called equalizer plates. Equalizer plates can be used to level loads when the legs of a sling are unequal. Plates are attached to the rigging with high-strength pins or bolts.

Inspect rigging plates and links before using them, and remove them from service if any of the following are present:

- Cracks in body
- Worn or elongated lifting eye
- Excessive rust or corrosion

1.2.4 Spreader and Equalizer Beams

Spreader beams (*Figure 13*) are used to support and protect long loads. These devices prevent the load from tipping, sliding, or bending. They decrease the possibility of a low sling angle and help prevent the sling from crushing the load. Equalizer beams are used to balance the load on sling

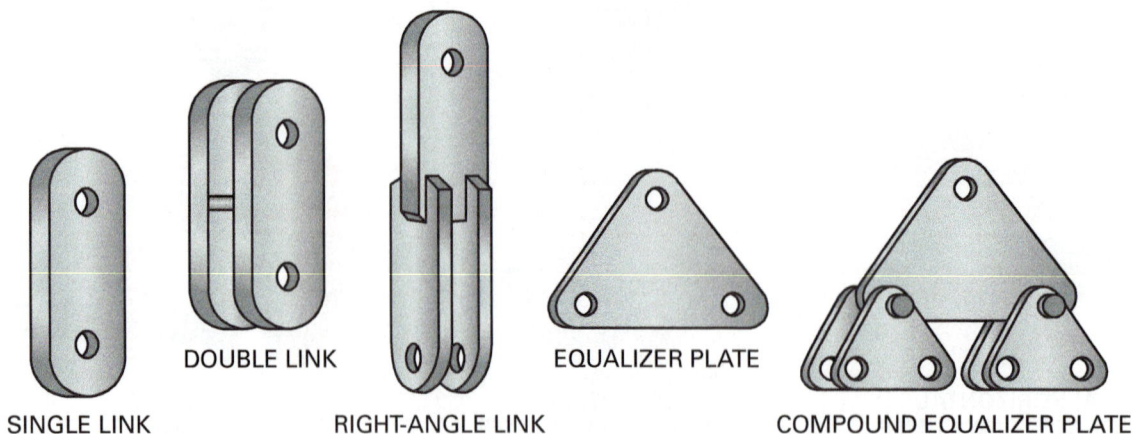

SINGLE LINK DOUBLE LINK RIGHT-ANGLE LINK EQUALIZER PLATE COMPOUND EQUALIZER PLATE

Figure 12 Rigging plates and links.

legs and to maintain equal loads on dual hoist lines when making multi-crane, or tandem, lifts.

Both types of beams are often fabricated to suit a specific application. They are commonly made of heavy pipe, I-beams, or other suitable material. Custom-fabricated spreader or equalizer beams must be designed by a qualified person (often an engineer) and have their capacity clearly stamped on the side. All such beams should be tested at 125 percent of their rated load. Information on any beams in use should be kept on file.

The rated load capacity of beams designed for use with multiple attachment points depends upon the distance between attachment points. For example, if the distance between the attachment points is doubled, the beam capacity is typically cut in half.

Before use, a spreader or equalizer beam should be inspected for the following:

Figure 13 Typical use of a spreader beam.

- Beam is clearly marked with the following information:
  - Manufacturer's name and address
  - Serial number
  - Device weight, if over 100 pounds (45 kg)
  - Rated load
  - *ASME Standard BTH-1* Design Category
  - *ASME Standard BTH-1* Service Class

Moving Materials Using Vacuum Principles

There are many devices used by cranes and other heavy equipment to grab, lift, and position construction materials. However, one method is a little more unusual than the others.

Vacuum lifters are equipped with vacuum pumps, usually powered by a small diesel engine, that draw out the air between the lift mechanism and the load. Irregular surfaces that cannot provide a tight seal are not candidates for vacuum lifting because integrity of the seal is essential for them to function. Vacuum lifters have been specifically designed for pipe, slabs, beams, and similar items. An advantage of this method is the elimination of a great deal of rigging hardware and the labor required to assemble it prior to a lift, saving both time and money. These devices and their design criteria are covered by *ASME Standard B30.20, Below-The-Hook Lifting Devices*, and *ASME Standard BTH-1, Design of Below-The-Hook Lifting Devices*. The power of a vacuum is surprising— rated load capacities for the strongest vacuum lifters exceed 20 tons (18 metric tons).

(A) PLATE LIFTER

(B) SLAB LIFTER

Figure Credit: Vaculift ™ , Inc. d.b.a. Vacuworx®

- Welds are free of cracks or other significant flaws.
- No cracks, nicks, gouges, or corrosion are present.
- Attachment points are not damaged or distorted.

Additional Resources

ASME Standard B30.5, Mobile and Locomotive Cranes. Current edition. New York, NY: American Society of Mechanical Engineers.

ASME Standard B30.10, Hooks. Current edition. New York, NY: American Society of Mechanical Engineers.

ASME Standard B30.20, Below-The-Hook Lifting Devices. Current edition. New York, NY: American Society of Mechanical Engineers.

ASME Standard BTH-1, Design of Below-The-Hook Lifting Devices. Current edition, New York, NY: American Society of Mechanical Engineers.

29 *CFR* 1926, Subpart CC, **www.ecfr.gov**

29 *CFR* 1926.251, **www.ecfr.gov**

29 *CFR* 1926.753, **www.ecfr.gov**

Mobile Crane Safety Manual (AEM MC-1407). 2014. Milwaukee, WI: Association of Equipment Manufacturers.

Willy's Signal Person and Master Rigger Handbook, Ted L. Blanton, Sr. Current edition. Altamonte Springs, FL: NorAm Productions, Inc.

North American Crane Bureau, Inc. website offers resources for products and training, **www.cranesafe.com**

1.0.0   Section Review

1. Which of the following statements about shackles is true?

   a. Shackles with pins worn more than 10 percent of their original size should be removed from service.
   b. The size of a shackle is based on its pin size.
   c. When connecting a shackle to a hook, the body of the shackle is hung on the hook.
   d. If a shackle is pulled at an angle rather than a straight pull, its rated load increases.

2. Which rigging devices require the *ASME Standard BTH-1* Design Category and Service Class on the labeling?

   a. Spreader beams
   b. Shackles
   c. Hooks
   d. Eyebolts

2.0.0 Slings and Hitches

Objective

Identify and describe various types of slings and sling hitches.

  a. Identify and describe wire-rope slings and their proper care.
  b. Identify and describe synthetic slings and their proper care.
  c. Identify and describe chain slings and their proper care.
  d. Explain the significance of sling angles and describe common hitches.
  e. Describe how to properly rig and handle piping materials and rebar.
  f. Identify and describe how to use taglines and knots for load control.
  g. Identify common rigging-related safety precautions.

Performance Tasks

  1. Inspect various types of rigging components and report on the condition and suitability for a task.
  2. Configure a sling to produce a single-wrap basket hitch.
  3. Configure a sling to produce a double-wrap basket hitch.
  4. Configure a sling to produce a single-wrap choker hitch.
  5. Configure a sling to produce a double-wrap choker hitch.
  6. Select the correct tagline for a specified application.
  7. Tie specific instructor-selected knots.

Trade Terms

Basket hitch: A common hitch made by passing a sling around a load or through a connection and attaching both sling eyes to the hoist line.

Bird caging: A deformation of wire rope that causes the strands or lays to separate and balloon outward like the vertical bars of a bird cage.

Choker hitch: A hitch made by passing a sling around the load, and then passing one eye of the sling through the other. The one eye is then connected to the hoist line, creating a choke-hold on the load.

Independent wire rope core (IWRC): Wire rope with a core consisting of wire rope, as opposed to a fiber or single-stranded core; considered to be the most durable for rigging applications.

Spur track: A relatively short branch leading from a primary railroad track to a destination for loading or unloading. A spur is typically connected to the main at its origin only (a dead end).

Tagline: A rope attached to a lifted load for the purpose of controlling load spinning and swinging, or used to stabilize and control suspended attachments.

Vertical hitch: A simple hitch that uses one end of a sling to connect to a point on the load and the opposite end to connect to the hoist line. Also known as a *straight-line hitch*.

Slings are available in a variety of configurations. Some of these include wire rope, synthetic web, and chain slings, which are all presented in this section. Some slings are made from natural fibers or synthetic rope, but these are not typically used for rigging applications involving mobile cranes. All types of slings are likely to need protection to ensure they are not damaged by sharp edges, protrusions, and other potential sources of damage. In many cases, the load itself also benefits from such protection.

Like hooks used in rigging, an ASME standard is devoted exclusively to slings. *ASME Standard B30.9, Slings*, provides a wealth of information about slings and their proper application. In addition, slings are also the topic of 29 *CFR* 1910.184.

2.1.0 Wire-Rope Slings

Wire-rope slings (*Figure 14*) are made of high-strength steel wires formed into strands and wrapped around a supporting core. They are lighter and easier to handle than chain slings, and can withstand substantial abuse and relatively high temperatures. However, because wire-rope slings can slip, the use of synthetic slings is often preferred. Wire-rope slings are still being used, so it is still important to learn the design, characteristics, applications, inspection, and maintenance of wire-rope slings.

2.1.1 Characteristics and Applications

Wire-rope slings usually consist of six strands, with each strand containing an average of 19 wires (written as 6 x 19), laid in a specific pattern

Figure 14 Typical wire-rope sling.

around a wire rope core. Cores other than wire are available, such as fiber cores, but independent wire rope core (IWRC) is the most common and considered the most durable. Wire rope will suffer the loss of only about 1 percent of its strength if a wire breaks. *Figure 15* shows examples of the most common wire-rope sling applications used in construction rigging.

Wire-rope slings should be protected where the slings are wrapped around the sharp edges of the object to be lifted (*Figure 16*). Of course, the load can often be damaged by the sling as well. Even if the edge of the load is a soft material that would not cut the sling, one or more individual wires may break if they are sharply bent. If wire-rope slings are kinked, the severe bending stress and displaced strands allow for unequal distribution of the live load, with some strands taking more than their normal load. Damage done to the rope by kinking is usually permanent, resulting in the disposal of many slings as well as some failures.

Protective material can include simple pieces of wood to separate the sling from the load. However, protective materials are also available in a variety of manufactured forms. *Figure 17* shows examples of some materials designed for the task. The plastic protectors shown here have magnets embedded in them, allowing them to remain in place on ferrous materials before the sling is applied. Others are made of metal alloys and are designed to slip around a wire-rope sling at any point along its length. Some type of protective material is often needed for any type of sling, depending upon the hitch used and the nature of the load being lifted. These protective materials are not used with wire rope only.

A special type of wire-rope sling, called a *braided-belt sling*, is made by braiding six or more small-diameter wire ropes together. This provides a sling with a wide, flat bearing surface of great strength and flexibility in all directions. They are especially well suited for a basket hitch or a choker hitch where sharp bends are encountered.

Rigging hardware placed into the eye of a sling must be appropriate to prevent failures and/ or sling damage. *ASME Standard 30.9*, Section 2.10.4(p) states that "an object in the eye of a (wire-rope) sling should not be wider than one half the length of the eye nor less than the nominal sling diameter." For example, if the eye of wire-rope sling is 6" (15 cm) long, then nothing wider than 3" (7.5 cm) should be placed through the eye. (Half of 6" is 3", or half of 15 cm is 7.5 cm.) The second portion refers to the minimal width of hardware placed in the sling eye; it should never be smaller in diameter than the wire-rope sling itself. Sling manufacturers provide information about the application of their products that should be followed as well.

2.1.2 Storage and Inspection

Store slings in a rack to keep them off the ground. The rack should be in an area free of moisture and away from acid, acid fumes, or extreme heat. Both moisture and fumes can lead to corrosion. Never let slings lie on the ground in areas where heavy machinery may run over them, or where they can become filled with sand and other abrasives internally.

> **WARNING!**
>
> Broken wires in a wire-rope sling are extremely sharp and can easily cut or puncture the skin. Always wear gloves when handling wire-rope slings and when inspecting them by running your hand along their length.

Slings should be regularly inspected for broken wires, kinks, rust, or damaged fittings. A visual inspection should be made before each use. Inspections at the beginning of each shift or work day are required by *ASME Standard B30.9*. Any slings found to be defective should be removed from service for repair or disposal. Repairs should not be attempted by anyone other than the manufacturer or other qualified party.

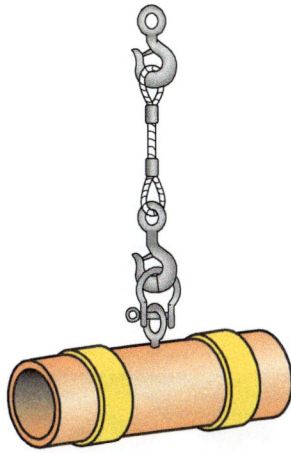

STRAIGHT VERTICAL **BASKET** **SINGLE CHOKER**

ALL OTHER HITCHES ARE A COMBINATION OR VARIATION OF THESE.
SEE BELOW:

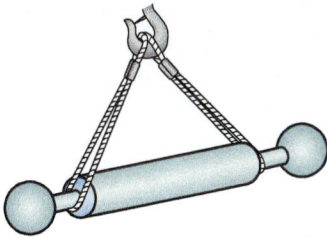

DOUBLE BASKET **DOUBLE WRAPPED BASKET** **DOUBLE CHOKER**

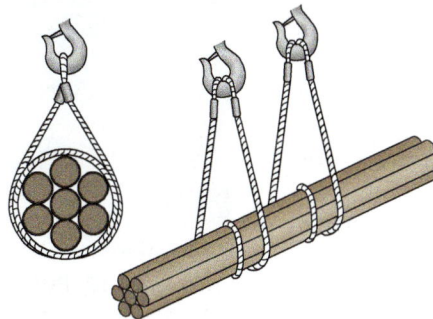

SINGLE DOUBLE WRAPPED CHOKER **PAIR OF DOUBLE WRAPPED CHOKERS**

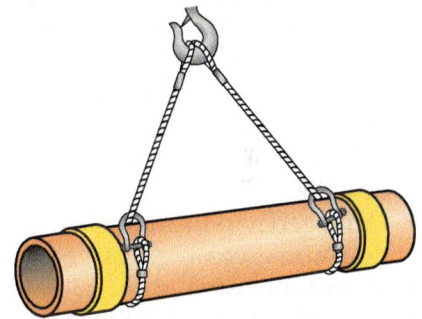

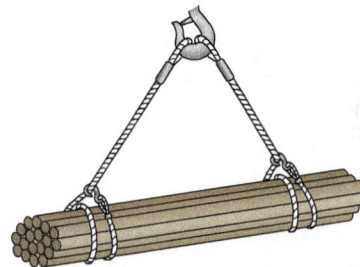

Figure 15 Examples of wire-rope sling applications.

Wire-rope slings must be properly tagged with specific information. This is required by both ASME and OSHA standards. *Figure 18* shows a typical wire-rope tag. A missing or illegible identification tag or data plate is a cause for sling disposal. Information that must be on the tag includes the following:

- Name or trademark of manufacturer or repair organization
- Rated load for at least one hitch and the angle upon which it is based

- Size/diameter
- Number of legs, if more than one

Wire-rope slings should also be removed from service if any of the following conditions are discovered; refer to *Figure 19* for examples:

- Localized abrasion or scraping that reduces the diameter of the sling by more than 5 percent of its original size
- Rope distortion, which includes kinking, crushing, and bird caging

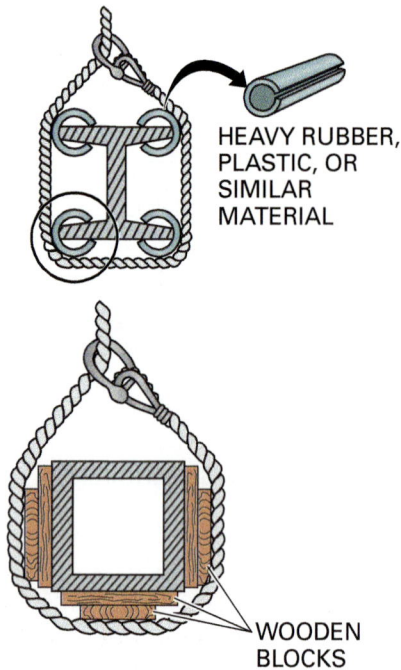

HEAVY RUBBER, PLASTIC, OR SIMILAR MATERIAL

WOODEN BLOCKS

Figure 16 Use of sling protection.

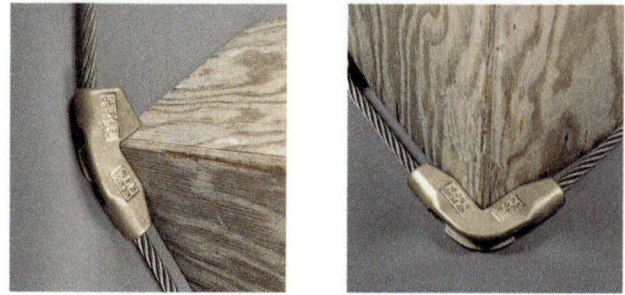

(A) UPPER AND LOWER WIRE-ROPE SADDLES

(B) PLASTIC SLING PROTECTORS FOR ALL SLING TYPES

Figure 17 Sling protection products.

- Evidence of heat damage, usually indicated by discoloration
- Damaged end fittings, such as cracks, deformation, and excessive wear
- Severe corrosion
- Any other type of damage that results in doubt about the integrity of the sling
- Broken wires

> **NOTE**
>
> *ASME Standard B30.9* does allow a specific number of broken wires depending on the rope design; consult the standard for details.

Sling Maintenance

Slings are required to have periodic inspections. *ASME Standard B30.9* requires that documentation of the most recent periodic inspection be maintained. Someone is responsible for maintaining that documentation as well as the slings themselves.

Most sling manufacturers now offer radio-frequency identification (RFID) chips permanently attached to new slings. Passive RFID chips called transponders are scanned by devices that receive the signal and synchronize important data with a database based on the embedded serial number. RFID tags are extremely durable and use Bluetooth technology; the tag can be located and information downloaded from a reasonable distance and without a direct line of sight to the receiver. The technology is not only used for slings, but for many other rigging components such as hooks and shackles.

Figure Credit: Lift-All Company, Inc.

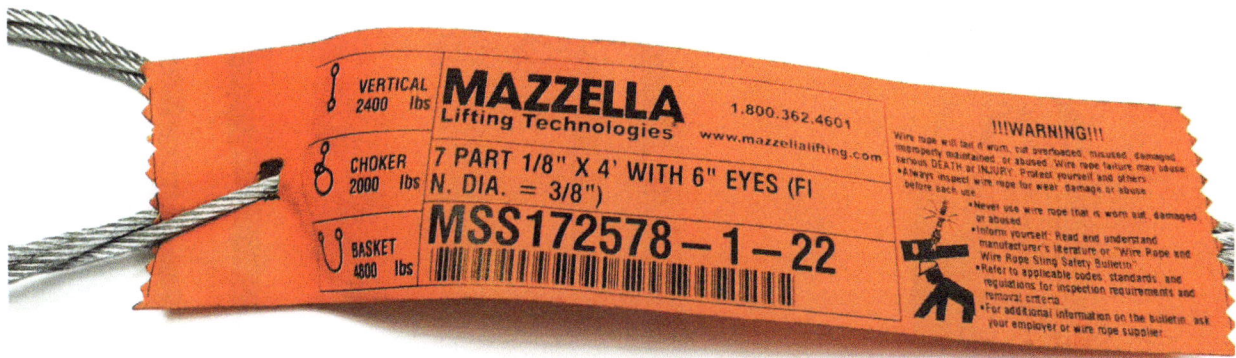

Figure 18 Wire-rope sling tag.

BROKEN WIRES

KINKING

BIRDCAGING

CRUSHING

CORROSION

Figure 19 Common types of wire-rope damage.

2.2.0 Synthetic Web Slings

Synthetic slings are widely used to lift loads, and they are especially suitable for easily damaged ones. Common types of synthetic slings, as well as guidelines for their storage and inspection, are outlined in the following sections.

2.2.1 Characteristics and Applications

Synthetic web slings commonly used for construction rigging are made of polyester, nylon, or other high-performance synthetic materials. While synthetic slings are useful in many situations, there are some applications and environments that are likely to damage them.

Synthetic web slings are available in a number of configurations, with common types shown in *Figure 20*. These include the following:

- *Endless slings* – Endless slings are also referred to as *continuous loop slings* or *grommet slings*. The ends of a piece of webbing are overlapped and sewn together to form an endless loop. They can be used for a vertical hitch, bridle hitch, choker hitch, or basket hitch.
- *Standard eye-and-eye slings* – Webbing in these slings is sewn to form a flat body with an eye on each end. The eye is in the same plane as the sling body. They are used for the same purposes as wire-rope slings of similar design.
- *Round slings* – Round slings are available in endless and eye-and-eye styles. They are generally made from a continuous length of polyester filament yarn covered by a woven sleeve. The eye-and-eye style of round sling simply has a sleeve wrapped around the two sides, forming eyes on the end.
- *Twisted eye-and-eye slings* – The eyes in twisted eye-and-eye slings are sewn at right angles to the plane of the sling body.

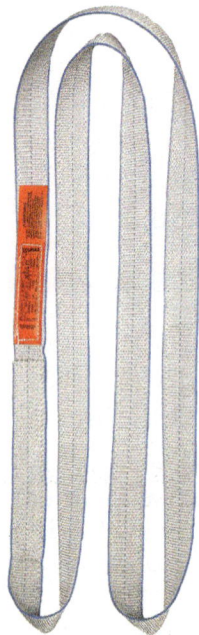

(A) ENDLESS

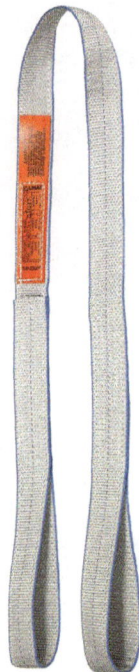

(B) STANDARD EYE-AND-EYE

(C) ENDLESS ROUND

(D) TWISTED EYE-AND-EYE

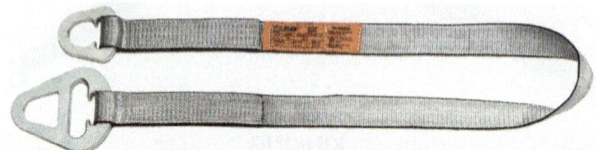

(E) SLING WITH TRIANGLE AND CHOKER HARDWARE

Figure 20 Examples of synthetic web slings.

- *Slings with attached hardware* – Web slings are also available with triangular and choker hardware end fittings.

Synthetic web slings can be cut by sharp edges and corners or damaged by excessive sling angles. A razor sharp edge is not required for damage to be done; a sling stressed with weight against a seemingly dull edge can result in failure. The same type of protection used with wire-rope slings must also be applied to web slings. Advantages of synthetic web slings include the following:

- Their texture and width reduce marring or scratching of polished, finished, or soft surfaces.
- They are less likely to crush fragile surfaces.
- They mold themselves to the shape of the load.
- Moisture and a variety of chemicals do not damage them.

- As long as they remain clean and dry, they are nonsparking and nonconductive. (Excessive soil can also make a sling conductive.)
- They minimize twisting of the load during lifting.
- They do not stain ornamental materials.
- They are lightweight and soft, making them easier and safer to handle.
- They stretch somewhat under load, allowing them to withstand shock.

ASME Standard B30.9, Section 5.10.4(p) provides the following guidance for fitting hardware into the eye of the sling: "An object in the eye of a sling should not be wider than one-third the length of the eye." For example, if the eye of a synthetic sling is 9" (23 cm) long, the maximum width of a hook or other object placed through the sling eye is 3" (7.5 cm). This is smaller than the maximum width allowed for wire-rope sling eyes, which is half the length of the eye.

2.2.2 Storage and Inspection

Nylon and polyester web slings should not be stored or allowed to contact objects at temperatures over 194°F (90°C) or below −40°F (−40°C). This temperature range is a requirement of *ASME Standard B30.9*; note that manufacturers may specify a more limited range for their products that would supersede the standard. It is unlikely that storage facilities would reach these temperatures, but the surface of a rigged load could in some cases. It is also important to limit long-term exposure to sunlight or ultraviolet (UV) light. Both can affect the strength of synthetic web slings. Sling manufacturers can assist with information related to sunlight and UV light exposure.

The strength of synthetic webbing slings can also be degraded by certain chemicals. This includes exposure in the form of solids, liquids, vapors or fumes. Consult the sling manufacturer for a specific list of the chemicals and their variations that can damage synthetic slings.

Like other types of slings, synthetic web slings should not be laid on the ground where they can be damaged or run over by heavy equipment. They also must be properly labeled with specific information. An unreadable or missing label is cause for removing a sling from service. However, a damaged label can be repaired by the manufacturer or other qualified party. When repairs are made, the repair organization must record their information on the label. Labels must include the following information:

- Name or trademark of manufacturer, or if repaired, the entity performing repairs
- Manufacturer's code or stock number
- Rated load for at least one hitch type and the angle upon which it is based
- Type of synthetic web material
- Number of legs, if more than one

Per *ASME Standard B30.9*, synthetic web slings should be removed from service if any of the following conditions are found during an inspection; refer to *Figure 21* for examples:

- Missing or illegible sling identification and rated load information
- Acid or caustic burns
- Melting or charring of any part of the sling
- Holes, tears, cuts, or snags
- Broken or worn stitching
- Excessive abrasion
- Knots in any part of the sling
- Discoloration and brittle or stiff areas on any part of the sling, which may indicate chemical or ultraviolet/sunlight damage has occurred
- For attached hooks, removal criteria as stated in *ASME Standard B30.10*
- For other attached rigging hardware, removal criteria as stated in *ASME Standard B30.26*
- Other conditions, including visible damage, that create doubt as to the continued use of the sling

2.3.0 Chain and Metal-Mesh Slings

Chain slings and metal-mesh slings are often used for lifts in high heat or rugged conditions. They are versatile because they can be adjusted over the center of gravity, and they are also very durable. However, because chain and metal-mesh are very heavy, they can be harder to inspect than other types.

2.3.1 Chain Slings

For some lifts, chains slings are more appropriate than wire-rope or web slings. For example, the use of chain slings is recommended when lifting rough castings that would quickly destroy wire or synthetic web slings. They are also used in high-heat applications or where wire-rope chokers are not suitable, and for dredging and other marine work because they withstand abrasion and corrosion better than wire rope. *Figure 22* shows some common configurations of chain slings and hooks.

Synthetic Slings and Rigging Incidents

Industrial Training International (ITI) compiled a webinar entitled "Rigging & Sling Failures: Case Studies & Solutions" in 2013. During ITI's research for the webinar, the results of which were supported by a poll taken of attendees during the webinar, it was found that over 80 percent of rigging accidents were related to synthetic slings. The incidents were generally due to cutting or severe abrasion resulting from a lack of sling protection used during lifting. As many rigging professionals have stated over the years, if you have a synthetic sling in your right hand, you should have sling protection in your left. All slings, regardless of their materials of construction, should be protected in use.

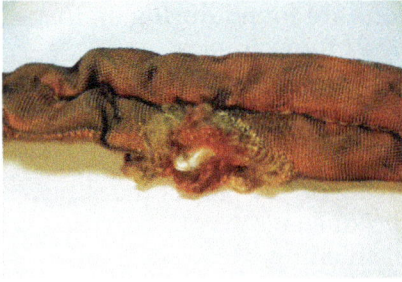

(A) JACKET AND WEB ABRASION

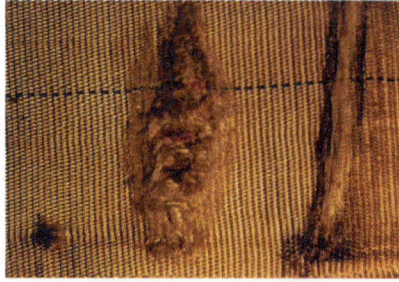

(B) JACKET AND WEB SLING ABRASION

(C) OUTER JACKET CUT

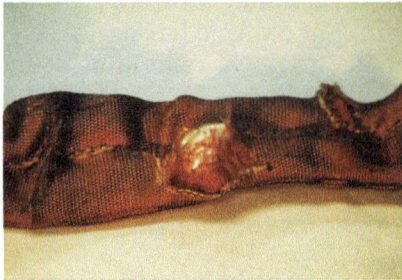

(D) INNER AND OUTER JACKET CUTS

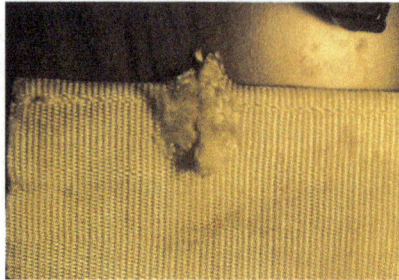

(E) CUT

(F) CUT WITH WARNING THREADS SHOWING

(G) BROKEN SPLICE OR STITCHING

(H) TENSILE DAMAGE

(I) OVERLOAD DAMAGE (TATTLE-TAILS PULLED IN)

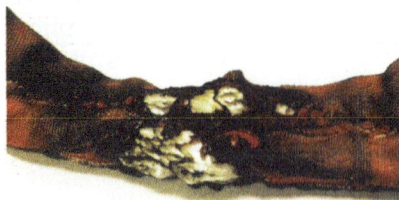

(J) SEVERE HEAT DAMAGE

Figure 21 Examples of synthetic-web sling damage.

SINGLE CHAIN
SLING WITH
SHAPED MASTER
LINK ON
EACH END

SINGLE CHAIN
SLING WITH
SHAPED MASTER
LINK AND
SLING HOOK

SINGLE CHAIN
SLING WITH
SHAPED MASTER
LINK AND
GRAB HOOK

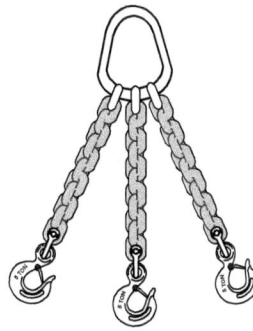

TRIPLE CHAIN
SLING WITH
SHAPED MASTER
LINK AND
SLING HOOKS

TRIPLE CHAIN
SLING WITH
SHAPED MASTER
LINK AND
GRAB HOOKS

SINGLE CHAIN
SLING WITH
SHAPED MASTER
LINK AND
FOUNDRY HOOK

SINGLE CHAIN
SLING WITH
SLING HOOK ON
EACH END

SINGLE CHAIN
SLING WITH
GRAB HOOK ON
EACH END

SINGLE CHAIN
SLING WITH
GRAB HOOK AND
SLING HOOK

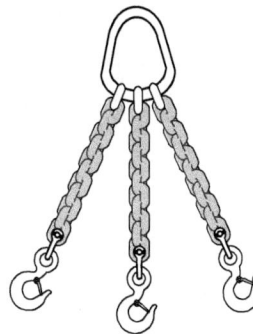

TRIPLE CHAIN
SLING WITH
SHAPED MASTER
LINK AND
FOUNDRY HOOKS

QUADRUPLE CHAIN
SLING WITH
QUADRUPLE
MASTER ASSEMBLY
AND SLING HOOKS

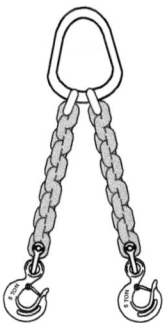

DOUBLE CHAIN
SLING WITH
SHAPED MASTER
LINK AND
SLING HOOKS

DOUBLE CHAIN
SLING WITH
SHAPED MASTER
LINK AND
GRAB HOOKS

DOUBLE CHAIN
SLING WITH
SHAPED MASTER
LINK AND
FOUNDRY HOOKS

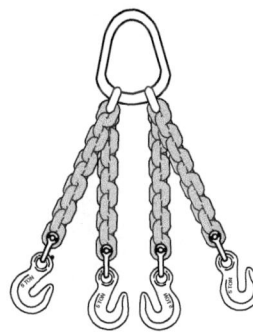

QUADRUPLE CHAIN
SLING WITH
QUADRUPLE
MASTER ASSEMBLY
AND GRAB HOOKS

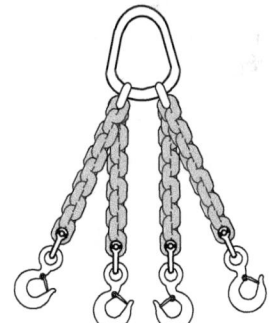

QUADRUPLE CHAIN
SLING WITH
QUADRUPLE
MASTER ASSEMBLY
AND FOUNDRY HOOKS

Figure 22 Common chain slings and hooks.

Chain links have two sides. Failure of either side causes the link to open and drop the load. Wire rope is frequently composed of as many as 114 individual wires, all of which must fail before the rope finally breaks. In other words, wire rope is more likely to experience a progressive failure than chain. Chains have less reserve strength and are more likely to fail quickly once the process begins.

Chains will stretch under excessive loading. This causes elongating and narrowing of the links until they bind on each other, giving visible warning. If overloading is severe, the chain will fail with less warning than a wire rope. When a chain link breaks, there is little or no warning.

2.3.2 Metal-Mesh Slings

Metal-mesh slings (*Figure 23*) are typically made of wire or chain mesh. They are similar in appearance and flexibility to web slings and are suited for some situations where other slings do not perform well. Metal-mesh slings have the following advantages:

- Resist abrasion and cutting
- Grip the load firmly without stretching
- Conform to irregular shapes
- Do not kink or tangle
- Can withstand high temperatures

ASME Standard B30.9 requires that the use of metal-mesh slings at temperatures above 550°F (288°C) requires manufacturer approval. These slings are available in several mesh sizes and can be coated with a variety of substances, such as rubber or plastic, to help protect the load. When used in high-temperature applications though, slings with coatings are not typically approved.

2.3.3 Storage and Inspection

Chain and metal-mesh slings must be stored inside a building or vehicle and hung on racks to reduce deterioration due to weather-related rust or corrosion. Never let chain slings lie on the ground in areas where heavy machinery can run over them.

Some manufacturers suggest lubrication of alloy chains while in use. However, slippery chains increase handling hazards. Chains coated with oil

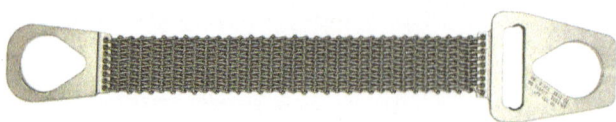

Figure 23 Metal-mesh sling.

or grease also attract dirt and grit that may cause abrasive wear. This is especially true of metal-mesh slings that are far more difficult to clean. Chain slings to be stored in exposed areas should be coated with a film of oil or grease for rust and corrosion protection.

Like all other slings, chain slings should be visually inspected before every lift. They should be removed from service if any of the following conditions are found during inspection:

- Missing or unreadable identification and/or rated load information
- Cracks or breaks
- Nicks, gouges, and excess wear
- Stretched, bent, twisted, or deformed links or end fittings; *ASME Standard B30.9*, Table 9-1.9.5.1 provides the minimum allowable thickness of any point on a link, based on the nominal size of the link.
- Evidence of overheating
- Excessive pitting or corrosion
- Lack of ability of chain or fittings to hinge freely
- Weld splatter
- Any other condition, including visible damage, that causes doubt about the continued use of the sling

Although metal-mesh slings are similar to chain slings, they are constructed in a very different way. Metal-mesh slings should be removed from service if any of the following conditions are found during inspection:

- Missing or unreadable identification tag
- A broken weld or brazed joint along the sling edge
- A broken wire in any part of the mesh
- A reduction in wire diameter of 25 percent due to abrasion, or 15 percent as the result of corrosion
- Lack of flexibility due to distortion of the mesh
- Distortion of the choker fitting so the depth of the slot is increased by 10 percent
- Distortion of either end fitting so the width of the eye opening is decreased by more than 10 percent
- Fittings that are pitted, corroded, cracked, bent, twisted, gouged, or broken
- A 15 percent reduction in the original cross-sectional area of metal at any point around
- Slings with individual spirals that are locked in place
- Any other condition, including visible damage, that causes doubt about the continued use of the sling

2.4.0 Sling Angles and Basic Hitches

Every sling, regardless of type and manufacturer, has a specified rated load capacity that should never be exceeded. Less obvious is that the load on a sling changes dramatically as the sling angle changes. What appears to be a light load for a sling can quickly become a very stressful load due to the additional stress placed on the sling as the sling angle changes. Sling angles and other significant factors in sling selection and use are explored in this section.

Since there are an infinite number of loads and load configurations, slings must be used in different ways to connect the load to the hoist line. An understanding of basic hitches and their performance characteristics help the crane operator understand how a reliable and properly made hitch should look.

2.4.1 Sling Capacity

Sling capacity depends on the sling material, sling construction, hitch configuration, number of slings, and the angle of the sling in the hitch used. This type of information, along with other relevant rigging information, is available from rigging equipment manufacturers and trade organizations in the form of easy-to-use pocket guides like the one shown in *Figure 24*. Sling capacity information is also provided in *ASME Standard B30.9* for various types of slings. *Table 1* is an example of a manufacturer's capacity table for wire-rope slings. Note that the capacity of a sling used in a simple basket hitch with a vertical pull is double the capacity for a straight vertical hitch. This is common, since two legs are being used to suspend the load instead of one. However, as the angle between the legs of the hitch widens, the rated load capacity of the sling is reduced. Remember that most anything you do with a sling that varies from a straight vertical pull is likely to negatively affect its rated load capacity. Choking a sling, for example, even for a vertical pull, significantly reduces the capacity of the sling.

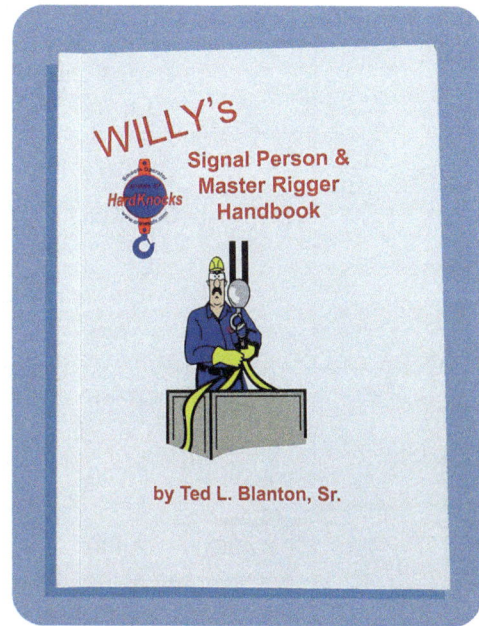

Figure 24 A rigging pocket guide.

2.4.2 Sling Angles

The angle formed by the legs of a sling, with respect to the horizontal plane, when tension is placed on the load is called the *sling angle*. The sling angle directly affects the tension applied to each sling. For this reason, maintaining acceptable sling angles is crucial to safe rigging. In addition, since the sling angle affects its rated load capacity, the rigger must ensure that even a sling at an acceptable angle remains within its rated load capacity.

To determine the effective weight placed on a sling at an angle, a sling angle factor is needed. The factor is applied to the actual load weight per sling to determine the tension on the sling. The proper way to calculate the sling angle factor is through some simple field measurements and math.

Figure 25 shows a load suspended from two slings in a vertical hitch. "L" represents the distance between the points of sling contact with the hook and the load. "H" represents the distance from where the sling contacts the hook down to

Convenient Rigging Information

Over the years, there have been quite a few rigging handbooks developed and marketed. Many are pocket-sized for the convenience of the rigger, allowing quick access to the needed information. As you might expect, rigging information has also gone electronic. Apps are now available from various manufacturers and trade organizations that are compatible with all popular smartphones, and more are sure to reach the market in the future. As OSHA and/or ASME standards change, it is much easier to update an app than it is to throw away and replace books.

Table 1 Wire-Rope Sling Capacity Table

| CLASS | SIZE (IN) | RATED CAPACITY - LBS* | | | | | | EYE DIMENSIONS (APPROXIMATE) | |
| | | VERTICAL | CHOKER** | BASKET HITCH | | | | WIDTH (IN) | LENGTH (IN) |
				(U)	30	60	90		
6 × 19 IWRC	¼	1,120	820	2,200	2,200	1,940	1,580	2	4
	⁵⁄₁₆	1,740	1,280	3,400	3,400	3,000	2,400	2½	5
	⅜	2,400	1,840	4,800	4,600	4,200	3,400	3	6
	⁷⁄₁₆	3,400	2,400	6,800	6,600	5,800	4,800	3½	7
	½	4,400	3,200	8,800	8,600	7,600	6,200	4	8
	⁹⁄₁₆	5,600	4,000	11,200	10,800	9,600	8,000	4½	9
	⅝	6,800	5,000	13,600	13,200	11,800	9,600	5	10
	¾	9,800	7,200	19,600	19,000	17,000	13,800	6	12
	⅞	13,200	9,600	26,000	26,000	22,000	18,600	7	14
	1	17,000	12,600	34,000	32,000	30,000	24,000	8	16
	1⅛	20,000	15,800	40,000	38,000	34,000	28,000	9	18
6 × 37 IWRC	1¼	26,000	19,400	52,000	50,000	46,000	36,000	10	20
	1⅜	30,000	24,000	60,000	58,000	52,000	42,000	11	22
	1½	36,000	28,000	72,000	70,000	62,000	50,000	12	24
	1⅝	42,000	32,000	84,000	82,000	72,000	60,000	13	26
	1¾	50,000	38,000	100,000	96,000	86,000	70,000	14	28
	2	64,000	48,000	128,000	124,000	110,000	90,000	16	32
	2¼	78,000	60,000	156,000	150,000	136,000	110,000	18	36
	2½	94,000	74,000	188,000	182,000	162,000	132,000	20	40

(The ¾ CHOKER value 7,200 is circled.)

* Rated capacities for unprotected eyes apply only when attachment is made over An object narrower than the natural width of the eye and apply for basket hitches only when the d/d ratio is 20 or greater, where d=diameter of curvature around which the body of the sling is bent, and d=nominal diameter of the rope.

** See choker hitch rated capacity adjustment chart.

an imaginary line that connects the two points of sling contact with the load. To determine the sling angle factor, the length (L) is divided by the height (H). The result is the factor to be applied to the weight of the load on each sling.

In the example shown in *Figure 25*, the height (H) is 74" (188 cm), and the length (L) is 86" (218 cm). (The length will always be longer than the height when the sling is at an angle.) The sling angle factor is calculated as follows:

US measure:
Sling angle factor = L ÷ H
Sling angle factor = 86" ÷ 74"
Sling angle factor = 1.162

Metric:
Sling angle factor = L ÷ H
Sling angle factor = 218.4 cm ÷ 187.9 cm
Sling angle factor = 1.162

The total weight of the load in this example is 2,000 lbs (907 kg). Since there are two slings, each sling must lift 1,000 lbs (454 kg). (2,000 lbs ÷ 2 slings = 1,000 lbs per sling.) Now apply the sling angle factor to determine the actual tension placed on each sling:

US measure:
Sling tension = load per sling ÷ sling factor
Sling tension = 1,000 lbs ÷ 1.162
Sling tension = 1,162 lbs

Metric:
Sling tension = load per sling ÷ sling factor
Sling tension = 453.5 kg ÷ 1.162
Sling tension = 527 kg

Figure 26 shows the effect of various sling angles on sling loading when 1,000 pounds (454 kg) of load are applied to each sling. Note that the tension on the slings is much higher when the legs are positioned at an angle of 30 degrees relative to the horizontal plane than when the legs are at an angle of 60 degrees. Optimum sling angles fall between 60 and 45 degrees to the horizontal plane. Angles of 30 degrees are occasionally required, but lesser angles are generally considered hazardous and unnecessary. At 30 degrees, the sling angle factor is 2.0, meaning that the tension on the sling is already double the actual load weight. If a lift plan leads to sling angles less than 30 degrees, changes in the rigging approach are likely needed.

These calculations for finding sling angle factor and tension help to determine whether slings are applied within their rated load capacity, and what that load actually is. However, they do not tell the rigger the sling angle. Tables have been developed that equate the sling angle factor to

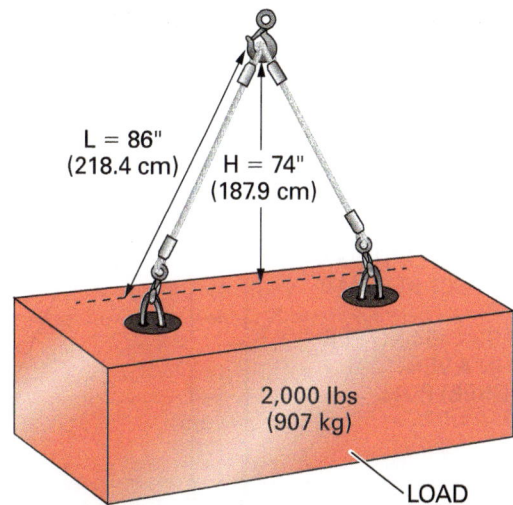

US MEASURE	METRIC
Sling angle factor = L ÷ H	Sling angle factor = L ÷ H
Sling angle factor = 86" ÷ 74"	Sling angle factor = 218.4 ÷ 187.9 cm
Sling angle factor = 1.162	Sling angle factor = 1.162

Figure 25 Determining the sling angle factor.

the angle. These tables can be used to determine the sling angle if the factor is known, or to determine the sling angle factor if the angle is known. *Table 2* is an example of such a table. Using the sling angle factor of 1.162 calculated in the previous example, it can be determined from the table that the sling angle is between 55 and 60 degrees. This is a very acceptable and safe sling angle for lifting.

2.4.3 Common Hitches

The way a sling is arranged to hold the load is referred to as a *hitch*. Hitches can be made using just the sling or by combining slings with connecting hardware. There are three basic types of hitches: vertical hitches, choker hitches, and basket hitches.

One of the most important parts of a rigger's job is making sure that the load is held securely. The type of hitch used depends on the nature of the load. For example, different hitches are used to secure a load of pipes, a concrete slab, or heavy machinery. Controlling the movement of the load once the lift is in progress is another extremely important part of the rigger's job. Therefore, the rigger must also consider the intended movement of the load when choosing a hitch. For example, some loads are lifted straight up and then lowered down to the same spot. Other loads may be lifted, turned 180 degrees in midair, and then set down in a completely different place. The

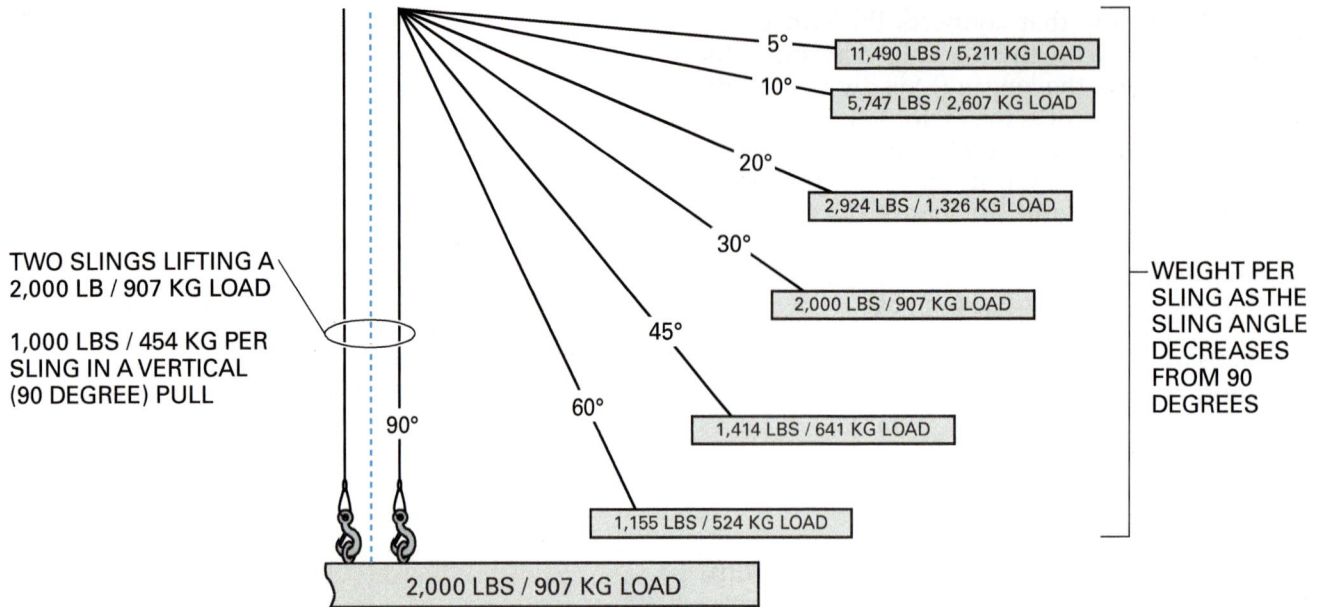

Figure 26 Effect of various sling angles.

Table 2 Sling Angle and Sling Angle Factor

Sling Angle	Sling Angle Factor
5	11.490
10	5.747
15	3.861
20	2.924
25	2.364
30	2.000
35	1.742
40	1.555
45	1.414
50	1.305
55	1.221
60	1.155
65	1.104
70	1.064
75	1.035
80	1.015
85	1.004
90	1.000

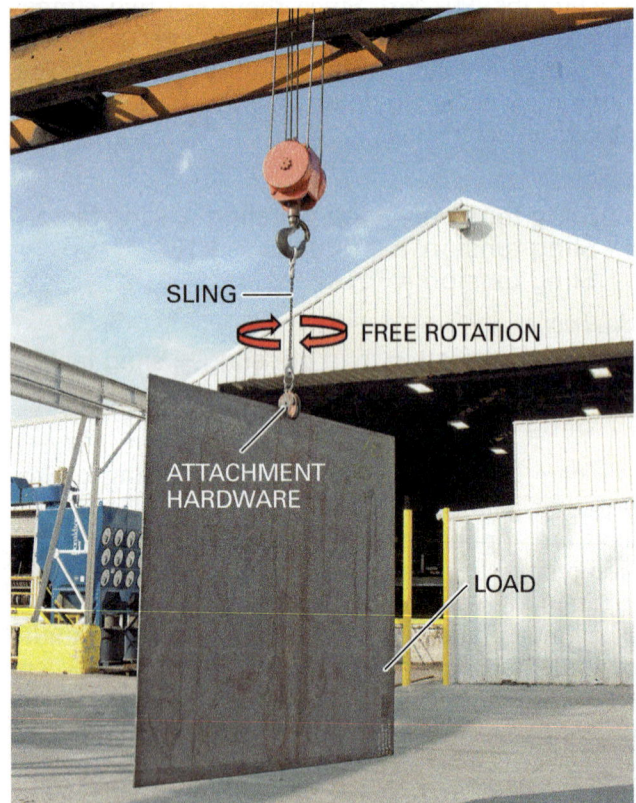

Figure 27 A vertical hitch using a plate clamp.

construction of the three basic types of hitches will be reviewed here.

The single vertical hitch (*Figure 27*) is used to lift a load straight up. With this hitch, some type of attachment hardware is needed to connect the sling to the load. The single vertical hitch allows the load to rotate freely. If you do not want the load to rotate freely, some method of load control must be used, such as a tagline.

Another version of the vertical hitch is the bridle hitch (*Figure 28*). The bridle hitch consists of two or more vertical hitches attached to the same hook, master link, or ring. This hitch allows the slings to be connected to the same load without the use of such devices such as spreader beams. Multiple-leg bridle hitches provide increased stability and balance for the load being lifted.

The Ten-Inch Rule

A alternate way to determine the sling angle factor in the field is done using a tape measure. First, measure the distance up from a horizontal line between the connection points, and find the point where the sling is exactly 10" above the line (as shown in the image). Then measure the distance from that point down the angled sling to the connection point. That line will be longer, since it travels at an angle. In the example shown here, that distance measures 14". Now simply place a decimal point between the 1 and the 4 to arrive at a sling angle factor of 1.4. If the factor is 2.0 or greater, the sling angle is excessive and applies too much stress to the slings. Change the rigging to reduce the sling angle.

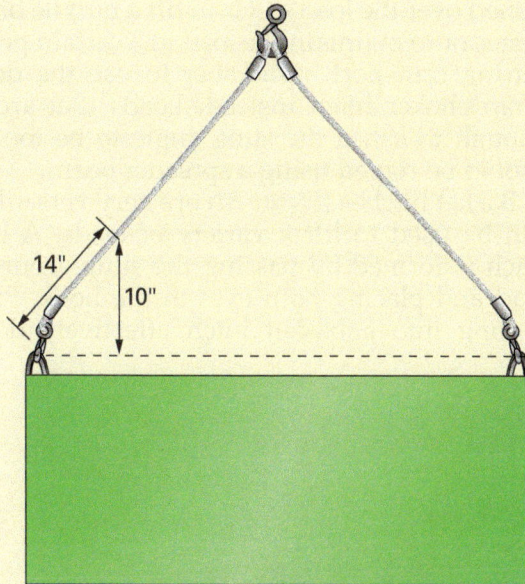

14"
10"

HORIZONTAL LINE BETWEEN
CONNECTION POINTS

- Find the point of the sling that is 10" above the horizontal line connecting the lift points.
- Measure along the sling down to the connecting point.
- Place a decimal point between the two numbers to determine the sling angle.

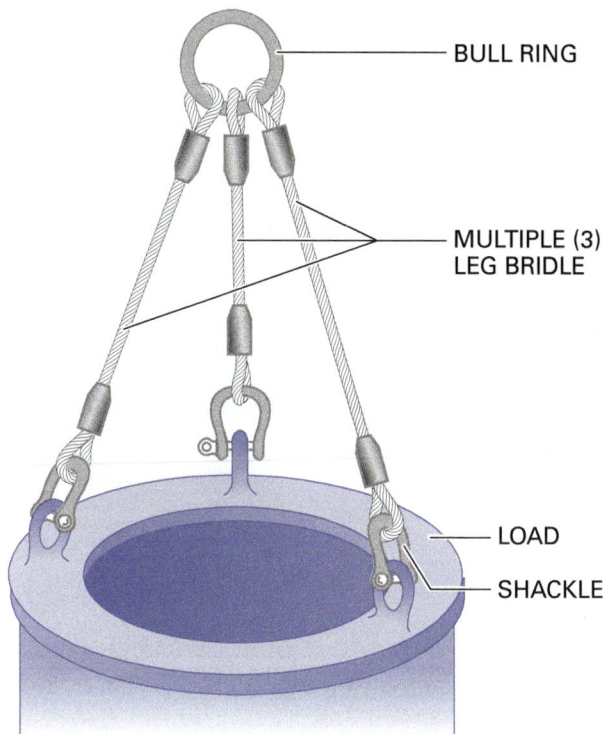

BULL RING

MULTIPLE (3)
LEG BRIDLE

LOAD

SHACKLE

Figure 28 A multi-leg bridle hitch.

However, it is important to note that a bridle hitch results in slings that are at an angle other than 90 degrees to the horizontal plane. Therefore, the stress applied to the sling is increased and must be accounted for in sling selection.

A choker hitch is often used when a load has no attachment points or when the attachment points are not practical for lifting (*Figure 29*). The hitch is made by wrapping the sling around the

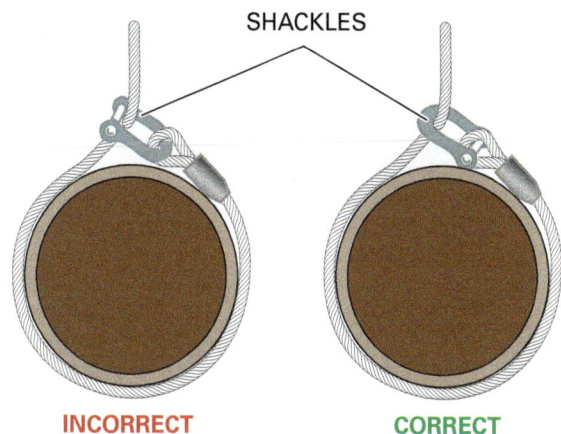

SHACKLES

INCORRECT   CORRECT

Figure 29 A choker hitch.

load and passing one eye of the sling through a shackle to form a constricting loop around the load. It is important that the shackle used in a choker hitch be oriented properly, as shown in *Figure 29*. It is also important to place a single choker hitch at the load's CG. Otherwise, the load will be unbalanced when lifted and will slip out of the hitch. The choker hitch affects the capacity of the sling, reducing it by a minimum of 25 percent. This reduction must be considered when choosing the proper sling.

A choker hitch does not grip the load as securely as the name implies. It is not recommended for loose bundles of materials because it tends to push loose items up and out of the choker. Many riggers use the choker hitch for loose bundles, mistakenly believing that forcing the choke down provides a tight grip. This actually increases the stress on the choked leg of the sling.

Instead, to gain gripping power, use a double-wrap choker hitch (*Figure 30*). The double-wrap choker uses the load weight to provide the constricting force, so there is no need to try and force the sling into a tighter choke. A double-wrap choker hitch is ideal for lifting bundles of items, such as pipes and structural steel. It will also keep the load in a certain position, which makes it ideal for equipment installation lifts.

When an item more than 12' (3.7 m) long is being rigged, the general rule is to use two choker hitches spaced far enough apart to provide the stability needed to transport the load. When two hitches are used, the hoist line should be positioned over the load's CG. To lift a bundle of loose items, or to maintain the load in a certain position during transport, remember to use the double-wrap choker hitch instead. Loads that are long enough to cause the sling angle to be too great should be rigged using a spreader beam.

Basket hitches (*Figure 31*) are very versatile and can be used to lift a variety of loads. A basket hitch is formed by passing the sling around the load and placing both eyes in the hook. Placing a sling into a basket hitch effectively doubles

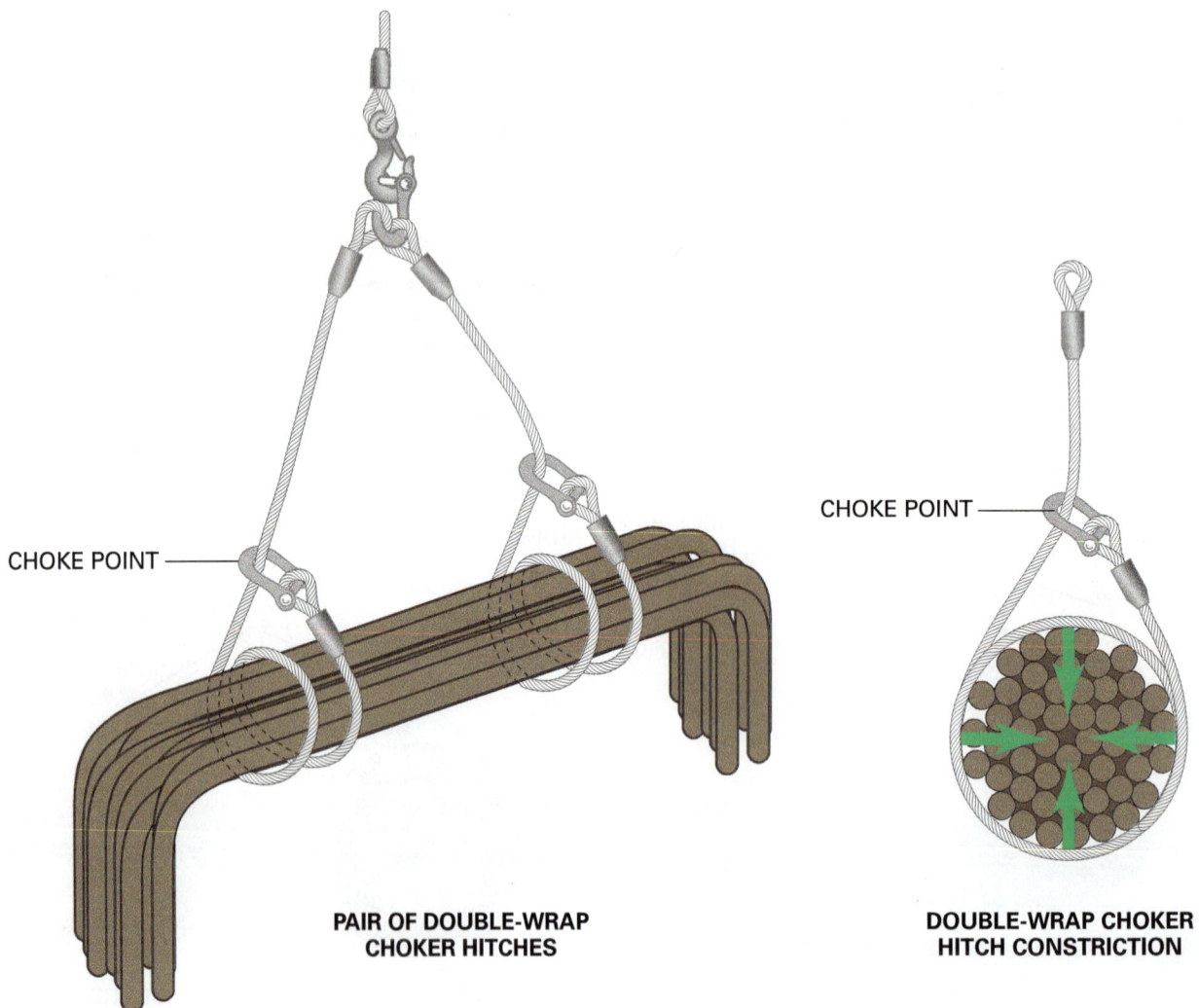

CHOKE POINT

CHOKE POINT

**PAIR OF DOUBLE-WRAP
CHOKER HITCHES**

**DOUBLE-WRAP CHOKER
HITCH CONSTRICTION**

Figure 30 A double-wrap choker hitch.

the capacity of the sling. This is because the basket hitch creates two sling legs from one sling. However, this does not provide secure control of the load.

The double-wrap basket hitch (*Figure 32*) combines the constricting power of the double-wrap choker hitch with the capacity advantages of a basket hitch. This means it is able to hold a larger load more tightly. The double-wrap basket hitch requires a considerably longer sling length than a double-wrap choker hitch, since both sling eyes must be connected to the crane hook. If it is necessary to join two or more slings together, the load must be in contact with the sling body only, not with the hardware used to join the slings. The double-wrap basket hitch provides support around the load. Just as with the double-wrap choker hitch, the load weight provides the constricting force for the hitch.

2.4.4 Finding the Load's Center of Gravity

The center of gravity (CG) of an object is the point around which the weight of the object is concentrated. The CG must be known when using any hitch in order to properly and safely position the load line over the load.

The CG of an object is the point where the weight times the length of the object on one side is equal to the weight times the length of the other side; the point in between these two sides is the CG. This is illustrated in *Figure 33*. When the two calculations are equal, the CG has been located. The symbol to identify the CG, also shown in *Figure 33*, will be seen on drawings for lifted components and equipment and sometimes on the load itself.

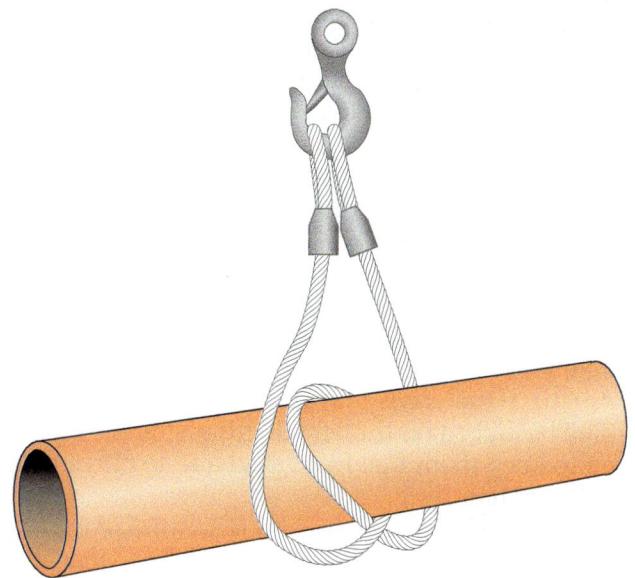

Figure 32 A double-wrap basket hitch.

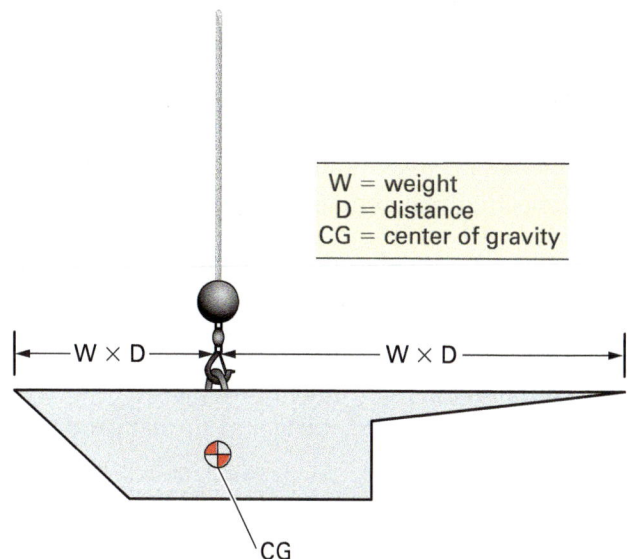

W = weight
D = distance
CG = center of gravity

$W \times D$ $W \times D$

CG

Figure 33 Calculating the center of gravity.

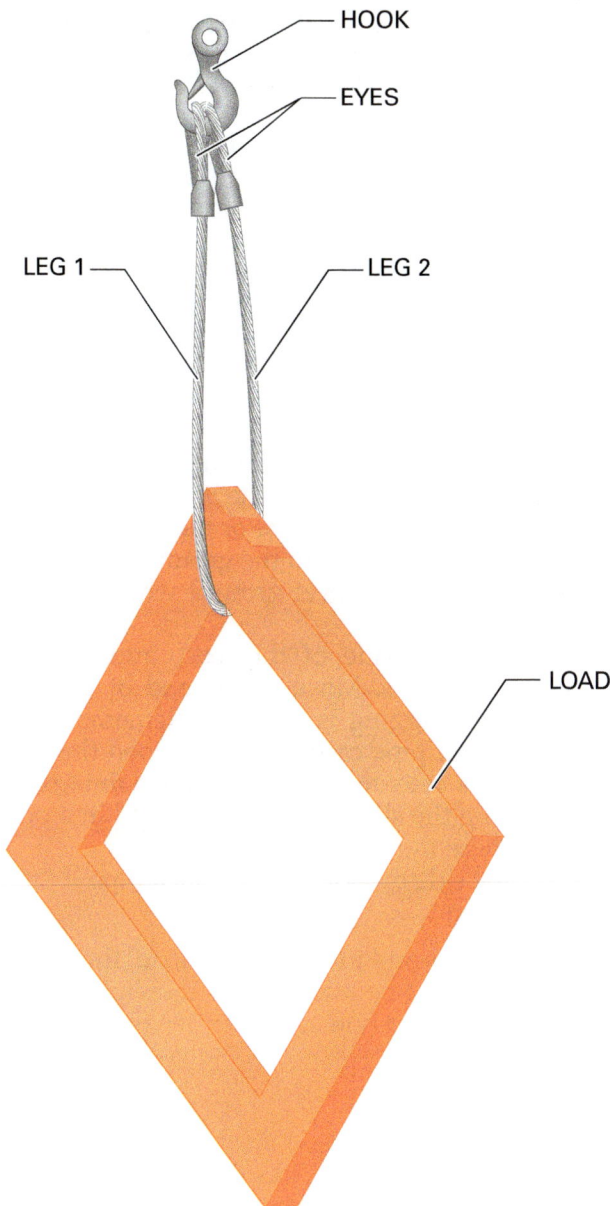

Figure 31 A simple basket hitch.

The simplest way of determining the CG of many objects is to experiment by lifting the object a few inches off the ground by a single point. If the object tilts, the single point by which the object is hanging is not directly over the CG. That means the lifting point needs to move in the same direction the load is tilting. When the object hangs level, the CG is vertically below the point from which the object is hung.

2.5.0 Rigging and Handling Piping Materials and Rebar

Reinforcing bars, piping, and similarly shaped construction materials are ordered from a supplier (usually the lowest bidder meeting the specification). Delivery of these materials is scheduled to coincide with the construction schedule as precisely as possible. Long-term storage of construction materials at the jobsite is not generally desirable.

These materials usually arrive at the job site on flatbed trucks or tractor trailers. If a spur track is available at the job site, shipments may be received on flatbed or gondola-style railroad cars. Where delivery of steel is scheduled to meet daily placement requirements, truckloads are delivered to the points of placement. Workers should be aware of safety factors that are necessary to achieve safe, efficient handling, storage, and hoisting of pipe and structural materials.

Trucks should be promptly unloaded to minimize jobsite traffic and potential safety hazards. Structural materials that are stored on the ground must be placed on timbers or other suitable blocking to keep them free from mud and to allow safe and easy handling.

Piping materials, rebar, and structural stock are normally stored by type and size first, then by length within each size if necessary. Materials such as pipe are usually of the same standard length unless sections or structural components have been precut or prefabricated before shipping. Tags are kept at the same end of each piece or bundle for easy identification. When a bundle is opened and part of the stock removed, the unit with the tag should remain with the bundle, and a new number reflecting the remaining quantity should be recorded. All mud and debris generally must be washed off before placement. The interior of piping materials should be kept clean and dry for most applications. For some applications, such as HVACR refrigerant piping, interior cleanliness is essential.

Use the following guidelines when rigging and handling a load of pipe, rebar, and similar products:

- Pipe of the same length should be lifted at the same time. If significantly different lengths of pipe must be lifted, lift each different length separately.
- Ensure that the open part of the hook is facing away from center when using slings with fixed or sliding hooks.
- Do not use carbon steel slings on stainless steel pipe to prevent cross-contamination of the metals. Carbon steel particles embedded in the stainless steel can rust and not only ruin the outer appearance, but also contaminate welds to the point of rejection. Use synthetic web slings on stainless steel and painted pipe.
- Lay the pipe on blocks so the slings can be worked on and off the load. Blocks should be made from hardwood and be thick enough to prevent the pipe from contacting uneven ground. Adequate ground clearance is also needed to slip a sling around the bundle. Pipe placed on blocks requires wedges to keep it from rolling off the blocking. Large diameter pipe requires larger chocks to keep it from rolling. In general, there should be 1" (2.5 cm) of chock per 1' (3 m) of diameter. For instance, a 42" (1,050 DN) pipe should have a chock that is about 3½" thick (9 cm) to prevent rolling. Note that a chock that is too thick can be pushed away by a rolling load.
- Use the hook and load line of the crane as the centerline with which to line up the load.
- As soon as the load clears the ground, check its orientation. If it is not level, the signal person should signal the operator to stop lifting the load, and guide the load back onto the blocks to adjust the position of the slings.
- Never stand underneath the load.
- Control all loads with one or more tag lines.
- Keep the load as low to the ground as possible.
- Store all pipe on level ground whenever possible.
- Stand to one side of the load, and guide it onto the blocks. Move to the end of the pipe before the rigging is released. Remember that a bundle of pipe can easily roll when the sling tension is released. Avoid pinch points when removing the sling, and do not stand in a position where the pipe may roll on you.
- Lay pipe side by side. Do not stack pipe if it can be avoided.

To select slings and properly rig a load, the weight of the load must be known. It is therefore beneficial to understand how to determine the weight of these materials.

2.5.1 Determining the Weight of Pipe

Every foot of pipe weighs a given amount, depending on the wall thickness and nominal size of the pipe. Therefore, if you know the weight per foot and the length, the total weight of the pipe can be determined.

Table 3 is an example of a weight chart for carbon steel pipe. Always be sure you consult a chart for the proper material as well as the correct size and weight. There are also separate charts for metric pipe sizes, which will report the weight as kilograms per meter rather than pounds per foot. Tables like the one shown here are available from a variety of sources online, including pipe distributors.

The numbers in the leftmost column represent the nominal pipe size in inches. The designations and numbers across the top represent wall thickness. The wall thickness of pipe is often designated by a schedule number; the numbers 10 through 160 across the top are schedule numbers. To use the table, find the point at which the nominal size and wall thickness, or schedule, of the pipe meet. The number in this block is the weight of one foot of pipe. To find the total weight, multiply the total number of feet by the weight per foot. Note that there is a significant difference in the weight of pipe made from other materials, such as PVC, copper, and stainless steel.

For example, to find the weight of five 10-foot lengths of 20" Schedule 40 carbon steel pipe, read right from the 20.0 in the Nominal Pipe Size column until reaching the Schedule 40 column. The number found there is 123.1. This is the weight, in pounds, of one foot of pipe. There is a total of 50' of pipe (5 lengths × 10' = 50'). Multiply the weight per foot from the table by the total length in feet to find the total weight of the pipe load (123.1 × 50' = 6,155 lbs).

Table 3 Carbon Steel Pipe Weights

Nominal Pipe Size (inches)	Wall Thickness								
	STD	XS	XXS	10	40	60	80	120	160
	Weight Per Foot in Pounds								
2.0	3.65	5.02	9.03	–	3.65	–	5.02	–	7.06
2.5	5.79	7.66	13.7	–	5.79	–	7.66	–	10.01
3.0	7.58	10.25	18.58	–	7.58	–	10.25	–	14.31
3.5	9.11	12.51	22.85	–	9.11	–	12.51	–	–
4.0	10.79	14.98	27.54	–	10.79	–	14.98	18.98	22.52
6.0	18.97	28.57	53.16	–	18.97	–	28.57	36.42	45.34
8.0	28.55	43.39	72.42	–	28.55	35.66	43.39	60.69	74.71
10.0	40.48	54.74	104.1	–	40.48	54.74	64.40	89.27	115.7
12.0	49.56	65.42	125.5	–	53.56	73.22	88.57	125.5	160.3
14.0	54.57	72.09	–	36.71	63.37	85.01	106.1	150.8	189.2
16.0	62.58	82.77	–	42.05	82.77	107.5	136.6	192.4	245.2
18.0	70.59	93.45	–	47.39	104.8	138.2	170.8	244.1	308.6
20.0	78.60	104.1	–	52.73	123.1	166.5	208.9	296.4	379.1
22.0	86.61	114.8	–	58.07	–	197.4	250.8	353.6	451.1
24.0	94.62	125.5	–	63.41	171.2	238.3	296.5	429.5	542.1
26.0	102.6	136.2	–	85.73	–	–	–	–	–
28.0	110.6	146.9	–	92.41	–	–	–	–	–
30.0	118.7	157.5	–	99.08	–	–	–	–	–
32.0	126.7	168.2	–	105.8	229.9	–	–	–	–
34.0	134.7	178.9	–	112.4	244.9	–	–	–	–
36.0	142.7	189.6	–	119.1	282.4	–	–	–	–
42.0	166.7	221.6	–	–	330.4	–	–	–	–

2.5.2 Determining the Weight of Rebar

The weight of rebar is determined the same way as the weight of pipe. *Table 4* shows a chart of rebar weight by bar size. This particular example shows the bar size and other characteristics in both imperial and metric units. Virtually all rebar is made from carbon steel, so it would be rare to encounter rebar made from another material. If that is the case however, the supplier would need to be contacted for the weight information.

Once the size and total length of rebar in a bundle is identified, the weight per foot (or per meter) is multiplied by the appropriate length to determine the weight. Always remember to work within the same system of units—kilograms per meter, or pounds per foot.

2.5.3 Rigging Valves

Rigging a valve correctly involves knowing where to place the sling, how to place the sling, what kind of sling to use, and what kind of valve is being lifted. Manufacturers of large valves often provide drawings to show how the part is best rigged for lifting. Very large valves and valve components may be equipped with factory-installed lifting lugs.

Synthetic slings are usually best for rigging all types of valves, since many are also painted and easily scarred by metal slings. Never use a carbon steel sling to rig a stainless steel valve to avoid surface contamination of the stainless steel.

To rig a valve like the one shown in *Figure 34*, place a synthetic sling on each side of the valve body between the bonnet and the flanges. If the handwheel remains attached, bring the slings up

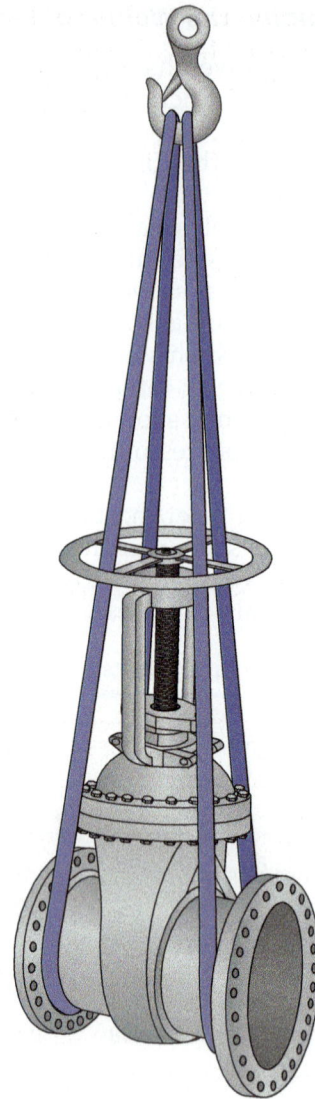

Figure 34 Rigging a valve.

Table 4 ASTM Standard Metric and US Measure Reinforcing Bar

Bar Size		Nominal Characteristics*					
		Diameter		Cross-Sectional Area		Weight	
Metric	[in-lb]	mm	[in]	mm	[in]	kg/m	[lbs/ft]
#10	[#3]	9.5	[0.375]	71	[0.11]	0.560	[0.376]
#13	[#4]	12.7	[0.500]	129	[0.20]	0.944	[0.668]
#16	[#5]	15.9	[0.625]	199	[0.31]	1.552	[1.043]
#19	[#6]	19.1	[0.750]	284	[0.44]	2.235	[1.502]
#22	[#7]	22.2	[0.875]	387	[0.60]	3.042	[2.044]
#25	[#8]	25.4	[1.000]	510	[0.79]	3.973	[2.670]
#29	[#9]	28.7	[1.128]	645	[1.00]	5.060	[3.400]
#32	[#10]	32.3	[1.270]	819	[1.27]	6.404	[4.303]
#36	[#11]	35.8	[1.410]	1006	[1.56]	7.907	[5.313]
#43	[#14]	43.0	[1.693]	1452	[2.25]	11.38	[7.65]
#57	[#18]	57.3	[2.257]	2581	[4.00]	20.24	[13.60]

*The equivalent nominal characteristics of inch-pound bars are the values enclosed within the brackets.

through the handwheel spokes so that the valve cannot tilt from front to back. Do not place a sling around the handwheel or through the valve bore. The handwheel is not built to support the weight of the valve. A valve rigged around the handwheel is unsafe, even if the valve is going to be moved a short distance. Placing a sling through the bore of the valve can destroy the inner workings, even if soft synthetic slings are used.

2.6.0 Taglines and Knots

Taglines are natural fiber or synthetic ropes used to control the load (*Figure 35*). The absence or improper use of taglines can turn a simple hoisting operation into a hazardous situation. Taglines are used to maintain lateral control and prevent spinning of a suspended load as the crane and/or boom move. Loads of all shapes and sizes are subject to some level of dynamic and wind forces once in the air.

When selecting a tagline, several factors need to be considered. Natural fiber, or manila, is notably weaker than synthetic fibers such as nylon, polyester, polypropylene, or polyethylene. Although any rope that is wet becomes an electrical conductor, natural fiber rope absorbs water readily and may remain wet enough to conduct electricity for a long time. Most (but not all) synthetic ropes do not absorb moisture. A nonconductive tagline should be used when working in the vicinity of power lines.

Synthetic rope is lighter than natural fiber and has a high strength-to-weight ratio. Its resistance to water and reduced electrical conductivity do give it a distinct safety advantage. However, synthetic ropes are more easily damaged by heat. Significant contact with surfaces at 150°F (66°C) can result in a loss of strength, and many synthetic rope materials begin to melt at 300°F (149°C).

The diameter of a tagline should be large enough so that it can be gripped well when wearing gloves. Rope with a diameter of ½" (12 mm) is common, but ¾" (19 mm) and 1" (25 mm) diameter rope is sometimes used on heavy loads or where the tagline must be extremely long. Taglines should never have knots or loops tied in them. Terminating hardware, such as snaps and carabiners (*Figure 36*), may be added to make an easy connection to some loads. This hardware does not need to be designed and rated for lifting purposes, but must be substantial enough to handle the stress.

WARNING!

Always wear gloves when using a tagline to control a load. The momentum of a moving load can cause the rope to slide through the hands unexpectedly, causing severe rope burns. Never wrap a tagline around an arm or leg in an attempt to stop a load's swing. Never place yourself between a fixed object and a suspended load. Manning a tagline is a significant responsibility that requires careful thought and attention to execute the task safely and prevent serious injury to personnel and/or damage to property.

(A) HEAVY-DUTY SNAP

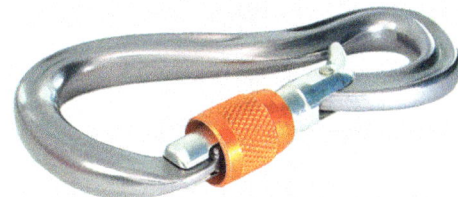

(B) CARABINER WITH LOCK

Figure 36 Rope hardware for quick connections.

Figure 35 Taglines in control of a load.

Taglines should be of sufficient length to allow control of the load from its original lift location until it is safely placed or until load control is transferred to other team members. Special consideration should be given to situations where a long tagline could interfere with the safe handling of loads, such as steel erection projects. Think through the lift and the material movement and consider where the tagline(s) will be and what obstructions might be encountered.

When working near power lines, there is always the hazard of electric shock or electrocution. An insulating link (*Figure 37*) can be added to taglines to help protect the user if the load or tagline contacts a power source.

Taglines should be attached to loads at a location that provides the best physical advantage in maintaining control. Long loads, for example, should have taglines attached as close to the ends as possible. The tagline should also be located in a place that allows personnel to access it for removal after the load is placed.

> **WARNING!**
>
> Avoid overcompensating during load control through exaggerated movements of the tagline; do not jerk the line. Taglines that are too long can be caught on objects or drag unnecessarily. Avoid tying knots in the line, as the potential for being caught on objects as it moves along the ground are significantly increased. Never tie the tagline to the load hook.

Figure 37 Insulating link.

To properly handle a tagline, you must determine the physical advantage intended. Consider the lift and which direction the load may be most likely to swing or rotate. With a long load, for example, try to maintain a position that is 90 degrees to the length of the object (*Figure 38*).

Whenever possible, keep yourself and the tagline in view of the crane operator. Stay alert. Do not become complacent during the lift. Be aware of the location of any excess rope and do not allow it to become fouled or entangled around your legs or on nearby objects.

Taglines should not to be used to pull or yank a load away from its natural vertical suspension. They also cannot be used in any way that results in them supporting or carrying any portion of the load. Large loads often require the use of multiple taglines. In these cases, tagline personnel must work as a team and coordinate their actions.

2.6.1 Knots

Knots used to attach taglines to loads should be tied properly to prevent slipping or accidental loosening, but they also must be easy to untie after the load is placed. Whenever possible, taglines should be of one continuous length and free of splices. If joining two taglines together becomes necessary, it is best done using the short-splice method. The short-splice method results in a knot diameter roughly twice that of the rope itself, and retains more rope strength than other methods. Larger knots tied in the middle of the taglines can sometimes create difficulties. However, the short splice can be a challenging knot to create, as it involves weaving of the individual rope strands. The short-splice method can be learned from numerous websites and internet videos if desired. In general, it is always best to avoid knots of any kind in a tagline and use a continuous length of rope instead.

Some recommended knots for rigging are the bowline and the clove hitch. The bowline (*Figure 39*) is used to form a secure loop in the end of a rope. It is sometimes called a *rescue knot* or the *king of knots* because it is reliable enough to be used for rescue work. It can be backed up with a second knot for extra security or tied twice, into a double bowline, for extra strength. A bowline does not slip or bind when under load, but it is easy to untie when there is no load.

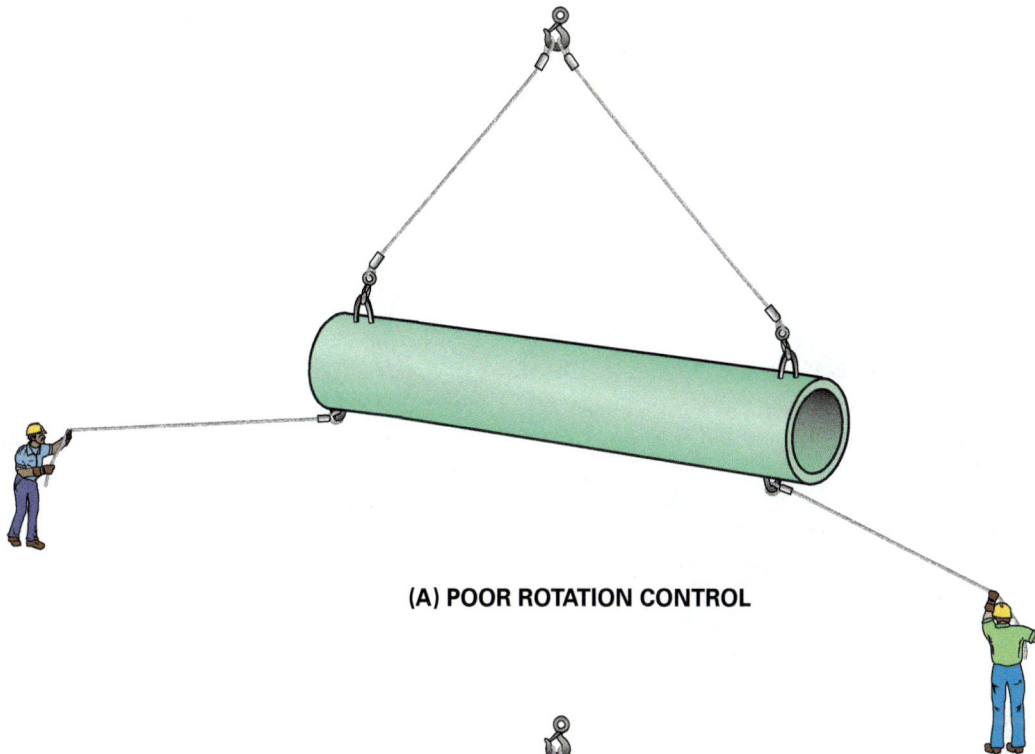

(A) POOR ROTATION CONTROL

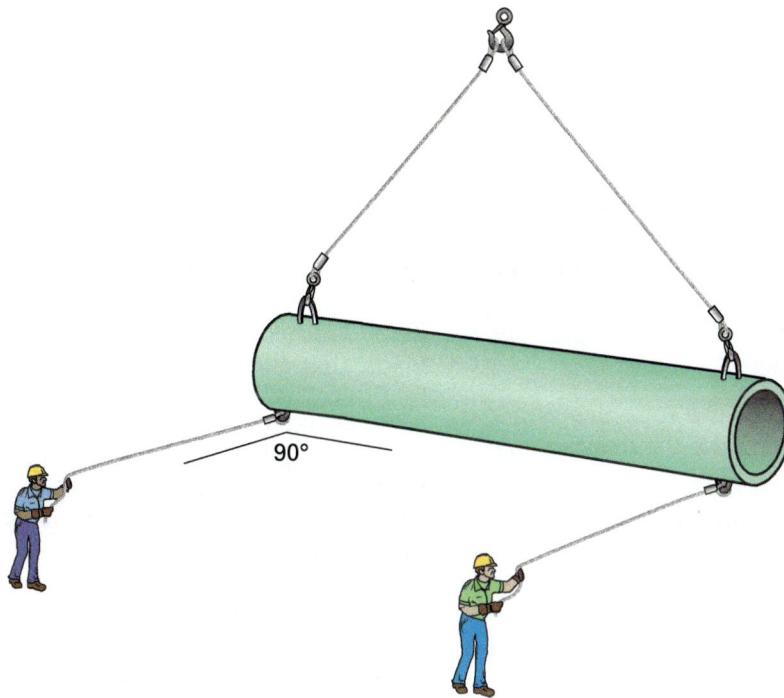

(B) BEST ROTATION CONTROL

Figure 38 Best position for controlling load rotation.

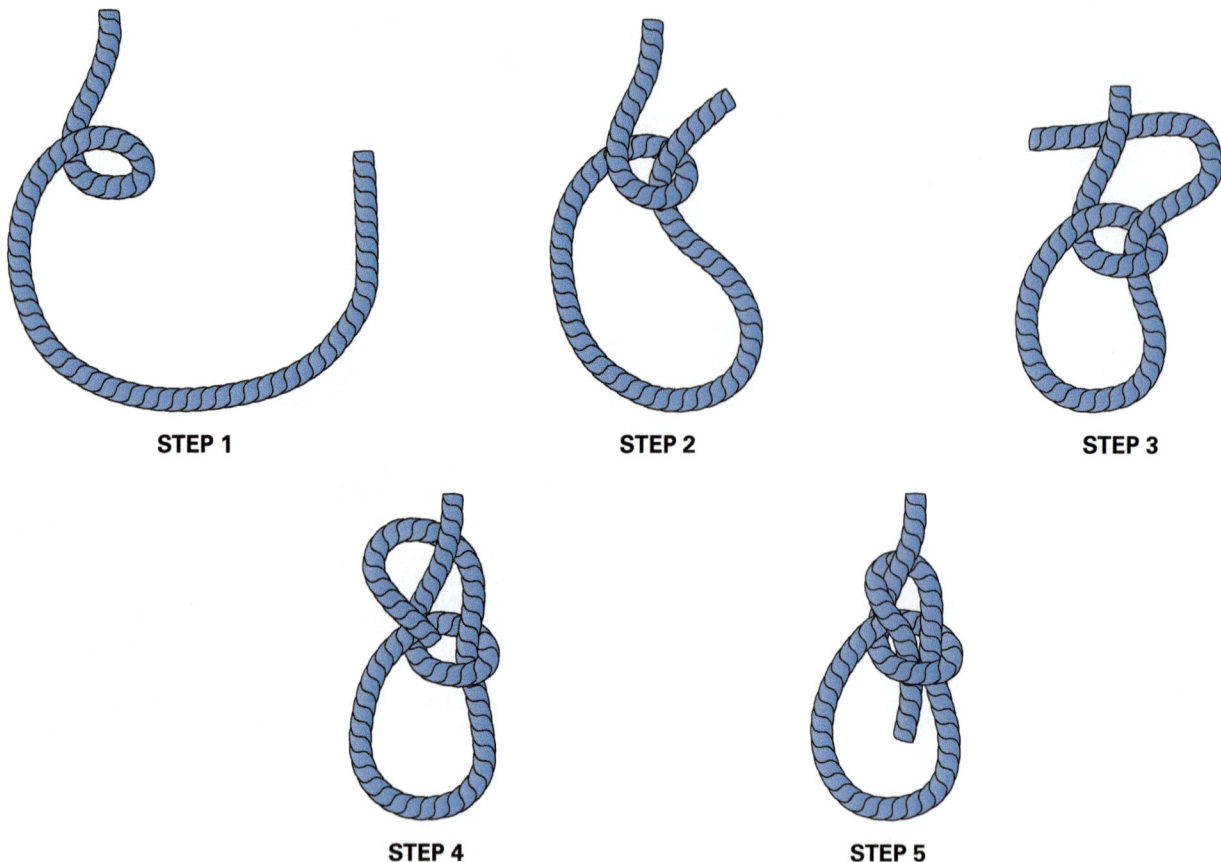

STEP 1 **STEP 2** **STEP 3**

STEP 4 **STEP 5**

Figure 39 Tying a bowline.

The bowline reduces the strength of the rope that it is tied in by as much as 40 percent. Bowline knots also require a long tail for reliability.

The following steps here, shown in *Figure 39*, can be used to tie a bowline:

Step 1 Form a small loop in the rope, leaving a long enough tail to make the desired size of main loop.

Step 2 Pass the tail of the rope through the small loop.

Step 3 Pass the tail under the standing end.

Step 4 Pass the tail back down through the small loop.

Step 5 Adjust the main loop to the required size and tighten the knot, making sure to maintain a sufficiently long tail.

NOTE

Some people use this saying to help them remember how to tie a bowline: "The rabbit comes out of his hole, around a tree, and back into the hole."

A half hitch (*Figure 40*) can be used to tie a rope around an object such as a rail, bar, post, or ring. It is commonly used for tasks such as suspending items from overhead beams and carrying light loads that have to be removed easily. The half hitch is also widely used in making other knots, such as the clove hitch. Because the half hitch is not very stable by itself, it is not suitable for heavy loads or tasks in which safety is paramount. It is often used to back up and secure another knot that has already been tied.

The following steps, shown in *Figure 40*, can be used to tie a half hitch:

Step 1 Form a loop around the object.

Step 2 Pass the tail of the rope around the standing part and through the loop.

Step 3 Tighten the hitch by pulling on the working end and the standing part of the rope simultaneously.

A clove hitch (*Figure 41*) is one of the most widely used general hitches. It is typically used to make a quick and secure tension knot on a fixed object that serves as an anchor, such as a post, pole, or beam. A clove hitch can also be used as the first knot when lashing items together.

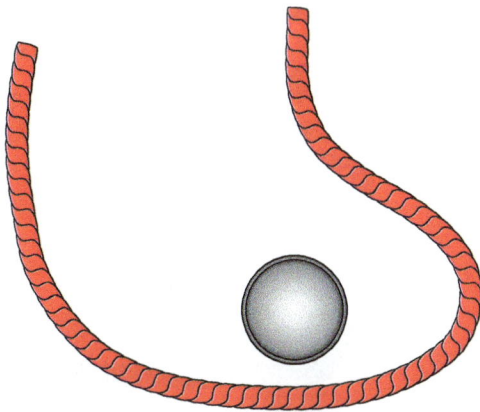

STEP 1

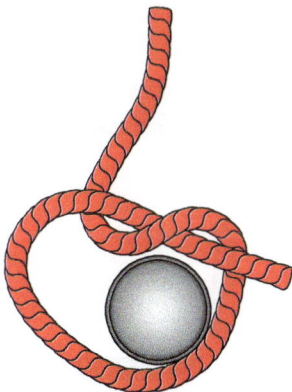

STEP 2

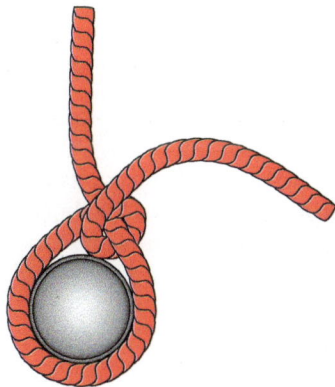

STEP 3

Figure 40 Tying a half hitch.

Because it is a tension knot, a clove hitch loosens when tension is removed from the rope. For these reasons, a clove hitch should not always be used as-is. Additional half hitches or an overhand safety knot should be added to make a clove hitch more secure, unless loosening when tension is released is intentional.

There are alternate techniques for tying a clove hitch. Regardless of the technique, the structure of the resulting knot consists of two half hitches made in opposite directions.

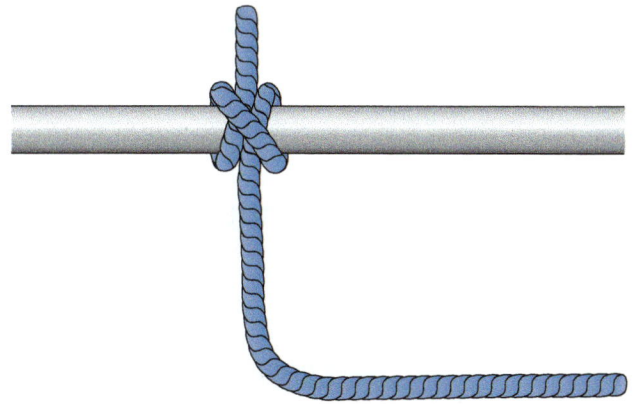

Figure 41 Clove hitch.

One common technique for tying a clove hitch that is used to attach a rope to a ring or upright structural component is to thread the end. This is called the *threading-the-end technique*. Use the following steps, shown in *Figure 42*, to tie a clove hitch using the threading-the-end technique:

Step 1 Pass the working end of the rope over the object.

Step 2 Pass the working end back over the standing part and then over the object. This forms a half hitch.

Step 3 Thread the working end back under itself and up through the loop to form a second half hitch.

Step 4 Pull both the working and standing ends evenly to tighten the hitch.

Another method for tying a clove hitch, the stacked-loops technique, allows the rope to be dropped quickly over a standing object, instead of tying the knot around the object. It also allows the hitch to be tied at any point in a rope, not just at an end. Use the following steps, shown in *Figure 43*, to tie a clove hitch the stacked-loops technique:

Step 1 Form two identical loops in a rope, one in the right hand and one in the left.

Step 2 Cross the loops one above the other to form a knot.

Step 3 Place the knot over the stake or post.

Step 4 Tighten the knot by pulling simultaneously on both ends of the rope. Note that the completed clove hitch consists of two half hitches stacked on each other.

Remember that an additional half hitch is often added for increased security.

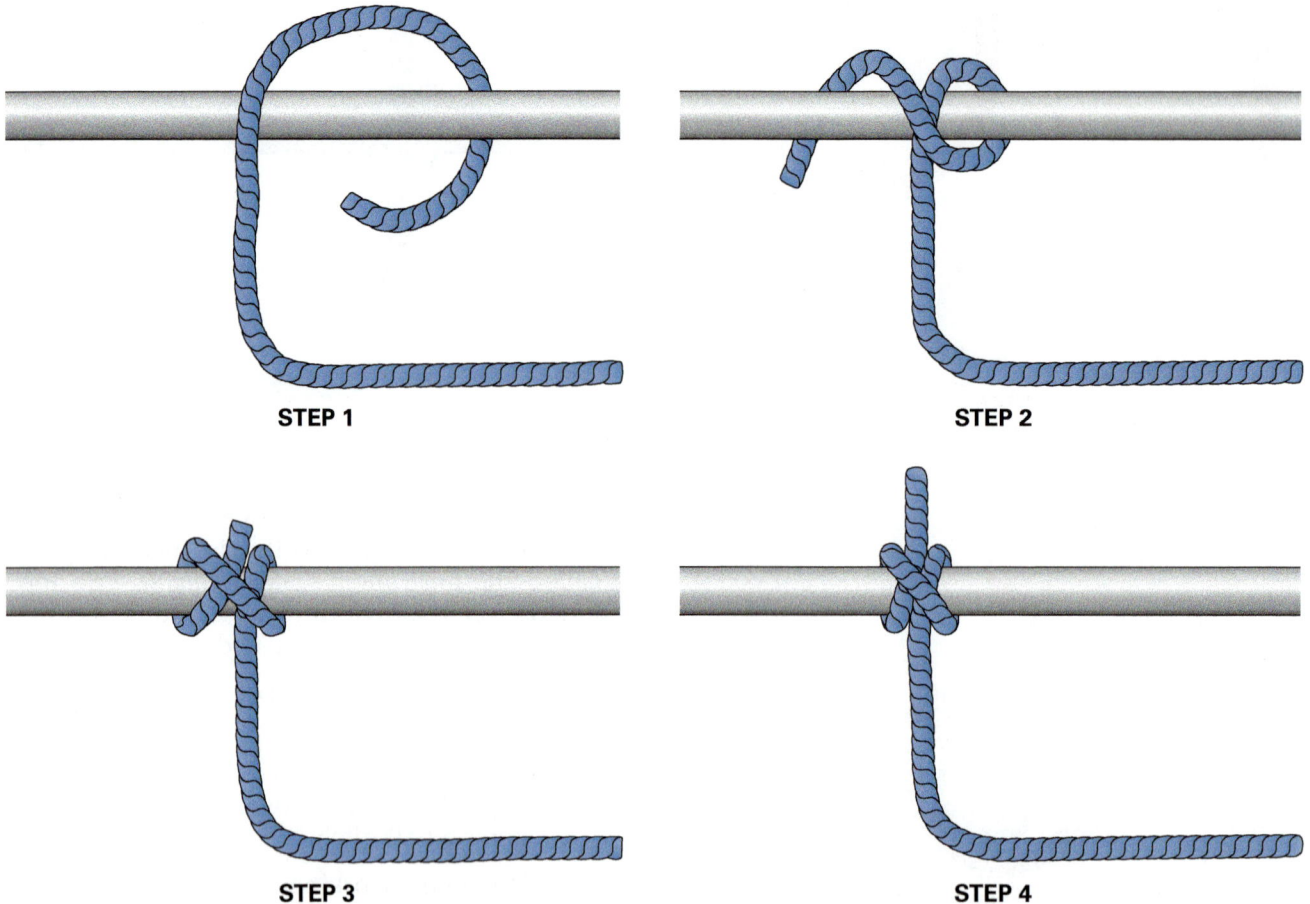

STEP 1   **STEP 2**

STEP 3   **STEP 4**

Figure 42 Threading-the-end technique for tying a clove hitch.

2.7.0 Rigging Safety Precautions

Workers on the job are responsible for their own safety and the safety of their fellow workers. Project and corporate management, in turn, has a responsibility to each worker. The responsibility of management and supervisors is to ensure that the workers who prepare and use the equipment, and those who work with or around it, are well trained in operating procedures and safety practices. Each worker is expected to put that training to good use. This section describes safety guidelines that are related to rigging.

> **NOTE**
>
> Safety guidelines pertaining to mobile cranes are covered in detail in NCCER Module 21106-18, "Crane Safety and Emergency Procedures" from *Mobile Crane Operations Level One*.

Did You Know?

The Ashley Book of Knots

Although *The Ashley Book of Knots* by Clifford W. Ashley is no longer in print, sailors, climbers, campers, macramé artists—anyone with more than a passing interest in knots—continue to find this book cited as the definitive work on the subject of knot tying. Originally published by Doubleday & Company in 1944, it contains the histories of and instructions for tying over 3,900 types of knots, accompanied by 7,000 pen-and-ink drawings.

Clifford Ashley was born in 1881, in New Bedford, Massachusetts. By trade he was an author and artist, but while sailing on many types of boats and researching knots he performed a wide variety of jobs. He spent 11 years writing and drawing the illustrations for his book. He died three years after the book's publication. Ashley continues to be regarded as a fine marine painter as well as one of the world's leading authorities on knot tying.

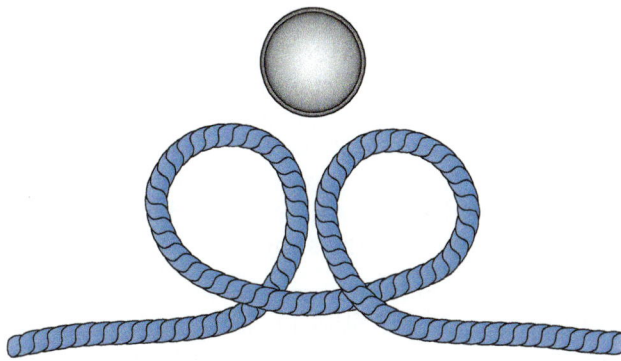

STEP 1

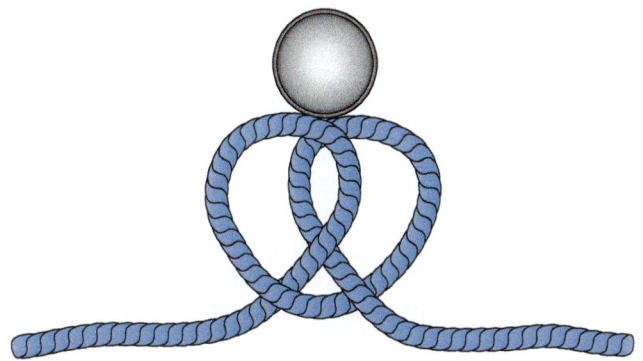

STEP 2

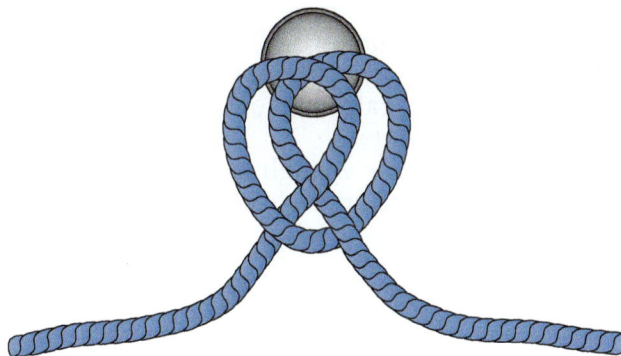

STEP 3

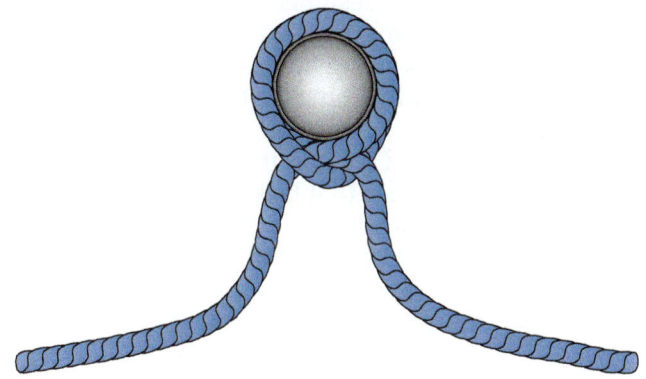

STEP 4

Figure 43 Stacked-loops technique for tying a clove hitch.

One very important rigging precaution is to determine the weight of every load before attempting to lift it. The weight of the load must obviously be within the capacity of the crane, but this is not the responsibility of the rigger. However, the load weight must be considered in the selection of the rigging components before the lift ever begins.

Unfortunately, accurately determining the weight is not always possible. This issue is addressed in 29 *CFR* 1926.1417, which requires that the crane operator verify the weight of the load in one of the following two ways:

- The weight can be determined from an industry-recognized source such as the manufacturer of the product(s), or by an industry-recognized calculation method. This latter method could be used, for example, to calculate the weight of an I-beam based on the weight per foot of such products.
- If the weight cannot be verified through the above methods, the operator may begin the lift using a load-weighing device, load-moment indicator, rated-capacity indicator, or rated-capacity limiter. This exercise is used to

determine if the lifted load is more than 75 percent of the crane's maximum rated capacity at the longest operating radius that the lift will require. If it is greater than 75 percent, the lift must be aborted until the actual weight can be verified through the first method.

It is equally important to rig the load so that it is stable and its center of gravity is below the hook. The personal safety of riggers and hoisting operators depends a lot on common sense. The following safety practices should be observed:

- Following the OSHA procedures from 29 *CFR* 1926.1417, determine the weight of loads before rigging whenever possible.
- Coordinate the rigging process as necessary with other personnel on the job site to minimize foot and vehicle traffic in the area.
- Know the rated load capacity of the equipment and tackle, and never exceed it. Remember that the rated load capacity of all hoisting and rigging equipment is based on ideal conditions. Normal working conditions are rarely ideal, so it is important to recognize factors that can reduce equipment capacity or increase the load imposed, such as sling angles.

- Examine all hardware, tackle, slings, and equipment before use. Remove defective components from service for evaluation and possible repair by others.
- Always wear gloves when handling slings, taglines, and similar rigging equipment.
- Never wrap hoist ropes around a load. Use only slings or other appropriate lifting devices.
- Ensure that all slings are made from the same material when using two or more slings on a single load.
- Never lift loads with one or two legs of a multi-leg sling until the unused slings are secured.
- Only use lifting beams for the purpose for which they were designed.
- All personnel must stand clear and not underneath loads being lifted and lowered, and when slings are being withdrawn from beneath the load.
- Never point-load a hook unless it has been specially designed for that task.
- Keep hands and feet away from pinch points as the slack is taken up by the crane.
- Use one or more taglines as necessary to keep the load under control.

- Immediately report defective equipment or rigging issues to the rigging supervisor or lift director, who must issue orders to proceed after safe conditions have been ensured.
- Prepare blocking at the load landing zone before the load arrives, rather than after it is suspended above. Allow the load to safely land before removing the slings.

Weather conditions can be a major factor in lifting operations. Stop hoisting and/or rigging operations when weather conditions such as the following present a hazard to property, workers, or bystanders:

- Winds exceeding crane manufacturer recommendations
- Lightning or thunder
- Poor visibility due to conditions such as darkness, dust, fog, rain, or snow, impairing view of rigger or hoist crew
- Temperature low enough to cause crane structures to fracture upon shock or impact

Crane Collapse in Manhattan

In early 2016, a lattice-boom crawler crane was attempting to lower its 565-foot (172-meter) boom and secure the crane due to high winds. Unfortunately, the wind gusts intensified and the boom came crashing into the streets of Manhattan. One bystander was killed and several others were injured. The crew had already begun clearing the streets of traffic to prepare to receive and secure the boom, minimizing the loss of life and injuries.

Figure Credit: © a katz/Shutterstock.com

Additional Resources

ASME Standard B30.5, Mobile and Locomotive Cranes. Current edition. New York, NY: American Society of Mechanical Engineers.

ASME Standard B30.9, Slings. Current edition. New York, NY: American Society of Mechanical Engineers.

ASME Standard B30.20, Below-The-Hook Lifting Devices. Current edition. New York, NY: American Society of Mechanical Engineers.

ASME Standard BTH-1, Design of Below-The-Hook Lifting Devices. Current edition, New York, NY: American Society of Mechanical Engineers.

29 *CFR* 1926, Subpart CC, **www.ecfr.gov**

29 *CFR* 1926.251, **www.ecfr.gov**

29 *CFR* 1926.753, **www.ecfr.gov**

Mobile Crane Operations Level One, NCCER. Third Edition. 2018. New York, NY: Pearson Education, Inc.

NCCER Module 00106-15, *Introduction to Basic Rigging*.

Mobile Crane Safety Manual (AEM MC-1407). 2014. Milwaukee, WI: Association of Equipment Manufacturers.

Willy's Signal Person and Master Rigger Handbook, Ted L. Blanton, Sr. Current edition. Altamonte Springs, FL: NorAm Productions, Inc.

Knots: The Complete Visual Guide, Des Pawson. First American Edition. 2012. New York, NY: DK Publishing.

North American Crane Bureau, Inc. website offers resources for products and training, **www.cranesafe.com**

2.0.0 Section Review

1. A sling with a wide, flat bearing surface that is especially well-suited for basket hitches where sharp bends are encountered is the _____.

   a. strand-core wire rope sling
   b. fiber-core wire rope sling
   c. braided-belt sling
   d. chain sling

2. If a sling is needed to lift a load with a polished surface, the best choice would be a _____.

   a. synthetic web sling
   b. chain sling
   c. metal-mesh sling
   d. wire-rope sling

3. Metal-mesh slings must have the approval of the manufacturer if they are to be used at temperatures above _____.

   a. 195°F (91°C)
   b. 375°F (191°C)
   c. 550°F (288°C)
   d. 750°F (399°C)

4. Refer to *Figure SR01*. Based on the measurements shown, what is the sling angle factor that should be applied to the weight of the lifted load on each sling?

   a. 0.841
   b. 1.194
   c. 1.213
   d. 1.414

5. Refer to *Table 2* in Section 2.4.2. Based on the generally accepted safe range of sling angles, is the sling angle associated with a sling angle factor of 1.194 within the acceptable range?

   a. Yes
   b. No

6. To prevent 24" pipe material (600 DN) from rolling, the chock needs to be _____.

   a. 1" thick
   b. 2" thick
   c. 3½" thick
   d. 5" thick

7. One disadvantage of synthetic rope compared to natural fiber is that synthetic rope _____.

   a. is significantly heavier
   b. readily conducts electricity
   c. has a low strength-to-weight ratio
   d. is more easily damaged by heat

8. The procedures for determining the weight of a load to be lifted are covered in _____.

   a. *29 CFR 1910.140*
   b. *ASME Standard B30.9*
   c. *29 CFR 1926.1417*
   d. *ASME Standard BTH-1*

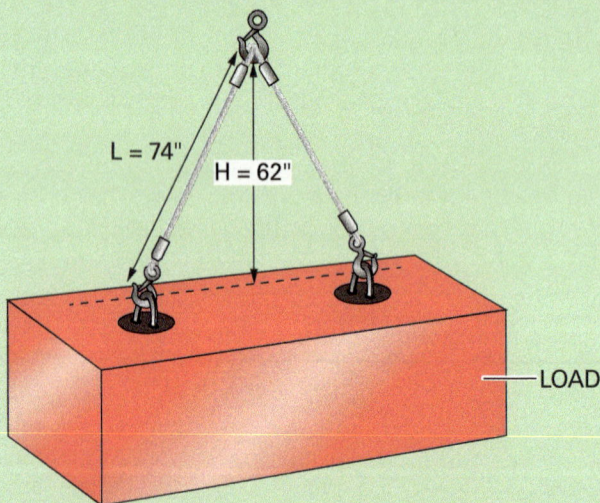

L = 74"
H = 62"
LOAD

Figure SR01

3.0.0 HOISTING EQUIPMENT AND JACKS

Objective

Identify and describe how to use various types of hoisting and jacking equipment.

a. Identify and describe how to use manual and powered hoisting equipment.
b. Identify and describe how to use jacks.

Performance Tasks

Select, inspect, and demonstrate the safe use of the following rigging equipment:

- Block and tackle
- Chain hoist
- Ratchet-lever hoist
- One or more types of jacks

Trade Terms

Gantry: A framed overhead structure supported by legs on each end, used to cross over obstructions. Gantries can be portable or permanent, providing support for hoisting equipment or raising and supporting lighting, cameras, and similar equipment.

Hauling line: The portion of a rope or chain on hoisting equipment that the operator uses to raise or lower the load. Also known as a hauling part.

Parts of line: The resulting number of lines that are supporting the load block when a line is reeved more than once.

R igging isn't only about connecting loads to cranes. Riggers are also involved in equipment movement and placement inside of buildings and other locations where a crane is either unnecessary or impractical. This section presents various types of hoisting and jacking equipment.

3.1.0 Manual and Powered Hoisting Equipment

Both manual and powered hoisting equipment are used when it is necessary to lift components into position, or raise one component to insert something beneath it. Common hoisting equipment used includes block and tackle rigs, chain hoists, and ratchet-lever hoists.

3.1.1 Block and Tackle

The block and tackle is the most basic lifting device. It is used to lift or pull light loads. A block consists of one or more sheaves (pulleys) fitted into a wood or metal frame with a hook attached to the top. The tackle is the line or rope and end attachments connected to the block. Some block and tackle rigs have a brake that holds the load once it is lifted and others do not. The types that do not have a brake require continuous pull on the hauling line, or the hauling line must be tied off to hold the load.

There are two types of block and tackle rigs: simple and compound. A simple block and tackle consists of one sheave and a single line (*Figure 44*). It is used to lift or pull very light loads. The hook is attached to the load, and the load is lifted by pulling the line. The load capacity of this type of block and tackle is equal to the capacity of the load line. The block must be attached to a building structure or other support by a method that provides adequate load capacity to support the load and the tackle. Adequate capacity and stability of the supporting structure is crucial and it must be evaluated carefully by qualified personnel.

A compound block and tackle (*Figure 45*) uses more than one block. It has an upper, fixed block that is attached to the building structure or other support and a lower, traveling block that is attached to the load. Each block may have one or more sheaves. The more sheaves the blocks have, the more parts of line the block and tackle has, and the higher the lifting capacity. The compound block and tackle multiplies the power applied to the rope, so a worker can lift a much heavier load than is possible with a simple block and tackle.

3.1.2 Chain Hoists

A chain hoist, also called a *chain fall*, is a very useful and commonly used by a number of crafts. Chain hoists should be used for straight, vertical lifts only—just like a crane. They may be damaged if used for angled lifts or horizontal pulls. Although chain hoists are more popular, generally more durable, and capable of lifting heavier loads, there are electric and pneumatic cable hoists as well.

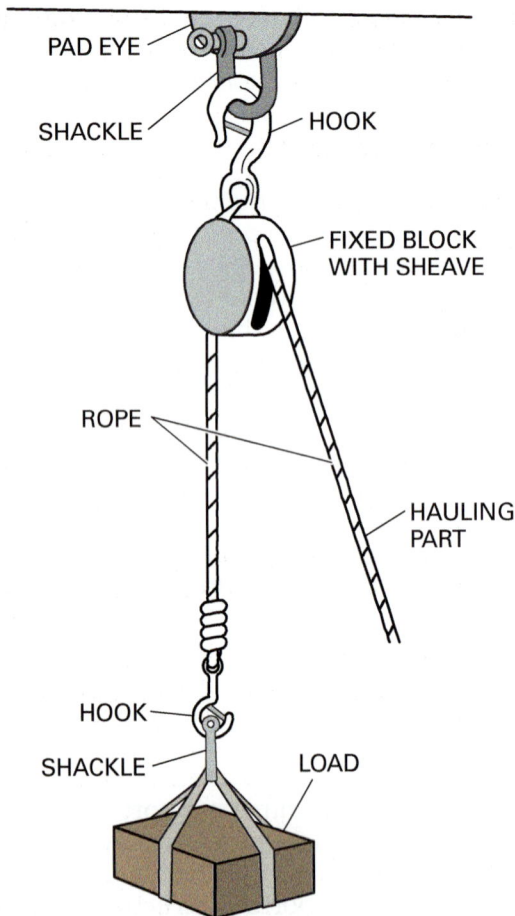

Figure 44 Simple block and tackle.

Figure 45 Compound block and tackle.

Chain hoists are standard equipment in many fabrication shops and rigging departments because they are dependable, portable, and easy to use. Some common types of chain hoists are the spur-geared (manual) chain and the electric (powered) chain.

The spur-geared chain hoist has two chains. An endless chain, the hand chain, drives a single pocketed sheave, which in turn drives a gear-reduction unit. The load chain is fitted to the gear-reduction unit and has a hook that attaches to the load. The gear-reduction unit provides the mechanical advantage, just as it does in many motor-driven conveyors and similar industrial systems. The spur-geared chain hoist is commonly used for significant loads and is convenient since it does not require a power source. Two common types are shown in *Figure 46*. The trolley-style chain hoist is designed to run on the rails of an I-beam, like the beam of a gantry.

A gantry (*Figure 47*) is often used indoors to hoist and move components. Gantries have maximum-load limitations just like chain hoists, but very large gantries with enormous capacities

Around the World

Block and Tackle History

Archimedes, a Greek mathematician, physicist, engineer, inventor, and astronomer, is credited with the development of the block and tackle around 250 BC. However, Archytas of Tarentum is thought to have developed the concept of a pulley—an essential component of the assembly to reduce friction. Block and tackle has been used in the construction of many structures that continue to fascinate us, including Stonehenge in the United Kingdom. Block and tackle systems are also considered to be the source of inspiration for modern cranes.

are used in shipping harbors and similar locations. These smaller versions are portable and very handy. A trolley-style chain hoist can be placed on the horizontal beam, or a separate trolley assembly can be purchased and a hook-type chain hoist can be attached to the trolley.

Electric chain hoists (*Figure 48*) work much like manual chain hoists, except that they have an electric motor instead of a pull chain to raise and lower the load. Electric chain hoists are faster and more efficient than manual chain hoists. Most are controlled by a pendant with pushbuttons, suspended from the hoist. There are a few battery-powered hoists on the market, but with limited capacity. As battery technology continues to advance, battery-operated hoists with more capacity may eventually be available. Most electric hoists operate on 120 and/or 240 VAC. Note that air-powered (pneumatic) chain hoists are also available.

Special care should be taken when raising a load with an electric chain hoist, since it is hard to tell how much force the electric motor is exerting. Motor sound may provide a clue. If the load gets caught on something immoveable, the chain hoist could be overloaded and damaged. A good electric chain hoist has reliable built-in overload protection that will open the motor circuit to prevent damage.

The load capacity of any chain hoist should be clearly marked on the data plate. The capacity should never be exceeded. Like a crane, it is necessary to know the weight of the load before beginning. Always use a sling to rig the load for a chain hoist if there is no single, dedicated lifting eye on the load. The chain of the hoist should never be wrapped around an object and used as a sling.

Like other lifting devices, chain hoists must be carefully inspected before each use. The chain must be checked to ensure that it has no significant defects. Before using a hoist, read the manufacturer's instructions and be aware of its limitations.

(A) HOOK-STYLE CHAIN HOIST

(B) TROLLEY-STYLE CHAIN HOIST

Figure 46 Manual chain hoists.

Figure 47 Portable gantry.

Figure 48 Electric chain hoist.

Follow these basic steps to select, inspect, use, and maintain a chain hoist:

Step 1 Select a chain hoist of adequate capacity to handle the load. Ensure that the overhead structure used to suspend the hoist also has adequate capacity.

Step 2 Inspect the load chain and hook to ensure that they are not excessively worn, bent, or deformed in any way.

Step 3 Inspect the sheaves to ensure that they are not bent or excessively worn.

Step 4 Ensure that the chain hoist has proper lubrication. Maintain per manufacturer's instructions.

Step 5 Hang the chain hoist from a suitable structure using the proper rigging. Provide an adequate power source for electric and pneumatic models, ensuring that it is routed to the hoist neatly and out of the way. The power cord or air hose should not be allowed to dangle straight down where it can interfere with the task.

Step 6 Position the hoist directly over the load. Lower the load hook, and connect it to the load using the proper rigging.

Step 7 Raise and place the load. To raise the load using a manual chain hoist, pull the hand chain. To raise the load using an electric chain hoist, press and hold the Up push-button on the handheld control.

Step 8 Disconnect the load hook from the load once it is in place.

Step 9 Remove the chain hoist from its support.

Step 10 Coil the chain so that it will not get tangled.

Step 11 Store the chain hoist in its proper place.

> **NOTE**
>
> For detailed information about chain hoist inspection and operation, consult *ASME Standard B30.16, Overhead Hoists (Underhung).*

3.1.3 *Ratchet-Lever Hoists and Come-Alongs*

Ratchet-lever hoists and come-alongs (*Figure 49*) are used for short pulls on heavy loads. You can see by the length of the chain in the figure that the ratchet-lever hoist cannot accommodate a long pull. The term come-along is widely used to identify both tools, but they are not the same thing. A come-along uses a cable, whereas a ratchet-lever hoist uses a chain. Linemen often use ratchet-lever hoists with a flat synthetic strap. Ratchet-lever hoists are often designed and rated for vertical lifts, but cable-type come-alongs can only be used to pull horizontally. Do not use a come-along or ratchet-lever hoist for lifting unless you are certain it is rated by the manufacturer to do so.

> **WARNING!**
>
> Never use a come-along for vertical lifts. They do not have the same safety-braking mechanisms as ratchet-lever hoists to prevent the load from slipping.

Come-alongs and ratchet-lever hoists are portable, easy to use, and available in varying capacities. Always read and follow the manufacturer's instructions. Follow these steps to select, inspect, use, and maintain a ratchet-lever hoist:

Step 1 Select a hoist of adequate capacity to handle the load. Ensure that it is rated for vertical lifting.

Step 2 Inspect the chain and hooks to ensure that they are not excessively worn, bent, or deformed in any way. Inspect the device overall for significant damage or signs of being overloaded.

Step 3 Hang the device from a suitable structure, and ensure that structure is adequate for the load.

Step 4 Turn the ratchet release to the mid position.

(A) RATCHET-LEVER HOIST

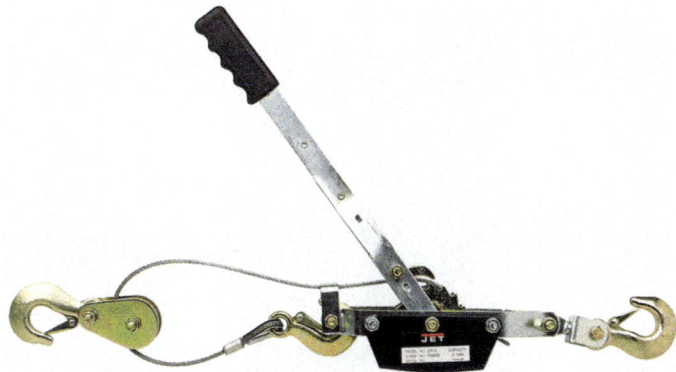

(B) COME-ALONG

Figure 49 A ratchet-lever hoist and a come-along.

Step 5 Position the hoist directly over the load. Pull the chain out enough to attach it to the load.

Step 6 Attach the load hook to the load, using the proper rigging.

Step 7 Turn the fast-wind handle to take the slack out of the chain.

Step 8 Turn the ratchet control to the Up position.

Step 9 Pump the ratchet handle to raise the load.

Step 10 Turn the ratchet control to the Down position.

Step 11 Pump the ratchet handle to lower the load until there is slack in the chain.

Step 12 Disconnect the hook from the load.

Step 13 Dismount and store the hoist in its proper place.

3.2.0 Jacks

A jack is a device used to raise or lower equipment. Jacks are also used to move heavy loads a short distance, with good control over the movement. The following are the three basic types of jacks:

- Ratchet
- Screw
- Hydraulic

3.2.1 Ratchet Jacks

The ratchet jack (*Figure 50*), also called a *railroad jack*, is used to raise loads under 25 tons. It uses the lever-and-fulcrum principle. The downward stroke of the lever raises the rack bar one notch at a time. A latching mechanism, called a pawl, automatically springs into position, holding the load and releasing the lever for the next lifting stroke. They can lift full jack capacity on the toe or on the cap. Ratchet jacks are rated by lifting capacity and the length of their stroke, or lifting distance.

3.2.2 Screw Jacks

Screw jacks (*Figure 51*) are used to lift heavier loads than ratchet jacks, but more slowly. A simple screw jack uses the screw-and-nut principle to lift the stem. A simple lever is placed into the hole and turned to raise or lower the threaded stem. For heavier loads and jacks, a gear-reduction unit is used to reduce the amount of power required to turn the nut.

3.2.3 Hydraulic Jacks

Hydraulic jacks (*Figure 52*) are operated by the pressure of pumped fluid. These jacks can lift a surprising amount of weight. Simple bottle jacks have a hand-operated lever to pump the jack and raise it. A bypass valve is opened to allow fluid to flow out of the cylinder and lower the jack.

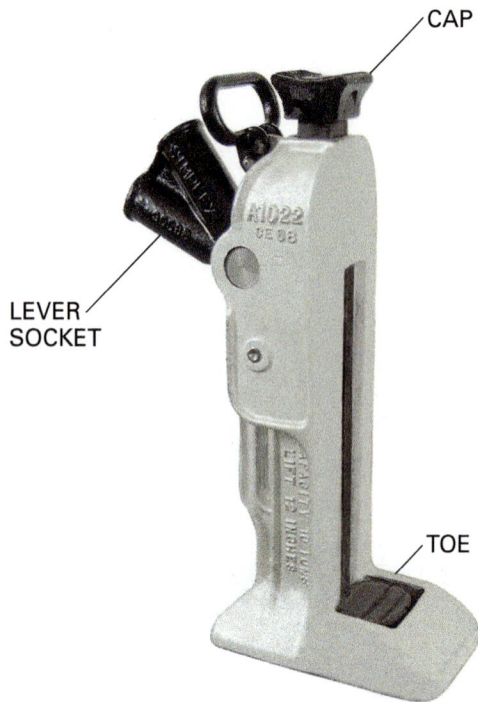

Figure 50 Ratchet jack.

Toe jacks are a different version of bottle jacks that provide a lifting toe at the base. The toe can be pushed under a load that is much closer to the floor, while a bottle jack requires a lot of clearance to fit under the load.

Jacks with higher lifting capacities use an external pump. The pump is connected to the jack by a hose and a quick-disconnect coupling. The pump can be hand-operated, foot-operated, electric, air-operated, or even gasoline-engine driven. It is not uncommon to see portable hydraulic jacks that are rated for 1,000 tons.

3.2.4 Jack Use and Maintenance

In time, a jack can become damaged or worn and fail under a load. To avoid such failures, all jacks should be carefully inspected before each use. Apply the following general guidelines when using jacks:

- Inspect jacks before using them to ensure that they are not damaged in any way. For jacks that use an external pump, also inspect the pump, hoses, and couplings carefully. The connecting hose is the most fragile part of the system and is subject to cuts and deep scarring during normal use.

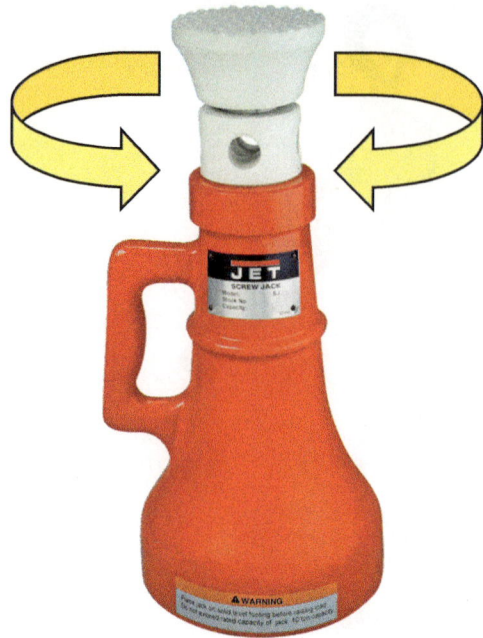

Figure 51 Screw jack.

- Thoroughly clean all hydraulic hose connectors before connecting them.
- Never exceed the load capacity of the jack.
- Use wood softeners when jacking against metal.
- Never place jacks directly on earth when lifting; provide a solid footing.
- Position jacks so the direction of force is perpendicular to the base and the surface of the load. Then raise the load evenly to prevent the load from shifting or falling.
- Use the proper jack handle, and remove it from the jack when it is not in use. Do not use extensions to the jack handles. If the added power of an extension is necessary, the wrong jack is being used.
- Never step on a jack handle to create additional force; use a foot-operated pump and an appropriate jack.
- Always apply blocking or cribbing under a raised load when jacking. Never leave a jack under a load without having the load blocked up so that it will not fall if the jack fails suddenly.
- Brace loads to prevent the jacks from tipping.
- Lash or block jacks when using them in a horizontal position to move an object.
- Never jack against any kind of roller or wheel.
- Match multiple jacks for uniform lifting of a single object.

(A) BOTTLE JACK

(C) LOW-PROFILE

(B) TOE JACK

(D) HYDRAULIC HAND PUMPS

Figure 52 Hydraulic jacks.

Additional Resources

29 *CFR* 1926.251, **www.ecfr.gov**

Willy's Signal Person and Master Rigger Handbook. Current edition. Ted L. Blanton, Sr. Altamonte Springs, FL: NorAm Productions, Inc.

North American Crane Bureau, Inc. website offers resources for products and training, **www.cranesafe.com**

3.0.0 Section Review

1. The mechanical advantage gained by using a spur-geared chain hoist is provided by _____.

   a. multiple parts of chain
   b. a trolley
   c. an operating lever
   d. the gear-reduction unit

2. Which of the following is a true statement about the use of jacks?

   a. If a load is difficult to lift, you can use your foot to operate the jack handle.
   b. Always apply blocking or cribbing under a raised load.
   c. The capacity of a hydraulic jack is only limited by the hydraulic pressure applied.
   d. A toe jack is just a different version of a screw jack.

SUMMARY

Selecting and setting up hoisting equipment, hooking cables to the load to be lifted or moved, and helping guide the load into position are all part of the rigging process. Performing this process safely and efficiently requires the rigger to properly select equipment, use it in the correct way, understand safety hazards, and know how to prevent accidents while working efficiently.

Riggers must have great respect for the hardware and equipment of their trade because their lives may depend on them functioning correctly.

Although they are not required to be rigging professionals, crane operators must also understand rigging practices and be able to distinguish between proper and improper methods for the safety of all jobsite personnel.

1. Threaded blind holes used with eyebolts should have a minimum depth of _____.

   a. 1¼ times the bolt diameter
   b. 1½ times the bolt diameter
   c. 1¾ times the bolt diameter
   d. 2 times the bolt diameter

2. A device used as a connection point for a load that has no reduction in rated load when an angular pull is applied is a _____.

   a. shouldered eyebolt
   b. swivel hoist ring
   c. rigging hook
   d. shoulderless eyebolt

3. Custom-fabricated spreader and equalizer beams are tested at _____.

   a. 125 percent of their rated load
   b. 200 percent of their rated load
   c. 250 percent of their rated load
   d. 350 percent of their rated load

4. To keep them in place while the sling is positioned, manufactured protective materials for slings may be equipped with _____.

   a. adhesive strips
   b. magnets
   c. hook-and-loop fasteners
   d. wooden edges

5. A wire rope sling should be removed from service if localized abrasion and scraping has reduced the diameter more than _____.

   a. 25 percent of the original rope diameter
   b. 15 percent of the original rope diameter
   c. 10 percent of the original rope diameter
   d. 5 percent of the original rope diameter

6. When the eyes of a synthetic web sling are sewn at right angles to the plane of the sling body, it is called a(n) _____.

   a. round sling
   b. endless sling
   c. standard eye-and-eye sling
   d. twisted eye-and-eye sling

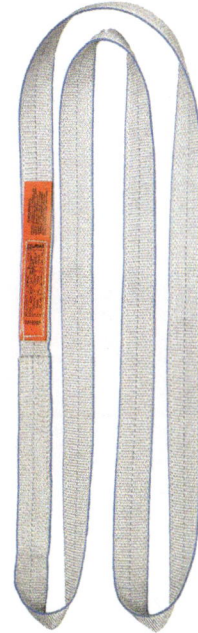

Figure RQ01

7. The type of sling shown in *Figure RQ01* is a(n) _____.

   a. standard eye-and-eye
   b. twisted eye-and-eye
   c. endless
   d. round

8. Which of the following is a disadvantage of chain slings when compared to wire-rope slings?

   a. Chain slings do not handle abrasion and corrosion as well as wire rope.
   b. Chain slings cannot be used in high-heat applications.
   c. Chain slings have less reserve strength and are more likely to fail without warning.
   d. Chain slings are more easily damaged by sharp corners than wire rope.

9. Which of the following slings is best suited for situations where the load is abrasive, hot, and/or tends to cut?

   a. Wire-rope sling
   b. Metal-mesh sling
   c. Synthetic web sling
   d. Endless grommet sling

10. Which of the following types of damage to a metal-mesh sling is allowed to be more extensive than the other before it is removed from service?

    a. Abrasion
    b. Corrosion

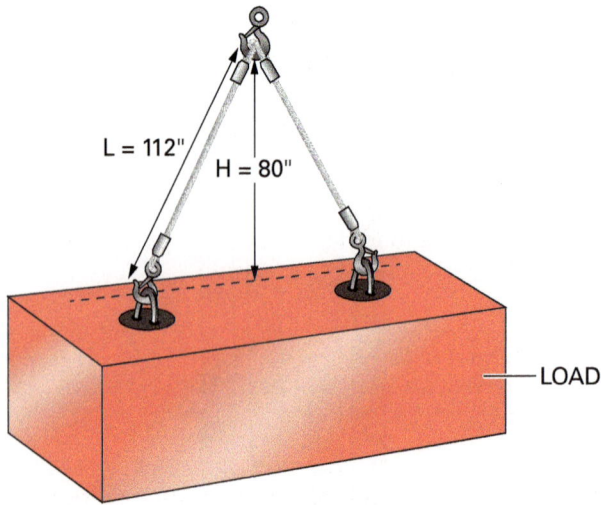

Figure RQ02

11. Refer to *Figure RQ02*. Based on the measurements shown, what is the sling angle factor that should be applied to the weight of the lifted load on each sling?

    a. 0.902
    b. 1.003
    c. 1.213
    d. 1.400

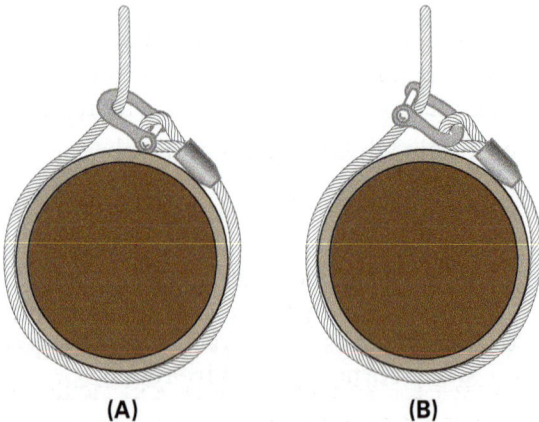

Figure RQ03

12. Refer to *Figure RQ03*. Which of the two drawings, A or B, represent the proper way to orient the shackle?

    a. Drawing A
    b. Drawing B

13. Regardless of how a clove hitch is actually tied, the result is the same as _____.

    a. two bowlines tied one after another
    b. the combination of a bowline and a half hitch
    c. two half hitches made in the same direction
    d. two half hitches made in opposite directions

14. When the weight of a load cannot be determined in other ways, it can be lifted slightly while monitoring the weight using instruments in the crane. The lift can proceed if the load is less than _____.

    a. 40 percent of the crane's capacity at the longest required operating radius
    b. 75 percent of the crane's capacity at the longest required operating radius
    c. 85 percent of the crane's capacity at the longest required operating radius
    d. 100 percent of the crane's capacity at the longest required operating radius

15. Which of the following is a correct statement?

    a. All come-alongs are rated and best used for vertical lifts.
    b. A ratchet-lever hoist can accommodate any length of chain.
    c. A come-along uses a cable while a ratchet-lever hoist uses a chain.
    d. Come-alongs and ratchet-lever hoists use the same type of braking mechanisms.

Trade Terms Quiz

Fill in the blank with the correct term that you learned from your study of this module.

1. Damage that occurs to wire rope identified by the separation of strands that then balloon outward like vertical bars is called _____.

2. The type of wire rope construction considered to be the most durable for rigging applications is the _____ type.

3. A(n) _____ is attached to a lifted load for the purpose of controlling load spinning and swinging.

4. Rigging devices used to distribute the weight of a load on multi-crane (tandem) lifts are called _____.

5. If a hole does not fully penetrate the material, resulting in a hole with a bottom, it is referred to as a(n) _____.

6. The portion of a hook directly below the center of the lifting eye is called the _____.

7. A common hitch made by passing a sling through a load or connection and attaching both sling eyes to the hoist line is called a(n) _____.

8. The _____ identifies the amount of stress required to bring a rigging component to its breaking point.

9. The terms *safe working load, working load limit,* and *rated capacity* are all synonyms for _____.

10. A simple hitch that uses one end of a sling to connect to a point on the load and the opposite end to connect to the hoist line is the _____.

11. The angle formed by the legs of a sling with respect to the horizontal plane when tension is placed on the rigging is referred to as the _____.

12. A(n) _____ is made by passing a sling around the load and then passing one eye of the sling through the other.

13. A(n) _____ is comprised of 2 or more single-leg hitches that is used for lifting objects equipped with lifting lugs or similar points of connection.

14. Rigging devices that are often used when the object being lifted is too long or large to be lifted from a single point, or when the use of slings around the load may crush the sides, are _____.

15. _____ have three or more holes in them and are used to level loads when sling lengths are not equal.

16. Plates with two holes, referred to as _____, are used as termination hardware to connect rigging to specific lift points.

17. A hook should always be positioned over a load's _____.

18. A major load of pipe or similar materials might be delivered to a site on a _____.

19. A chain hoist or ratchet lever hoist used for lifting can be suspended under a _____.

20. The _____ have a lot to do with the lifting assistance provided by a block and tackle.

21. To raise a load using a block and tackle, the user must pull on the _____.

Trade Terms

Basket hitch	Equalizer beams	Minimum breaking	Sling angle
Bird caging	Equalizer plates	strength (MBS)	Spreader beams
Blind hole	Gantry	Parts of line	Spur track
Bridle hitch	Hauling line	Rated load	Tagline
Center of gravity (CG)	Independent wire rope	Rigging links	Vertical hitch
Choker hitch	core (IWRC)	Saddle	

Harold "Ed" Burke
Rigging Training Specialist,
Mammoet USA South

Please give a brief synopsis of your construction career and your current position.

I began working as a rigger with Mammoet in 1997. I had never been around cranes before, so I was green, for sure. I was fortunate, though, to learn and grow under the guidance of some of the best riggers in the industry—people like Dean and Mark Bell, Jeff Bland, and Jason Marks, just to name a few of those I respect and owe a debt of gratitude. I've been a part of some great projects across the United States and abroad, such as The Netherlands, Belgium, Italy, Taiwan, and Chile. In October 2013, I was approached by our safety director and asked if I would consider serving as a trainer for Mammoet USA. I accepted and I'm proud to still be a part of it.

How did you get started in the construction industry?

I had been doing general industry work since high school, with some time working in residential construction. I was fortunate to get an opportunity with Mammoet through a connection there. When I was hired, the same person told me, "I got you the job—you'll have to keep it."

Who or what inspired you to enter the construction industry?

Honestly, it was exactly what I needed at that point in my life. I was looking to better myself. When I started with Mammoet, I made more money for the remaining eight months of the year than in any full year before that. The earning potential was a key factor, but I really needed some positive changes in my life, and the construction industry surely provided that.

How has training in construction impacted your life and career? What types of training have you completed?

When I got into crane and rigging in the late 1990s, there was not so much emphasis on formal training. Most training happened through mentoring on the job. The industry has changed dramatically in the 20 years since I became involved. I really believe quality training contributes to a safer and more efficient work force. Knowledge truly is power; I say that all the time.

Why do you think credentials are important in the construction industry?

I believe valid credentials provide benchmarks, showing where an individual's skill is in relation to an accepted standard of competency. Credentials provide a glimpse of their knowledge foundation on a given topic or craft.

What do you enjoy most about your career?

The people, without a doubt. I have had the pleasure of working with some of the best people across the world from my days in the field. I continue to work with some of the best in my training environment today. I wouldn't be an effective educator without the people who have been a part of my career and growth as a craft professional. If you surround yourself with great professionals, you have a much better chance of becoming one yourself.

Would you recommend construction as a career to others? Why?

Absolutely! Why, you ask? Not everyone is suited for the college experience or for a career in an office. The earning potential of young men and women graduating from high school who are willing to work hard, learn, and be an asset to their chosen craft is really fantastic. They can start to build real wealth immediately instead of taking on crippling college-loan debt.

What advice would you give to someone who is new to the construction industry?

I see new hires all the time, and those that are also new to the industry. What I tell them is this: Show up early, stay late, do more than is expected, be the first to volunteer, learn all you can by keeping your ears open and your mouth closed, and do a good enough job to convince your supervisor he can't do the next job without you. You must build a reputation for yourself, and it must be a good one.

How do you define craftsmanship?

Many would say that it is skill in a particular craft, and I agree with that. But there is a noticeable level of pride and passion evident in the work and attitude of a true craftsman. You really have to love what you do to be happy and productive, and you pour yourself into it to be a craftsman. That's my take on it.

Trade Terms Introduced in This Module

Basket hitch: A common hitch made by passing a sling around a load or through a connection and attaching both sling eyes to the hoist line.

Bird caging: A deformation of wire rope that causes the strands or lays to separate and balloon outward like the vertical bars of a bird cage.

Blind hole: A hole that does not penetrate the material completely, leaving a hole with a bottom.

Bridle hitch: A type of hitch comprised of 2 or more single-leg hitches, used for lifting objects equipped with lifting lugs or other points of connection.

Center of gravity (CG): The point at which the entire weight of an object is considered to be concentrated, such that supporting the object at this specific point would result in its remaining balanced in position.

Choker hitch: A hitch made by passing a sling around the load, and then passing one eye of the sling through the other. The one eye is then connected to the hoist line, creating a choke-hold on the load.

Equalizer beams: Beams used to distribute the load weight on multi-crane lifts. The beam attaches to the load below, with two or more cranes attached to lifting eyes on the top.

Equalizer plates: A type of rigging plate that has three or more holes, used to level loads when sling lengths are unequal.

Gantry: A framed overhead structure supported by legs on each end, used to cross over obstructions. Gantries can be portable or permanent, providing support for hoisting equipment or raising and supporting lighting, cameras, and similar equipment.

Independent wire rope core (IWRC): Wire rope with a core consisting of wire rope, as opposed to a fiber or single-stranded core; considered to be the most durable for rigging applications.

Hauling line: The portion of a rope or chain on hoisting equipment that the operator uses to raise or lower the load. Also known as a *hauling part*.

Minimum breaking strength (MBS): The amount of stress required to bring a rigging component to its breaking point. The MBS is a factor in determining a components' rated load capacity.

Parts of line: The resulting number of lines that are supporting the load block when a line is reeved more than once.

Rated load: The maximum working load permitted by a component manufacturer under a specific set of conditions. Alternate names for rated load include *working load limit* (WLL), *rated capacity*, and *safe working load* (SWL).

Rigging links: Links or plates with two holes used as termination hardware to appropriate lifting points.

Saddle: The portion of a hook directly below the center of the lifting eye.

Sling angle: The angle formed by the legs of a sling with respect to the horizontal plane when tension is placed on the rigging.

Spreader beams: Beams or bars used to distribute the load of a lift across more than one point to increase stability. Spreader beams are often used when the object being lifted is too long or large to be lifted from a single point, or when the use of slings around the load may crush the sides.

Spur track: A relatively short branch leading from a primary railroad track to a destination for loading or unloading. A spur is typically connected to the main at its origin only (a dead end).

Tagline: A rope attached to a lifted load for the purpose of controlling load spinning and swinging, or used to stabilize and control suspended attachments.

Vertical hitch: A simple hitch that uses one end of a sling to connect to a point on the load and the opposite end to connect to the hoist line. Also known as a *straight-line hitch*.

Additional Resources

This module presents thorough resources for task training. The following reference material is recommended for further study.

ASME Standard B30.5, Mobile and Locomotive Cranes. Current edition. New York, NY: American Society of Mechanical Engineers.

ASME Standard B30.9, Slings. Current edition. New York, NY: American Society of Mechanical Engineers.

ASME Standard B30.10, Hooks. Current edition. New York, NY: American Society of Mechanical Engineers.

ASME Standard B30.16, Overhead Hoists (Underhung). Current edition. New York, NY: American Society of Mechanical Engineers.

ASME Standard B30.20, Below-The-Hook Lifting Devices. Current edition. New York, NY: American Society of Mechanical Engineers.

ASME Standard BTH-1, Design of Below-The-Hook Lifting Devices. Current edition, New York, NY: American Society of Mechanical Engineers.

29 *CFR* 1926, Subpart CC, **www.ecfr.gov**

29 *CFR* 1926.251, **www.ecfr.gov**

29 *CFR* 1926.753, **www.ecfr.gov**

Mobile Crane Operations Level One, NCCER. Third Edition. 2018. New York, NY: Pearson Education, Inc.

NCCER Module 00106-15, *Introduction to Basic Rigging*.

Mobile Crane Safety Manual (AEM MC-1407). 2014. Milwaukee, WI: Association of Equipment Manufacturers.

Willy's Signal Person and Master Rigger Handbook, Ted L. Blanton, Sr. Current edition. Altamonte Springs, FL: NorAm Productions, Inc.

Knots: The Complete Visual Guide, Des Pawson. First American Edition. 2012. New York, NY: DK Publishing.

The following websites offer resources for products and training:

Occupational Safety and Health Administration (OSHA), **www.osha.gov**

Electronic Code of Federal Regulations, **www.ecfr.gov**

North American Crane Bureau, Inc. website offers resources for products and training, **www.cranesafe.com**

Figure Credits

Link-Belt Construction Equipment Company, Module opener

Columbus McKinnon Corporation, Figures 1, 4A, 4C, 46, 48, 49A

Konecranes Americas, Inc., SA01

Courtesy of The Crosby Group LLC, Figures 4B, 27

J. C. Renfroe & Sons, Figure 9

Vaculift™, Inc. d.b.a. Vacuworx®, SA02

Lift-All Company, Inc., Figures 14, SA03, 19, 20, 23, Review Question Figure 1, Exam Figure 1

Lift-It Manufacturing Co., Inc., Figure 17A

Linton Rigging Gear Supplies LLC - www.lrgsupplies.com, Figure 17B

Mazzella Companies, Figure 18

Ed Gloninger, Figure 21

© Donvictorio/Dreamstime.com, Figure 35

© iStock.com/krungchingpixs, Figure 36A

© iStock.com/Cajarima01, Figure 36B

Insulatus Company, Inc., Figure 37

© a katz/Shutterstock.com, SA05

© Steve Norman/Shutterstock.com, Figure 45

Vestil Manufacturing, Figure 47

Walter Meier Manufacturing Americas, Figures 49B, 51, 52A, 52B, Exam Figure 3

Photos courtesy of Enerpac, Figures 50, 52C, 52D

Section Review Answer Key

	Answer	Section Reference	Objective
Section One			
	1. a	1.1.2	1a
	2. a	1.2.4	1b
Section Two			
	1. c	2.1.1	2a
	2. a	2.2.1	2b
	3. c	2.3.2	2c
	4. b	2.4.2	2d
	5. a	2.4.2	2d
	6. b	2.5.0	2e
	7. d	2.6.0	2f
	8. c	2.7.0	2g
Section Three			
	1. d	3.1.2	3a
	2. b	3.2.4	3b

Section Review Calculations

2.0.0 SECTION REVIEW

Question 4

Divide the length (L) by the height (H) to determine the sling angle factor:

Sling angle factor = L ÷ H
Sling angle factor = 74" ÷ 62"
Sling angle factor = 1.194

The sling angle factor is **1.194**.

NCCER CURRICULA — USER UPDATE

NCCER makes every effort to keep its textbooks up-to-date and free of technical errors. We appreciate your help in this process. If you find an error, a typographical mistake, or an inaccuracy in NCCER's curricula, please fill out this form (or a photocopy), or complete the online form at **www.nccer.org/olf**. Be sure to include the exact module ID number, page number, a detailed description, and your recommended correction. Your input will be brought to the attention of the Authoring Team. Thank you for your assistance.

Instructors – If you have an idea for improving this textbook, or have found that additional materials were necessary to teach this module effectively, please let us know so that we may present your suggestions to the Authoring Team.

NCCER Product Development and Revision
13614 Progress Blvd., Alachua, FL 32615

Email: curriculum@nccer.org
Online: www.nccer.org/olf

❏ Trainee Guide   ❏ Lesson Plans   ❏ Exam   ❏ PowerPoints   Other _____

Craft / Level: _____   Copyright Date: _____

Module ID Number / Title: _____

Section Number(s): _____

Description: _____

_____

_____

_____

_____

Recommended Correction: _____

_____

_____

_____

_____

Your Name: _____

Address: _____

_____

Email: _____   Phone: _____

21106-18

Crane Safety and Emergency Procedures

OVERVIEW

Cranes are used to accomplish very important tasks in various construction and industrial settings. When working with or near cranes, safety is always the highest priority. Crane operators and other members of the lift team must embrace their responsibility as the manager of a powerful machine that can both accomplish great things and destroy property and lives. Thousands of successful crane operations occur each day without incident; all of the lifts in the future can end the same way. The goal of this module is to present a wide variety of safety information related to crane operation and prepare lift team members for their role in a safe workplace.

Module Ten

Trainees with successful module completions may be eligible for credentialing through the NCCER Registry. To learn more, go to **www.nccer.org** or contact us at 1.888.622.3720. Our website has information on the latest product releases and training, as well as online versions of our *Cornerstone* magazine and Pearson's product catalog.

Your feedback is welcome. You may email your comments to **curriculum@nccer.org**, send general comments and inquiries to **info@nccer.org**, or fill in the User Update form at the back of this module.

This information is general in nature and intended for training purposes only. Actual performance of activities described in this manual requires compliance with all applicable operating, service, maintenance, and safety procedures under the direction of qualified personnel. References in this manual to patented or proprietary devices do not constitute a recommendation of their use.

Objectives

When you have completed this module, you will be able to do the following:

1. Identify relevant OSHA and ASME standards and general crane safety considerations.
   a. Identify safety standards relevant to mobile cranes and their operation.
   b. Identify general safety considerations for mobile crane operation.
2. Identify mobile-crane operation considerations related to specific applications and explain how to respond to various incidents.
   a. Describe the purpose of pre-lift meetings and identify the topics of discussion.
   b. Identify safety considerations related to power lines.
   c. Identify safety considerations related to weather conditions.
   d. Describe safety considerations related to specific crane functions and how to respond to various incidents.

Performance Tasks

This is a knowledge-based module; there are no Performance Tasks.

Trade Terms

Avoidance zone
Competent person
Critical lift
High-voltage proximity warning device
Insulating link
Minimum clearance distance

Prohibited zone
Recloser
Shock loading
Standards
Standard lift

Industry Recognized Credentials

If you are training through an NCCER-accredited sponsor, you may be eligible for credentials from NCCER's Registry. The ID number for this module is 21106-18. Note that this module may have been used in other NCCER curricula and may apply to other level completions. Contact NCCER's Registry at 888.622.3720 or go to **www.nccer.org** for more information.

Contents

Figures and Tables

1.0.0 CRANE SAFETY

Objective

Identify relevant OSHA and ASME standards and general crane safety considerations.
   a. Identify safety standards relevant to mobile cranes and their operation.
   b. Identify general safety considerations for mobile crane operation.

Trade Terms

Competent person: As defined by OSHA, an individual who is capable of identifying existing and predictable hazards in the surroundings or working conditions which are unsanitary, hazardous, or dangerous to employees, and who has the authorization to take prompt corrective measures to eliminate such hazards.

Shock loading: A sudden, dramatically increased load imposed on a crane and rigging, usually as the result of momentum from the load that occurs due to swinging side-to-side, dropping the load and then stopping it suddenly, and similar actions that create momentum.

Standards: As defined by OSHA, statements that require conditions, or the adoption or use of one or more practices, means, methods, operations, or processes, that are reasonably necessary or appropriate to provide safe or healthful employment and places of employment. Standards developed by some organizations are voluntary in nature, while OSHA standards and those they incorporate by reference are enforceable by law.

Equipment can be damaged and people can be severely injured or killed in crane accidents. Lives, careers, and companies can all be lost as the result of a crane accident. This module provides common safety guidelines for the operation of mobile cranes. In addition, it provides an overview of situations that can occur on the jobsite and the steps to take in response to various situations.

Injuries and fatalities related to crane operations happen for a variety of reasons. The vast majority of incidents are preventable. Many incidents fall into the category of struck by / caught between incidents. One of the leading causes of fatalities in crane operations over the years has been electrocution. These are not cases of electrical problems developing in the cranes; they are electrocutions resulting from power line contact. For this reason, there are a number of OSHA directives related to operations around power lines that must be followed without compromise.

Because working on or around mobile cranes can be dangerous (*Figure 1*), members of a lift team must understand that the first responsibility on the job is safety. This responsibility includes personal safety, the safety of others, and the safety of the equipment and materials on the jobsite. One must know the safety requirements for each jobsite and be aware of any unique hazards that may be associated with the work. Accidents can happen anywhere at any time, but they can be prevented when safety is always the first priority.

1.1.0 Safety Standards

The mobile crane industry is very large and complex. The possibility of major damage and loss of life demands that the industry be monitored. There are several groups that monitor and regulate crane operations. The three primary organizations to be introduced here are the following:

- The Occupational Safety and Health Administration (OSHA)
- The American Society of Mechanical Engineers (ASME)
- The American National Standards Institute (ANSI)

ASME and ANSI are nonprofit professional organizations. The principal difference in the two is that ASME does the bulk of the technical research and offers guidance (standards) based on this research and engineering studies. ANSI is more focused on embracing and supporting standards of their choosing through a specific set of procedures to gain consensus; ANSI does not develop standards independently. Both have an international component as well as a domestic function and provide guidance that serves to reduce or eliminate hazards and the resulting accidents. OSHA, on the other hand, is an office of the US Government, established under the Occupational Safety and Health Act of 1970.

There are several terms often used in discussions of publications from these organizations that should be defined, beginning with the term *standard*. A standard, as it applies to the crane industry, outlines specific conditions, or the adoption or use of one or more practices and methods, necessary or appropriate to provide safe, functional products and/or places of employment. Some standards may be referred to as *national consensus standards*. This means several things:

Figure 1 Crane accidents can result in property damage, serious injuries, and fatalities.

- The standard has been adopted by one or more nationally recognized organizations under a specific set of procedures whereby it can be determined that persons affected by it have reached substantial agreement on its adoption.
- The standard was created in a manner that afforded an opportunity for diverse views to be considered.
- The standard has been designated as a national consensus standard by the US Secretary or the Assistant Secretary of Labor, after consultation with other appropriate federal agencies.

OSHA and ASME standards frequently refer to other standards, often informing the reader that those standards are incorporated by reference. This means that the standards referred to must also be considered and recognized as if they were a part of the text. This helps to reduce the duplication of effort to develop a similar piece of work.

ASME develops many standards, some of which apply to cranes and the crane industry. However, ASME and ANSI do not have the power to enforce standards. ANSI selects standards from organizations like ASME and works to gain consensus and acceptance on a national or international scale. However, ANSI standards are considered voluntary and are not written as laws or regulations, since ANSI is also a nonprofit organization.

OSHA, however, is different. OSHA was created under federal law, and employers are required to follow the standards they develop, adopt from others, or incorporate by reference.

OSHA often creates their own enforceable standards, some of which are based on the work of organizations like ASME and ANSI.

In some cases, OSHA may simply adopt or incorporate the standards of other respected organizations by reference. A good example can be found in this statement from 29 *CFR* 1926.1433(b): "Mobile (including crawler and truck) and locomotive cranes manufactured on or after November 8, 2010 must meet the following portions of *ASME B30.5-2004* (incorporated by reference, *see* §1926.6)..." This section of the OSHA standards goes on to list specific passages from the ASME standard that apply. Incorporation by reference into an OSHA standard effectively makes those standards enforceable as well.

It is very important to understand that OSHA standards are not just suggestions; they are enforceable laws. Employers can be punished for failing to comply with OSHA standards. As an employee, you have a responsibility to your employer, as well as to yourself, to follow the OSHA standards. Remember that the standards of other organizations that are incorporated by reference into OSHA standards and are made mandatory by their language also become legally enforceable.

1.1.1 Crane-Industry Safety Standards

Mobile crane operations are governed primarily by several standards. It is important to note that all standards related to crane operations typically contain safety information. That is their priority and reason for existence. Even standards such as *ASME Standard B30.10, Hooks* are based on safety. Consider that the standards related to hooks and their design and fabrication are created to ensure that they are safe and reliable when used properly. Of course, this is also true of OSHA standards, all of which exist to support OSHA's mission of safety in the workplace.

29 *CFR* 1926, Subpart CC, *Cranes and Derricks in Construction* is arguably the most important set of standards in the crane industry. This is especially true since OSHA standards are enforceable by law. Subpart CC includes 29 *CFR* 1926.1400 through 1926.1442, plus a listing of other standards that are incorporated by reference. Topics covered in this standard include, but are not limited to, the following:

- Ground conditions for crane support
- Assembly and disassembly of equipment and attachments
- Powerline safety
- Equipment inspections
- Wire rope inspection, selection, and installation
- Required safety devices and operational aids
- Crane operation
- Signaling
- Work area control
- Operator, signal person, and crane maintenance personnel qualifications
- Hoisting personnel
- Equipment modifications
- Crane operation and safety when used on floating barges

ASME Standard B30.5, Mobile and Locomotive Cranes is the most important ASME standard relevant to crane operators. Note that some portions of this standard have been incorporated into the OSHA standards by reference. Again, this means they are not simply suggestions, but are legally enforceable. Topics covered in this standard include, but are not limited to, the following:

- *Personnel competence* – "Persons performing the functions identified in this Volume shall meet the applicable qualifying criteria stated in this Volume and shall, through education, training, experience, skill, and physical fitness, as necessary, be competent and capable to perform the functions as determined by the employer or employer's representative."
- *Crane construction and characteristics* – Items covered here include: crane load ratings; boom hoists and telescoping boom mechanisms; crane travel; controls; cabs; and structural performance.
- *Inspection, testing, and maintenance* – This section covers the crane as well as the inspection and replacement of the wire ropes.
- *Operation* – The specific qualifications of crane operators are provided here, including the physical requirements as well as those related to testing. The specific requirements are covered in NCCER Module 21101-18, "Orientation to the Trade," from *Mobile Crane Operations Level One*. The role and responsibilities of each individual that is part of a typical lift crew, including the operator, are also outlined. Following this information, the standard addresses a wide variety of common crane movements, such as attaching, lifting, and swinging the load, and provides safety guidelines specific to each action. Crane hand signals are also found in the Operation section of the standard.

> **NOTE**
>
> All of the topics listed above have not been incorporated by reference into the OSHA standards. For a list of ASME and ANSI standards that have been incorporated by reference and the OSHA standards they affect, see 29 *CFR* 1926.6.

Another important ASME standard is *ASME Standard P30.1, Planning for Load Handling Activities*. The standard documents lift-planning considerations that extend beyond cranes to other load-handling equipment as well. Guidance is divided into two categories—Standard Lift Plans and Critical Lift Plans— based on the degree of exposure to hazards. Lift planning will be covered in detail in *Mobile Crane Operations Level Three*.

Throughout this module, standards are referenced where appropriate. Note that this text attempts to present the standards as accurately

Glossaries

Some OSHA and ASME standards contain glossaries to clearly define important terms used in the text. Although every trade has its own verbiage that changes over the years, it is helpful to be familiar with the definition of terms as the standard-setting organizations see them. Many of the OSHA and ASME standards that apply to the mobile crane industry have a glossary at the beginning to ensure that the meaning of a given term is not misunderstood or misapplied.

as possible, but does not present all relevant standards or the requirements they contain. It is the responsibility of crane operators, riggers, signal persons, and all other members of a lift team to directly review and follow the appropriate standards. Requests for interpretations and clarifications of the various standards can be addressed directly to OSHA, ASME, or other issuing authority.

1.2.0 Mobile Crane Safety Considerations

Mobile-crane operators must be aware of the unavoidable hazards associated with the trade. Lifted loads will be moved above and around other workers, and such loads represent an extreme hazard to workers in the area (*Figure 2*). You may also work during inclement weather conditions where wind, slippery surfaces, and other hazards exist. When working near mobile cranes, look up and be mindful of the hazards above and around you, but do not forget the potential hazards that exist on the ground.

As a result of the industry's efforts and losses experienced by employers, construction-trade contractors have made the development of a safety culture in the organization a priority. However, it isn't just about complying with the laws—most employers truly care about the lives of their employees and their families, and developing a safety culture on the job supports their concerns for employee safety and welfare.

Safety consciousness and helping to build a culture that promotes safety from within is extremely important. The earning ability of injured employees may be reduced or eliminated for the rest of their lives. The number of injured employees can be significantly reduced if each employee is committed to safety awareness and exhibits

that attitude in their daily work. Full participation in the employer's safety program is a matter of personal responsibility. Making safety the first priority is the key to reducing accidents, injuries, and fatalities on the job. Most accidents can be avoided, because most result from human error. Show that you are a team player by helping to establish and support a safety culture within your organization and on the jobsite.

1.2.1 Personal Protection

Hard hats, safety shoes, safety glasses, and barricaded cranes to discourage personnel entry into the area are among the personnel-protection requirements for almost every jobsite (*Figure 3*). Gloves are also required in many cases, especially when working with rigging equipment such as wire rope. Other personal protective equipment (PPE) may be required at specific jobsites, such as those that produce hazardous chemicals that could be released to the environment. It is essential that every worker be familiar with the requirements of each individual jobsite, and embrace those requirements consistently.

1.2.2 Basic Rigging Safety

Riggers and other members of lift teams must be capable of selecting suitable rigging and lifting equipment, as well as directing the movement of the crane to assure the safety of all personnel and the load itself. All rigging operations must be planned, supervised, and accomplished by qualified and competent personnel. [OSHA defines competent person in 29 *CFR* 1926.32(f).]

One very important rigging requirement is to determine the weight of all loads before attempting to rig and lift them. Crane operators must

Figure 2 Lifting and positioning large loads is hazardous work.

Figure 3 Wearing the correct PPE is a crucial first step toward a safe working environment.

know the weight of the load to ensure it is within the rated load capacity of the crane under the circumstances of the lift. Riggers, however, must also know the weight of the load to ensure that the rigging equipment and techniques used are suitable for the task.

The following rigging-related safety practices should be followed at all times:

- Determine the weight of the load before rigging. If this is not possible, the load can be lifted slightly while the crane operator monitors the instrumentation and determines the weight. If it does not exceed 75 percent of the crane's rated load capacity at the operating radius required, the lift can be made. If it does exceed that value, the load must be set down and the weight reevaluated per 29 *CFR* 1926.1417(o)(i). Note that rigging components or techniques may need to be changed based on any new weight information that is discovered.
- Ensure that the appropriate rigging equipment and components are available. Using an inappropriate piece of equipment due to an equipment shortage can lead to rigging failures. Know the rated load capacity of the rigging equipment and never exceed the limit.
- Ensure that the rigging equipment has been properly inspected and is in good working condition. Remove any damaged or defective equipment from service.
- Always maintain the manufacturer's information for the rigging equipment in an easily accessible location. The literature provides information on hitch configurations, lift angles, and similar information that may be needed as the rigging process proceeds.
- Recognize factors in the lift that can reduce rigging equipment capacity. Remember that the rated load capacity of all hoisting and rigging equipment is based on ideal conditions; lifts often involve conditions that are less than ideal.
- Use proper padding and protection to protect slings as well as the surface of the load.
- Never place loads on the tip of the hook, where it is weakest and most likely to fail.

Gloves for Everyone

There is a time and place in every construction trade for gloves. Injure your hands, and you have damaged the most versatile and important construction tool you will ever own. Craft professionals are observed every day doing tasks without gloves where it is clear that the protection is needed. There has been a long history of workers rejecting gloves for any number of reasons. Many workers have rejected the use of gloves in years past because they felt too restrictive and awkward.

There are more work gloves on the market today than ever before. Today's gloves are miles ahead of the work gloves of old that fit poorly, were unnecessarily bulky and clumsy, and were constructed only of relatively simple materials such as leather and canvas. Although wearing gloves on the job may seem awkward at first, there is a glove out there that fits you well and offers essential protection for your hands without restricting movement. If you work with the right gloves for a while, you will soon feel naked without them. Look for reasons to wear them instead of reasons to reject them, and find your pair of gloves.

- Observe the area where loads will be placed and ensure that it is clear of obstructions and properly prepared for load placement. Preparing the target area while the load is suspended represents poor planning. The practice of lowering the load just above the landing zone and then placing needed blocking is hazardous since the riggers are forced to work beneath the load. Riggers need to think ahead of the crane.
- When serving as part of the rigging team, do not assume that the crane operator and other members of the lift team see the same potential hazards that you do. They may also see something that you do not see. Discuss the lift prior to beginning and share any concerns about obstructions, power lines, and other hazards with the rest of the team. Also listen to and consider any concerns that other members of the team may share. Every lift offers its own unique hazards and obstructions (*Figure 4*).
- Rig the load and connect it to the crane with the center of gravity directly below the hook.
- Always consider where your fingers, hands, and feet are in relation to pinch points. Wear appropriate gloves when handling rigging equipment.
- Never ride a load or the hook.
- Remain outside the load's fall zone at all times unless it is required to guide or receive a load.
- If there is ever any doubt about the reliability or arrangement of the rigging or you observe something unexpected, stop the lift, lower the load, and report it to lift supervision.

Figure 4 A complex jobsite with multiple cranes.

1.2.3 Pre-Lift Considerations

Careful planning, detailed inspections, and timely maintenance help prevent accidents. Crane operators must demonstrate their attention to detail as they prepare for a lift as well as during the lift. Prior to operating a crane, the operator should accomplish the following:

- Determine if there are any locally established restrictions placed on crane operations, such as traffic considerations or time restrictions for noise abatement.
- Ensure that a complete operating manual is in the crane cab. The manual should remain with the crane at all times.
- Accurately determine the weight of the load. Regardless of the perceived accuracy, begin every lift slowly to ensure there are no surprises in the load weight.
- Confirm that load charts in one form or another are readily available for use and review them. Refer to the load charts for every lift and always remain within the capabilities of the crane.
- Determine the deductions to be made from the rated load capacity due to attachments or other crane-related factors.
- Look for documentation of recent crane inspections, as well as any deficiency-correction statements, and review the results.
- Examine the site and ensure that it is suitable to support the crane. Ask questions and seek information about the presence of underground utilities such as gas, oil, electrical, and telephone lines; sewage and drainage piping; and underground tanks. Also ask if the area has been recently excavated. Determine if the lifting operation is limited in some way by stability or structural concerns.
- Evaluate the weather conditions and be familiar with the wind speed limitations of the crane.
- Determine how close the crane or load path may be to power lines throughout the lift and whether the clearance is sufficient.
- Confirm that the crane boom is assembled correctly or extends as designed.
- Determine the hoist line pull and the maximum permissible line pull.
- Ensure there is a safe path to move the crane around, if point-to-point movement on site is necessary.
- Make sure that the crane can rotate unobstructed in the required quadrants for the planned lift.
- Complete the required daily pre-start inspection.
- Ensure that the crane is level before lifting.

1.2.4 Load-Handling Safety

The safe and effective control of the load involves the strict observance of load-handling safety requirements by the entire lift team, including the crane operator. This includes making sure that the swing path of the crane upperworks remains clear of personnel and obstructions any time the crane is in operation (*Figure 5*). Barricades or other visual barriers are required. Also keep the path of any planned load movement clear. Many people tend to watch the load when it is in motion, which prevents them from watching for hazards on the ground.

Here are a few additional precautions related to crane operation and load handling to consider in every lift:

- Consult and follow all applicable safety standards (29 *CFR* 1926.1402 and *ASME Standard B30.5*) and manufacturer guidelines to properly stabilize mobile cranes. Most modern cranes are equipped with outriggers. To be properly stabilized, the outriggers must be used to relieve the weight from the tires of most truck cranes (*Figure 6*). Crawler cranes can usually operate directly from their tracks, but they may also use outriggers in certain conditions.
- When computing equipment loads, the blocks, hooks, slings, equalizer beams, lifting components, and other equipment below the hook must also be taken into consideration. Crane load ratings only extend to the hook.
- Avoid allowing a suspended load to swing more than necessary. This subjects the equipment to additional side loading that can cause a failure of a component or tip the crane. Keep the load directly below the boom.
- Crane operators must avoid snatching or stopping the descent of a suspended load suddenly. Rapid acceleration and deceleration results in

Figure 6 Extended and lowered outriggers.

shock loading, greatly increasing the stress on equipment and rigging.
- Physical control of the load beyond the ability of the crane operator may be required. Tag lines are used to limit the unwanted movement of the load as it reacts to the motion of the crane, wind, or other external influences. They are also used to allow the controlled rotation of the load for final positioning in the landing zone (*Figure 7*). Tag lines are attached after the rigger verifies that the load is balanced.

1.2.5 Signaling

Topics of signaling and communication with a crane operator are covered in 29 *CFR* 1926.1419 through 1926.1422. According to these standards, a signal person is required in the following situations:

- When the points of operation, including the load travel path or the area in the vicinity of the load and its landing place, are not in full view of the crane operator
- When the crane will travel and the view in the direction of travel is obstructed
- Any time the crane operator or workers handling the load determine that it is necessary

Voice signals can be used as well as hand signals. When voice signals are used, the crane operator, signal person, and lift director must all agree on the voice signals to be used. 29 *CFR* 1926.1421 requires that voice signals contain three elements, provided in the following order: a function with direction, such as hoist up or boom left; the distance and/or speed of the function; and a command to stop the function. If electronic devices such as radios are to be used, they must be tested at the site before beginning the operation.

Figure 5 Crane barricading is required.

Figure 7 Using taglines.

The crane operator's version of any radio or telephone used must allow for hands-free reception.

Using hand signals to communicate with a crane operator is also very common. There are established hand signals used for communicating load navigation directions. The required hand signals are referred to as the Standard Method in 29 *CFR* 1926.1419(c)(1) and they are pictured in the appendix of the publication. In addition, 29 *CFR* 1926.1422 requires that hand signal charts be posted in a conspicuous location near the lift operation.

Standard hand signals, when used correctly and known by both parties, provide the needed information to the crane operator. Nonstandard hand signals may be developed and used by a lift team when standard hand signals are not feasible for some reason or the use and operation of a

crane attachment is not provided for in the standard hand-signal set.

Serving in the role of a signal person requires qualification, as outlined in 29 *CFR* 1926.1428. However, it is important to note that any member of a lift team that becomes aware of an issue that affects safety is authorized to display or speak the Stop or the Emergency Stop signals (*Figure 8*). Crane operators are required to obey these two signals regardless of their source. To build flexibility into lift teams, it is not unusual for a crane operator or rigger to also seek certification as a signal person.

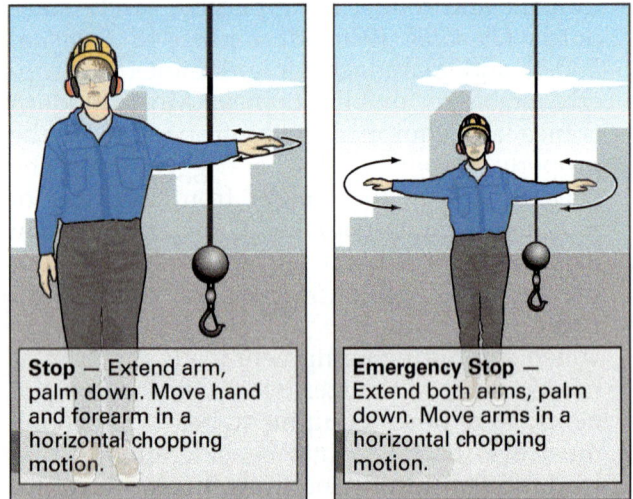

Stop — Extend arm, palm down. Move hand and forearm in a horizontal chopping motion.

Emergency Stop — Extend both arms, palm down. Move arms in a horizontal chopping motion.

Figure 8 The Stop and Emergency Stop hand signals.

Additional Resources

ASME Standard B30.5, Mobile and Locomotive Cranes. Current edition. New York, NY: American Society of Mechanical Engineers.

ASME Standard B30.20, Below-The-Hook Lifting Devices. Current edition. New York, NY: American Society of Mechanical Engineers.

ASME Standard P30.1, Planning for Load Handling Activities. Current edition. New York, NY: American Society of Mechanical Engineers.

29 *CFR* 1926, Subpart CC, **www.ecfr.gov**

29 *CFR* 1926.251, **www.ecfr.gov**

29 *CFR* 1926.753, **www.ecfr.gov**

Mobile Crane Safety Manual (AEM MC-1407). 2014. Milwaukee, WI: Association of Equipment Manufacturers.

The following websites offer resources for products and training:

American National Standards Institute (ANSI), **www.ansi.org**

The American Society of Mechanical Engineers (ASME), **www.asme.org**

Occupational Safety and Health Administration (OSHA), **www.osha.gov**

North American Crane Bureau, Inc., **www.cranesafe.com**

Electronic Code of Federal Regulations, **www.ecfr.gov**

1.0.0 Section Review

1. Which of the following statements is *not* a required characteristic of a national consensus standard?

   a. The standard has been adopted by one or more nationally recognized organizations under a specific set of procedures.
   b. The standard was created in a manner that afforded an opportunity for diverse views to be considered.
   c. The standard has been adopted or incorporated through reference by the American National Standards Institute (ANSI).
   d. The standard has been designated as a national consensus standard by the US Secretary or the Assistant Secretary of Labor.

2. Positioning a load just above a landing zone and then placing any needed blocking or support for the load is _____.

   a. considered the safest way to do it
   b. considered hazardous
   c. required by law
   d. a function of the load owner

2.0.0 SITE SAFETY AND EMERGENCIES

Objective

Identify mobile-crane safety considerations related to specific applications and explain how to respond to various incidents.

a. Describe the purpose of pre-lift meetings and identify the topics of discussion.
b. Identify safety considerations related to power lines.
c. Identify safety considerations related to weather conditions.
d. Describe safety considerations related to specific crane functions and how to respond to various incidents.

Trade Terms

Avoidance zone: An area both above and below one or more power lines that is defined by the outer perimeter of the prohibited zone. As the name implies, any part of the crane should avoid this area whenever possible, and may not enter the area except under special circumstances.

Critical lift: As defined in ASME Standard B30.5, a hoisting or lifting operation that has been determined to present an increased level of risk beyond normal lifting activities. For example, increased risk may relate to personnel injury, damage to property, interruption of plant production, delays in schedule, release of hazards to the environment, or other significant factors.

High-voltage proximity warning device: An early-warning device that senses the electric fields created by high-voltage power lines and alerts the crane operator and/or the lift team to the hazard.

Insulating link: An electrical insulating device used on the crane hook to protect workers in contact with the load from the danger of electrocution in the event the crane contacts a power line. The link can also provide some level of protection for the crane if the load alone contacts a power line.

Minimum clearance distance: The OSHA-required distance that cranes, load lines, and loads must maintain from energized power lines. This OSHA term is synonymous with the ASME term prohibited zone.

Prohibited zone: An area of specific dimensions, based on the voltage of a power line(s) that no part of the crane is allowed to enter during normal operations. Special considerations and preparations are required if the crane's task must place any part of it within the prohibited zone. The prohibited zone is a term used by ASME that is synonymous with the term minimum clearance distance used by OSHA.

Recloser: A device that functions much like a circuit breaker, or in conjunction with a circuit breaker, in power distribution and transmission systems that automatically recloses the circuit after a fault has been detected and the circuit has been opened. Reclosers allow the power system to be re-energized quickly after a transient (temporary) condition, such as a tree limb falling across power lines and then falling to the ground, has occurred. If the fault reoccurs upon closure, the circuit will typically remain open until the situation has been addressed by power line workers or operators.

Standard lift: A lift that can be accomplished through standard procedures, allowing load-handling and lift team personnel to execute it using common methods, materials, and equipment.

One of the ways to avoid accidents and incidents in crane operations is to discuss the plan in detail in a pre-lift meeting. However, in spite of such meetings and a consistent focus on safety, accidents and equipment failures can and do happen. This section focuses on some of the common environmental hazards that crane operators encounter and how to respond to a variety of emergency situations.

One of the keys to a successful response to an emergency is the crane operator's intimate knowledge of the equipment and its controls. There is rarely time for research and a great deal of thought when an operator is confronted with an emergency situation. It is far better to consider how you should react to a given problem throughout the process in order to be well prepared for an unexpected event.

2.1.0 Pre-Lift Meetings

One of the best ways to avoid incidents and hazards is to plan fir the lift carefully. *ASME Standard P30.1, Planning for Load Handling Activities* provides guidance in lift planning. The first requirement is to evaluate a load-handling activity and place it in a category based on the following characteristics:

> **NOTE**
> Pre-lift planning is discussed in detail in NCCER Module 21304-18, "Lift Planning," from *Mobile Crane Operations Level Three.*

- The potential hazard to people that the operation represents
- Hazards that exist in close proximity to the operation
- The complexity of the activity
- The potential for problems that may be caused by the weather or other environmental conditions
- The capacity and ability of the load-handling equipment to cope with the stresses involved
- The potential for an adverse commercial impact, such as the loss of a unique or irreplaceable load, or the costly delay of a major project
- Site requirements that are unique, such as the effect on roadways or other infrastructure

The standard does not limit the evaluation process to these areas; other areas of concern can also be factors. Documentation of this evaluation process is not required, but it is certainly a good idea to do so. Once the evaluation is complete, the activity is then placed in one of two major categories—a standard lift or a critical lift.

A standard lift, per *ASME Standard B30.5*, is one that "can be accomplished through standard procedures, and that the load-handling activity personnel can execute using common methods, materials, and equipment." A critical lift, again per the ASME standard, is one that has been evaluated and it has been determined that the activity "exceeds standard lift plan criteria and requires additional planning, procedures, or methods to mitigate the greater risk." 29 *CFR* 1926.751, *Steel Erection*, defines a critical lift as "a lift that (1) exceeds 75 percent of the rated capacity of the crane or derrick, or (2) requires the use of more than one crane or derrick." Although the term is not defined in 29 *CFR* 1926, Subpart *CC*, both of the above definitions are widely applied in the industry. In fact, both the OSHA and ASME definitions are applied by most organizations. A load that is over 75 percent of the rated capacity becomes a critical lift automatically, based on the OSHA standard, but other conditions can also lead to a lift being labeled as critical. These other conditions represent the influence of the ASME definition.

2.1.1 Standard Lift Planning

A standard lift plan can be written or verbal. There is no OSHA requirement for a standard lift plan to be documented, although its documentation is a good idea. However, many employers require both the evaluation process and the standard lift plan to be documented. A standard lift plan should address the following:

- The load, its center of gravity, and available points of attachment
- Confirmation that the load is within the crane's rated load capacity
- Rigging
- Movement of the crane and/or the load
- The personnel required
- Site conditions such as weather, crane support and ground conditions, and utilities
- Communication method
- Site control of non-essential personnel and pedestrians
- Contingency plans
- Emergency action plans
- Equipment inspection during repetitive processes

Any time the operation is not going according to plan, the operation should stop and the situation evaluated. Any changes should be clearly communicated to all members of the lift team.

2.1.2 Critical Lift Planning

Unlike standard lift plans, critical lift plans are required to be in written form. An example of a critical lift planning worksheet is available in the *Appendix* for review. *ASME Standard P30.1* provides examples and templates for the evaluation process as well as for lift planning.

Essentially, a critical lift plan addresses the same topics as a standard lift plan. However, each topic is considered at a deeper level. Since the lift has been classified as critical, there is at least one element of the lift that requires additional planning and possibly a deviation from normal procedures. This information must be carefully considered and documented in the plan.

The lift director typically schedules a pre-lift meeting to construct, discuss, and review the details of the plan and ensure all personnel involved understand their role. The lift director has overall responsibility for the lift from start to finish. He or she ensures that all the appropriate preparations have been made, the plan is executed as scheduled, and that the lift is stopped if it is not going as planned. Once the activity is stopped for any reason, only the lift director can restart it. Upon completion, a post-lift review is common to assess the process and determine what can be done better in the future. Any recommendations are shared with the lift team and others involved in the process.

2.2.0 Working Around Power Lines

Operating mobile cranes where they can become electrified by power lines is an extremely hazardous practice, although it is sometimes necessary. Work must be performed so that there is no possibility of the crane, load line, or load contacting an energized component and becoming a conductive path. Contact with high-voltage power sources is a major cause of fatalities associated with crane operations. However, these accidents can be prevented.

It is important to note that contact between the crane and power line is not always necessary to initiate an incident. Due to the high voltage carried by some power lines, electricity can jump across an open gap and create a sustainable arc between the energized power line and any path to ground—in this case, a metal crane. Moist air masses allow a larger gap to be crossed. Establishing an arc across a gap is the principle on which automotive spark plugs are based.

Surrounding every energized power line is an area referred to as the prohibited zone (*Figure 9*). The prohibited zone is an area around an energized power line that no part of a crane, boom, load, or load line is allowed to enter. The extent of the prohibited zone is shown in *Table 1* for various line voltages. The table reflects the clearance requirements established by OSHA in 29 *CFR* 1926.1408, Table A, and 1926.1411, Table T. They are identical to the values in tables provided in *ASME Standard B30.5* at the time of this writing (2017). OSHA uses different terminology however, using the term minimum clearance distance in place of prohibited zone. The line voltage determines how much clearance is required. Note that there are different values for cranes in operation versus those that are in transit with no load and the boom lowered.

Figure 9 also shows an avoidance zone. The avoidance zone is the area above and below the prohibited zone, defined by imaginary vertical lines. The distance of the vertical lines from the power lines is determined by the outer edge of the prohibited zone. The avoidance zone exists due to the increased probability of accidental power line contact when working in the area. The prohibited zone, or minimum clearance distance, extends away from the power lines in all directions, so there is a prohibited area above and below the lines as well as to their left and right. The avoidance zone is above and below the prohibited zone.

Note that the term *avoidance zone* is not a term used by ASME or OSHA, but it is commonly used to identify these hazardous areas. OSHA does not address the area above power lines that is beyond the minimum clearance distance. However, 29 *CFR* 1926.1408(d) does address crane operation in the avoidance zone below the power lines. Cranes cannot be used to operate under energized power lines (which is within the avoidance zone) unless any part of the machine or load cannot physically reach the prohibited zone. Although the crane operator may have no intention of fully extending a boom to accomplish the desired task, it could be extended and therefore poses too great a risk.

However, there are times when a task must be done inside the avoidance zone, and sometimes

---

Did You Know?

Power Transmission and Distribution—What's the Difference?

Power transmission lines carry very high-voltage power from the point of power generation to the numerous general locations served by the system. The power carried by transmission lines is extremely hazardous; physical contact with a common transmission line is not required for serious injury or electrocution. The voltage is sufficient for the power to establish an arc across an air gap to any nearby grounded conductor. The higher the voltage, the larger the gap that the arc can cross. To use this power, it must be transformed to a much lower voltage. This cannot be done with simple pole-mounted transformers and the limited protection features this approach offers. Transmission lines are generally routed to substations for voltage reduction where a great deal more control and safety features are available. The high voltage allows the conductor size to be relatively small.

Power distribution lines carry high-voltage power as well, but not as high as transmission lines. Power distribution lines generally carry power from a substation to our homes and places of business. The voltage applied to a common distribution line is still quite dangerous and deadly, as many squirrels and similar creatures have discovered. The voltage on distribution lines must also be transformed to a usable level, usually using a pole- or ground-mounted transformer. Large industrial users, however, often have a substation of their own that intercepts power from a transmission line or substantial distribution line and transforms the power to the various voltages needed by the facility.

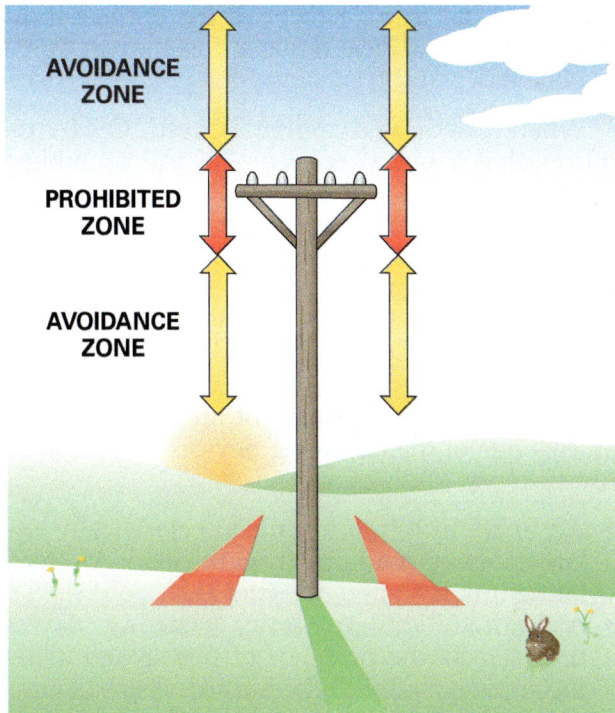

Figure 9 The prohibited zone and the avoidance zone.

Table 1 Minimum Clearance Distance for High-Voltage Power Lines

CRANE IN OPERATION	
POWER LINE VOLTAGE IN	MINIMUM CLEARANCE IN FEET (METERS)
0 to 50	10 (3.05)
50 to 200	15 (4.60)
200 to 350	20 (6.10)
350 to 500	25 (7.62)
500 to 750	35 (10.67)
750 to 1,000	45 (13.72)
Over 1,000	Distance established by the utility owner / operator or registered professional engineer who is a qualified person in power transmission and distribution

CRANE IN TRANSIT (with no load and the boom or mast lowered) [1]	
POWER LINE (kV)	MINIMUM CLEARANCE IN FEET (METERS)
0 to 0.75	4 (1.22)
0.75 to 50	6 (1.83)
50 to 345	10 (3.05)
345 to 750	16 (4.87)
750 to 1,000	20 (6.10)
Over 1,000	Distance established by the utility owner / operator or registered professional engineer who is a qualified person in power transmission and distribution

(1) Environmental conditions such as fog, smoke, or precipitation may require increased clearances.

within the prohibited zone as well. 29 *CFR* 1926.1410 addresses these situations and outlines the requirements that must be met. Some of these situations are addressed in the sections that follow. However, it is important to read and study the OSHA standards whenever work in the vicinity of power lines is planned, especially when they are energized.

Before operations begin near power lines, the owner of the power lines or an authorized representative must be notified and provided with all relevant information. In addition, the cooperation of the owner must be requested. Any overhead line must be considered to be electrically energized unless and until the owner or utility confirms that it is not energized, per 29 *CFR* 1926.1408(e).

There are four scenarios to consider when operating a mobile crane near power lines:

- Power lines de-energized and grounded
- Power lines energized and the crane operating near the prohibited zone
- Power lines energized and the crane operating within the prohibited zone
- Crane in transit with no load and the boom lowered

Each of these scenarios will be discussed further in the sections that follow.

2.2.1 Near De-Energized, Grounded Power Lines

Working in the vicinity of power lines that have been de-energized and grounded is always the preferred situation, since the vast majority of the hazard has been removed. To ensure power lines have in fact been de-energized, the following steps must be taken:

- The utility or owner of the power lines must be contacted first. When possible, they will de-energize the lines.
- The lines must be visibly grounded to avoid electrical feedback and be appropriately marked at the jobsite location (*Figure 10*).
- A qualified representative of the power line owner or the electrical utility must be on site to verify that the first two steps have been completed and that the lines are not energized.

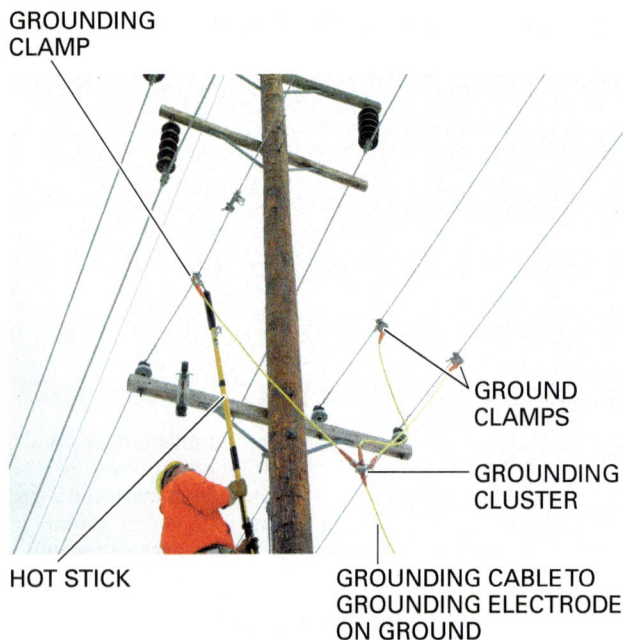

GROUNDING CLAMP

GROUND CLAMPS

GROUNDING CLUSTER

HOT STICK

GROUNDING CABLE TO GROUNDING ELECTRODE ON GROUND

Figure 10 A worker installs temporary grounds on power distribution lines.

Of course, power lines and other lines such as those providing cable and telephone service still represent a significant hazard and an obstruction that can easily become entangled in the boom or hoist lines. Even when they are de-energized, all overhead lines must be treated with great respect.

2.2.2 Near Energized Lines and Near Prohibited Zone

When work must be conducted in the vicinity of the prohibited zone, specific precautions must be taken to ensure that the crane and load do not enter the hazardous areas. Per 29 *CFR* 1926.1408:

- A planning meeting must be held with the lift team and other involved workers to review the plan to avoid encroachment.
- Nonconductive tag lines must be used.
- A visual aid must be erected to aid the crane operator in determining where the prohibited zone boundary begins. For voltages up to 350 kilovolts (kV), or 350,000 volts, the lift team can choose whether to use the minimum clearance distances from the OSHA tables, or to simply use the listed minimum clearance dimension for 350kV lines of 20 feet (6.1 meters). The visual aid must be elevated and can be fabricated several different ways. A suspended warning line with flags attached is one example. If the crane operator cannot clearly see the visual aid, a dedicated spotter must be used to monitor the crane, load line, and load to ensure no part enters the hazardous areas. This individual is in addition to any signal person that may be

part of the team. However, the spotter must have the qualifications required of a signal person.

When a dedicated spotter is used, OSHA requires that a visual reference of the prohibited zone be provided to the spotter. This can be done by placing visible lines on the ground, aligned with the edge of the zone; placing a line of small posts or stanchions in a row; or using existing points of reference as a sight line, such as a sign post behind the spotter and a distant object that aligns with the edges of the zones. If the use of a spotter is chosen instead of an operator's aid, at least one of the following devices must also be used:

- An insulating link installed to electrically isolate the crane load line from the load. An insulating link is shown in *Figure 11*.
- A high-voltage proximity warning device that detects the presence of power and alerts the crane operator to the hazard. These devices sense the electric fields around power lines to determine their presence. The alert feature is usually a series of lights—green, yellow, and red—that indicates how close the hazard is to the instrument. The components of the system are shown in *Figure 12*.
- A device that automatically warns an operator to stop movement of the boom or load line.
- A device that automatically limits the movement of the crane.

> **CAUTION**
>
> Many high-voltage proximity warning devices used are unable to sense the presence of direct current (DC) power. Although DC power lines are rare in the United States, they are used in a few power transmission systems, and are more likely to be encountered in Europe, South America, and Asia. DC power is also often associated with power generated through wind and solar systems. When working near energized power lines and related systems, be sure that the equipped proximity alarm system is capable of detecting the power in use.

Keep in mind that power lines tend to move and sway with the wind. Essentially, wind causes the prohibited zone to be in motion. The anticipated horizontal and/or vertical movement of power lines due to the wind must be added to the clearance distances in OSHA's tables. The utility or owner's representative must be consulted to determine the specific distance to be added for a given situation.

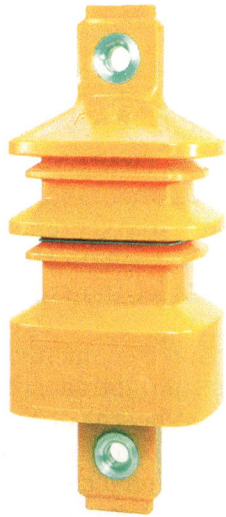

Figure 11 An insulating link.

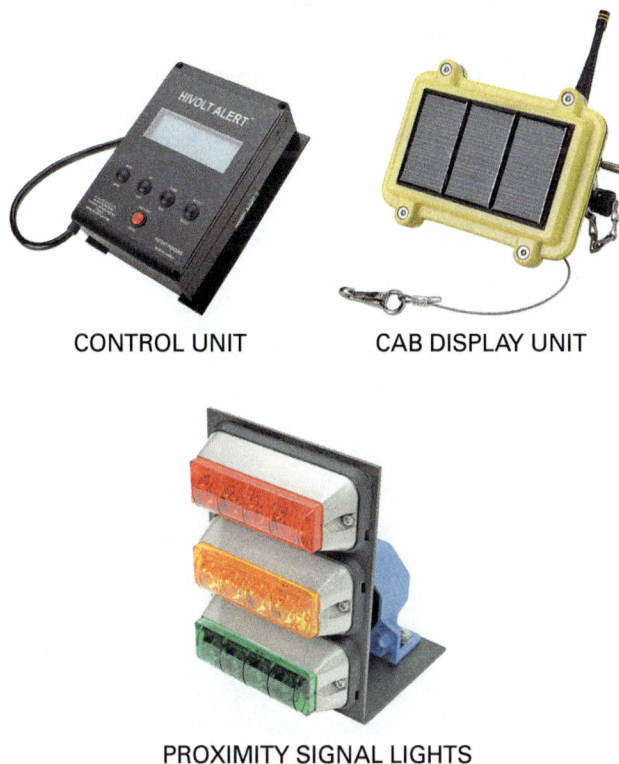

CONTROL UNIT   CAB DISPLAY UNIT

PROXIMITY SIGNAL LIGHTS

Figure 12 High-voltage proximity warning system.

2.2.3 Within Prohibited Zone and Power Lines Energized

Crane operations can be performed within the prohibited zone if the task is absolutely necessary and the crane employer can show that the task cannot be done in any other practical and safe way. Such work places the crane and lift team in close proximity to a major hazard. 29 *CFR* 1926.1410 provides the majority of the guidance for this situation.

All of the steps associated with operating a crane within reach of the prohibited zone are also required in this case. In this case, the required

planning meeting must include the utility operator or owner's representative who is a qualified person in the field of electrical power transmission and distribution. The procedures and techniques that help avoid an incident are determined and documented. The completed documentation must be readily available at the jobsite. Per the OSHA standard, the resulting plans and procedures must include the following:

- Any device such as a recloser that can automatically re-energize a power line after a fault occurs must be disabled if possible.
- A dedicated spotter that can communicate with the crane operator directly and is provided with the aforementioned visual references must be in place.
- An elevated warning line to act as a visual aid for the crane operator must be erected.
- An insulating link must be installed to isolate the load from the load line. No worker other than the crane operator can be allowed to contact any part of the crane or load line above the insulating link.
- If any of the rigging devices, such as slings, will be within the prohibited zone, they must be nonconductive. Tag lines must also be nonconductive.
- A perimeter at least 10 feet (3 meters) away from the crane must be established with barricades to prevent workers and others from getting too close to the crane. If there are structures around the crane that prevent placing barricades that far away, they must be placed as far away as possible. All persons must be kept away from the crane and the work area except those that are essential to the task.
- The crane and any other involved equipment must be properly grounded.
- Power line insulating hose and/or blankets (*Figure 13*) must be applied by the utility if such products are available for the voltage of the lines. Note that such products cannot provide effective protection when the voltage is very high.

Figure 13 Installing insulating line hose and blankets on power lines.

In addition to the initial planning meeting, the utility or owner's representative must also meet with the work team(s) at the site to review the procedures. 29 *CFR* 1926.1410(h) directs that involved employers, as well as the utility/owner's representative, together identify an individual that will be responsible for implementing the procedures that have been developed. The individual has the authority to stop work at any time for safety reasons. If the procedures are not working out as planned, the process stops until new procedures can be developed and implemented.

OSHA also requires that all crane operators and crew members involved with lifts in the prohibited zone of power lines must be specifically trained. 29 *CFR* 1926.1408(g) provides a list of the training topics to be covered.

Note that the guidance regarding work around power lines in *ASME Standard B30.5* is slightly different from the OSHA directives. In this case, the related ASME standards have not been incorporated by reference into the OSHA standard, but the ASME standard contains valuable guidance that should be followed regardless.

2.2.4 Crane Transit with No Load and Boom Lowered

While in transit with no load and the boom lowered, the minimum clearance as specified in *Table 1* must be maintained. You will recall that the lower half of the table provides a separate set of clearance requirements for cranes in transit. Consider however, that a crane bouncing along rough terrain or crossing a rise under a power line can become taller than its specified height. The condition can be momentary, but a moment is all it takes to make contact with a power line. When moving around the site, the effect of speed and terrain on the height of the crane must be considered when evaluating the minimum clearance distance. Additional clearance may be in order to accommodate these factors.

2.2.5 Power Line Contact Emergency Procedures

In spite of everyone's best efforts, something has gone wrong and the boom or load has made contact with an energized power line. Now what? First and foremost, the crane operator must not panic.

Did You Know?

Power Lines That Re-Energize Automatically

At some point in your life, you have likely experienced a power outage in your home or place of business. Sometimes the power goes off, but returns a moment later. Sometimes it quickly goes off again and stays off for a significant period of time, while on other occasions it may return a second time and stay on. Did you ever wonder what's going on at those moments?

Lots of things can disturb power lines and cause what is known as a fault to occur. For example, a wet tree limb can fall across the lines. This causes a short circuit and the current flow becomes extraordinary. The power line must be de-energized when this happens to prevent serious damage to the system or its many components. In many cases though, the fault is temporary – the limb simply falls to the ground after making initial contact.

Many power distribution systems are protected by a special type of circuit breaker called a recloser (shown here). When a fault is sensed, the recloser opens like a circuit breaker, but is also programmed to close the circuit again after a very short interval. If the fault has cleared itself, power is restored and the event is over. If not, a fault is again detected and the recloser opens the circuit. Reclosers may be programmed for one or more attempts to clear the fault, depending upon the utility's policies and the type of customer being served by the circuit. Leaving a recloser active on a circuit near a crane lift could cause a serious incident such as the boom or other object touching the power line to go from bad to much worse very quickly, as it may try repeatedly to re-energize the line.

If power line contact occurs, first try to gently reverse the action that caused the contact. Snatching or grabbing at the controls in response can cause the load to swing out of control, making matters worse. Side loading or another condition that upsets the balance of the crane can then occur, causing the crane to tip. If the contact between the crane and the power line can be broken, the immediate danger to the operator and lift team has been resolved. However, a power outage may have resulted, and the crane may have sustained damage. The crane must be carefully inspected for damage caused by the electrical contact. Wire rope should be replaced if it touches an energized line since the arc is easily sufficient to melt and/or badly scorch the rope. Arcing can occur in a number of other areas in the crane as well. Assuming all is well because the crane still functions is a dangerous practice.

If there is no immediate sign of fire or explosion, remain inside the cab. The operator is usually safest inside the cab at this moment (often safer than any other team member). The crane operator is at the same electrical potential as the equipment and is not in the path of the power as it seeks a path to ground. Note, however, that this is not true when operating a boom truck with standing controls. In this case, the operator might be standing on the ground while in contact with the controls or the frame of the vehicle. This creates a very dangerous situation for the crane operator. For cranes with cabs, however, unless an extreme emergency such as an explosion or fire involving the crane presents itself, operators should remain in the cab and avoid touching the ground.

If you must exit the crane due to fire or an explosion, try to jump off from the lowest point of the crane. As you jump, make no further contact with the crane in an attempt to stabilize yourself. If contact with the crane is made after your feet contact the ground, you become a path to ground and create a complete circuit. Land on the ground with your feet close together and make no further contact with the crane. Do not exit the crane one foot at a time while holding onto the crane.

While moving away from the crane, do not run or take long strides. Instead, shuffle your feet along in very small steps (about six inches, or 15 cm) or hop away with your feet together until you are a safe distance from the crane. High-voltage current transmitted from the power lines through the crane to the ground energizes the ground around the crane (*Figure 14*). As the distance from the crane increases, the voltage and difference in electrical potential decreases. The rate of the decrease varies depending on the resistance of

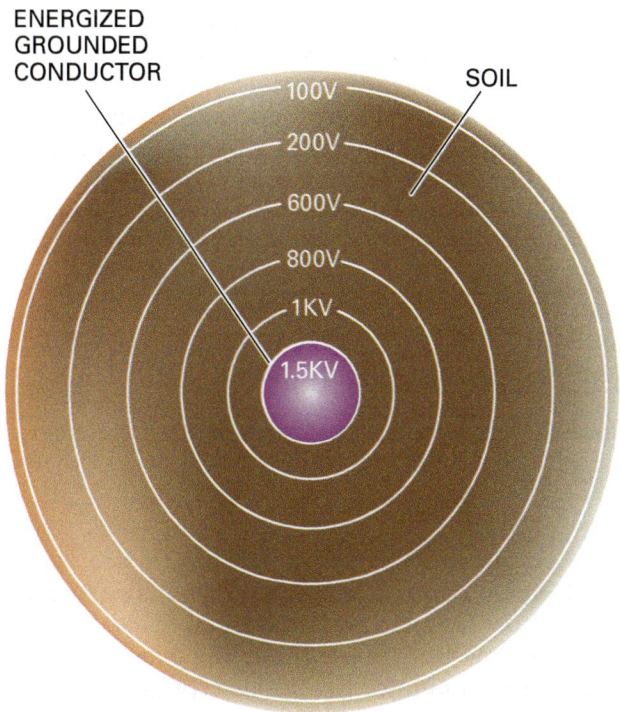

Figure 14 Voltage applied to the ground diminishes with distance.

the surrounding soil. If you take large steps, it is possible for one foot to be in a high-voltage area and the other to be in a lower-voltage area. This increases the difference in electric potential between the two feet, possibly initiating the flow of electricity through the body.

> **WARNING!**
> Do not casually step down from an electrically energized crane. Both feet need to touch the ground at the same time to minimize the potential for serious injury or death. Also, if an electrical circuit is created by making contact with the crane and the ground at the same time, the possibility of electrocution is high.

Instruct all other personnel on the site to stay away from the crane and anything connected to it. If they are close to the crane, instruct them to move away in short steps as well.

If you can remain in the crane, wait until the electrical authorities de-energize the circuit and confirm that the crane is no longer energized. It is likely that a substation circuit breaker or a recloser detected the fault and has opened to a point where it must be manually reset. However, this cannot be assumed until qualified personnel provide that information through testing.

Report every incident involving contact with power lines to the electrical authority and safety officer. If there was ever a chance of such an incident occurring due to work within the prohibited zone, utility personnel should have already been on site. As soon as possible, ensure that someone on the site has called for medical assistance or any other help needed on the scene. Lifts are made with cranes in the vicinity of power lines on a daily basis without incident. Take every precaution to ensure your encounters with work in this area are safe and successful as well.

2.3.0 Hazardous Weather

Mobile crane operators work outdoors. Under certain environmental conditions, such as extremely hot or cold weather or in high winds, work can become uncomfortable and maybe dangerously so. For example, snow and rain can have a dramatic effect on the weight of the load and on ground compaction. During the winter, the tires, outriggers, and crawlers can freeze to the ground. This may lead the operator to the false conclusion that the crane is on stable ground. As weight is then added during the lifting operation, an outrigger

float, tire, or crawler track may sink into mushy ground below the frozen surface. Heavy rain can also cause the ground under the crane to become unstable. The crane set-up site must be carefully evaluated to ensure that there is sufficient stability, including the condition of the soil below the surface.

High winds and lightning represent significant hazards on the jobsite (*Figure 15*). Both must be taken seriously. Crane operators must be prepared to respond appropriately to weather changes in order to avoid accidents and injuries. Fortunately, it is relatively rare for high winds or lightning to arrive without at least being reported as a possibility in weather reports. As a general rule, the lift team and crane operators have time to react appropriately. The site supervisor or lift director (possibly the same individual, depending on the lift characteristics) is responsible for ensuring that factors such as wind, heavy rain, fog, and the soil conditions that change as a result of weather are properly considered and addressed.

Figure 15 Wind and lightning hazards.

2.3.1 Wind

High winds typically start out as less dramatic gusts. Operators must be keenly aware of changing wind speeds. The crane boom should be down and stowed before the wind becomes too strong, not after the threshold has been passed. Keep in mind that the wind speed can be dramatically higher at the tip of the boom than it is at ground level. The operator must end and secure crane operations as soon as possible when the wind speed is increasing. This involves placing the boom in the lowest possible position and securing the crane. However, since wind speed affects capacity, a lift may have to be postponed due to a loss of crane capacity at wind speeds that are still acceptable for operation in general.

> **CAUTION**
>
> The wind chart shown in *Table 2* is for a specific crane model. The information provided cannot be applied to all cranes. Always check and follow the manufacturer's wind chart speed for the specific crane in use.

Crane operators should be familiar with and follow the crane manufacturer's guidance related to wind. Their guidelines will differ from model to model. It is important to have the correct information for the specific crane in use. *Table 2* provides an example of a crane manufacturer's wind chart, showing how various wind speeds affect the rated load capacity of the crane. Note that the rated load capacity shown on the load charts is valid through wind speeds of 20 mph (32.2 kph). The chart also shows different capacity deductions for boom lengths less than or greater than 250 feet (76 meters). These reductions are applied to the load chart being used. This particular

crane cannot be operated at all when the wind speed is above 45 mph (72.4 kph).

It is also important to point out that wind affects the load as much as the crane. It is not unusual for a crane to have the capacity to lift a given load at a wind speed of 30 mph (78.3 kph). But many loads have a very low weight-to-surface area ratio. A sheet pile is a good example. With a great deal of surface area and limited weight, a sheet pile is easily blown around by winds of 30 mph. In some cases, even though the crane is capable of making the lift, the wind's effect on the load has to be a significant factor in the decision. Perhaps additional tagline personnel can be put on the job, but the safer thing to do is wait until the wind speed has dropped to a more acceptable level. In this example, the wind hazard is not about the crane—it is about the load, the workers that must handle it, and other equipment or property in the area that could be damaged by it.

It is important to make this point regarding changing wind speeds. Assume that you are lifting a load with a wind speed of 25 mph. If the wind speed chart in *Table 2* applied, the rated load capacity of the crane must therefore be reduced by 20 percent. Now assume that the load represents 50 percent of the normal rated-load capacity, so the load weight now represents 70 percent of the capacity; still within the parameters of a standard lift plan. However, if the wind speed increases just 6 mph to 31 mph, the operation must be stopped and the crane secured. Although the crane itself can operate at this wind speed, the reduction of capacity is now 40 percent, and the load weight now represents 90 percent of the crane's capacity. Under these conditions, the lift becomes a critical lift that requires a documented lift plan. The alternative is to simply wait until

Table 2. Example of a Wind Speed Chart

Boom and Boom + Jib Lengths up to 250'	
Description	**Allowable Windspeeds in Miles Per Hour (mph)**
Boom and Boom + Jib Lengths Greater than to 250'	
1. Normal Lifting Operation. (See Capacity Charts.)	0–20 mph
2. Reduced Operation. Capacities must be reduced by 20%.	21–30 mph
3. Reduced Operation. Capacities must be reduced by 40%.	31–40 mph
4. Reduced Operation. Capacities must be reduced by 70%.	41–45 mph
5. No Operation. Store attachment on ground.	Over 45 mph
1. Normal Lifting Operation. (See Capacity Charts.)	0–20 mph
2. Reduced Operation. Capacities must be reduced by 35%.	21–30 mph
3. Reduced Operation. Capacities must be reduced by 60%.	31–40 mph
4. Reduced Operation. Capacities must be reduced by 70%.	41–45 mph
5. No Operation. Store attachment on ground.	Over 45 mph

the wind speed is lower. This type of scenario and the related decision-making process is repeated daily in the crane industry.

2.3.2 Lightning

Because crane booms extend so high and are made of metal, they are easy targets for lightning. Operators must be constantly aware of this threat. Lightning can usually be detected when it is several miles away. As a general rule of thumb, sound travels near the ground about 1 mile in 5 seconds, or about 1 kilometer in 3 seconds. Therefore, if there is a five-second delay between the flash of lightning and the sound of thunder, the lightning strike was roughly one mile away. Be aware, however, that successive lightning strikes can touch down up to 8 miles apart. That means once you hear thunder or see lightning, it is close enough for the next strike to present a hazard.

In some high-risk areas, local proximity sensors provide warnings when lightning strikes occur within a 20-mile radius. Once a warning is given, lightning is spotted, or thunder is heard, the crane operator must secure crane operations as soon as practical.

There have been many cases when the warning signs have not been taken seriously enough. Even if the above common rule of thumb were completely accurate, the process provides no information allowing one to determine when and where the next lightning strike will occur. If lightning is seen or thunder is heard, it is time to secure the crane and ignore the math.

Crane operators and the rest of the lift team must pay attention to the signs of thunderstorms and other weather events developing. It is best not to attempt operations that require a significant amount of time, such as a lengthy concrete pour, when there is a possibility that the operation may have to stop before it is complete. Doing so encourages the team to rush or remain in operation longer than it should once the warning signs are evident, in an attempt to complete the task. Both responses raise the potential for an accident.

Once crane operations have been shut down, all personnel should seek indoor shelter away from the crane. Even with the boom in the lowest position, it may be taller than surrounding structures and could still be a target for lightning strikes. Always wait a minimum of 30 minutes from the last instance of lightning or thunder before resuming work.

If lightning strikes a crane, a thorough inspection of the crane will be required. If lightning strikes the wire rope, for example, it may be damaged beyond safe use. All electrical systems need to be tested before the crane is returned to service, in addition to a thorough visual inspection.

2.4.0 Other Operational Safety Topics and Incidents

There are several other issues that affect the safety of a crane operation, as well as specific incidents to which a crane operator may need to respond. These issues are presented in the sections that follow.

2.4.1 Manufacturer's Requirements and Guidance

To operate a mobile crane safely, the operator must use the manufacturer's data and documentation provided for the specific crane in use. These manuals provide information on required startup checks and periodic inspections, as well as inspection guidelines. These manuals also provide many safety precautions and restrictions of use. Ignorance of any of these requirements or precautions is hazardous to the safe operation of the crane and could make the operator liable if an accident should occur. Operators should always read and follow the manufacturer's instructions. Crane manuals also provide information related to certain types of equipment failures and error messages that may present themselves. The manufacturer has reasons for any specific responses they provide. Follow the manufacturer's recommendations in all such cases.

2.4.2 Moving Cranes Safely

During the course of a job, cranes and other heavy equipment are moved to, around, and away from the site. Many accidents and injuries happen during the movement of heavy equipment. It is important to be especially safety conscious whenever equipment is moving.

Always follow these guidelines when driving equipment on public roads:

- Know and obey all state and local laws.
- Secure all attachments and loose gear.
- Use proper warning signs and flags per state and federal Department of Transportation (DOT) requirements.
- Drive slowly and never speed.
- Allow extra time to enter traffic.
- Stay in the extreme right lane on multi-lane highways.
- Travel with your lights on, day or night.
- Be aware of the crane's turning radius.
- Turn cautiously; allow for extensions or attachments and for structural clearances. Some

equipment is top-heavy and will tip over if a turn is made too fast.

- Be aware of the crane's stopping distance. Due to their size and weight, cranes can develop a great deal of momentum. Be especially careful when driving downhill.

When driving on the job site, follow these guidelines:

- Never drive a machine on a job site, in a congested area, or around people without a spotter or flagger to guide you. The spotter or flagger is responsible for determining and controlling the driver's speed.
- Be sure everyone is in the clear while backing up, hooking up, or moving attachments. When backing, allow a few moments for the back-up alarm to announce your intention before putting the crane in motion.
- If you cannot see your area clearly from the operator's seat and have no spotter, dismount and examine the site for possible hazards before proceeding.
- Wait for an all-clear signal from spotters before moving.
- Signal a forward move with two blasts of the horn; signal a reverse move with three blasts of the horn.

- Yield the right-of-way to moving equipment on haul roads and in pits.
- Maintain a safe distance from all other vehicles.
- When moving, keep the crane in gear at all times; never coast.
- Maintain a speed consistent with ground conditions.
- Pass only when necessary; use caution.
- Watch for overhead electrical power lines and ensure you have sufficient clearance. Refer to the lower portion of *Table 1*.
- Watch for flags indicating buried utilities (*Figure 16*).

2.4.3 Using Cranes to Lift Personnel

Although using a crane to hoist personnel is generally discouraged, it can be done safely with the correct equipment (*Figure 17*) and procedures. There are many personnel-platform styles to choose from, and they can be custom-made by several vendors to suit unique needs. Using a crane to lift personnel is prohibited by *29 CFR 1926.1431*, unless the employer can demonstrate that the erection and use of a more conventional means to access an area is more hazardous than using the crane. When it is allowed, a personnel platform that meets the requirements of the

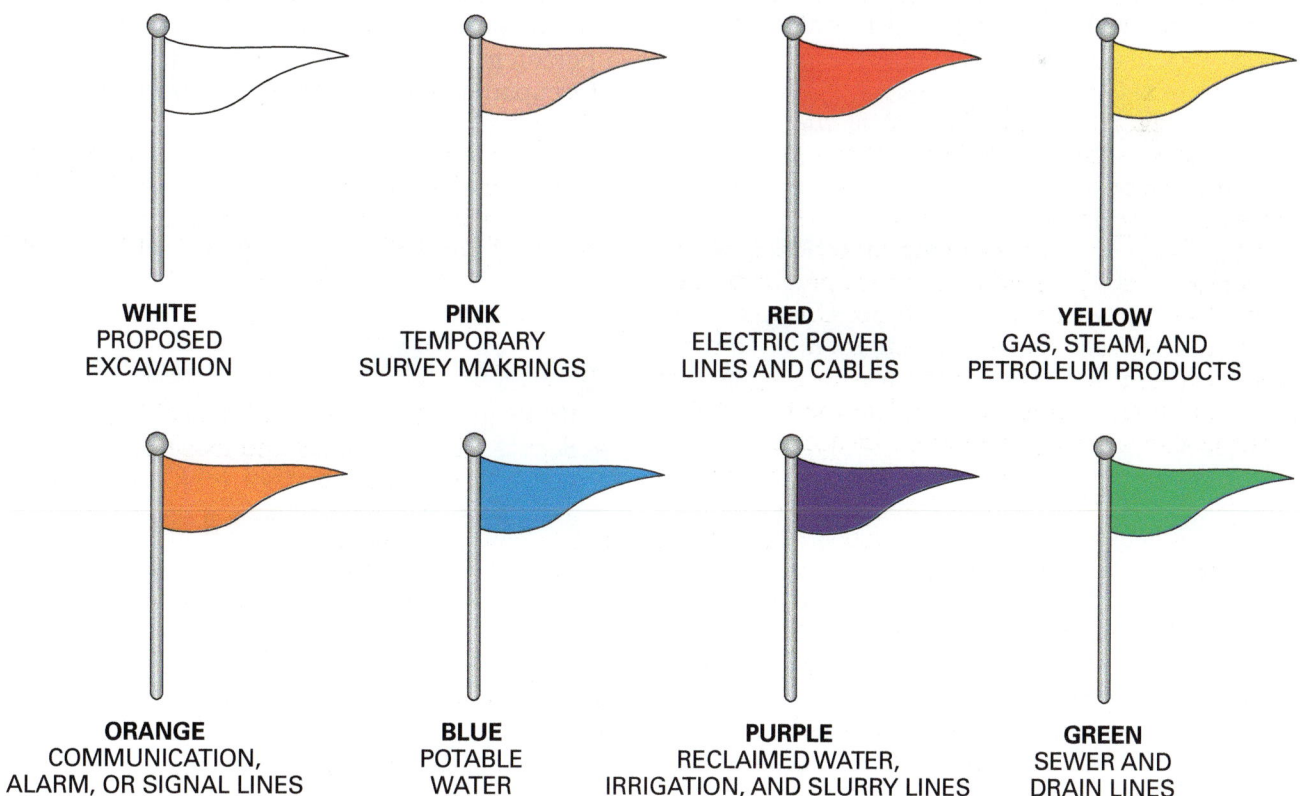

WHITE
PROPOSED
EXCAVATION

PINK
TEMPORARY
SURVEY MAKRINGS

RED
ELECTRIC POWER
LINES AND CABLES

YELLOW
GAS, STEAM, AND
PETROLEUM PRODUCTS

ORANGE
COMMUNICATION,
ALARM, OR SIGNAL LINES

BLUE
POTABLE
WATER

PURPLE
RECLAIMED WATER,
IRRIGATION, AND SLURRY LINES

GREEN
SEWER AND
DRAIN LINES

Figure 16 Flag colors used to identify underground utilities and related areas.

OSHA standard must be used (with some special exceptions).

ASME Standard B30.23, Personnel Lifting Systems, is devoted exclusively to the topic of hoisting personnel and the equipment requirements, as the name implies. Note, however, that this ASME standard has not been incorporated by reference into the OSHA standards, but the two standards do have a lot of similarities. Some requirements found in the standards include the following:

- The crane must be level within 1 percent, and outriggers must be used if the crane has them.
- The load cannot be more than 50 percent of the rated load capacity of the crane for the configuration and operating radius.
- If the personnel platform is in a stationary working position, the primary and secondary boom and vehicle braking systems and locking features must be engaged.
- The load line hoist drum must be equipped with a system that regulates lowering speed. If the crane has the capability to allow free-fall of the load line, it cannot be used for hoisting personnel.
- A boom angle indicator, a boom hoist limiting device, and anti-two-blocking devices are all required. If a luffing jib is in use, a jib angle indicator and a jib hoist limiting device are also required.
- Trial lifts, using an equal or greater weight than the expected load of the personnel and

their equipment, must be made. The crane must lift the load and move it to each location that the personnel will need to access. Trial lifts must continue to be conducted at the beginning of each shift, any time the crane is moved to a new location, and whenever the lift route changes in a way that adds new hazardous factors to the task.

- Proof testing is also required. The platform and rigging must be tested at 125 percent of the platform's rated load capacity and then inspected by a competent person. This testing must be done at each new jobsite and after any repairs or alterations have been performed on the personnel platform. Proof testing can be done in conjunction with the trial lifts. Some manufacturers of personnel platforms have developed a simple system of attaching a weight that equals 125 percent of the capacity to the bottom of the platform for this purpose (*Figure 18*).
- Unless the personnel platform is equipped with crane controls, the operator must remain at the controls in the cab at all times.
- Personnel being lifted must remain in contact with either the crane operator or the signal person (if used) at all times.
- If the wind speed exceeds 20 mph (32.2 kph), a qualified person must determine if it is safe to lift personnel for the required task, or whether the lift should be ended or postponed. Other weather issues may also prompt a qualified person to stop or postpone the lift.
- Occupants of the personnel platform must be equipped with personal fall-arrest equipment, with the lanyard attached to a structural member of the platform.
- Any other lift lines on the crane may not be used for lifting other items while lifting

Figure 17 Enclosed round personnel platform.

Figure 18 Personnel platform with detachable weight for proof testing.

personnel. Pile-driving operations are an exception to this rule.

- Unless the task directly involves work on a power line, hoisted personnel cannot be placed within 20 feet (6.1 meters) of lines up to a voltage of 350 kV, or within 50 feet (15.2 meters) of power lines over 350 kV.

Note that there are a number of other requirements, especially for special situations such as pile driving, lifting personnel in and out of drilled shafts, and transferring personnel to the site of a task in a marine environment. The above list does not represent all of the requirements and conditions found in the OSHA or ASME standards. If hoisting personnel is part of your work schedule, it is important to review the requirements that apply to your particular situation.

2.4.4 Incidents During Lifting Operations

Mechanical malfunctions or lapses in judgement during a lift can be very serious. If an equipment failure or operator error causes the operating radius to increase unexpectedly, the crane can tip or the structure could collapse. Loads can also be dropped during a mechanical malfunction. A sudden loss of load on the crane can cause a whiplash effect that causes the crane to tip or the boom to fail. The chance of these types of incidents occurring in modern cranes is greatly reduced because of system redundancies and safety backups. However, failures do happen, so the operator must stay alert at all times.

If a mechanical problem occurs, the operator should attempt to lower the load immediately. Next, the operator should secure the crane, tag the controls indicating the crane is out of service, and report the problem. The crane should not be operated until it is checked and repaired if necessary, by a qualified technician.

Carelessness by the operator can lead to accidents other than those associated with overloading the crane. These incidents include the following:

- *Striking the boom* – The operator must never allow the boom to strike any structure or load. Even what seems to be mild contact can dent, bow, or bend the lower boom chords and compromise the integrity of the boom. A serious incident may result in total boom collapse. If the boom touches or rests on another structure, the boom-loading changes from a compression force to a bending force. The boom is very strong in compression but weak in bending. If the boom, mast, or jib is struck or damaged, stop the lift and leave the boom where it is, un-

less it is creating a new significant hazard by remaining in position. The load on the boom increases as the boom is lowered. As a result, a damaged boom or boom suspension could collapse during the lowering process. A second crane may be required to help lower a significantly damaged boom. The site supervisor or lift director will generally make this decision.

- *Backward collapse of a boom* – When operating near the minimum radius, with the boom at its highest angle, boom down as you set the load down. This will compensate for the tendency of the boom to move or jump back against the boom stops when the load is released, especially if the load on the boom is relieved too quickly. This action occurs because of the elasticity in the boom and boom hoist systems (*Figure 19*), and it can result in a backward collapse of the boom.

Another factor that may cause the backward collapse of a boom or upset the balance of the crane is high winds. Consult and follow the wind speed charts for the specific crane in use. Other factors that may contribute to boom collapse include the following:

- Continuing to pull on the hoist line after two-blocking has occurred if the crane is not equipped with a functional anti-two-blocking device
- Starting or stopping a swing suddenly if the boom is at a high angle
- Sudden forward movement of a crane that can send the boom over backward if it is being carried at a high boom angle
- Snubbing the hook block to the boom foot, then pulling it up tight
- Instability in certain positions when the crane is traveling on an incline

- *Two-blocking* – Two-blocking refers to a situation that results in the load block or hook assembly contacting an upper load block (if equipped) or the boom-point sheave assembly. Damage can occur to the sheaves, block, and/or wire ropes. However, if the load block makes contact and the operator continues to wind the drum, the crane is essentially pulling against itself. This can result in serious damage and boom failure. The devices shown in *Figure 20* offer protection against two-blocking. The dangling weight that encircles the hoist rope is attached to a switch. If the load of the weight is removed from the switch due to the load block contacting and raising it up, the switch opens and stops the hoist drum.

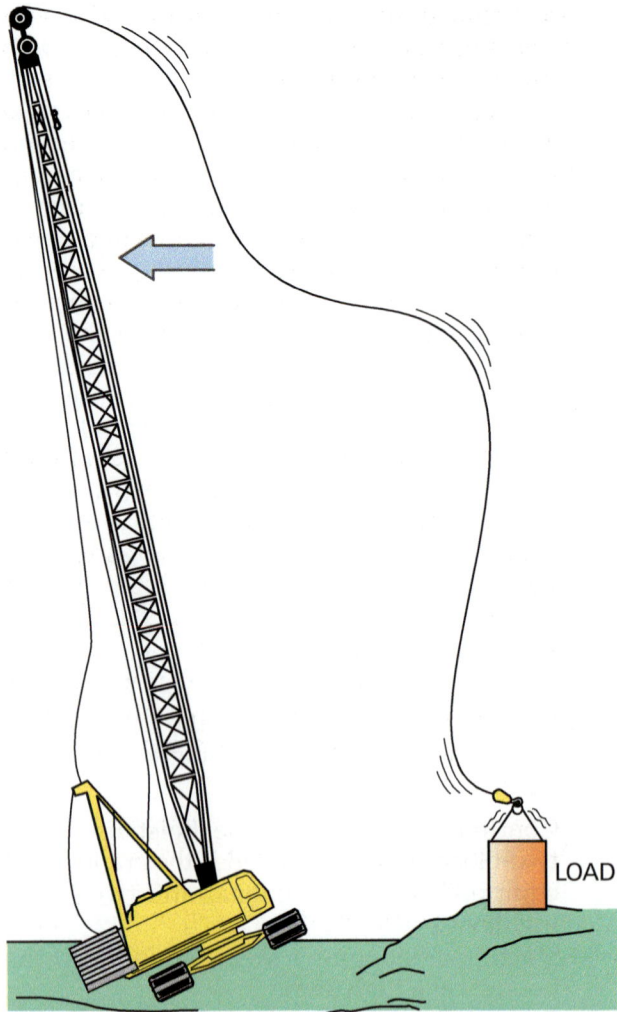

Figure 19 Crane response to the sudden release of a load.

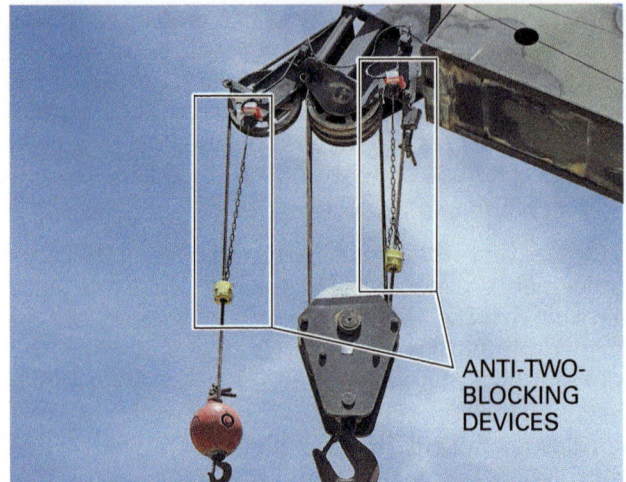

Figure 20 Anti-two-blocking devices.

In *Figure 20*, note the red devices that are inserted into the anti-two-block switches near the sheaves. These devices can be inserted into the switch to disable the anti-two-blocking safety feature. This may be necessary when stowing a mobile crane and preparing it for travel, so that the load block can be drawn up closer to the sheaves. They must not remain in place when the crane begins normal operations.

Crane operators must avoid two-blocking by being attentive to the position of the load block and boom. Do not rely on anti-two-blocking devices as control devices. They are there for safety purposes only.

2.4.5 Fire

The operator's judgment is crucial in determining the correct response to fire. The preferred first response is to cease crane operation, lower the load if practical, and secure the crane. In all cases of fire, evacuate the area even if the load cannot be lowered or the crane secured. After emergency services have been notified, a qualified individual may judge if the fire can be combated with a fire extinguisher. A fire extinguisher can be successful at fighting a small fire in its beginning stage, but a fire can get out of control very quickly. The operator must keep in mind that the highest priority is preventing loss of life or injury. Do not be overconfident in your ability to control a fire. Even trained firefighters using the best equipment can be overwhelmed and injured by fires.

According to 29 *CFR* 1926.1433(d)(6), all mobile cranes are required to have an accessible fire extinguisher on the equipment. *ASME Standard B30.5* indicates that it should be a minimum of a 10BC-rated portable fire extinguisher (*Figure 21*). Note that this has not been incorporated by reference into the OSHA standards. The operator should be trained in its use. This type of extinguisher is designed to combat Class B (flammable liquids) and Class C (electrical) fires. A BC extinguisher is typically charged with dry chemicals. The number 10 is an indicator that the extinguisher should be sufficient to fight a fire that covers 10 square feet (0.9 square meters). The effective range of this type of extinguisher is generally 5 to 20 feet (1.5 to 6 meters).

The extinguisher is constructed so that the extinguishing agent becomes pressurized when the pin is pulled and the handle is compressed. Remember to use the PASS method to fight a fire with a fire extinguisher:

- **P**ull the pin from the handle, breaking the tamper seal.
- **A**im the nozzle at the base of the fire while 8–10 feet (2.5–3 meters) away.
- **S**queeze the discharge handle.
- **S**weep the nozzle back and forth at the base of the flames.

ASME standards require that refueling of the crane be done when the engine is not running and in the absence of smoking, open flames, or any other sources of ignition. If refueling is being done with a portable container, the container must be a safety-type fuel can equipped with an automatic-closing cap and a flame arrester.

The best way to prevent a fire is to make sure the three elements needed for fire—fuel, oxygen, and a source of ignition—are never present in the same place at the same time. Oxygen in the atmosphere is impossible to eliminate, leaving only the fuel source and a source of ignition within human control. Here are some basic safety guidelines for fire prevention:

- Always operate in a well-ventilated area, especially when flammable materials such as shellac, lacquer, paint stripper, or construction adhesives are in use.
- Never smoke, strike a lighter, or light a match when you are working with flammable materials.
- Dispose or store oily rags only in approved, self-closing metal containers. Store other combustible materials only in approved containers.
- Know where to find additional fire extinguishers, what kind of extinguisher to use for different kinds of fires, and how to use the extinguishers when called upon.
- Make sure all extinguishers are fully charged. Never remove the tag from an extinguisher; it shows the date the extinguisher was last serviced and inspected.
- Keep open fuel containers away from any sources of sparks, fire, or extreme heat.
- Do not fill a gasoline or diesel fuel container while it is resting on a truck bed liner or other

ungrounded surface. The flow of fuel creates static electricity that can ignite the fuel if the container is not grounded.

- Be prepared at all times to call on professional firefighters, either by using the 911 system or through a direct call to the nearest fire department, even for a small fire. If you are in unfamiliar territory, ask for the location and contact information of the nearest fire department. Time permitting, this call is best made before a personal attempt to fight the fire begins.

2.4.6 Accident Investigations

Lifting operations must always be conducted in the safest manner possible to prevent accidents. However, should an accident occur, the operator should follow the required emergency procedures. Your employer is also likely to have specific procedures to be followed in the event of an accident. The crane operator must have an understanding of the type of information that may be requested on a typical accident investigation checklist. A great deal of information is typically required, including the following:

- Company name and mailing address
- Person receiving the report (name, title, and phone number)
- Investigator's name, title, and company
- Date and time of investigation
- Location and exact time of accident
- Description of the equipment involved, including the manufacturer, model, serial number, age of machine, unit number, and configuration
- Summary of the accident as recalled by witness(es)
- List of people who will make formal statements, including witnesses to the events from one hour before the accident to the end of the accident sequence, all personnel who perform maintenance on the involved equipment, and all personnel involved in determining and planning the lift
- Sketches of the accident scene to scale with as much detail as possible
- Photographs from as many angles as possible
- Weather conditions at the time of the accident
- Ground conditions
- Boom length
- Operating radius
- Actual load weight including load block, sling, and boom attachments
- Background information on the crane operator, riggers, and other members of the lift team – name, age, training, and experience

Figure 21 10BC-rated fire extinguisher.

- Summary of the accident and sequence of events
- Any conclusions based on findings

Normally, a safety officer or inspector from the company responsible for the jobsite and/or crane will conduct the accident investigation and complete the required forms. After the investigation is completed, the report is filed by the responsible company with the proper state and federal authorities.

If the accident results in one or more fatalities that are immediate or occur within 30 days of the incident, the accident must be reported to OSHA within eight hours. If the accident causes inpatient hospitalization, amputation, or the loss of an eye, it must be reported to OSHA within 24 hours. One minor exception relates to the timing of the hospitalization, amputation, or eye loss: If it occurs more than 24 hours after the accident, it is not required to be reported. After a report is submitted, investigators from OSHA and/or the National Institute for Occupational Safety and Health (NIOSH) will also conduct investigations into the accident.

Employers may require the crane operator and others involved in the incident to complete controlled substance and/or alcohol testing immediately following the event.

Additional Resources

ASME Standard B30.5, Mobile and Locomotive Cranes. Current edition. New York, NY: American Society of Mechanical Engineers.

ASME Standard B30.23, Personnel Lifting Systems. Current edition. New York, NY: American Society of Mechanical Engineers.

ASME Standard P30.1, Planning for Load Handling Activities. Current edition. New York, NY: American Society of Mechanical Engineers.

OSHA Standard 1926, Subpart CC, Cranes and Derricks in Construction. **www.ecfr.gov**

Mobile Crane Safety Manual. 2014. Milwaukee, WI: Association of Equipment Manufacturers.

The following websites offer resources for products and training:

American National Standards Institute (ANSI), **www.ansi.org**

The American Society of Mechanical Engineers (ASME), **www.asme.org**

Occupational Safety and Health Administration (OSHA), **www.osha.gov**

North American Crane Bureau, Inc., **www.cranesafe.com**

Electronic Code of Federal Regulations, **www.ecfr.gov**

2.0.0 Section Review

1. Which of the following characteristics would most likely place a lift in the category of a critical lift?

   a. The load weight exceeds 60 percent of the crane's rated load capacity at the required operating radius.
   b. Winds at the time of the lift are forecast to be as high as 15 mph.
   c. A load must be lifted between two occupied buildings that are very close together.
   d. The boom of a hydraulic crane will need to be fully extended.

2. The minimum clearance to be maintained from power lines carrying up to 50 kilovolts (kV) when the crane is operating nearby is _____.

   a. 10 feet (3.05 meters)
   b. 15 feet (4.60 meters)
   c. 20 feet (6.10 meters)
   d. 25 feet (7.62 meters)

3. Which of the following lift factors is most likely to prompt a decision to stop a lift due to wind, even if the load weight is well within the limitations for the situation and the crane in use?

   a. The load has to be lifted over 25 feet up (7.6 meters).
   b. The load is primarily constructed of wood.
   c. The load has a high weight-to-surface area ratio.
   d. The load has a low weight-to-surface area ratio.

4. When driving a mobile crane, the crane operator signals a forward move with _____.

   a. one horn blast
   b. two horn blasts
   c. one forward hand signal
   d. two forward hand signals

5. A crane accident must be reported to OSHA within eight hours if the accident results in _____.

   a. injuries to two or more workers
   b. one or more fatalities
   c. a fire inside the crane cab
   d. property damage exceeding $5,000

SUMMARY

The size of a crane and the huge loads it handles expose the crane operator and anyone in the vicinity of the crane to the potential for serious injury and even death. However, crane operations are carried out thousands of times a day without incident. Mobile crane operators must be skilled, knowledgeable, and mentally and physically fit to reduce the potential for accidents. Other members of a lift team must be equally fit for their role. There are many rules and guidelines that apply to crane operations. The vast majority of accidents occur when those rules aren't followed.

Mobile crane operators carry a great deal of responsibility on their shoulders. It is imperative that they embrace safety as the number priority on every job, every day. This module is just the beginning of what you will learn about safety as you progress in your training. Decide now to be a part of establishing and supporting a safety culture within your organization, and commit yourself to the task of learning how to keep yourself and your co-workers safe in the most effective way possible.

1. Which of the following is the most important ASME standard for crane operators?
   a. *ASME Standard B30.20*
   b. *ASME Standard P30.1*
   c. *ASME Standard B30.5*
   d. *ASME Standard B30.23*

2. Information that must be kept in the crane cab at all times includes _____.
   a. the complete operating manual
   b. a copy of *ASME Standard B30.5*
   c. a copy of 29CFR1910.333
   d. the title to the crane, confirming the owner

3. Pre-lift planning guidance is provided by _____.
   a. *ASME Standard B30.5*
   b. *ASME Standard P30.1*
   a. 29CFR1926.1431
   b. 29CFR1910.180

4. The prohibited zone is defined as _____.
   a. any quadrant the crane is not designed to work in according to the manufacturer's guidance
   b. an area around an energized power line that no part of the crane or load should enter
   c. the length and operating radius at which the crane cannot operate if it is making a lift on uneven soil
   d. a circle defined by the outermost point of the crane's outriggers when extended

5. If a visual aid to identify the prohibited zone is not provided for the crane operator when operating near energized power lines, the crane operator must be in constant contact with a(n) _____.
   a. professional engineer qualified in power distribution and transmission
   b. designated representative of the electrical utility or power line owner
   c. dedicated spotter whose sole responsibility is to monitor the required clearance
   d. site supervisor and the local fire department

6. If it becomes necessary to jump off a crane that has contacted a power line, you should keep your feet _____.
   a. as far apart as possible
   b. about one foot apart, side-by-side
   c. about one foot apart, one in front of the other
   d. close together as you land

7. The rate of voltage decrease in the ground surrounding a crane after it has contacted a power line varies depending on the _____.
   a. type of boom
   b. length of the boom
   c. voltage of the power line
   d. resistance of the soil

8. The wind speed at which the operation of a given crane must be limited or halted is determined by _____.
   a. the crane manufacturer
   b. the crane operator
   c. OSHA
   d. crane manufacturer

9. When personnel are being hoisted on a personnel platform, the total weight of the load, basket, and rigging may not exceed _____.
   a. 100 percent of the crane's rated load capacity
   b. 75 percent of the crane's rated load capacity
   c. 50 percent of the crane's rated load capacity
   d. 25 percent of the crane's rated load capacity

10. If the boom, mast, or gantry is struck or damaged, the operator should first _____.
    a. lower the boom
    b. raise the boom
    c. bring the boom to the ground
    d. stop the lift

Trade Terms Quiz

Fill in the blank with the correct term that you learned from your study of this module.

1. A(n) _____ presents an increased level of risk beyond normal lifting activities.

2. Although the ones developed by many organizations are voluntary in nature, the _____ established by OSHA are enforceable by law.

3. An individual that can identify existing and predictable hazards and has the authority to take prompt corrective measures is referred to by OSHA as a(n) _____.

4. As a general rule, cranes are not allowed to handle materials on the ground beneath power lines because this area is part of the _____.

5. The area of specific dimensions that surrounds energized power lines that no part of a crane is allowed to enter is called the _____.

6. A device that senses the electric fields created by power lines and alerts the crane operator to the danger is a(n) _____.

7. A rigger or crane operator can use a(n) _____ to help protect workers in contact with a load from the danger of electrocution in the event the crane contacts a powerline.

8. When a lift can be accomplished using common procedures, methods, and equipment, it is referred to as a(n) _____.

9. If a crane operator lowers a load quickly and then brings it to a sudden stop in the air, the crane and rigging will experience _____.

10. OSHA refers to the space around energized power lines that no part of a crane or load can enter as the _____.

11. A brief power outage that is followed by power being quickly restored is probably the result of a _____ doing its job.

Trade Terms

Avoidance zone
Competent person
Critical lift
High-voltage proximity warning

device
Insulating link
Minimum clearance distance
Prohibited zone

Recloser
Shock loading
Standards
Standard lift

EXAMPLE—CRITICAL LIFT PLANNING WORKSHEET

POWER CONSTRUCTION

CRITICAL LIFT PLANNING WORKSHEET

PROJECT: _____   SUBCONTRACTOR: _____

COMPETENT PERSON: _____   QUALIFIED RIGGER: _____

IS OPERATING ENGINEER CITY OF CHICAGO CERTIFIED?   ☐ YES   ☐ NO   ☐ N/A

IF MULTIPLE CRANES ARE REQUIRED FOR THE LIFT, A SEPARATE WORKSHEET IS REQUIRED FOR EACH CRANE.

CRANE INFORMATION

CRANE OWNER / SUPPLIER: _____

BOOM TYPE: ☐ TELESCOPING BOOM   ☐ LATTICE BOOM

CRANE BASE: ☐ ON RUBBER TIRE   ☐ OUTRIGGERS   ☐ CRAWLER   ☐ AIRCRAFT

BOOM LENGTH: _____

JIB LENGTH: _____

COUNTERWEIGHT: _____

CAPACITY OF CONFIGURATION: _____

ANNUAL CERTIFICATION DATE: _____

LOAD DATA & RIGGING

WHAT IS BEING HOISTED? (TYPE OF MATERIAL / PRODUCT) _____

HOW WILL THE LOAD BE HOISTED? (RIGGING CONFIGURATION) _____

WILL ENGINEERED PICK POINTS BE UTILIZED?   ☐ YES   ☐ NO

WHAT TYPE(S) OF RIGGING IS NEEDED? _____

WHO IS PROVIDING THE RIGGING? _____

HAS THE RIGGING BEEN INSPECTED?   ☐ YES   ☐ NO

TAG LINES UTILIZED (IF NOT, WHY)   ☐ YES   ☐ NO

WEIGHT OF LOAD: _____

RIGGING WEIGHT: _____

BLOCK & LINE WEIGHT: _____

TOTAL LOAD WEIGHT: (RIGGING + BLOCK & LINE + LOAD WEIGHT) _____

IS LOAD GREATER THAN 75% OF CHART?   ☐ YES   ☐ NO

Critical Lift Planning Worksheet   Revised October 2012

Figure A01A Critical Lift Planning Worksheet (1 of 2)

POWER CONSTRUCTION   **CRITICAL LIFT PLANNING WORKSHEET**

COMMUNICATIONS & FALL PROTECTION

WHAT TYPE OF COMMUNICATIONS WILL BE USED? ☐ HAND SIGNALS   ☐ HARD LINE   ☐ 2 WAY RADIO   ☐ OTHER: _____

HAND SIGNALS MUST BE POSTED / CELL PHONES ARE NOT APPROVED METHOD

IDENTIFY SIGNAL PERSON: _____

IS FALL PROTECTION REQUIRED FOR SIGNALPERSON? ☐ YES ☐ NO   IF YES, WHAT METHODS WILL BE UTILIZED? _____

SITE CONSTRAINTS & SOIL CONDITIONS

ARE OVERHEAD POWER LINES / OBSTRUCTIONS PRESENT: ☐ YES ☐ NO   IF YES, IDENTIFY LOCATIONS: _____

PRECAUTIONS FOR OVERHEAD POWERLINES / OBSTRUCTIONS: _____

PRECAUTIONS FOR OVERHEAD PROTECTION; PROTECTION OF OCCUPIED SPACES AND PEDESTRIANS: ☐ YES ☐ NO   IF YES, WHAT IS PLAN: _____

GROUND CONDITIONS: ☐ ACCEPTABLE   ☐ NOT ACCEPTABLE

EXPLAIN REQUIRED ACTION TO CORRECT: _____

OUTRIGGER PLACEMENT (ATTACH LOAD CHART): ☐ FULL EXTENSION   ☐ HALF EXTENSION   ☐ OTHER

IS THE CRANE RATED FOR THIS CONFIGURATION? ☐ YES ☐ NO

WILL OUTRIGGERS BE PLACED ON / NEAR SHORING OR OPEN EXCAVATION? ☐ YES ☐ NO

IF YES, IS THE SHORING DESIGNED TO HANDLE THE IMPOSED LOAD: ☐ YES ☐ NO ☐ UNKNOWN (IF NO OR UNKNOWN, CONTACT ENGINEER)

WILL THE OUTRIGGERS BE PLACED ON, OVER, OR NEARLY OVER THE TOP OF UNDERGROUND UTILITIES: ☐ YES ☐ NO

IF YES, WHAT PRECAUTIONS WILL BE TAKEN: _____

IS LIFT BEING MADE BY AIRCRAFT? ☐ YES ☐ NO (IF YES, REFER TO APPENDIX C OF POWER'S CRANE POLICY)

SUBMITTAL

SUBMITTED BY: _____

REVIEWED BY (POWER REPRESENTATIVE): _____

DATE: _____   *THIS FORM DOES NOT REPLACE POWER'S MOBILE CRANE CHECKLIST*
A SEPARATE MCCL NEEDS TO BE COMPLETED WHEN THE CRANE ARRIVES ON SITE

Critical Lift Planning Worksheet   Revised October 2012

Figure A01B Critical Lift Planning Worksheet (2 of 2)

Trade Terms Introduced in This Module

Avoidance zone: An area both above and below one or more power lines that is defined by the outer perimeter of the prohibited zone. As the name implies, any part of the crane should avoid this area whenever possible, and may not enter the area except under special circumstances.

Competent person: As defined by OSHA, an individual who is capable of identifying existing and predictable hazards in the surroundings or working conditions which are unsanitary, hazardous, or dangerous to employees, and who has the authorization to take prompt corrective measures to eliminate such hazards.

Critical lift: As defined in ASME Standard B30.5, a hoisting or lifting operation that has been determined to present an increased level of risk beyond normal lifting activities. For example, increased risk may relate to personnel injury, damage to property, interruption of plant production, delays in schedule, release of hazards to the environment, or other significant factors.

High-voltage proximity warning device: An early-warning device that senses the electric fields created by high-voltage power lines and alerts the crane operator and/or the lift team to the hazard.

Insulating link: An electrical insulating device used on the crane hook to protect workers in contact with the load from the danger of electrocution in the event the crane contacts a powerline. The link can also provide some level of protection for the crane if the load alone contacts a power line.

Minimum clearance distance: The OSHA-required distance that cranes, load lines, and loads must maintain from energized power lines. This OSHA term is synonymous with the ASME term prohibited zone.

Prohibited zone: An area of specific dimensions, based on the voltage of a power line(s) that no part of the crane is allowed to enter during normal operations. Special considerations and preparations are required if the crane's task must place any part of it within the prohibited zone. The prohibited zone is a term used by ASME that is synonymous with the term minimum clearance distance used by OSHA.

Recloser: A device that functions much like a circuit breaker, or in conjunction with a circuit breaker, in power distribution and transmission systems that automatically recloses the circuit after a fault has been detected and the circuit has been opened. Reclosers allow the power system to be re-energized quickly after a transient (temporary) condition, such as a tree limb falling across power lines and then falling to the ground, has occurred. If the fault reoccurs upon closure, the circuit will typically remain open until the situation has been addressed by power line workers or operators.

Shock loading: A sudden, dramatically increased load imposed on a crane and rigging, usually as the result of momentum from the load that occurs due to swinging side-to-side, dropping the load and then stopping it suddenly, and similar actions that create momentum.

Standards: As defined by OSHA, statements that require conditions, or the adoption or use of one or more practices, means, methods, operations, or processes, that are reasonably necessary or appropriate to provide safe or healthful employment and places of employment. Standards developed by some organizations are voluntary in nature, while OSHA standards and those they incorporate by reference are enforceable by law.

Standard lift: A lift that can be accomplished through standard procedures, allowing load-handling and lift team personnel to execute it using common methods, materials, and equipment.

Additional Resources

This module is intended as a thorough resource for task training. The following reference works are suggested for further study.

ASME Standard B30.5, Mobile and Locomotive Cranes. Current edition. New York, NY: American Society of Mechanical Engineers.

ASME Standard B30.20, Below-The-Hook Lifting Devices. Current edition. New York, NY: American Society of Mechanical Engineers.

ASME Standard B30.23, Personnel Lifting Systems. Current edition. New York, NY: American Society of Mechanical Engineers.

ASME Standard P30.1, Planning for Load Handling Activities. Current edition. New York, NY: American Society of Mechanical Engineers.

29 CFR 1926, Subpart C, **www.ecfr.gov**

29 CFR 1926.251, **www.ecfr.gov**

29 CFR 1926.753, **www.ecfr.gov**

Mobile Crane Safety Manual (AEM MC-1407). 2014. Milwaukee, WI: Association of Equipment Manufacturers.

The following websites offer resources for products and training:

American National Standards Institute (ANSI), **www.ansi.org**

The American Society of Mechanical Engineers (ASME), **www.asme.org**

Occupational Safety and Health Administration (OSHA), **www.osha.gov**

North American Crane Bureau, Inc., **www.cranesafe.com**

Electronic Code of Federal Regulations, **www.ecfr.gov**

Figure Credits

Link-Belt Construction Equipment Company, Module opener, Figures 4, 6, Table 2

© Gary Whitton/Shutterstock.com, Figure 1

Carolina Bridge Co., Figures 2, 3

Mechanix Wear, SA01

© Donvictorio/Dreamstime.com, Figure 7

Salisbury Electrical Safety, Figure 10

Insulatus Company, Inc., Figure 11

Atlas Polar Company Ltd., Figure 12

Topaz Publications, Inc., Figures 13, 20, SA02

Lifting Technologies, Figures 17, 18

Courtesy of Amerex Corp., Figure 21

Power Construction Company, LLC, Appendix

Section Review Answer Key

Answer	Section Reference	Objective
Section One		
1. c	1.1.0	1a
2. b	1.2.2	1b
Section Two		
1. c	2.1.0	2a
2. a	2.2.0; Table 1	2b
3. d	2.3.1	2c
4. b	2.4.2	2d
5. b	2.4.6	2d

NCCER CURRICULA — USER UPDATE

NCCER makes every effort to keep its textbooks up-to-date and free of technical errors. We appreciate your help in this process. If you find an error, a typographical mistake, or an inaccuracy in NCCER's curricula, please fill out this form (or a photocopy), or complete the online form at **www.nccer.org/olf**. Be sure to include the exact module ID number, page number, a detailed description, and your recommended correction. Your input will be brought to the attention of the Authoring Team. Thank you for your assistance.

Instructors – If you have an idea for improving this textbook, or have found that additional materials were necessary to teach this module effectively, please let us know so that we may present your suggestions to the Authoring Team.

NCCER Product Development and Revision
13614 Progress Blvd., Alachua, FL 32615

Email: curriculum@nccer.org
Online: www.nccer.org/olf

❏ Trainee Guide   ❏ Lesson Plans   ❏ Exam   ❏ PowerPoints   Other _____

Craft / Level: _____   Copyright Date: _____

Module ID Number / Title: _____

Section Number(s): _____

Description: _____

_____

_____

_____

_____

Recommended Correction: _____

_____

_____

_____

_____

Your Name: _____

Address: _____

_____

Email: _____   Phone: _____

21102-18

Basic Principles of Cranes

OVERVIEW

Mobile cranes range from small, simple models to large, mechanically and electrically complex pieces of equipment. In addition, cranes that appear similar often differ in their operation and configuration. A fundamental understanding of crane types and how they move from place to place is important to riggers and signal persons as well as to crane operators. Cranes also have different types of lifting booms, each of which has its own strengths and weaknesses. This module presents specific features of various cranes and booms, and introduces the basic principles of lifting and leverage.

Module Eleven

Trainees with successful module completions may be eligible for credentialing through the NCCER Registry. To learn more, go to **www.nccer.org** or contact us at 1.888.622.3720. Our website has information on the latest product releases and training, as well as online versions of our *Cornerstone* magazine and Pearson's product catalog.

Your feedback is welcome. You may email your comments to **curriculum@nccer.org**, send general comments and inquiries to **info@nccer.org**, or fill in the User Update form at the back of this module.

This information is general in nature and intended for training purposes only. Actual performance of activities described in this manual requires compliance with all applicable operating, service, maintenance, and safety procedures under the direction of qualified personnel. References in this manual to patented or proprietary devices do not constitute a recommendation of their use.

Objectives

When you have completed this module, you will be able to do the following:

1. Identify and describe various types of cranes and crane components.
   a. Identify and describe mobile cranes based on their means of travel.
   b. Identify and describe various types of crane booms.
   c. Identify and describe common crane attachments and accessories.
   d. Describe common crane instrumentation and safety devices.
   e. Identify and describe various crane reeving patterns.
2. Identify factors related to lifting capacity and explain their significance.
   a. Explain the significance of ground conditions and a level surface.
   b. Describe the bearing surface and explain how to determine the required blocking.
   c. Define and describe the significance of the center of gravity and the quadrants of operation.
   d. Describe the significance of boom length, angle, operating radius, and elevation.
   e. Explain how to use a load chart and understand the basic concepts of critical lifts.

Performance Tasks

Under the supervision of your instructor, you should be able to do the following:

1. Verify the boom length of a telescopic- and/or lattice-boom crane using manufacturer's data or a measuring tape.
2. Measure the operating radius of a telescopic- and/or lattice-boom crane using a measuring tape.
3. Calculate the amount of blocking needed for the outrigger of a specific crane.
4. Verify that a crane is level.

Trade Terms

Anti-two-blocking device	Duty cycle	Jib backstay	Parts of line
Backfill	Dynamic loads	Jib forestay	Pendants
Base mounting	Effective weight	Jib mast	Quadrant of operation
Base section	Floats	Lattice boom	Reach
Block and tackle	Grapples	Leads	Reeving
Blocking	Gross capacity	Leverage	Ring gear drive
Boom torque	Hardpan	Load moment	Sheave
Carbody	Headache ball	Lowboy	Swallow
Carrier	Hoist drum	Luffing	Tipping fulcrum
Center of gravity (CG)	Hoist reeving	Luffing jib	Upperworks
Check valve	Hydraulic motors	Net capacity	Wheelbase
Counterweights	Idlers	Non-ferrous	Whip line
Crane mat	Impact loads	Open-throat boom	
Crawler frames	Interpolation	Operating radius	
Critical lift	Jib	Outriggers	

Industry Recognized Credentials

If you are training through an NCCER-accredited sponsor, you may be eligible for credentials from NCCER's Registry. The ID number for this module is 21102-18. Note that this module may have been used in other NCCER curricula and may apply to other level completions. Contact NCCER's Registry at 888.622.3720 or go to **www.nccer.org** for more information.

Contents

Figures and Tables

1.0.0 INTRODUCTION TO MOBILE CRANES

Objective

Identify and describe various types of cranes and crane components.

    a. Identify and describe mobile cranes based on their means of travel.
    b. Identify and describe various types of crane booms.
    c. Identify and describe common crane attachments and accessories.
    d. Describe common crane instrumentation and safety devices.
    e. Identify and describe various crane reeving patterns.

Trade Terms

Anti-two-blocking device: Two-blocking refers to a condition in which the lower load block or hook assembly comes in contact with the boom tip, boom tip sheave assembly or any other component above it as it is being raised. If this occurs, continuing to apply lifting power to the cable can result is serious equipment damage and/or failure of the hoist line. An anti-two-blocking device, therefore, prevents this condition from occurring.

Base mounting: A crawler crane assembly consisting primarily of the carbody, ring gear drive, crawler frames, and tracks.

Base section: The lowest portion of a telescopic boom that houses the other telescopic sections but does not extend.

Block and tackle: A system of two or more pulleys, which form a block, with a rope or cable threaded between them, reducing the force needed to lift or pull heavy loads.

Blocking: Wood or a similar material used under outrigger floats to support and distribute loads to the ground. Also referred to as *cribbing*.

Boom torque: A twisting force applied to the crane boom, typically resulting from imbalanced reeving of the boom tip sheave assembly ropes.

Carbody: The part of a crawler-crane base mounting that carries the rotating upperworks.

Carrier: The base of a wheeled crane that provides crane movement and supports the upperworks.

Check valve: A valve designed to allow flow in one direction but closes as necessary to prevent flow reversal.

Counterweights: Weights added to the crane, usually on the end opposite the boom, to help counter the weight of the load and improve stability.

Crane mat: A portable platform, typically made of large wooden timbers bolted together, used to support and spread the weight of a crane over a larger ground area.

Crawler frames: Crane assemblies comprised of the crawler tracks, track idlers, and track power sources of a crawler crane. Also called *tread members* or *track assemblies*.

Duty cycle: An expression of equipment use over time. In the case of mobile cranes, an 8-, 16-, or 24-hour rating expressed as a percentage.

Grapples: Devices used to pick up bulk items, containers, rocks, trees and tree limbs, etc. Grapples typically have several jaws that operate like fingers to pick up material, using mechanical or hydraulic power.

Headache ball: A heavy round weight often attached to a load line to provide sufficient weight to allow the load line to unspool from the drum when there is no live load. Larger versions of headache balls are used to swing into structures to demolish them.

Hoist drum: A drum is a cylindrical component around which a rope is wound. The hoist drum is used to wind or unwind the rope for hoisting or lowering the load; the part of a crane that spools and unspools the lifting line.

Hoist reeving: The reeving pattern applied to the hoist sheaves. Single- or multiple-line hoist reeving is used for whip, boom, and jib lines.

Telescopic boom: A crane boom that extends and retracts in sections that slide in and out, powered by hydraulic pressure.

Hydraulic motors: Motors powered by hydraulic pressure provided by an external pump. Hydraulic motors are often used to power the tracks of crawler cranes, instead of complex drive systems connected directly to the diesel engine.

Idlers: Pulleys, wheels, or rollers that do not transmit power, but guide or place tension on a belt or crawler-crane track.

Jibs: Extensions attached to the boom point to provide added boom length for reaching and lifting loads. Jibs may be in line with the boom, offset to another angle, or adjustable to a variety of angles. A jib is sometimes referred to as a *fly*.

Jib backstay: A piece of standing rigging that is routed from the jib mast back to the main boom to help support the jib.

Jib forestay: A piece of standing rigging that is routed from the far tip of the jib back to the jib mast, holding the tip of the jib up.

Jib mast: A structure mounted on the main boom that provides a fixed distance for the point of connection of the jib forestay and jib backstay. Also referred to as a *jib strut*.

Lattice boom: A boom constructed of steel angles or tubing to create a relatively lightweight but strong, rigid structure.

Leads: Steel structures that provide support for a pile hammer and help to align and position the hammer with the pile to be driven. The hammer can travel up or down in the leads as necessary.

Load moment: The force applied to the crane by the load; the leverage of the load, opposing the leverage of the crane. The load moment is calculated by multiplying the gross load weight by the horizontal distance from the tipping fulcrum to the center of gravity of the suspended load. The load moment is usually reported to the operator as a percentage of the crane's capacity at the present set of conditions. As those conditions change, such as the boom angle, the load moment changes as well.

Lowboy: A trailer with a low frame for transporting very tall or heavy loads. A typical lowboy has two drops in deck height: one right after the gooseneck connecting it to the tractor, and one right before the wheels. This allows the trailer deck to be extremely low compared with common trailers.

Luffing: Changing a boom angle by varying the length of the suspension ropes.

Luffing jib: A jib mounted on the end of a boom that can be positioned at different angles relative to the main boom.

Non-ferrous: Having no iron. Ferrous metals, such as steel, contain iron and are magnetic as a result.

Open-throat boom: A lattice boom with an opening in the boom structure near the far end, allowing the hoist lines to drop through the boom rather than over the end of the boom.

Outriggers: Extendable or fixed members attached to a crane base that rest on ground supports at the outer end to stabilize and support the crane.

Parts of line: When a line is reeved more than once, the resulting number of lines that are supporting the load block.

Pendants: Ropes or strands of a specified length with fixed end connections, used to support a lattice boom or boom components. According to 29 *CFR* 1926.1401, a pendant may also consist of a solid bar.

Reach: The combined operating height and radius of a boom, or the combination of boom and jib.

Reeving: A method often used to multiply the pulling or lifting capability by using wire rope routed through multiple pulleys or sheaves a number of times.

Ring gear drive: Sometimes referred to as the swing circle. An assembly that provides the point of attachment and pivot point for the upperworks of a crane. The ring gear is typically driven by hydraulic pressure, allowing the upperworks to rotate on a set of bearings that reduce friction and transfer the weight of the upperworks (and any load) to the carbody.

Sheaves: Wheels that have a groove for a belt, rope, or cable to run in. The terms *sheave* and *pulley* are often used interchangeably.

Swallow: The space between the sheave and the frame of a block, through which the rope is passed.

Upperworks: A term that refers to the assembly of components above the ring gear drive; the rotating collection of components on top of the base mounting or carrier; may also be referred to as the *house*, or as the *superstructure* as defined in 29 *CFR* 1926.1401.

Wheelbase: The distance between the front and rear axles of a vehicle.

Whip line: A secondary hoisting rope usually of lower capacity than that provided by the main hoisting system. When a whip line exists, it is typically out at the tip of a jib, while the main hoist line is closer to the crane and operated from the tip of the main boom.

This section describes the capabilities and limitations of various crane types used in construction and related applications. Jobsite environmental conditions and how these conditions determine the loads that can be safely lifted are also discussed. Even cranes of the same basic type differ greatly from one manufacturer to another, so operators must always be prepared to familiarize themselves with the information supplied by the manufacturer for the specific crane in use.

Crane selection for a particular task depends upon the technical requirements of crane capacity, reach, site clearances, and site conditions. Economic factors may also dictate the type of crane to be used. The most significant differences between mobile cranes revolve around two distinct characteristics:

• How they travel from one place to another
• The type of boom equipped

1.1.0 Approaches to Travel

As the name implies, mobile cranes are those that have a means of propulsion. A crane does not need to be fast, nor does it need to be drivable on the highway to be considered mobile. Although many mobile cranes can be driven on the highway, the mobility implied by their name is more about their ability to move around in general. The vast majority of mobile cranes are powered by diesel engines due to the amount of torque they develop and their durability. Diesel engines are especially suited for mobile cranes, just as they are for large trucks. A variety of crane types, based on their means of movement, are presented in the sections that follow.

1.1.1 Crawler Cranes

Crawler cranes are available in a variety of configurations. *Figure 1* shows a typical crawler crane with a lattice boom. Crawler cranes are also available with a telescopic boom. Crawler cranes are so named because they move slowly on tracks that provide superior traction and stability. The track assemblies are called crawler frames or *tread members*. The hoist drum and ring gear drive in this model are powered by hydraulic motors that provide very precise control. The unit shown has a diesel engine driving hydraulic pumps that in turn drive the hydraulic motors. The hoist drum manages the wire rope used for lifting, and the ring gear drive rotates and positions the upper portion of the crane.

The motors that power the tracks, placing the crane in motion, are also hydraulic (*Figure 2*). A number of hydraulic pumps often exist in a single crane. The counterweights can be lowered all the way to the ground using hydraulic actuators. The operator cab shown here can tilt up to 20 degrees to provide better visibility of the boom and/or load. The tracks of very large crawler cranes may be powered by a drive shaft from the engine instead.

The lifting and rotating assembly, called the upperworks, is mounted on top of the crawler assembly. The complete assembly beneath the

(A) CRAWLER CRANE

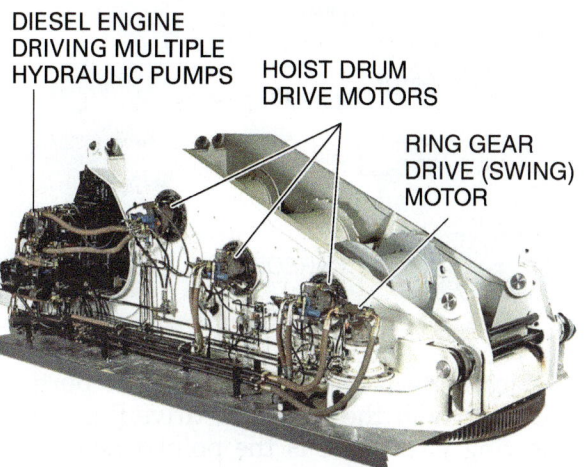

(B) HYDRAULIC HOIST DRUM AND RING GEAR POWER UNIT

Figure 1 Typical crawler crane.

upperworks is referred to as the base mounting. The base mounting of the crawler crane transmits loads imposed on the upperworks down to the ground. To accomplish this effectively, the base needs to be extremely stiff. The base is stiffened by using complex castings or heavy weldments. The base mounting, shown in *Figure 3*, consists primarily of the carbody, ring gear drive or swing circle, crawler frames, tracks, and the propelling mechanism. Note that the base of a crawler crane should not be called a carrier. (Only wheeled crane bases are called *carriers*.)

HYDRAULIC MOTOR

Figure 2 Hydraulic drive motor.

UPPER MECHANISMS
MOUNT ONTO THE
RING GEAR DRIVE

RIGHT
CRAWLER
FRAME

LEFT
CRAWLER
FRAME

CARBODY

Figure 3 Crawler base mounting.

A carbody is the central or main portion of the base, with the ring gear drive fixed on top. The ring gear drive is the point where the upperworks is joined to the base mounting. The ring gear drive transmits the loads imposed on the upperworks to the base, and must also serve as a low-resistance surface for the swinging motion. Crawler frames are mounted on each side of the carbody. The crawler frames hold the crawler track motors, idlers, and crawler tracks (*Figure 4*).

Most crawler cranes are powered by an engine located in the upperworks. Older cranes use a mechanical system of drive shafts to couple power from the engine to the crawler tracks. Newer machines have hydraulic motors mounted directly to the crawler frames to propel the unit. Instead of complex drive arrangements, hydraulic tubing from the pump(s) is routed to the crawler motors. However, some larger machines have separate engines mounted on the base mounting

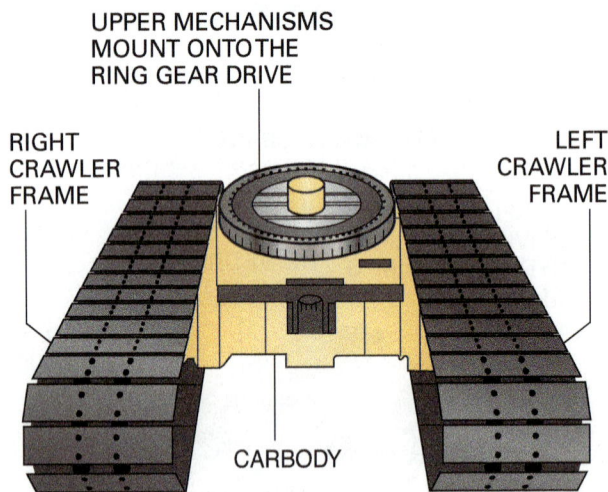

IDLERS

Figure 4 Crawler track and track idlers.

to provide power to the crawler tracks via mechanical drive shafts.

The width between the crawler tracks affects the stability of the crawler crane. The greater the width between the tracks, the greater the lifting capacity and stability. However, transporting cranes with widely spaced tracks can be difficult. This has led to the development of crawlers with extendable/retractable crawler frames. The cranes are transported while the tracks are retracted, making the crane as narrow as possible. When they arrive at the site, the tracks are extended to their full width to achieve increased stability. Most crawler frames are designed to extend or retract without assistance, while others require jacks, blocking, and other equipment to get the job done. Others may be equipped with removable crawler frames to further reduce the crane width and allow easier transportation by trailer (*Figure 5*). Lowboy trailers and railcars are commonly used to transport these cranes. Crawler cranes are definitely not designed to travel on the open road under their own power.

When comparing crawler cranes to other cranes with equal lifting capacity, crawler cranes generally offer lower rental rates than other cranes, but their transit and erection costs are higher. If the nature of the work and the ground conditions make it necessary to operate a machine from a crane mat and/or from a number of different positions, siting costs must also be considered. The cost of putting a truck crane in place is usually far less than that of placing and leveling a supportive mat for a crawler. However, when power and endurance are critical characteristics, a crawler crane is the best choice.

Figure 5 Crawler crane disassembled and loaded for transport.

1.1.2 Wheeled Truck Cranes

Wheeled truck cranes are available in configurations similar to crawler cranes; they are available with both lattice and telescopic booms. A truck crane carrier is analogous to a crawler crane's base mounting, and should not be confused with an ordinary commercial truck chassis. Truck crane carriers are designed and manufactured with a completely different priority than commercial trucks. Some of the larger units may have nine axles or more.

Truck crane carriers may be designed for moving around the job site only, traveling on the highway and smooth ground, or both. To accommodate these extremes of travel in a single crane, transmissions with more than 30 forward gears, as well as special creeping gears, are available. The creeping gears are used for very slow movement at the job site.

Highway speeds can range from 35 to 70 mph (48 to 113 kph). Job site ramps up to approximately 40 percent grade can generally be climbed by a wheeled truck crane with no load. Brakes are designed to hold position on similar grades. *Figure 6* and *Figure 7* show typical truck cranes. Note that some large truck cranes must have their counterweights removed before they are driven on the highway due to handling and highway-loading concerns.

Some wheeled truck cranes that are designed to be superior in off-road situations. Known as rough-terrain cranes (*Figure 8*), they are often found on construction sites due to their versatility, maneuverability, and ability to handle the

Figure 6 Telescopic-boom truck crane with rotating controls.

Figure 7 Lattice-boom truck crane with rotating controls.

Figure 8 Rough-terrain crane with rotating controls.

terrain. Rough-terrain cranes are usually dual-axle carriers with telescopic booms and a relatively short wheelbase. They typically have larger tires and more versatile steering capabilities than common wheeled truck cranes. Steering options include both two-wheel and all-wheel steering capabilities. Two-wheel and all-wheel drive capabilities are also available. Having only two axles and a short wheelbase allows the crane to turn sharply and take full advantage of the four-wheel steering capability.

Many rough-terrain cranes have a single cab. The crane is both driven and operated from the same cab. The cab may swing with the boom (a crane with rotating controls, referred to as a *swing cab*) or remain fixed (a crane with fixed or stationary controls, referred to as a *fixed cab*). Rough-terrain cranes generally travel at less than highway speeds (about 30 mph, or 48 kph) due to their gear ratios, and their short wheelbase delivers a poor highway ride. As a result, they are usually transported to the job site on low-boy trailers. Heavier lifts are made over the front of the crane, as that is its position of greatest lift capacity.

There are two small crane types that might be considered a subset of rough-terrain cranes. Pick-and-carry cranes are often based on rough-terrain designs. A true pick-and-carry crane, also called a *cherry picker*, does not have outriggers, as it is designed to lift a load and travel (at low speeds) with it suspended. However, they are generally designed for highway travel from site to site. Another small crane type is the carry-deck crane, which is considered an American version of the pick-and-carry crane. The carry-deck crane has a small deck that can accommodate a load as it is moved from one place to another. While some are suitable for highway driving, others are not. They can be designed with rough-terrain

characteristics, or be designed for shipping/receiving, fabrication shop, or similar duties where rough terrain characteristics are not important.

There is one additional wheeled truck crane type that fills the void between standard models that excel on the highway and those that excel in rough territory—the all-terrain crane (*Figure 9*). All-terrain cranes can be considered a hybrid between standard wheeled truck cranes and rough-terrain cranes. This type of crane is quite capable of highway travel. At the same time, it is also designed to traverse reasonably rough terrain, within limits.

All-terrain cranes typically have more axles and are available in greater capacities than rough-terrain models. For example, the largest rough-terrain crane presently offered by one company has a nominal capacity of 135 tons (122 metric tons); the largest all-terrain model offered by the same company is rated at 450 tons (408 metric tons). All-terrain cranes are not generally as maneuverable in tight quarters as rough-terrain models because their length and additional axles limit maneuverability. They also may not be equipped with the same steering options that are available on rough-terrain models. However, they do offer performance on the highway as well as on a rough job site, filling the needs of many users. All-terrain cranes generally set-up quick and have long boom capabilities with additional jibs for added reach. These features make them a favorite with crane rental services.

As a general rule, wheeled truck cranes do not offer the stability of a crawler crane. As a result, the vast majority of them are equipped with outriggers. In spite of its ability to maneuver in challenging terrain, no mobile crane is designed to be used on unstable ground or when it is not level. Outriggers are designed to provide stability and a means to level the crane. Lift capacity and safety are both increased as a result. The outriggers should remove the weight of the crane and the lifted load from the wheels and tires altogether. *Figure 10* shows the two most common types of hydraulic outriggers. Cantilever outriggers are used on smaller cranes, since their reach is limited; larger cranes require outriggers of greater strength and able to provide a wider stance. Boom trucks and other small cranes may have outriggers that must be deployed and positioned manually.

1.1.3 Railcar-Mounted Cranes

Railcar-mounted cranes (*Figure 11*) can also have either a telescopic or a lattice boom. The two types of booms have different weight capacities and lift heights. Like lattice-boom crawler cranes, the lattice boom railcar-mounted crane has greater

Figure 9 All-terrain crane.

weight capacity and height potential than a telescopic-boom version. However, the telescopic boom offers greater versatility and flexibility, as well as quick setup, when that is an important consideration. Self-propelled railcar-mounted cranes are used primarily for railroad track and bridge repair, railyard maintenance, loading and unloading rail freight, and to clear train derailments.

1.2.0 Crane Booms

Cranes not only differ by means of travel; they also differ in the types of booms they employ. In addition, there are special accessories that can be attached to, or suspended under, the crane boom.

1.2.1 Telescopic Booms

Telescopic-boom cranes are among the most widely used types of mobile cranes. This is partially because of their ease of setup and teardown. Hydraulic power for a wide variety of applications expanded dramatically in the 1950s, making the telescopic boom possible. The simplest telescopic-boom crane is the boom truck (*Figure 12*). Boom trucks are commonly used to transport and deliver or pick up materials. The boom is operated from outside the cab, in a standing position. Note that the deck area of the crane can be used to move materials from one place to another, and then be unloaded by the crane.

Telescopic booms are sectional, as shown in *Figure 13*. The crane shown here has a total of four boom sections, one of which serves as the

housing for the other three. A headache ball is often attached to the load line to provide sufficient weight to allow the load line to unspool from the drum when there is no live load. Telescopic booms allow the operator to change the length of the boom at any time by extending or retracting the boom sections, even while the crane is loaded. Hydraulic actuators, often hidden from view, extend and retract the individual sections. The base section—the lower portion of the boom that does not extend—is raised up and down by one or more hydraulic actuators. Operators must be very careful when extending the boom of a crane, as the crane's capacity changes dynamically as the boom length changes. This increases the possibility of the crane tipping over.

1.2.2 Lattice Booms

Lattice booms can be much longer than most telescopic booms. However, lattice-boom cranes are more time-consuming to prepare for a lift, and to prepare for travel. The site preparation and erection of lattice-boom cranes may take anywhere from a day for a smaller crane to possibly weeks for very large cranes. The length of setup time depends on the length of the boom and any attachments to be used.

Newer lattice-boom cranes are powered by computer-controlled systems. Older models are very mechanical in their control, using complex linkage. Computer-controlled models have redundant fail-safe devices to assure operational safety. These devices consist of many components, including: automatic braking systems,

(A) CROSSBEAM OUTRIGGERS

(B) CANTILEVER OUTRIGGERS

Figure 10 Hydraulic outriggers.

load-moment measuring devices, mechanical-system monitoring devices, and function-lockout systems that stop operation when a system fault occurs. Such controls provide for better safety, control, and accuracy than older friction-operated models.

As noted earlier, lattice booms are available on both crawler and wheeled truck cranes. Lattice-boom crawler cranes are specifically designed for heavy-duty service. The reliability and versatility of the lattice-boom crawler crane makes it a widely applied crane design. Besides lifting heavy loads to great heights, these cranes are also ideal for applications with a high duty cycle. The duty cycle refers to the consistency and repetitive nature of the work over a period of time. *Figure 14* shows an example of a lattice-boom crawler crane. The load block shown connects and moves up and down with the load, riding on the crane reeving.

Lattice-boom truck cranes provide the mobility of a truck crane with the extreme lifting capacity of a lattice-boom crane. Depending on the size of the crane and its gross vehicle weight, some components such as boom sections, counterweights, and outriggers usually have to be removed before highway travel to meet local, state, and federal weight and physical size restrictions. *Figure 15* shows a typical lattice-boom truck crane. Many parts of the crane and boom are identified as well. The identified parts of the lattice-boom assembly are the same for crawler cranes.

There are a number of possible configurations for lattice booms, as shown in *Figure 16*. Lattice booms that are sectional (very common) begin with a boom base—the section that attaches to the upperworks of the crane. Additional sections can be added to increase boom length. A lattice boom is supported by a network of wire ropes or lines called pendants. The boom hoist reeving and boom pendants are used to raise and lower the boom. If the crane is rigged with a jib, it is supported by a jib forestay and a jib backstay. Note that a jib may also be referred to by some in the industry as a *fly*. A luffing jib is raised and lowered with jib hoist reeving and luffing jib pendants, as shown in *Figure 15*. The process of adjusting the jib angle is called luffing. Both fixed

Inventor of the Hydraulic Crane

Sir William Armstrong (1810–1900) is generally credited with development of the first hydraulic crane. A scientist and inventor, he was responsible for the first home in England to be powered by hydroelectricity. He is also considered the father of modern artillery.

Armstrong was inspired to create a hydraulic crane (in this case, the powering fluid was water) as part of a water-piping project to move water from reservoirs to distant homes in Newcastle. The water pressure was more than needed in areas of lower elevation. He proposed to use the excess water pressure to power a crane for unloading ships more effectively. The experiment proved extremely successful and more cranes were added to further increase the efficiency of the harbor.

Figure 11 Railcar-mounted crane working in the yard.

CRANE CONTROLS

Figure 12 Telescopic-boom truck with stand-up control.

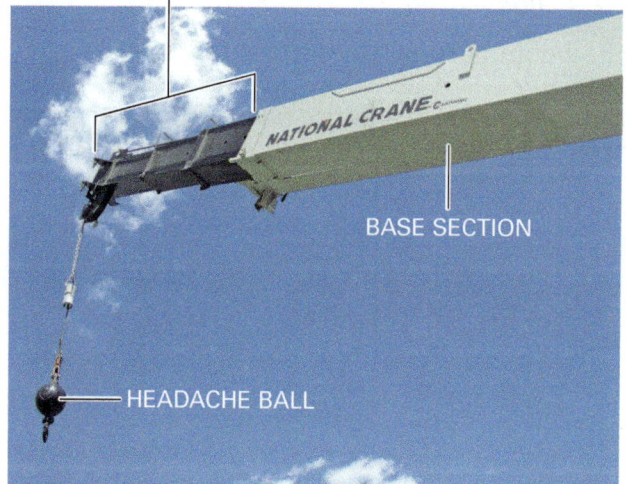

EXTENDABLE SECTIONS

BASE SECTION

HEADACHE BALL

Figure 13 Sections of a telescopic boom.

and luffing jibs require a jib mast, or jib strut, to support the forestay/backstay or the jib hoist reeving. The load hoist line and its reeving are used to actually lift the load. Note that some of the captions shown in *Figure 16* refer to an open-throat boom. This refers to an opening in the boom structure near the tip. The boom tip is open on the bottom side and is designed to provide

clearance for multi-part reevings at a high boom angle when the main sheaves are positioned along the centerline of the boom. When the main sheaves are offset below the center line of the boom, the need for an open-throat boom is eliminated. Many new lattice booms manufactured at present are built with the main sheaves offset to eliminate the open-throat design.

LOAD
BLOCK

Figure 14 Lattice-boom crawler crane.

1.2.3 Jibs

The boom is sometimes extended by adding a jib (*Figure 17*). A jib can be added to either a telescopic boom or to a lattice boom. The jib may be fixed in position to the boom, or it may be a luffing jib. A luffing jib allows the angle between the main boom and jib to be adjusted. An adjustable jib angle provides more flexibility in the lifting arrangement, allowing the crane to reach areas that may not be accessible otherwise. Standards require all jibs to have a mechanical stop that will prevent the jib from being positioned more than 5 degrees above the center line of the main boom.

The lifting capacity of a jib declines as the jib moves closer to a horizontal position. *Figure 18* demonstrates the reduction of capacity at different jib angles.

1.3.0 Crane Attachments

Various attachments have been developed to enable a crane to perform specific tasks beyond simple lifting. These attachments allow cranes to be used for tasks such as excavation, concrete pouring, scrap or debris removal, and loose-material lifting. Refer to *ASME Standard B30.20* for below-the-hook lifting device requirements.

1.3.1 Specialty Buckets

Cranes are often used to pick up and move materials such as dirt, concrete, or even water. Specialty buckets designed for these tasks include clamshell buckets, concrete buckets, and drag buckets.

Clamshell buckets (*Figure 19*) are hinged at the top center, allowing the bottom of the bucket to open and close like a clamshell. Digging or dredging clamshell buckets are built with counter weights, levers, and teeth that enable the bucket to open up and dig sharply into the ground as it is lowered. As it is lifted, the clamshell closes and

Duty-Cycle Work

You may hear the terms *duty-cycle crane* and *lift crane* used at times. A duty-cycle crane is generally considered to be a crane designed for long-term, repetitive work that may not approach the maximum lift capacity of a crane. Lift cranes need to be lightweight and maneuverable enough to get to where the load is. A duty-cycle crane generally works in a single area, perhaps for years at a time. Duty-cycle cranes generally have larger engines and dissipate heat more efficiently. Duty-cycle cranes also need faster hoist/fall speeds and hydraulic systems with individual pumps dedicated to a task. Lift cranes often use one or more hydraulic pumps that provide pressure to a manifold that supplies all of the cranes systems with pressure as needed. Many duty-cycle cranes are fixed in position, but mobile cranes are also needed for duty-cycle applications. The dredging activity shown here is a good example of duty-cycle work for a mobile crane.

Figure Credit: Liebherr USA, Co.

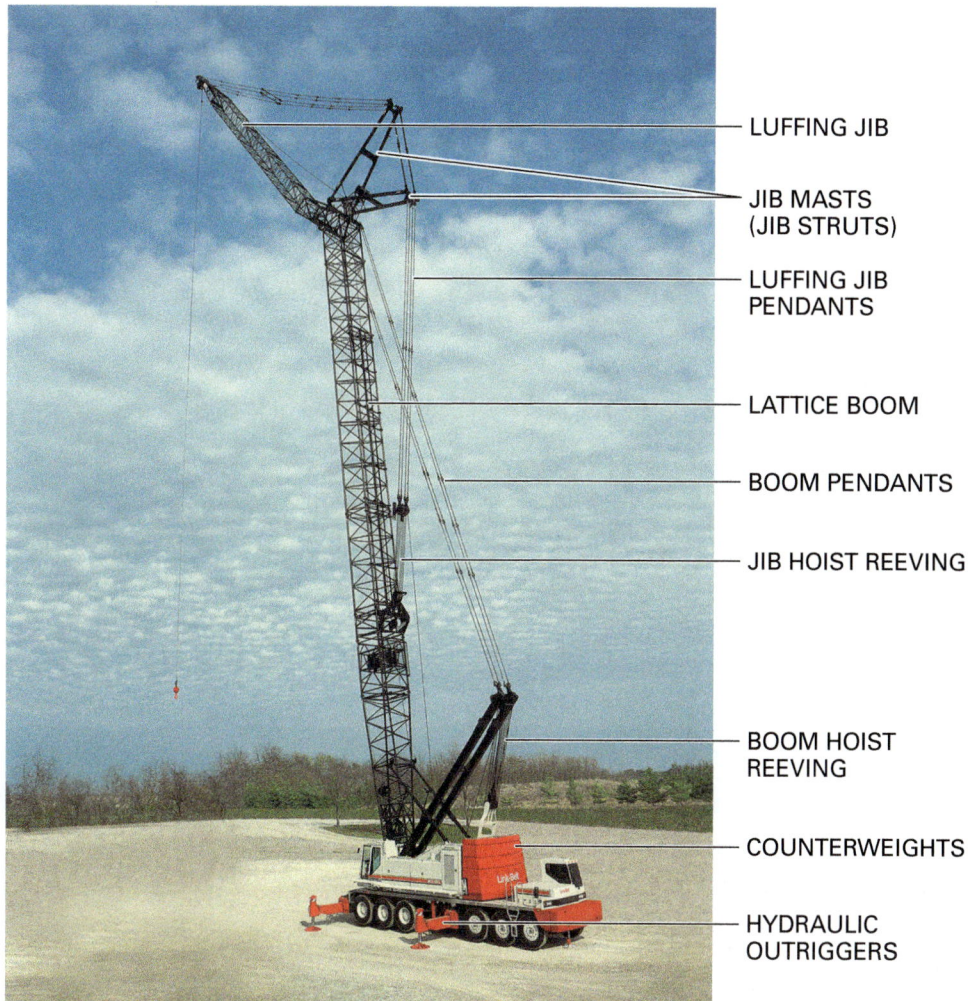

LUFFING JIB

JIB MASTS
(JIB STRUTS)

LUFFING JIB
PENDANTS

LATTICE BOOM

BOOM PENDANTS

JIB HOIST REEVING

BOOM HOIST
REEVING

COUNTERWEIGHTS

HYDRAULIC
OUTRIGGERS

Figure 15 Lattice-boom truck crane.

removes a bucket load of material. Rehandling clamshell buckets operate on the same principle as digging clamshell buckets but, instead of digging, are used to move free-flowing stockpile materials such as coal, fertilizer, or wood chips. Clamshell buckets for cranes are available with single, two-, and three-rope operating configurations to match the control capabilities of the crane.

Concrete buckets (*Figure 20*) are used to lift loads of cement up to a pour site. Concrete buckets may lift as much as 8 cubic yards (6 cubic meters) to workers well above ground level. They may be used to lift and dispense cement under the complete control of the crane operator, or the bucket may be controlled by another party on the site using a remote control.

Drag buckets are used in dredging, mining, or material-handling operations. They can also handle large loads of 8 cubic yards (6 cubic meters) or

more. Drag buckets are equipped with teeth and are pulled across a surface by an auxiliary winch on a crane. When full, they are lifted by the crane, and the contents are typically dispensed into a truck or railcar, or onto a barge.

Self-dumping bins are also popular for moving loose or dry, granular materials. The type shown in *Figure 21* is not controlled or tipped through a rope or cable. The crane operator merely sets the bin down on a firm surface and then moves the boom slightly to allow a latch to open on the bin lifting arm. As the bin is then lifted with the latch open, the contents are dumped in a controlled manner. The latch can be reset by again placing the bin on a firm surface and moving the boom slightly. This is a simple design that allows a crane operator to function more independently and efficiently.

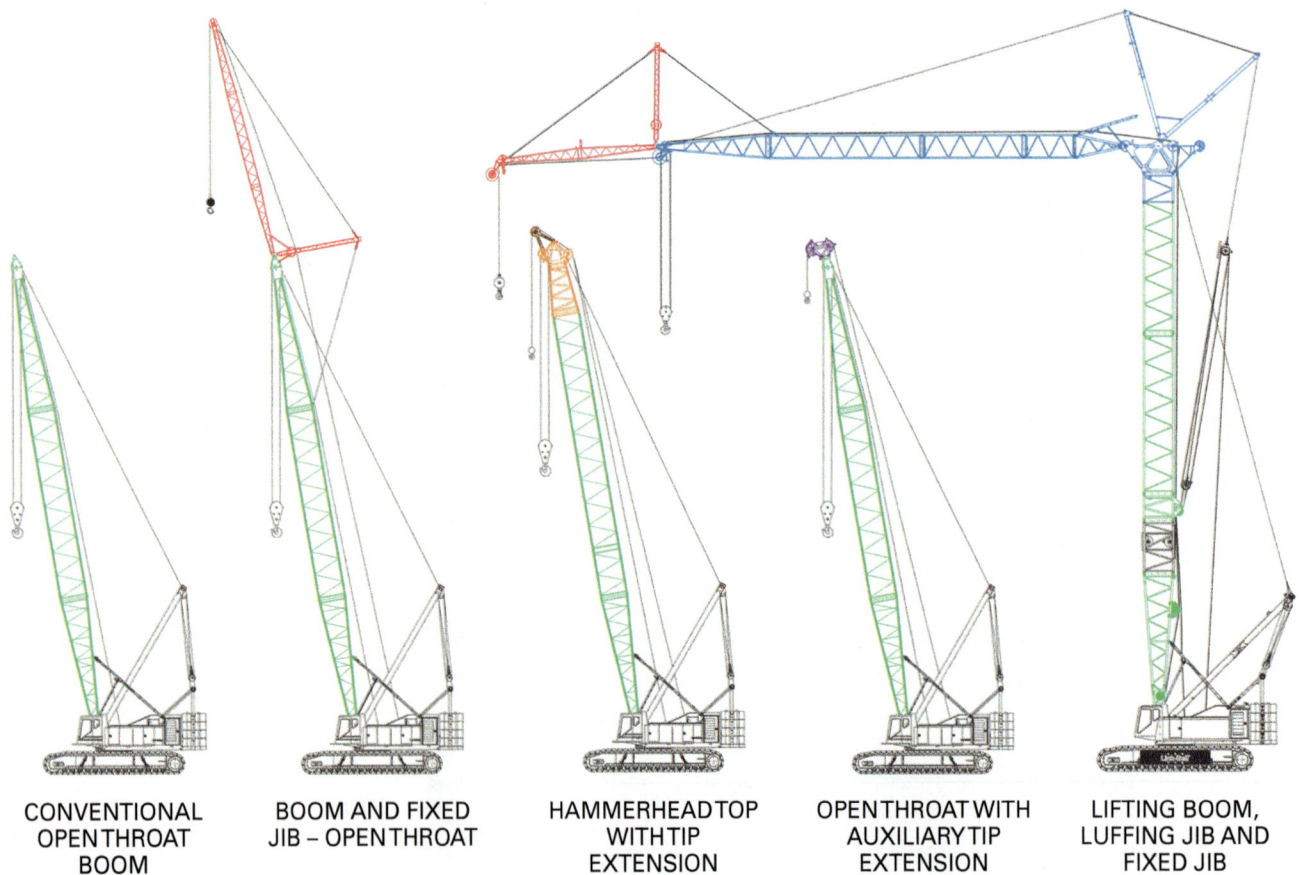

Figure 16 Examples of lattice-boom configurations on crawler cranes.

| CONVENTIONAL OPEN THROAT BOOM | BOOM AND FIXED JIB – OPEN THROAT | HAMMERHEAD TOP WITH TIP EXTENSION | OPEN THROAT WITH AUXILIARY TIP EXTENSION | LIFTING BOOM, LUFFING JIB AND FIXED JIB |

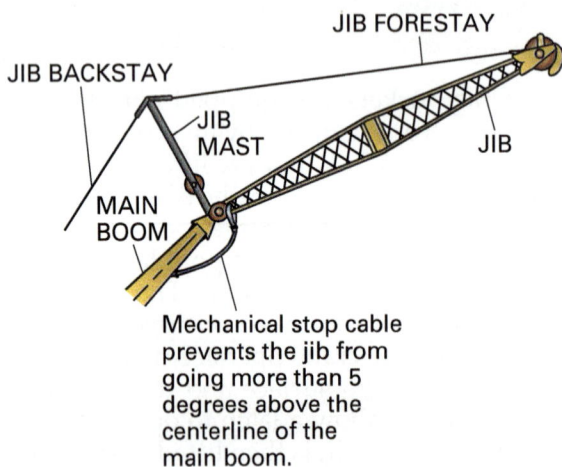

Figure 17 Typical jib.

1.3.2 Grapples and Magnets

Grapples and magnets are used to pick up and move materials that are irregular in size and shape such as scrap metal or tree branches. Grapples (*Figure 22*) usually have from two to six tines that interlock, acting much like fingers to grab material. The style of grapple shown here is sometimes referred to as an orange-peel grapple. Grapples can be specifically designed for scrap metal, rocks, logs, or brush.

Magnets are used to pick up magnetic metal objects. *Figure 23* shows an electromagnet designed for crane use. The crane operator is usually in control of the power supply to the electromagnet. The magnet is energized to pick up a load of metal, and de-energized to drop the load. Remember that magnets do not work on nonferrous materials such as aluminum, brass, and copper.

1.3.3 Special Lifting Devices

Cranes can be fitted with clamps and similar devices designed to easily lift and move specific and unique loads. Clamps can be attached to a crane to move stacks of bricks and blocks, concrete highway barriers, pipe, and many other objects. Product manufacturers will also build custom clamps and hooks for specific tasks. *Figure 24* shows several types of specialty accessories. Small clamping devices are also available, designed to clamp onto and lift sheet goods such as steel plate and sheet piles.

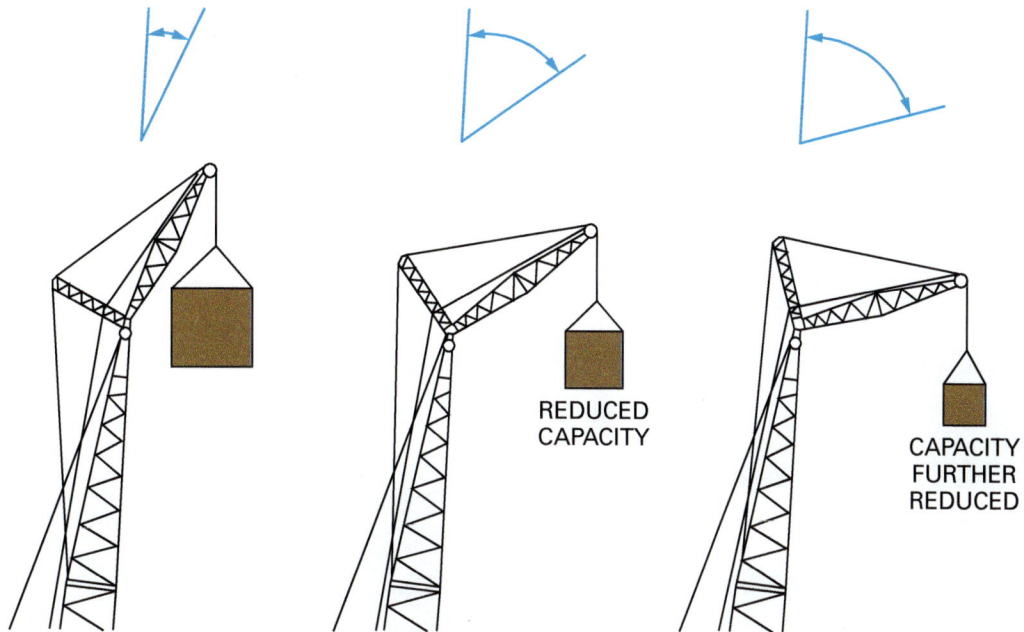

LIFTING CAPACITY DECLINES AS THE JIB IS MOVED CLOSER TO HORIZONTAL

Figure 18 Reduction of capacity as the jib is lowered.

The Spiderman of Cranes

Sometimes you need lifting capability in places where a crane cannot be driven, such as inside a building or many miles into the desert. This mini-crawler, with tracks tucked underneath the body, can get to the spot. Once in place, the spider-like legs are extended, lifting the rig off of its tracks and creating a stable lifting platform. Manufactured under the name of SpyderCrane®, these cranes are extremely versatile. Glaziers installing glass in tall buildings, for example, use special suction-cup accessories that allow the crane to pick up large sheets of glass and position them in the wall opening for installation from inside the building. There are also used for remote military and civil operations, where they can be flown in and dropped. There is no cab; they are remotely controlled.

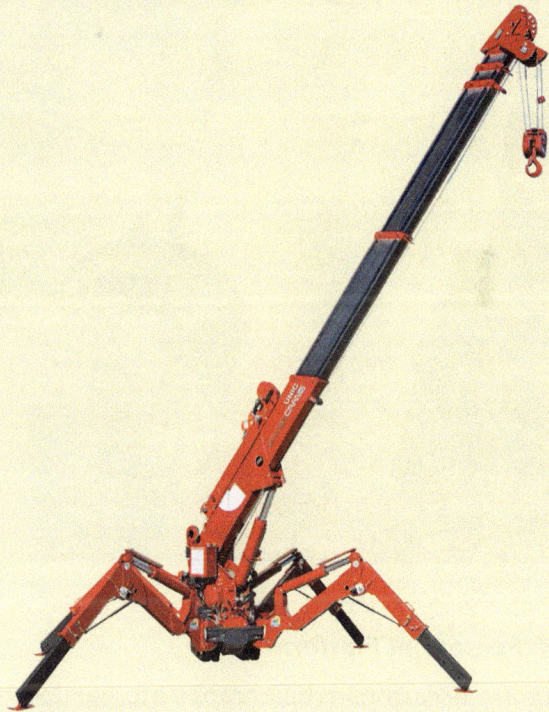

Figure Credit: Smiley Lifting Solutions - www.spydercrane.com

Figure 19 Clamshell bucket.

Figure 21 Self-dumping bin.

Figure 20 Concrete bucket.

1.3.4 Personnel Platforms

There are occasions when cranes are needed to lift personnel. 29 *CFR* 1926.1431 provides guidance in this area. The standard specifies that lifting by crane is only allowed if any other means to do so represents a greater hazard, or if there is no alternative. It also provides guidance in crane operation, the construction of the personnel platform, and other relevant areas. *ASME Standard B30.23, Personnel Lifting Systems,* is also dedicated to this topic.

Figure 22 Grapple.

Figure 25 shows a personnel platform that meets the OSHA and ASME requirements for a personnel platform. Platforms are available in a number of shapes and sizes. Some have no roof, while others have a fixed or removable roof section. Those with roof sections provide a greater margin of safety.

1.3.5 Pile Drivers

Cranes, especially lattice-boom crawler cranes, are an essential part of heavy pile-driving operations. Pile driving is heavy, repetitive work, imposing a great deal of wear and tear on the equipment. Lattice-boom crawler cranes are uniquely suited for the task.

There are two primary types of pile drivers typically managed by cranes: pile hammers and vibratory pile drivers. Pile hammers, either diesel-powered or hydraulic, represent the traditional pile-driving equipment class. They generally require a set of leads that support and position the hammer and pile, allowing the hammer to move freely up and down as the pile is driven. *Figure 26* shows a diesel-powered pile hammer and a pile in a set of swinging leads. Pile hammers like this

Figure 23 Crane electromagnet.

are almost exclusively handled by lattice boom crawler cranes.

Vibratory pile drivers use high-frequency vibration and their own weight to drive piles. They are best suited for driving sheet piles, but they can also drive concrete and wooden piles in favorable soil conditions, as shown in *Figure 27*. The vibration is created by an external hydraulic power source. Although lattice-boom crawler cranes are often used, vibratory units can be used with telescopic-boom cranes as well. A limited amount of vibration and stress is transmitted to the crane through the ropes.

1.3.6 Counterweights

The tipping force created by the weight of the boom and the lifted load and must be countered or the crane will tip over. Large counterweights are mounted to the backs of crane upperworks to balance the loads, increasing lift capacity. *Figure 28* shows a heavy-lift crane fitted with counterweights at the back of the upperworks to directly counterbalance the expected load. The weights are modular, so that weight can be added or removed incrementally. Counterweights can also be added to the crawler frames, as shown, to provide additional counterbalance for lifting over the side.

1.4.0 Safety Devices and Operational Aids

Prior to any mobile crane movement or operation, the operator should note the location of all the operating gauges and check their current readings. Instruments and gauges should be located in clear view of the operator, by design. The operator must always be aware of the status of the crane and keep a constant eye on the controls as well as the load. The first indication of a crane system failure is likely to come to an operator's attention through the instrumentation. *Table 1* describes the most basic instruments and gauges related to a crane's power plant. These instruments are similar to those provided in today's cars.

29 *CFR* 1926.1415 and 1926.1416 list the safety devices and operational aids that are required on mobile cranes. Of course, many more such devices are often applied to cranes; the OSHA standards outline the minimum requirements. The list of safety devices required by 29 *CFR* 1926.1415 includes the following items:

- A crane level indicator, either built into the equipment or readily available on it
- Boom stops, except for derricks and telescopic booms
- Jib stops, except for derricks
- A brake-locking feature on cranes equipped with foot-pedal brakes
- An integral holding device or hydraulic **check valve** for hydraulic outrigger and stabilizer jacks
- Rail clamps and stops on railcar-mounted cranes
- A working horn

The standard requires that all the devices above be in proper working order. If one or more of these devices fail to operate properly during a lift, the operation must be brought to a safe halt.

1.4.1 Operational Aids

29 *CFR* 1926.1416 outlines the operational aids required for mobile cranes. Like safety devices, the specified aids must be in proper working order; their failure during a lift requires the operator to bring the operation to a safe halt. However, the standard does allow specific periods of time for repairs to be made as long as specific alternative measures are taken to replace the function of the aid. For example, if a boom hoist limiting device has failed, the standard lists three acceptable alternative practices to be used while repairs are planned and executed.

(A) HIGHWAY BARRIER CLAMP

(B) PIPE CLAMP

(C) PIPE HOOK

Figure 24 Special lifting devices.

Diesel Fuel

Diesel engines have long been preferred for heavy, hard-working motor vehicles of all types. Diesel fuel however, produced significant amounts of pollutants for many years. The sulfur content of diesel fuel has been one of the primary concerns. Nearly 100 percent of the sulfur in fuel is emitted in the form of sulfur oxides, primarily sulfur dioxide (SO_2). SO_2 causes respiratory problems and reacts with other airborne chemicals to form sulfate particles that collect in the lungs.

The US Environmental Protection Agency (EPA) has been working to reduce the pollutants that originate from diesel fuel through the issuance of quality standards. Although the fuel has certainly risen in cost, the sulfur content of diesel fuel has been reduced to a tiny fraction of what it once was. Today's diesel fuel is referred to as ultra-low sulfur diesel, or ULSD, and contains less than 0.5 percent of the sulfur that it contained in the 1980s. The emission standards for diesel engines have also been steadily improved to reduce their impact on the environment.

Figure 25 Personnel lifting platform.

Figure 26 Pile hammer in leads.

OSHA has placed the required list of operational aids into two designated categories: Category I and Category II. Category I devices must be repaired within seven days; Category II devices must be repaired within 30 days. The list of Category I operational aids includes the following:

- Boom hoist limiting device
- Luffing jib limiting device
- Anti-two-blocking device

Category II operational aids include the following:

- Boom angle or radius indicator
- Jib angle indicator (if a luffing jib is installed)
- Boom length indicator (for cranes with a telescopic boom)
- Load-weighing device
- Outrigger/stabilizer position monitor (only required for recently manufactured equipment)
- Hoist drum rotation indicator (for hoist drums not visible to the operator; only required for recently manufactured equipment)

There are a number of exceptions included in the standards, such as those related to the age of the crane. For additional details on these exceptions, consult the listed OSHA standards (found in the *Code of Federal Regulations* [*CFR*]) directly.

Newer cranes may be equipped with sophisticated monitoring panels, providing the operator with a great deal of information and control in real time. Such panels and systems fall into the broad category of human-machine interfaces (HMIs). HMIs also include joysticks and similar electronic controls. Crane manufacturers often have specific names for their HMI systems, such as the Electronic Crane Operating System (ECOS) from Manitowoc and the LICCON System from Liebherr.

Graphical displays and color are extremely useful features that keep the operator constantly informed of the crane's capacity and load moment. Load charts are built into the interface, and information about the lift can be programmed into the system during the planning stages.

Touch-screen control is also becoming more common. *Figure 29* is an example of one HMI panel and some of the functions they typically provide. Remember however, that every crane brand and model is different, and new features are consistently being introduced. Documenting all of the control options available in today's cranes is virtually impossible. Each crane operator must study and understand each individual crane's features and controls.

Figure 27 Vibratory pile driver at work.

Figure 28 A lattice-boom crawler crane with counterweights.

UPPERWORKS
COUNTERWEIGHTS

CRAWLER FRAME
COUNTERWEIGHTS

Table 1 Common Crane Instruments and Gauges

WATER TEMPERATURE GAUGE	This gauge displays the current engine coolant temperature of water-cooled engines. It monitors the coolant through a sensor unit located in the engine block. This gauge is not present on air-cooled engines.
ENGINE OIL PRESSURE GAUGE	This gauge lets the operator know the status of the crane engine oil pressure.
VOLTMETER (BATTERY VOLTAGE)	Prior to starting the crane but with the key in the ON position, this gauge displays the crane battery's state of charge. With the crane running, this gauge shows the voltage output of the alternator.
FUEL GAUGE	With the ignition in either the ON or ACC position, this gauge indicates how much fuel is in the crane engine fuel tank.
TACHOMETER	This instrument indicates the crane engine's rotations per minute (rpm).

Figure 29 Crane operation monitoring panel.

« HMI PANEL FOR CRANE OPERATION MONITORING –
COMMON FUNCTIONS:

- Graphic representation of machine configuration
- Graphical step-by-step machine set-up
- Boom length & angle
- Jib length & angle
- Load on hook
- Rated load
- Load radius
- Tip height
- Anti-two block warning & function limiters
- Operating mode
- Audio/visual warning when the load on hook is within a preset percentage of the crane's rated load
- Audio/visual warning and limits functions when the load on hook is at a preset percentage of the crane's rated load
- Operator settable alarms

Note that all warning plates, both inside and outside the cab, should be kept readable and in place. They should be considered an integral part of the equipment, and should not be defaced or removed.

1.5.0 Crane Reeving

Reeving for cranes refers to the routing of ropes or cables through the swallow and around the sheaves of a block. Mobile cranes use the same single- and multiple-part reeving patterns as manually powered block and tackle arrangements use to multiply power.

When a single hoist line is reeved for a lift, the line must be run on the center sheave, or on the sheave next to the center when an even number of sheaves exists. Reeving the line on a sheave at or near one side of the boom causes boom torque that could damage the boom. The twisting of the boom also causes the hoist line to rub against the side of the sheave, causing excessive wear to both the sheave and the line. Figure 30 demonstrates the effect of boom torque. Single-line reeving

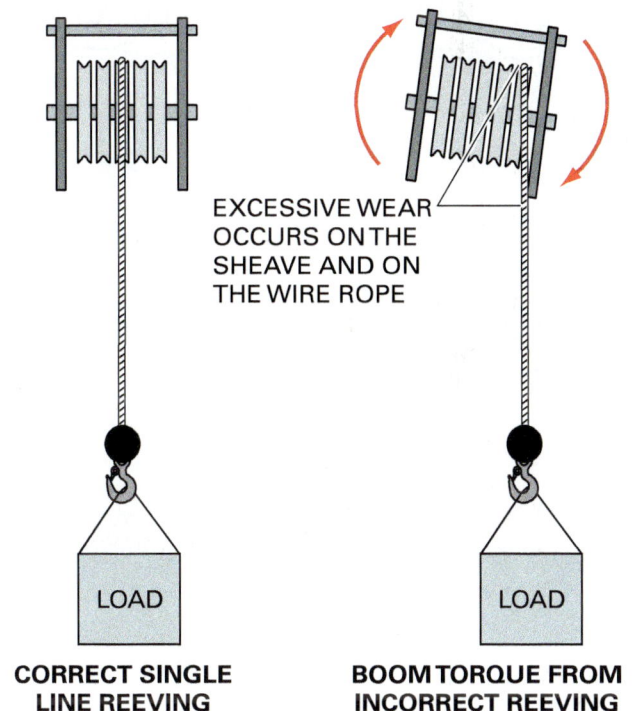

EXCESSIVE WEAR OCCURS ON THE SHEAVE AND ON THE WIRE ROPE

LOAD

LOAD

CORRECT SINGLE LINE REEVING

BOOM TORQUE FROM INCORRECT REEVING

Figure 30 Effect of incorrect single-line reeving.

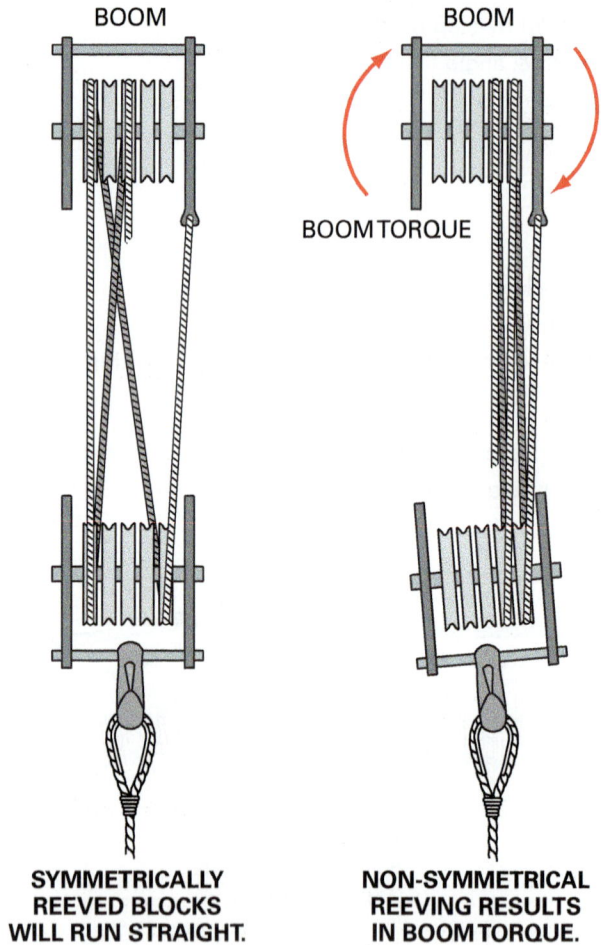

ON MULTIPLE-LINE SYSTEMS, BOOM TORQUE IS MINIMIZED IF THE PARTS OF THE LINE ARE DISTRIBUTED ON EITHER SIDE OF THE BOOM CENTER LINE.

BOOM

BOOM

BOOM TORQUE

SYMMETRICALLY REEVED BLOCKS WILL RUN STRAIGHT.

NON-SYMMETRICAL REEVING RESULTS IN BOOM TORQUE.

Figure 31 Balanced and unbalanced reeving.

may be used on smaller cranes or boom trucks, and for the whip line on larger cranes.

It is also important to reeve the boom and load block symmetrically for multiple-line lifts. If the load block is not reeved symmetrically and balanced, it will not run straight and will cause the same type of wear associated with boom torque. *Figure 31* shows balanced and unbalanced reeving for multiple-line lifts.

When a line is reeved more than once, the resulting number of lines that are supporting the load block are referred to as parts of line. *Figure 32* shows the correct reeving patterns for three- to five-part reeving. *Figure 33* shows the correct reeving patterns for six- to eight-part reeving. Very large cranes may require even more parts. Using these patterns will result in balanced reeving and smooth-running blocks with minimal boom torque. Notice that the hoist line terminates at the boom for even-numbered parts of line, but terminates at the load block for odd-numbered parts of line. Additional information on wire rope and the reeving process is provided in NCCER Module 21204-18, "Wire Rope" from *Mobile Crane Operations Level Two.*

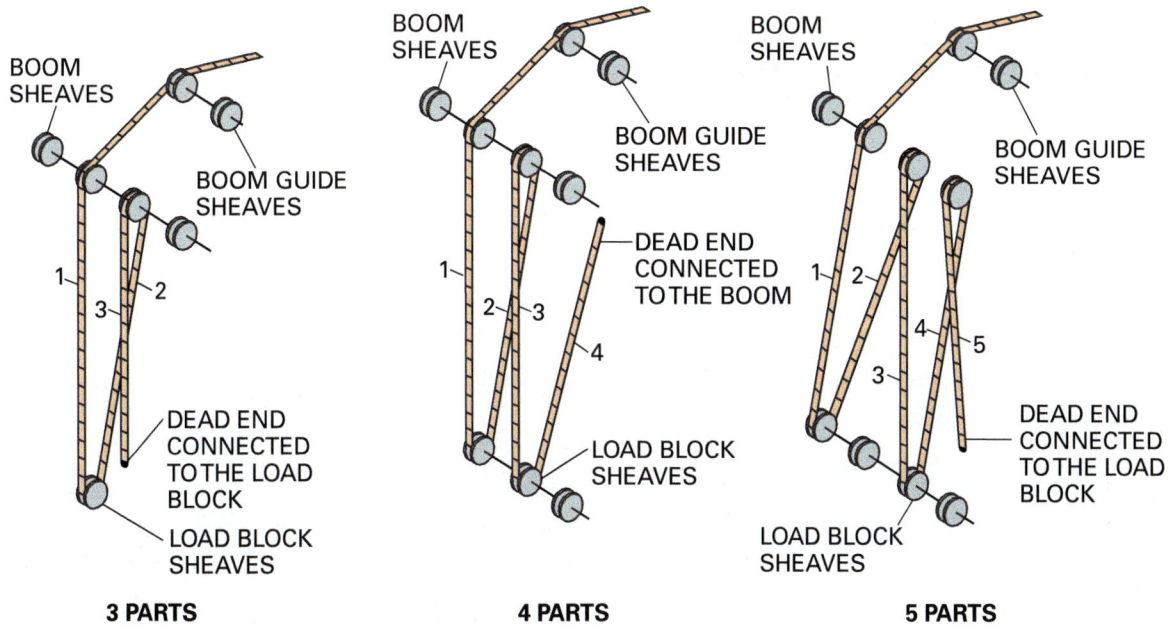

3 PARTS

4 PARTS

5 PARTS

Figure 32 Three-, four-, and five-part reeving.

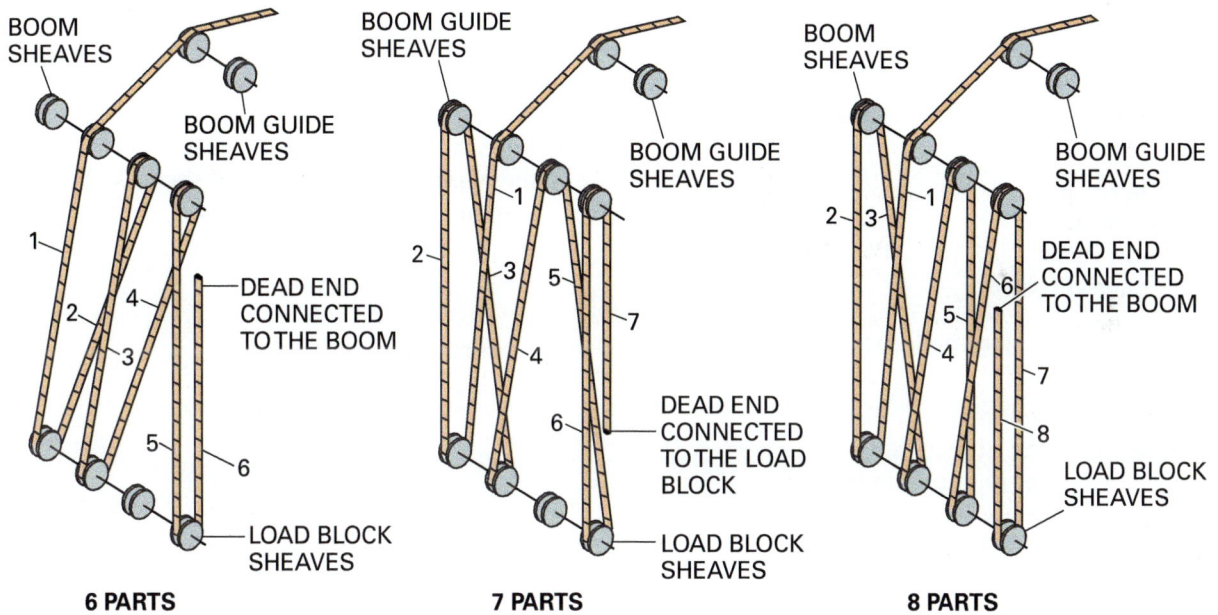

6 PARTS

7 PARTS

8 PARTS

Figure 33 Six-, seven-, and eight-part reeving.

Additional Resources

ASME Standard B30.5, Mobile and Locomotive Cranes. Current edition. New York, NY: American Society of Mechanical Engineers.

29 *CFR* 1926.1400, **www.ecfr.gov**

Mobile Crane Safety Manual (AEM MC-1407). 2014. Milwaukee, WI: Association of Equipment Manufacturers.

Cranes: Design, Practice, and Maintenance, Ing J. Verschoof. Second Edition. 2002. Hoboken, NJ: John Wiley and Sons, Inc.

IPT's Crane and Rigging Handbook, Ronald G. Garby. Current Edition. Spruce Grove, Alberta, Canada: IPT Publishing and Training Ltd.

North American Crane Bureau, Inc. website offers resources for products and training, **www.cranesafe.com**

1.0.0 Section Review

1. A crane with rotating controls has _____.
   a. an operator's cab that swings with the boom
   b. an operator's cab that remains stationary as the crane rotates
   c. a control console operated from a standing position
   d. controls mounted directly on the boom

2. If a jib can be adjusted to change its angle in relation to the main boom, it is called a(n) _____.
   a. lattice jib
   b. hydraulic jib
   c. luffing jib
   d. open-throat jib

3. A rehandling clamshell bucket is designed to _____.
   a. lift large volumes of water or oil
   b. dig into and lift hard soil
   c. pick up tree stumps and boulders
   d. pick up free-flowing material

4. Which of the following is a safety device required by OSHA standards?
   a. A brake-locking feature on cranes with foot-pedal braking
   b. A boom angle indicator
   c. A jib angle indicator
   d. An electronic crane operating system (ECOS)

5. When a single hoist line is reeved for a lift and there are an even number of sheeves, the hoist line must be run on the sheave _____.
   a. nearest the crane
   b. nearest the boom
   c. next to the center
   d. furthest from the load

SECTION TWO

2.0.0 BASIC LIFTING CONCEPTS

Objective

Identify factors related to lifting capacity and explain their significance.
a. Explain the significance of ground conditions and a level surface.
b. Describe the bearing surface and explain how to determine the required blocking.
c. Define and describe the significance of the center of gravity and the quadrants of operation.
d. Describe the significance of boom length, angle, operating radius, and elevation.
e. Explain how to use a load chart and understand the basic concepts of critical lifts.

Performance Tasks

1. Verify the boom length of a telescopic- and/or lattice-boom crane using manufacturer's data or a measuring tape.
2. Measure the operating radius of a telescopic- and/or lattice-boom crane using a measuring tape.
3. Calculate the amount of blocking needed for the outrigger of a specific crane.
4. Verify that a crane is level.

Trade Terms

Backfill: Soil and rock used to level an area or fill voids, such as the perimeter of building foundations or trenches. Areas with fresh backfill may not be stable enough to support a crane.

Center of gravity (CG): The point at which the entire weight of an object is considered to be concentrated, such that supporting the object at this specific point would result in its remaining balanced in position.

Critical lift: Defined in *ASME Standard B30.5* as a hoisting or lifting operation that has been determined to present an increased level of risk beyond normal lifting activities. For example, increased risk may relate to personnel injury, damage to property, interruption of plant production, delays in schedule, release of hazards to the environment, or other significant factors.

Dynamic loads: A load on a structure (in this case, a crane) that is not constant, but consistently changing as the result of one or more changes in various factors. Also referred to as shock loading, significant dynamic loads can be applied to a crane through abrupt motions and lifting a load from its support too quickly.

Effective weight: The weight of an accessory such as a boom extension or jib that reflects the effect of its weight on the lift, usually based on its position, rather than its actual weight. For example, a jib folded and stored on the main boom will have different effective weight than when it is installed on the main boom tip.

Floats: The portion of outriggers that touches the ground; the feet of the outriggers.

Gross capacity: The total amount a crane can safely lift under a given set of conditions. The gross capacity includes but is not limited to the load block, ropes, and rigging as well the primary load.

Hardpan: A hard, compacted layer of subsoil, usually with a major clay component.

Impact load: The dynamic effect on a stationary or mobile body as imparted by the forcible contact of another moving body or the sudden stop of a fall.

Interpolation: The process of estimating or calculating unknown values between two known values.

Leverage: The mechanical advantage in power gained by using a lever.

Net capacity: The weight of the item(s) that can be lifted by the crane; the gross capacity of a crane minus all noted capacity deductions.

Operating radius: The distance from the center of the boom's mounting point (usually the ring gear drive) to the center of gravity of the load.

Quadrant of operation: The direction of the boom relative to the base mounting or carrier body.

Tipping fulcrum: The point of crane contact with the ground where it would pivot if it were to tip over; the fulcrum of the leverage applied by the load. Depending on the attitude and type of crane, the tipping fulcrum may be the edge of one crawler assembly, one or more outriggers, or similar locations.

The lifting capacities of mobile cranes are affected by a wide range of conditions, from the type of soil beneath them to the height of the boom tip. These conditions include but are not limited to the following:

- Condition of the ground
- Crane being positioned within 1 degree of level
- Type and size of the bearing surfaces
- Type of crane (wheeled versus crawler)
- Type of boom
- The crane's capacity and actual configuration
- The center of gravity (CG) as the lift progresses
- Quadrant of operation
- Boom length, boom angle, and the resulting operating radius
- Swing out, side loading, and dynamic loads

Each of these factors will be explored further as you progress through the section.

2.1.0 Ground Conditions and Leveling

Ground conditions are a major factor in the ability of any crane to perform a lifting task safely. Even a small crane with a light load can overturn if the ground beneath gives way. In addition, the crane must be in a level position to ensure stability. It is important to note that load charts are only valid if the crane is level. Load charts assume a level condition.

2.1.1 Ground Conditions

Ground conditions around a job site can vary widely. Some areas may have been graded down to a hardpan, while excavations or depressions may have been filled but not compacted. Mobile cranes may encounter many different conditions just moving from one side of the site to the other. The ground beneath the crane must be able to support all the loads that are placed on it. This includes the weight of the crane, the load, and all rigging. In addition, there are impact loads that occur and dynamic loads that result from swinging, hoisting, lowering, and traveling. *Table 2* shows the weight-bearing capacities of various soil types.

Table 2 Bearing Capacities of Various Soil Types

SOIL	CAPACITY	
Cemented sand & gravel	135 PSI	931 kPa
Sand & gravel compact	110 PSI	758 kPa
Sand, coarse to medium compact	60 PSI	414 kPa
Sand, fine to silty compact	54 PSI	372 kPa
Clay compact	54 PSI	372 kPa
Silt compact	40 PSI	276 kPa

The backfill adjacent to new buildings is often not thoroughly compacted and cannot provide proper support for cranes. Other dangerous areas include the edges of excavations. These areas may give way beneath the crane as it approaches to place piping or tanks into the excavation. This is also dangerous for any workers who may be in the trench. Existing sewer and water lines pose hazards as well. Excavations or grading may remove enough covering soil so that the weight or vibration of a crane traveling or working above the line could cause it to collapse. *Figure 34* shows examples of each of these ground condition hazards.

For working near excavations, any crane component bearing weight (such as an outrigger) must be positioned at least 1.5 times the depth of the excavation away from the edge. For example, if a trench is 4 feet (1.2 meters) deep, the outrigger floats should be no closer than 6 feet (1.8 meters) from the edge. For crawler cranes, the leading edge of the track should also be positioned at least this far away. Different soil conditions may require that this distance be increased.

WARNING!

The clearance distance calculation for excavations provided here, although common, is not sufficient for every situation. Ensure that all factors related to the required clearance have been considered before positioning a crane or any other heavy equipment near an excavation. The unexpected failure of an excavation wall while a crane is positioned nearby can lead to fatalities in addition to serious personal injuries and/or property damage.

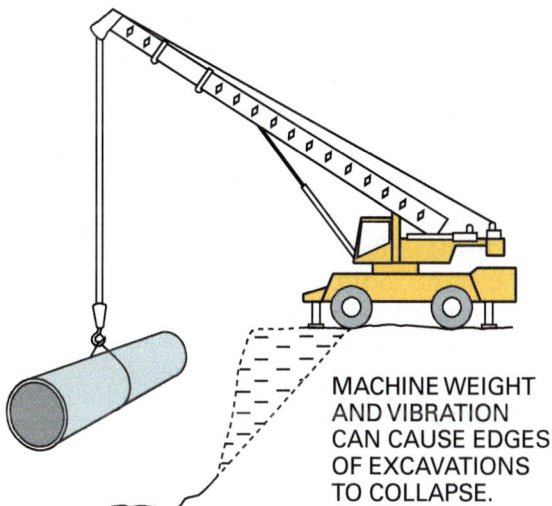

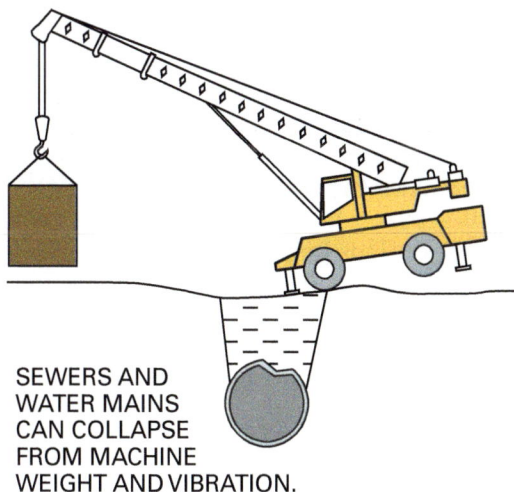

2.1.2 Leveling

Most load chart ratings require the crane to be perfectly level or level within a one-percent grade (0.57 degrees). This applies to all cranes including pick-and-carry models. When it is impossible to level a given crane, the manufacturer can sometimes provide a load chart that specifies how far off level the crane may be and the effect on load capacity. For example, a crane on a barge will be accompanied by a barge chart that considers the position of the barge in the water.

When a crane is not perfectly level, the following three problems can arise:

- The boom can be side loaded.
- The boom angle changes when the crane swings, which in turn changes the radius.
- The crane becomes more difficult to operate.

Table 3 demonstrates how capacity is reduced due to boom side-loading when the crane is not perfectly level.

An example would be found on the load chart for a crane with a 200-foot boom at an angle of 70 degrees that is out of level by just 1 degree (1.75 percent). Swinging the boom from the high side to the low side changes the operating radius by 6.5 feet (2 meters). With the crane out of level, the boom or counterweight will also try to swing downhill, complicating control.

The operator must know what the crane manufacturer considers a level condition, and then adjust the outriggers or level the supporting surface to meet those requirements.

The small target levels often provided by the crane manufacturers may not be accurate enough to confidently level the crane and meet the requirements of the load chart. On the other hand, some cranes are equipped with accurate electronic levels that can be read directly on the crane

UNCOMPACTED
BACKFILL AROUND
A NEW STRUCTURE

MACHINE WEIGHT
AND VIBRATION
CAN CAUSE EDGES
OF EXCAVATIONS
TO COLLAPSE.

SEWERS AND
WATER MAINS
CAN COLLAPSE
FROM MACHINE
WEIGHT AND VIBRATION.

Figure 34 Examples of ground condition hazards.

Table 3 Crane Capacity Decrease Due to an Off-Level Condition

CRANE CAPACITY DECREASE DUE TO OFF-LEVEL CONDITION			
Boom length or Radius	1.75% or 1 degree	3.5% or 2 degrees	5.25% or 3 degrees
Short boom, minimum radius	10%	20%	30%
Short boom, maximum radius	8%	15%	20%
Long boom, minimum radius	30%	41%	50%
Long boom, maximum radius	5%	10%	15%

instrument panel. When the equipped level is unreliable or the accuracy needs to be checked, the following two methods can be used to level the crane accurately:

- Place a carpenter's level in two different positions, 90 degrees apart, on a sturdy part of the crane, such as on top of the ring gear. The longer the level, the more accurate the reading. The bubble of the level should be between the lines in both positions.
- Boom up to the highest boom angle possible for the crane configuration. At two positions, 90 degrees apart, look straight ahead at the boom and determine if the ball is hanging straight in-line with the boom (plumb). If it is, the crane is reasonably level. In this case, the headache ball or load block is being used as a plumb bob.

2.2.0 Bearing Surface and Ground Pressure

The bearing surface refers to the points where the crane makes ground contact during a lift. It is a shortened form of the term loadbearing surface. The weight of both the crane and the load is concentrated on the bearing surfaces.

Crane bases are different and each is designed to place weight on various types of bearing surfaces. The three primary types of bearing surfaces used by cranes are the following:

- Outriggers
- Rubber (sitting directly on the tires, like a pick-and-carry crane)
- Crawler tracks

For a crane to be considered on its outriggers, the outriggers must be fully extended and the crane tires must be relieved of all weight. Some crane manufacturers supply capacity ratings based on partial outrigger extension, but that represents an exception, not the rule. If the outriggers are not fully extended, the tires are bearing weight, or both, the capacity chart for the crane on outriggers will be invalid.

The term *on rubber* refers to a wheeled truck crane being operated without any outriggers extended. The crane is being operated with its wheels and tires as the sole method of support and stability. Pick-and-carry cranes operate this way in most cases. Lifting capacity is significantly reduced as a result, but maximum lifting capacity is often not a priority for them.

Crawler cranes typically have capacity ratings for both crawlers extended and crawlers retracted when that feature is available. Extending the tracks provides a broader base and therefore greater capacity and stability.

The crane base in use has a great deal to do with ground pressure. For crawler cranes, the weight is distributed across the total surface area of the tracks in contact with the ground. When on rubber, the weight is distributed to the numerous small patches of area where the tires contact the ground. When outriggers are in use, the weight is being distributed across the total area of the floats in contact with the ground.

When the crane is sitting idle, the weight is distributed among the various bearing surfaces. The weight distribution is not likely to be uniform, but relatively so. Some bearing surfaces, such as those at the opposite end from the boom near the counterweights, may be carrying extra weight that will be balanced by the load. Once lifting begins, the weight distribution changes. Ground pressure increases most at the bearing surfaces nearest the load. The ground pressure at an outrigger can instantly become much greater than that under a track on similar lifts, due to the smaller bearing surface offered by the outrigger. *Figure 35* shows the relative ground pressures of track- and outrigger-supported cranes with the load supported over different areas of the crane. The darker red areas represent greater pressure. If the crane is not properly supported, one or more of these bearing surfaces can sink, allowing the crane to tilt away from level. As it moves away from a level condition, the lift capacity is dropping and stability is being compromised.

2.2.1 Mats and Blocking

To compensate for ground conditions and the changing ground pressures, crawler cranes are often supported on crane mats that are made of steel or wood. Wooden mats can be made from 8- to 12-inch (20- to 30-cm) square timbers that have been bolted together at uniform intervals. These mats, fabricated in manageable sections, are then combined to make a mat that extends at least 2 feet (0.6 meters) beyond the ends and edges of the crane tracks. Once the crane is positioned, timber blocking is secured to the mat at the ends and sides of the tracks. This helps prevent the crane from sliding easily in any direction. *Figure 36* shows a typical mat arrangement for a crawler crane.

For wheeled cranes, the load on the outrigger floats must also be spread across the ground surface. The material used to transfer and spread loads is called *blocking* or *cribbing*. Crane manufacturers must make outrigger floats small and light enough so that one person can move them, since

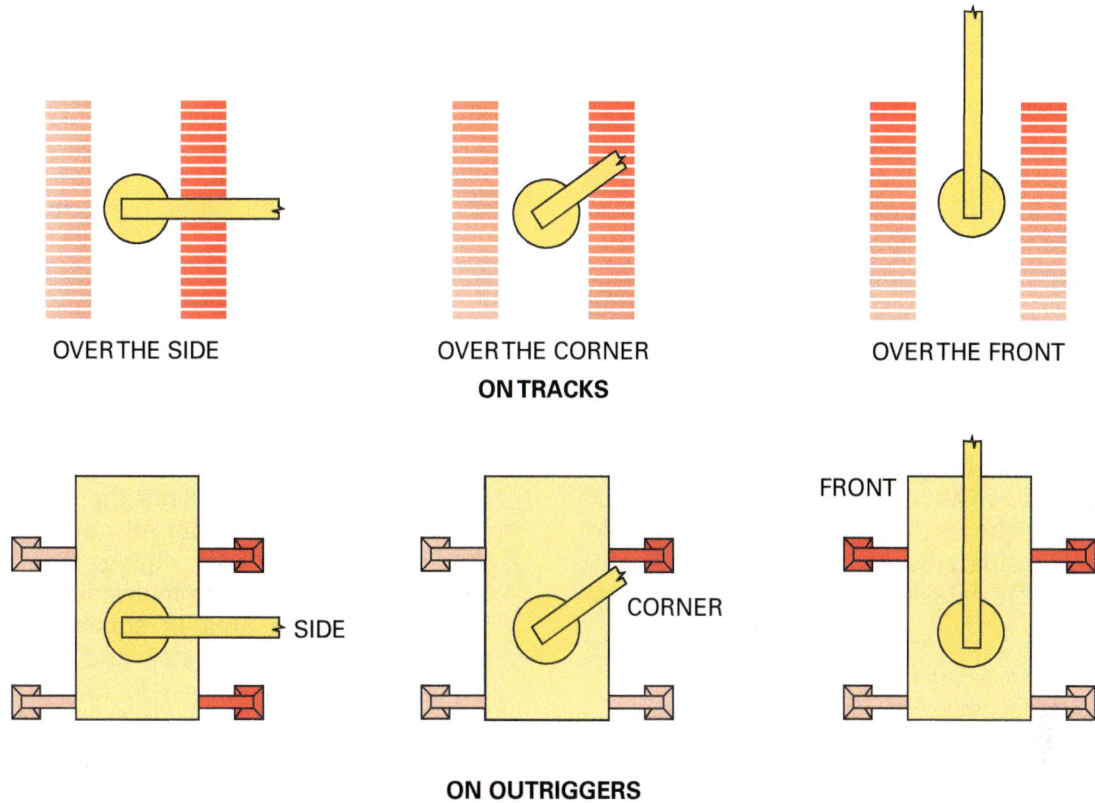

OVER THE SIDE   OVER THE CORNER   OVER THE FRONT

ON TRACKS

SIDE

CORNER

FRONT

ON OUTRIGGERS

Figure 35 Relative ground pressures applied by track- and outrigger-supported cranes.

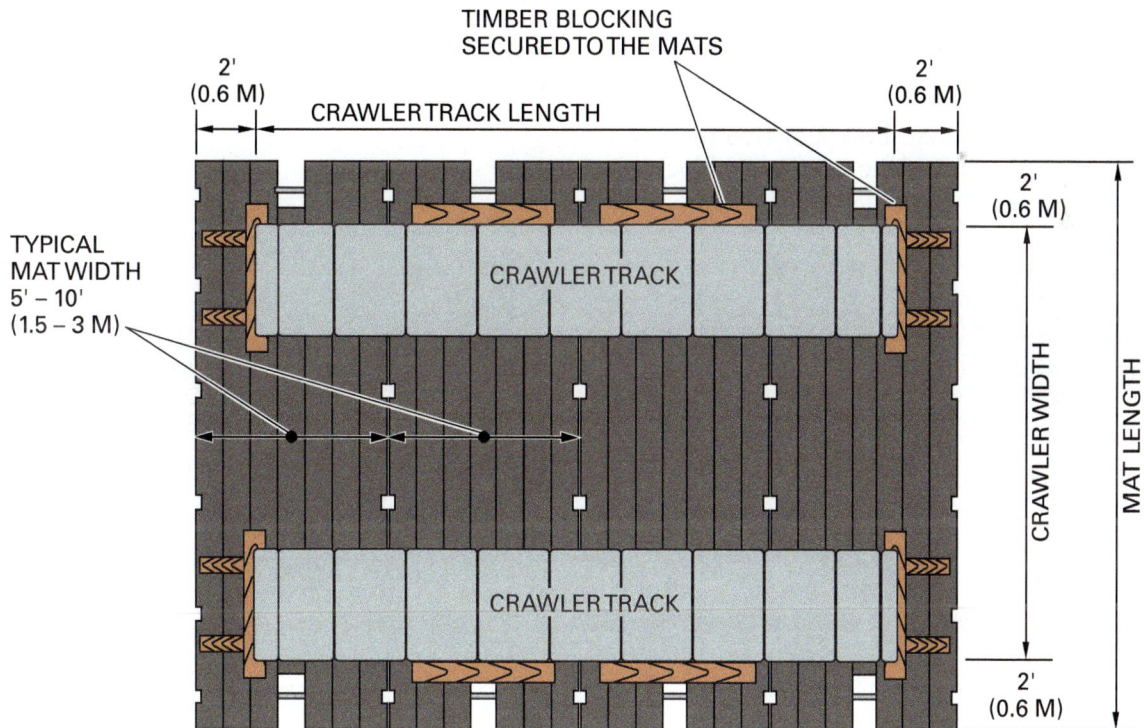

TIMBER BLOCKING
SECURED TO THE MATS

2'
(0.6 M)

CRAWLER TRACK LENGTH

2'
(0.6 M)

2'
(0.6 M)

TYPICAL
MAT WIDTH
5' – 10'
(1.5 – 3 M)

CRAWLER TRACK

CRAWLER WIDTH

MAT LENGTH

CRAWLER TRACK

2'
(0.6 M)

Figure 36 Typical crawler-crane mat arrangement.

they are often detached from the outrigger for travel. This limits the size of the float somewhat.

The blocking beneath the float must be made from a suitable material such as a hardwood. Softwoods are not generally acceptable. Heavier cranes often require blocking that is stacked to increase the total thickness. When wooden blocking is stacked, the lumber in each layer should be positioned at 90-degree angles to each other. See *Figure 37* for several examples of correct and incorrect blocking. Although blocking is often made of wood, it is not a requirement. Blocking products are also fabricated from man-made materials and marketed by various manufacturers.

Never assume the supporting surface will support the load without blocking unless specific technical data shows that it will. Blocking should always be considered a requirement unless specific and reliable data is presented to the contrary.

> **WARNING!**
>
> Concrete masonry units (CMUs) should never be used for crane blocking. Blocking should always be placed under the outrigger float, and never under the outrigger beam, inboard of the float.

Spreading the load means the surface area of the blocking should cover as much ground surface as possible, within reason. The area of the blocking, to do this effectively, must be larger than the area of the float. Either of two methods can be used to determine the minimum blocking area for average soil:

- *Method 1* – Method 1 is sometimes referred to as the *Rule of Three*. Multiply the area of the crane float by 3 to determine the required area of the blocking (float area × 3 = blocking area). Note that the units used can be imperial or metric. The same unit, such as square feet or square meters, used to determine the float area will be the unit of the result. For example, if square meters are used for the area of the float, the result will also represent square meters.

- *Method 2* – Method 2 is sometimes referred to as the *Rule of Five*. Divide the capacity of the crane (in tons) by 5 to determine the required area of the blocking in square feet [Crane Capacity (tons) ÷ 5 = blocking area (square feet)]. The result represents the total amount of blocking needed under all outriggers. Note that you cannot easily change the imperial units in this equation to metric units. It is best to determine the area in square feet, using tons for the crane capacity, and then convert the result area to a metric unit if desired. This method is generally preferred for smaller cranes that have relatively small floats.

> **WARNING!**
>
> These calculations are for average soil, which will not provide enough support for all soil conditions.

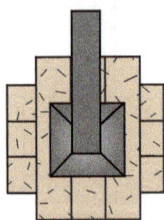

CORRECT

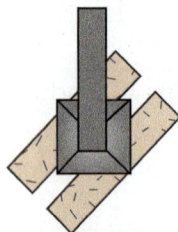

SPAN BLOCKING THAT LEAVES A GAP BETWEEN PIECES IS NOT ACCEPTABLE. BLOCKING SHOULD ALSO NOT BE POSITIONED IN A WAY THAT ALLOWS ANY PART OF THE FLOAT TO OVERHANG THE EDGES.

BE SURE BLOCKING IS STABLE.

INCORRECT

NEVER BLOCK UNDER THE OUTRIGGER BEAMS.

INCORRECT

Figure 37 Correct and incorrect use of outrigger blocking.

Manufactured Blocking

Not all crane blocking and mats are made by the crane owner. Those that are made by the owner are usually made of wood. However, there are a number of manufacturers that specialize in the fabrication and sale of crane support materials. Although some manufacturers do make and sell wooden blocking and mats, others use materials that are more durable.

The crane blocking shown here can be made to order for a given crane to ensure it is properly sized. The fibrous material used is far more weather resistant than wood, which tends to crack and split over time, as well as absorb water. The material is said to have the strength of steel blocking, but weighs 70 percent less than steel. The manufacturer also makes smaller pads using a thermoplastic material that offers the same basic advantages. Crane blocking and mat manufacturers will also assist in the engineering and selection process.

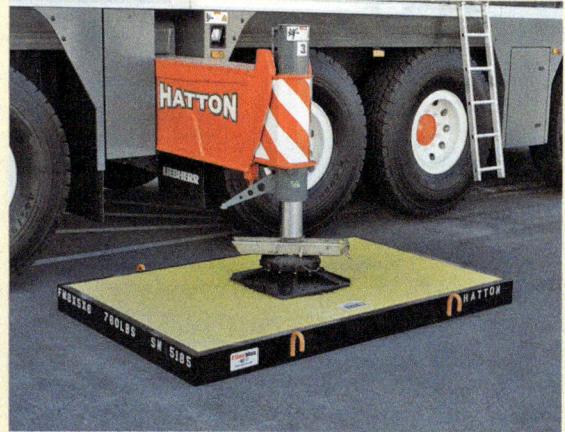

Figure Credit: DICA Outrigger Pads

An example calculation for each method is provided below. Unless there are unique circumstances, the results from either method show the minimum size of the blocking for a single outrigger. There is no reason that the blocking can't be larger; indeed, larger is better. For this reason, it is best to always round the results of any calculation up when rounding is needed.

Example for Method 1 (Square Floats):

What is the minimum blocking area needed for a 27-inch square float?

The blocking area can be calculated as follows:

Step 1 Determine the area of the float:

$$27" \times 27" = 729 \text{ in}^2$$

Step 2 Multiply by 3 to determine the required blocking area:

$$729 \text{ in}^2 \times 3 = 2{,}187 \text{ in}^2$$

Step 3 If needed, the result can be converted to square feet or any other unit you find convenient to use. The result has been rounded up to the nearest tenth:

$$2{,}187 \text{ in}^2 \div 144 \text{ in}^2/\text{ft}^2 = 15.2 \text{ ft}^2$$

> **NOTE**
> When rounding blocking area calculations, always round up (not down) to ensure your result is not below the minimum blocking area.

There are several ways this information can be applied. For example, if square blocking is desired, finding the square root of 15.2 ft² will give you the dimensions of the square. Since the square root of 15.2 is approximately 3.9, then the blocking should be at least 3.9' × 3.9'. Note that this is very close to a 4-foot square.

Many outrigger floats are round rather than square. For round floats, you will need to determine the area of the circle. To find the area of a circle, the radius of the circle (which is half its diameter) is squared and then multiplied by the constant *pi*, represented by the symbol π. Rounded to the nearest hundredth, pi has a value of 3.14. The equation for determining the area of a circle appears as:

$$\text{Area} = \pi r^2$$

Example for Method 1 (Round Floats):

What is the minimum size of the blocking needed for a round float that is 50 cm in diameter?

The blocking area can be calculated as follows:

Step 1 Determine the radius of the round float:
$$\text{Radius} = \text{diameter} \div 2$$
$$\text{Radius} = 50 \text{ cm} \div 2$$
$$\text{Radius} = 25 \text{ cm}$$

Step 2 Determine the area of the float:

$$\text{Area} = \pi r^2$$
$$\text{Area} = 3.14 \times (25 \text{ cm})^2$$
$$\text{Area} = 3.14 \times 625 \text{ cm}^2$$
$$\text{Area} = 1{,}963 \text{ cm}^2$$

Step 3 Multiply by 3 to determine the required blocking area:

$$1{,}963 \text{ cm}^2 \times 3 =$$
$$5{,}889 \text{ cm}^2 \text{ of blocking surface area required}$$

Step 4 Determine the diameter of the round blocking needed, using the area. The same equation can be used to first find the radius of the round blocking:

$$\text{Area} = \pi r^2$$
$$5{,}889 \text{ cm}^2 = \pi r^2$$
$$(5{,}889 \text{ cm}^2 \div 3.14) = (3.14 \times r^2) \div 3.14$$
$$1{,}876 \text{ cm}^2 = r^2$$
$$\sqrt{1{,}876 \text{ cm}^2} = \sqrt{r^2}$$
$$43.3 \text{ cm} = r$$

Step 5 Now determine the diameter of the round blocking needed by simply multiplying the radius by 2, as follows:

$$43.3 \text{ cm} \times 2 = 86.6 \text{ cm}$$

Therefore, a section of round blocking that is 86.6 cm in diameter is the minimum needed for the 50-cm round float.

Note that the blocking can be square even if the float is round. Square blocking is easier to construct from lumber. Again, you can determine the size of the square by finding the square root of the required area. Since the area needed is 5,889 cm², use a calculator to find the square root. The square root of 5,889 is 76.7. Therefore, a square section of blocking 77 cm × 77 cm will also work for the 50-cm round float.

Example for Method 2:
What is the blocking area required for a crane with a capacity of 75 tons?
The blocking area can be calculated as follows:

Step 1 Divide the capacity of the crane (in tons) by five, which represents the number of square feet of blocking area per ton of weight. Remember that the Rule of Five only works with square feet:

$$75 \text{ tons} \div 5 \text{ tons/ft}^2 = 15 \text{ ft}^2$$

Step 2 If desired, the result can be converted to square meters or any other unit you find convenient to use. To convert square feet to square meters, for example, multiply square feet by the conversion factor of 0.093 m²/ft²:

$$15 \text{ ft}^2 \times 0.093 \text{ m}^2/\text{ft}^2 = 1.4 \text{ m}^2$$

Since 15 ft² represents the total amount of blocking needed for the crane, each of four outrigger pads would require 3.75 ft² of blocking beneath them. Therefore, a 2' × 2' square section of blocking under each pad would work nicely and provide a bit of extra area.

When tasked with constructing blocking that must cover a given area, lumber sizes are a factor. First, remember that lumber, such as a 4 × 4 timber, is not made to its nominal size. The actual dimensions of a common 4 × 4 are 3.5" × 3.5". This must be taken into consideration as the actual size of the blocking is planned. Square blocking does not need to be perfectly square. However, it is important to ensure that the final dimensions result in a section of blocking that does not allow any part of the float to hang over the edge, and that has an area no less than the calculated area.

> **WARNING!**
> These calculations are for average soil, which will not provide enough support for all soil conditions.

2.3.0 Center of Gravity and Operating Quadrants

Mobile crane operators must have a clear understanding of a crane's center of gravity and its tipping fulcrum. The location of and relationship between these two points determines if a crane will stay upright or tip over. These concepts are important to all members of a lift team, such as signal persons and riggers who work in the vicinity of the crane and directly support the effort.

2.3.1 Center of Gravity

The center of gravity can be defined as the point at which the entire weight of an object is considered to be concentrated. Supporting an entire object at its precise center of gravity results in the object remaining balanced in position. The location of a crane's center of gravity in relation to its tipping fulcrum directly affects its leverage and stability. During a lift, the crane's center of gravity must remain between the tipping fulcrum and the other point of crane support (tracks, tires, or outriggers opposite the load). If the center of gravity moves outside of this area toward the load, the crane will tip over.

To understand the relationship between a crane's center of gravity and stability, it is first necessary to understand the principle of leverage. Leverage can be illustrated with a simple teeter-totter (*Figure 38*). When the weight of the heavy load, multiplied by the distance (X) from its center of gravity to the tipping fulcrum, is equal to the weight of the lighter load multiplied by the distance (Y) from its center of gravity to the tipping fulcrum, a condition of balance has been reached. This relationship is further illustrated with a long lever bent upwards in *Figure 39*.

Another example shows the relationship with the lighter load being suspended below the long lever in *Figure 40*. This figure represents the profile of a crane. Note that in all three figures, the loads will remain balanced as long as the load weights remain the same and the horizontal distances X and Y remain the same.

2.3.2 Quadrant of Operation

The center of gravity usually changes depending on the crane's quadrant of operation. When the crane boom shifts from one quadrant of operation, such as over the side, to another quadrant, the center of gravity shifts closer to, or farther away from, the tipping fulcrum. This, in turn, may increase or decrease the crane's leverage. For a rough-terrain crane, maximum lifting capacity (with the greatest stability margin) is accessed when the crane boom is operated directly over the front. Maximum lifting capacity for other wheeled cranes is accessed when the boom is over the rear.

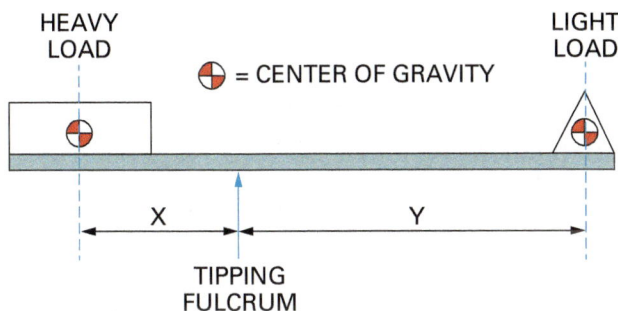

Figure 38 Teeter-totter demonstrating leverage.

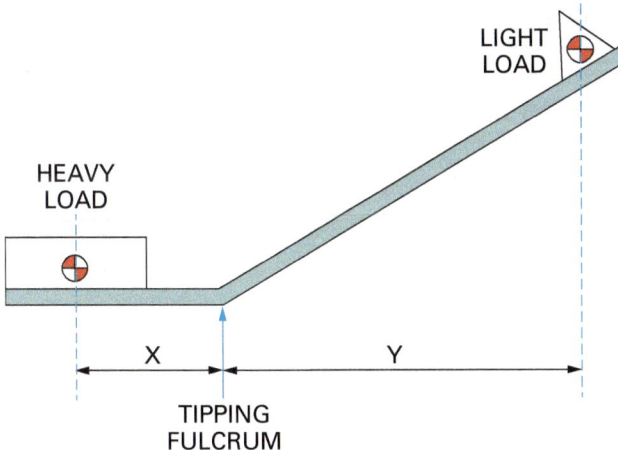

Figure 39 Leverage demonstrated on a bent lever.

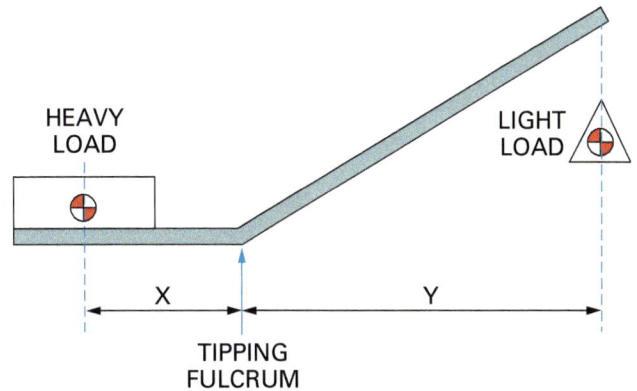

Figure 40 Crane-shaped lever.

The individual quadrants of operation are typically defined as follows:

- 360 degrees
- Over the front
- Defined arc over the front
- Over the rear
- Over the side

Figure 41 shows examples of defined quadrant boundaries for rough-terrain, wheeled truck, and crawler cranes. Note that the quadrant boundaries differ from crane to crane; always check the manufacturer's load charts and operating manual for the applicable quadrant boundaries.

Load charts are available for each quadrant of operation for which the crane is approved. The capacity for each quadrant can vary widely. Lifting from some quadrants may be prohibited entirely. This means that the lift and the position of the crane must be considered carefully beforehand. In addition, it means that operators must be cautious anytime the load must swing from one quadrant to another. In most situations, rotating the crane with a suspended load necessary in order to move the load.

If there is no load chart provided for a given quadrant, handling a load in that quadrant is not approved. This restriction means that no lifts, without exception, can be made from that quadrant, and loads cannot be swung into or through that quadrant. Most truck cranes (other than rough-terrain models) do not allow lifting over the front of the crane. Some are equipped with special bumper outriggers up front, allowing over-the-front operations. However, these are the exception rather than the rule. Check the charts available with each crane you operate to determine the approved operating quadrants and their boundaries. Load charts often provide pictures of the quadrant areas for clarity.

LIFTING AREAS ON RUBBER

OVER FRONT
QUADRANT

OVER SIDE
QUADRANT

OVER SIDE
QUADRANT

OVER REAR
QUADRANT

LIFTING AREAS ON OUTRIGGERS

OVER FRONT
QUADRANT

OVER SIDE
QUADRANT

OVER SIDE
QUADRANT

OVER REAR
QUADRANT

ROUGH-TERRAIN QUADRANTS

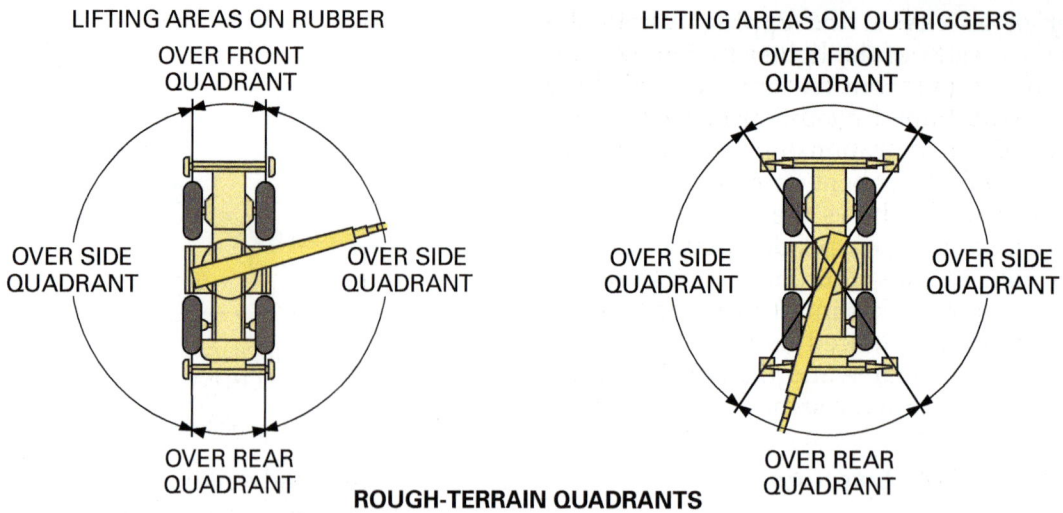

NOTE THAT LIFTING
FROM RUBBER IS A
RARE OPTION FOR
WHEELED CRANES
OTHER THAN
ROUGH-TERRAIN
MODELS.

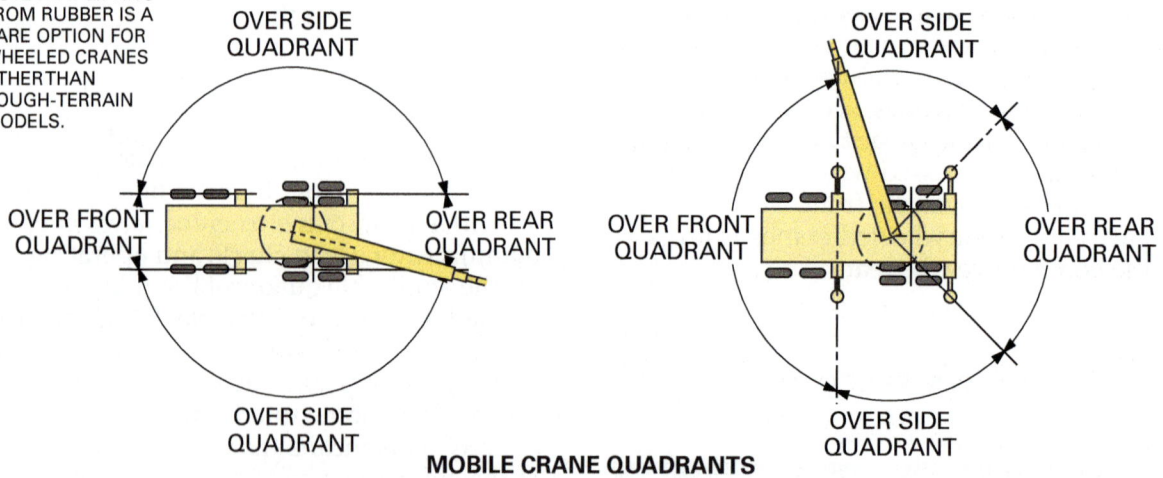

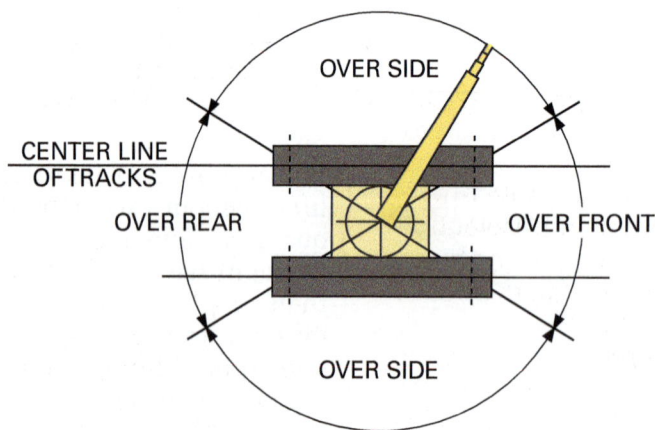

OVER SIDE
QUADRANT

OVER FRONT
QUADRANT

OVER REAR
QUADRANT

OVER SIDE
QUADRANT

OVER SIDE
QUADRANT

OVER FRONT
QUADRANT

OVER REAR
QUADRANT

OVER SIDE
QUADRANT

MOBILE CRANE QUADRANTS

OVER SIDE

CENTER LINE
OF TRACKS

OVER REAR

OVER FRONT

OVER SIDE

CRAWLER CRANE QUADRANTS

Figure 41 Quadrant definition drawings for rough-terrain, wheeled truck, and crawler cranes.

As explained earlier, the crane's stability depends on the relationship between the load leverage and the leverage of the crane. The crane leverage is determined by the distance between the crane's center of gravity and the tipping fulcrum. The load leverage is determined by the distance between the load's center of gravity and the same tipping fulcrum. Remember that the tipping fulcrum changes as a load swings from one quadrant to another, changing the relationship between it and the center of gravity.

In addition to changes in capacity and stability that occur as the boom moves from one quadrant to another, the leverage applied by the load changes as the angle of the boom changes. When the boom is near vertical, the leverage of the load is relatively small. As the boom is lowered towards a horizontal position, the load leverage increases due to the increased horizontal distance between it and the tipping fulcrum. Note that the safe lifting capacity of the crane is being reduced as the boom is lowered, as a direct result of this change in leverage. The crane remains stable as long as the crane's leverage is greater than the load's leverage.

The change in leverage that occurs as a boom is lowered is illustrated in *Figure 42*. Pay close attention to the changes in the length of the Y line, representing load leverage, as the boom drops toward horizontal. Once a crane starts to tip over, the crane leverage will rapidly decrease further due to the reduced distance to the tipping fulcrum (X_2), and tipping will accelerate.

The crane leverage for a wheeled crane (other than most rough-terrain models) in different quadrants varies more widely than a crawler crane. This is especially true for crawler cranes that can extend their crawler frames to increase stability. This is due to the large changes in the distance between the center of gravity of the crane and the tipping fulcrum of wheeled cranes.

The greatest leverage, and thus the greatest lifting capacity, is developed when the boom is over the rear of a wheeled crane. Even with the outriggers extended, the crane leverage with the boom over the side is less than over the rear, due to the shorter distance to the tipping fulcrum.

The least capacity is with the boom over the front, because the crane leverage is at its least and the weight of the cab and front drive train adds to the leverage of the load. Lifts over the front are not permitted with most wheeled truck cranes. *Figure 43* shows how the crane leverage changes for each position. Note that the crane leverage, represented by the line X, gets shorter from the top drawing to the bottom drawing.

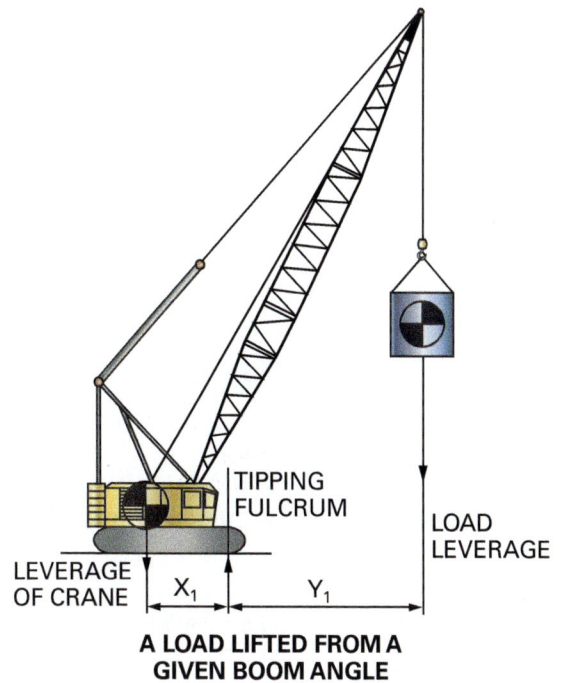

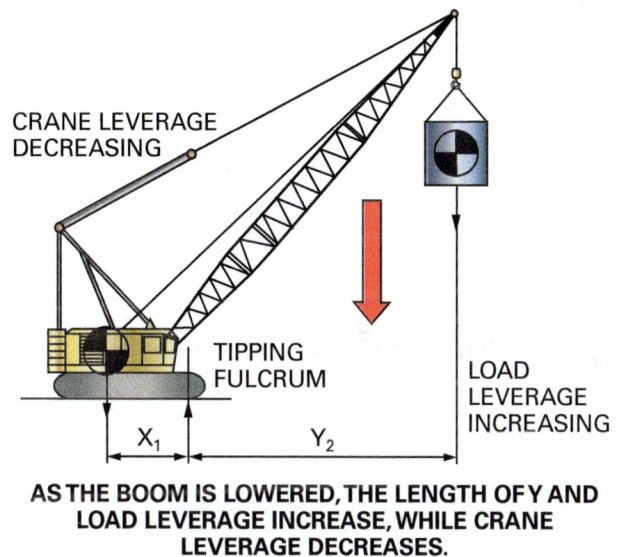

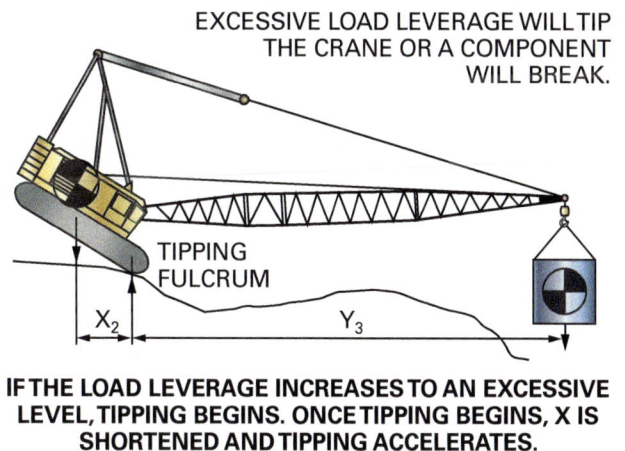

A LOAD LIFTED FROM A GIVEN BOOM ANGLE

AS THE BOOM IS LOWERED, THE LENGTH OF Y AND LOAD LEVERAGE INCREASE, WHILE CRANE LEVERAGE DECREASES.

IF THE LOAD LEVERAGE INCREASES TO AN EXCESSIVE LEVEL, TIPPING BEGINS. ONCE TIPPING BEGINS, X IS SHORTENED AND TIPPING ACCELERATES.

Figure 42 Relative tipping forces as a boom is lowered.

MOST CAPACITY
OVER THE REAR

X₁

TIPPING
FULCRUM

TIPPING FULCRUM OVER THE REAR

LESS CAPACITY
OVER THE SIDE

X₂

TIPPING
FULCRUM

TIPPING FULCRUM OVER THE SIDE

LEAST CAPACITY OVER THE FRONT
(Not permitted on most units.)

X₃

TIPPING
FULCRUM

TIPPING FULCRUM OVER THE FRONT

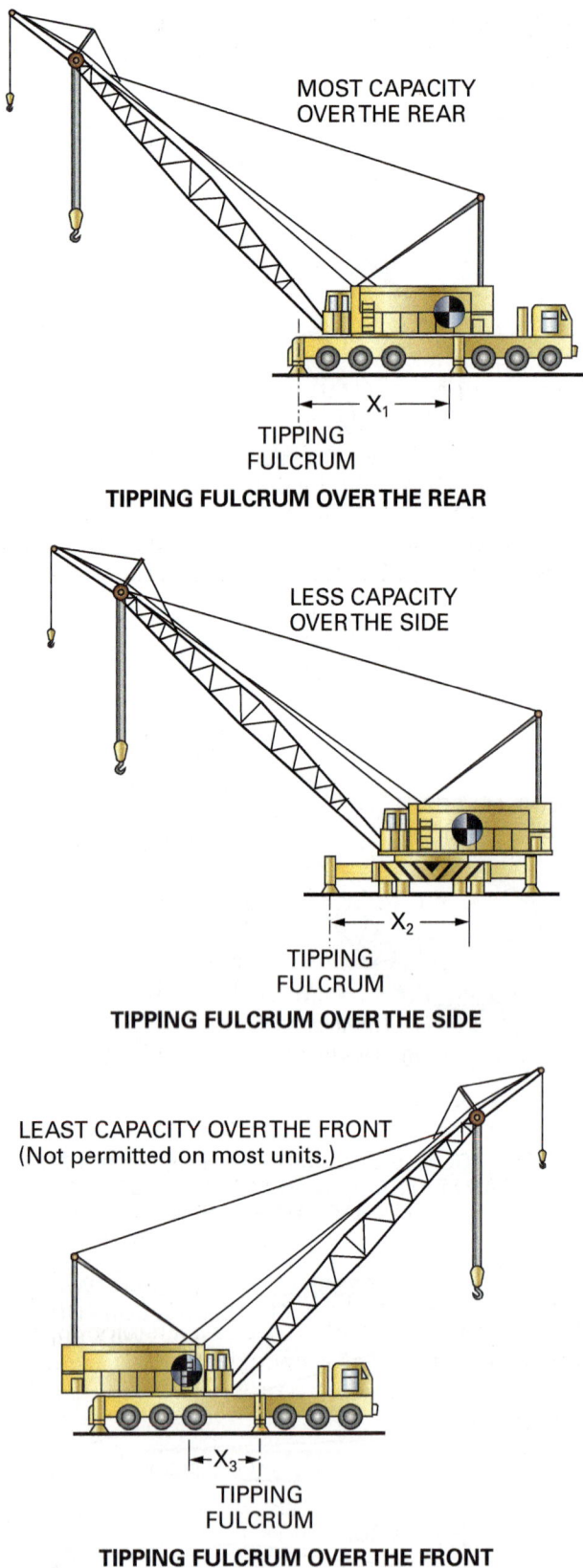

Figure 43 Tipping fulcrums for various quadrants.

To avoid tipping, cranes are rated with a built-in safety factor. The maximum rated capacity of any crane represents a percentage of the tipping load. *Table 4* shows the typical rated capacities of cranes as a percentage of their tipping load. Although these values are commonly used, they do not represent a standard shared by all manufacturers. It is best to know how the load capacity of the specific crane you are operating was determined.

2.4.0 Boom Position and Characteristics

Boom length, boom angle, operating radius, and boom-point elevation each have an effect on the capacity of a given crane. These characteristics are all within the control of the operator and are constantly changing (dynamic) as the boom position changes. *Figure 44* shows each of these measurements.

Boom length is the distance measured from the center of the boom hinge pin to the center of the sheaves at the boom tip. Boom length includes the jib and/or other boom extension when equipped. Boom length directly affects both crane capacity and reach, horizontally and vertically.

The boom angle is the angle existing between the horizontal plane and the center line of the boom in its present position. Generally, the smaller the angle (closer to the horizontal plane), the lower the capacity of the crane. As the boom angle increases, so does the crane's capacity. Smaller angles also place additional stress on the boom.

When a lattice boom is lowered to a smaller angle, it begins to sag in the middle from its own weight. This sag increases the pull on the pendants. These forces can exceed the strength

Table 4 Capacities of Cranes as a Percentage of Their Tipping Load

TYPE OF CRANE	CAPACITY
Locomotive	85%
Crawler	75%
Wheeled truck crane	
on outriggers	85%
on tires	75%
Commercial boom truck	
on stabilizers	85%

* Check your crane. The capacities shown above are not used by all manufacturers.

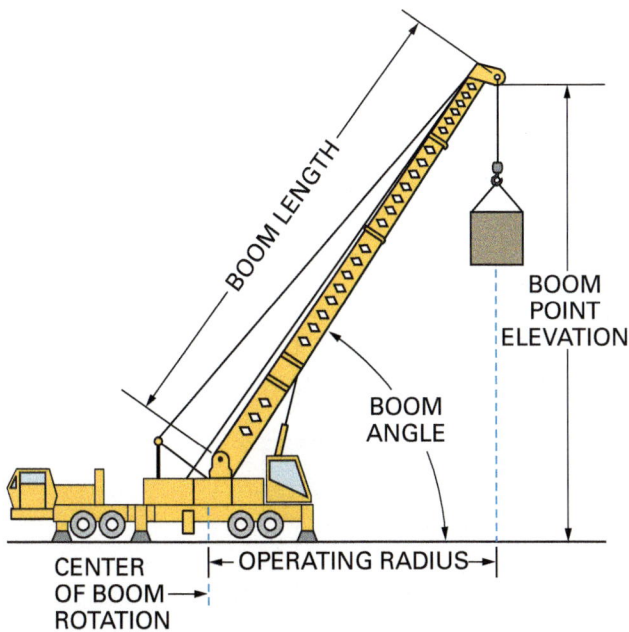

Figure 44 Boom length, boom angle, operating radius, and boom-point elevation.

of the lattice boom structure. Helper cranes are often needed to assist in assembling long booms to prevent lattice failure. With a telescopic boom, the forces of the lower angle typically result in the bending of one or more boom sections. The degree of bending depends on how far the individual sections are extended. There is less stress when all of the sections are retracted, even when the boom is in the horizontal position (a 0-degree boom angle). When the sections are fully extended, a great deal of stress is placed on the end of each section.

The operating radius is defined as the horizontal distance from the axis of boom rotation to the center of the load block with the load suspended. As the operating radius increases, capacity decreases. Cranes have both minimum and maximum operating-radius specifications. These specifications are directly dependent on boom length. Unless a longer boom is installed, the only way to increase the operating radius of the crane is to lower the boom. It makes sense then, that the crane's capacity decreases as the operating radius increases.

It is not unusual for the operating radius to increase as the load is lifted and the weight transfers to the crane, due to the deflection of the boom and crane structure (*Figure 45*). If the crane is lifting from rubber, tire compression nearest the tipping fulcrum will also cause the operating radius to increase as the crane leans slightly toward the load.

A third factor is related to the compression of the supporting surface. Although the crane may be perfectly level before the load is lifted,

compression of the earth directly beneath the tipping fulcrum (where most of the load is concentrated) will also cause the crane to lean toward the load. Although generally minor, any change in the operating radius as the lift progresses can become a major factor in a lift that is at or near the crane's capacity. In addition, the combined effect of all three of these factors in a single lift can interrupt what appears to be a safe and simple lift that was well within the crane's capacity during the planning stages.

Boom-point elevation is the vertical distance from the ground to the tip of the boom, boom extension, or jib point sheave. This dimension changes with the boom length (on telescopic booms) and the boom angle. It is used to determine the ability of the crane to reach a certain height and/or to determine clearances from power lines, buildings, or other hazards in the operating area.

2.4.1 Swing Out, Side Loading, and Dynamic Loads

The movement of the load in relation to the crane boom can cause additional stresses on the boom, resulting in failure. Swing out, side loading, and dynamic loads can cause potentially dangerous stress on the boom.

If the crane or boom is moved rapidly with a suspended load, the load will not initially follow the boom. When the swing begins, the load will lag behind the boom. As the load begins to swing to catch up, the centrifugal force of the swing will cause the load to swing out beyond the boom in the opposite direction. When the boom stops moving, the load will continue to swing. Each swing causes strain on the equipment and rigging that can result in failure. More importantly, swing out may also result in the load striking workers or equipment that were originally thought to be well outside the operating radius. For this reason, any movement of the crane or boom with a suspended load must be done gently and slowly. *Figure 46* demonstrates swing out.

Boom side-loading occurs when the load is positioned to one side of the boom due to the following factors:

- An off-level crane setup
- Rapid movement
- Sudden starting or stopping of boom rotation (swing out)
- Dragging of a load along the ground
- Lifting without first positioning the boom tip directly over the load
- Improper reeving

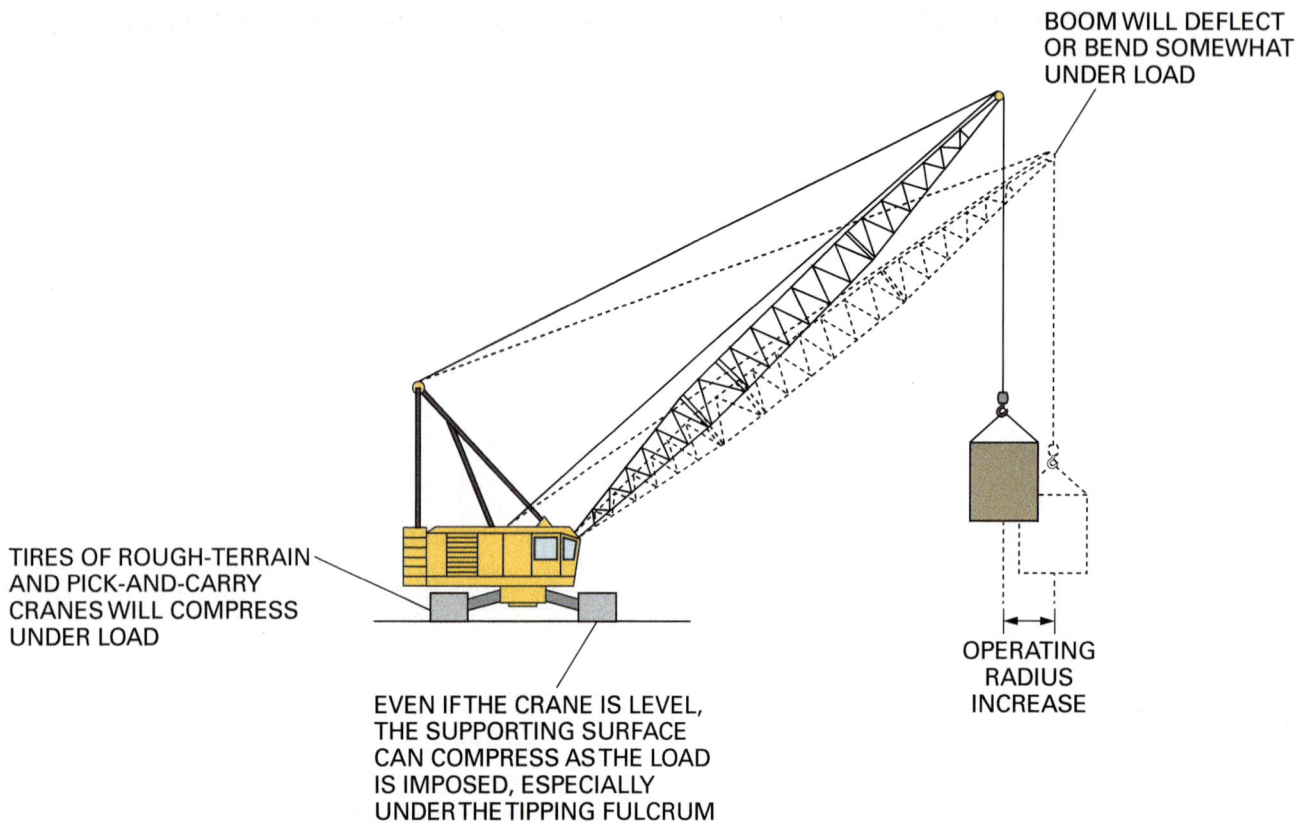

BOOM WILL DEFLECT OR BEND SOMEWHAT UNDER LOAD

TIRES OF ROUGH-TERRAIN AND PICK-AND-CARRY CRANES WILL COMPRESS UNDER LOAD

OPERATING RADIUS INCREASE

EVEN IF THE CRANE IS LEVEL, THE SUPPORTING SURFACE CAN COMPRESS AS THE LOAD IS IMPOSED, ESPECIALLY UNDER THE TIPPING FULCRUM

Figure 45 Increased operating radius due to various factors.

Dynamic loading, also called *shock loading*, refers to any changing stress placed on the crane by actions such as hoisting, lowering, stopping, or swinging the crane, particularly when these actions are sudden. When a crane is at rest with a suspended load on a firm, uncompromising surface and there is no wind, the load is said to be *static*. That is, a certain load is imposed on the crane structure and it remains relatively constant. However, the lightest breeze or any crane movement whatsoever can cause the load to fluctuate, making it *dynamic*. Sudden or rapid movements can easily increase the load by 30 to 100 percent; therefore, such movements must be avoided. Swing out and boom side-loading are two ways that dynamic loads are generated in a crane. It is important to note, however, that not all sources of dynamic loads are within the control of the operator. Wind, for example, is an uncontrollable factor.

2.5.0 Load Charts

A crane's load chart is as important to the mobile crane operator as the crane itself. It is a critical piece of the equipment. *ASME Standard B30.5, Section 5-1.1* says that "the crane manufacturer shall provide load rating charts and information for all crane configurations for which lifting is permit-

ted." Many manufacturers physically attach the load chart in a readily accessible location in the cab of the crane. Load charts are also built into the crane's software in models with a graphical display. The load chart indicates the maximum capacity of the crane under every permissible configuration. The understanding and use of load charts by crane operators and lift planners is critical to the safe operation of the crane.

The following are three basic load chart configurations. Note that there are many possible configurations for each crane, and each requires a different load chart. However, most charts fall into one of these three broad categories:

- Lifting from the boom with no extensions or jibs installed
- Lifting from the main load line with a boom extension and/or jib installed
- Lifting from the boom extension or jib

Remember that the advertised maximum capacity of a crane is calculated with the shortest boom option and lifting at the minimum operating radius. Do not think that just because a crane is rated at 50 tons (45.4 metric tons), it can lift that load in all situations. The advertised maximum capacity is likely more valuable in marketing efforts than it is to the crane operator. Experienced crane operators know that the load chart determines what

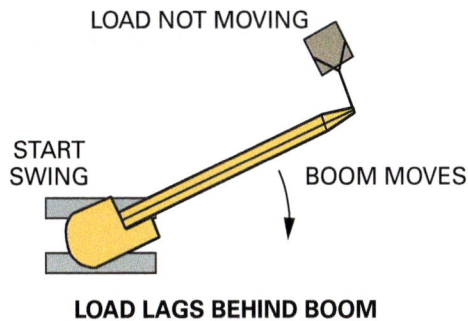

LOAD LAGS BEHIND BOOM

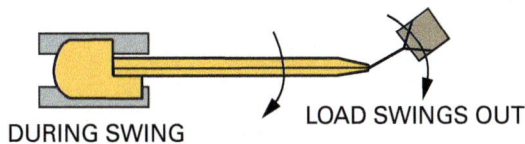

RAPID SWING CAUSES LOAD RADIUS TO INCREASE

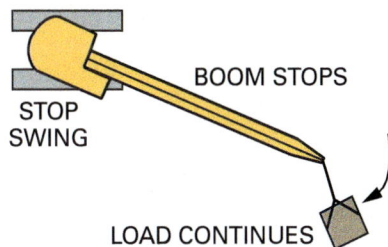

BOOM STOPS – LOAD CONTINUES TO SWING

Figure 46 Operating sequence that leads to swing out.

the crane is capable of in a given set of conditions, and that the advertised maximum capacity is substantially higher than its actual capacity in most situations and configurations.

> **WARNING!**
>
> Load charts are based on a level crane that is properly supported on an appropriately prepared surface. Any deviation from the conditions that the load chart is based on, including an out-of-level position, can result in property damage and personal injury.

Each crane manufacturer has their own way of displaying a load chart, but all contain the same type of information, which includes the following:

- Type of crane base
- Type of crane configuration
- Quadrant of operation
- Length of boom
- Angle of boom
- Load radius

- Allowed deductions for additional equipment from the gross capacity to arrive at the net capacity

Figure 47 shows a sample working-range diagram for a National Crane Series NBT40 boom truck with a five-section, 142-foot (43.3 m) boom and a 26-foot (7.9 m) folding jib. *Figure 48* shows the matching load chart. Notice the heavy, dark line on the load chart, separating the top from the bottom. The capacities listed above this line are based on the structural strength of the boom extension as the weakest link, rather than on the tipping force. Weights below the line are based on the tipping force of the load at the designated angle and boom length. Other manufacturers may identify those values by shading the capacity numbers in the appropriate area, or placing an asterisk beside them.

Many cranes also have range diagrams and load charts in metric units, depending upon the distribution of that particular model. Metric data is not available for this particular crane model. Note that the load chart reports the gross capacity for a given set of conditions. The net capacity is calculated from this figure to determine the maximum weight of the load to be lifted.

Many cranes are designed to be used with different styles of boom tops, such as hammer-head, tapered top, boom extensions, or jibs. Each style of top has its own capacity chart or deductions for the extensions and jibs. This particular range diagram and load chart are based on the jib being in its stowed position. When it is stowed rather than removed, it adds to the weight managed by the boom.

Boom angle, operating radius, and boom length are critical factors in calculating the capacity of a crane. If, in determining loads from a chart, the chart increments do not precisely match the lift plan, use the next longer boom length, operating radius, or the next lower boom angle on the chart. For example, a lift requires a boom length of 105 feet (32 meters) and an operating radius of 60 feet (18.3 meters). Referring to the chart shown in *Figure 48*, note that there is no listing for a 105-foot boom across the top. The next longest boom length, 115 feet (35 meters), must be used. This shows that the maximum allowable load is 6,200 pounds when the outriggers are set and the jib is stowed. For this particular crane, this capacity is also valid around the entire 360-degree circle, rather than only in one or two quadrants. The point here however, is to avoid trying to calculate a value between the 101-foot boom and the 115-foot boom. This is known as interpolation, and it is not allowed. Trying to interpolate values can result in an unsafe lift.

Boom deflection not shown

Height from ground in feet

Boom length and extension in feet

— 26' Ext.

— 142

80°

70°

60°

— 128

50°

— 115

— 101

40°

— 88

30°

— 74

20°

— 61

— 47

10°

— 34

81° Max boom angle

0°

Axis of rotation

170 160 150 140 130 120 110 100 90 80 70 60 50 40 30 20 10

Operating radius in feet from axis of rotation

(6'–9")

(8'–0")

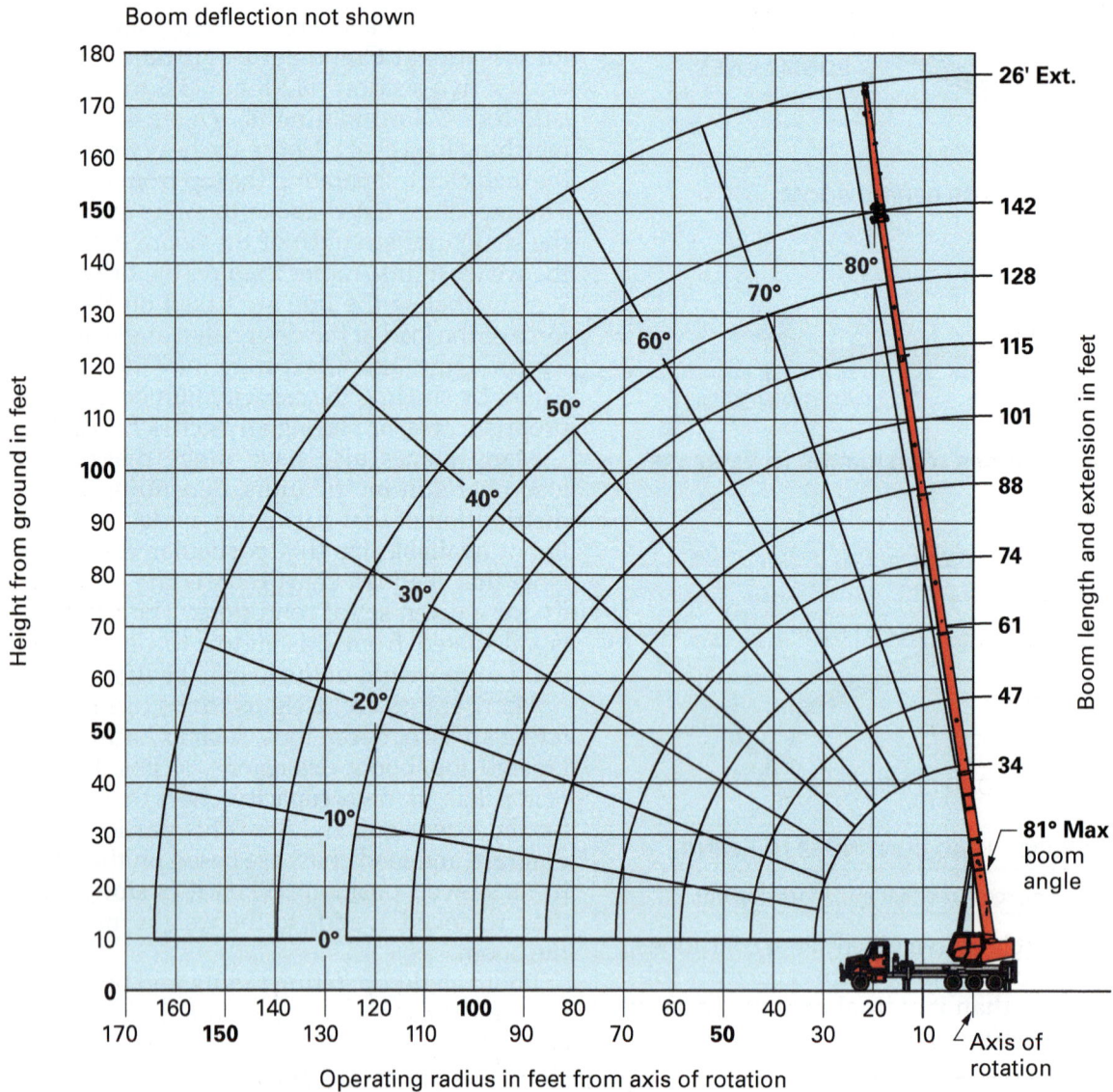

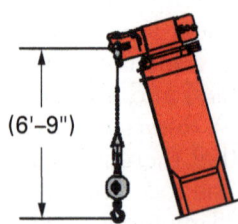

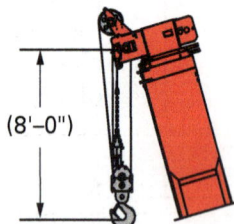

Dimensions are for largest furnished hookblock and headache ball with anti-two block activated.

*Drawing is to show the physical reach of the machine. Always refer to load chart to see what portions of this range are structurally and stability limited.

THIS CHART IS ONLY A GUIDE AND SHOULD NOT BE USED TO OPERATE THE CRANE.
The individual crane's load chart, operating instructions and other instructional plates must be read and understood prior to operating the crane.

Figure 47 Sample range diagram.

Radius in feet	#02										Radius in feet	#03
	Main boom length in feet											
	34	47-A	61-B	74-C	88-D	101-E	115-F	128-G	142			
7	79,475 (74.9)										33	4000 (80)
8	74,475 (73.1)											
10	65,975 (69.4)	39,600 (75.6)									50	3800 (75)
12	54,475 (65.7)	39,600 (73.1)	39,700 (77.4)									
15	42,475 (59.7)	39,600 (69.2)	37,700 (74.5)	33,750 (77.7)							65	3200 (70)
20	30,225 (48.9)	31,000 (62.3)	31,500 (69.5)	29,750 (73.7)	22,850 (76.7)	17,200 (78.8)					78	2450 (65)
25	22,725 (35.7)	23,450 (55)	23,950 (64.2)	24,250 (69.5)	20,500 (73.4)	15,550 (75.9)	12,850 (78.3)					
30	17,475 (13.5)	18,400 (46.9)	18,900 (58.8)	19,200 (65.2)	18,550 (70)	14,100 (73.1)	12,000 (75.8)	9900 (78)	7875 (79.5)		90	1800 (60)
35		14,750 (37.5)	15,250 (52.9)	15,550 (60.7)	15,800 (66.4)	13,000 (70.1)	11,000 (73.5)	9400 (75.8)	7475 (77.7)		101	1250 (55)
40		11,800 (25.2)	12,350 (46.6)	12,650 (56)	12,900 (62.6)	12,000 (67.1)	10,250 (71)	8900 (73.7)	7325 (75.9)			
45			10,000 (40.1)	10,300 (51.5)	10,550 (59.1)	10,700 (64.2)	9600 (68.4)	8400 (71.4)	7075 (74)		112	650 (50)
50			8100 (31.8)	8400 (46.2)	8600 (55)	8800 (60.8)	8950 (65.7)	7900 (69.1)	6675 (72)		Min. boom angle for indicated length (no load)	48°
55			6550 (20.6)	6900 (40.3)	7100 (50.8)	7250 (57.3)	7450 (62.8)	7450 (66.7)	6425 (70)			
60				5650 (33.6)	5850 (46.3)	6000 (53.7)	6200 (59.7)	6350 (64.1)	6075 (67.9)		Max. boom length at 0° boom angle (no load)	88 ft
65				4600 (25.4)	4850 (41.4)	5000 (49.9)	5150 (56.6)	5300 (61.4)	5425 (65.6)			
70				3750 (12.6)	4000 (35.9)	4100 (45.9)	4300 (53.3)	4400 (58.6)	4525 (63.1)			80027138
75					3250 (29.6)	3350 (41.6)	3550 (49.9)	3650 (55.7)	3775 (60.6)			
80					2600 (21.6)	2700 (36.9)	2850 (46.4)	2950 (52.7)	3125 (58)			
85					2100 (7.2)	2100 (31.5)	2250 (42.6)	2350 (49.6)	2475 (55.3)			
90						1650 (25.1)	1350 (38.5)	1850 (46.3)	1925 (52.6)			
95						1250 (16.5)	950 (34)	1350 (42.8)	1475 (49.7)			
100							600 (28.8)	950 (39.2)	1075 (46.7)			
105							500 (22.6)	650 (35.1)	675 (43.6)			
110									525 (41.6)			
Minimum boom angle (°) for indicated length (no load)						0	22.5	35	43.4			
Maximum boom length (ft) at 0° boom angle (no load)						101						

NOTE: Loads displayed in pounds. ( ) Boom angles are in degrees.
#LMI operating code. Refer to LMI manual for operating instructions.

Lifting capacities at zero degree boom angle						
Boom angle	Main boom length in feet					
	34	47-A	61-B	74-C	88-D	101-E
0°	16,825 (31.5)	9550 (44.5)	5600 (58.5)	3550 (71.5)	2000 (85.5)	900 (98.5)

NOTE: ( ) Reference radii in feet.   80027135

NOTE: Loads displayed in pounds
( ) Boom angles are in degrees.
#LMI operating code. Refer to LMI manual for operating instructions.

Boom extension capacity notes:
1. All capacities above the bold line are based on structural strength of boom extension.
2. 26 ft extension length may be used for single line lifting service
3. Radii listed are for a fully extended boom with the boom extension erected. For main boom lengths less than fully extended, the rated loads are determined by boom angle. For boom angles not shown, use the rating of the next lower angle.
   Warning: Operation of this machine with heavier loads than the capacities listed is strictly prohibited. Machine tipping with boom extension occurs rapidly and without advance warning.
4. Boom angle is the angle above or below horizontal of the longitudinal axis of the boom base section after lifting rated load.
5. Capacities listed are with outriggers properly extended and vertical jacks set.
6. When lifting over the main boom nose with 26 ft extension erected, the outriggers must be fully extended or 50% (17.5 ft) spread.

THIS CHART IS ONLY A GUIDE AND SHOULD NOT BE USED TO OPERATE THE CRANE.
The individual crane's load chart, operating instructions and other instructional plates must be read and understood prior to operating the crane.

Figure 48 Sample load chart.

Note that this section merely introduces load charts and doesn't represent a full presentation of the topic. Additional information on load charts is presented in future modules.

2.5.1 Capacity with No Attachments

All cranes have their greatest lifting capacity when they have no attachments. This is because attachments add to the lifted load. To calculate the capacity with no attachments, use the following steps:

Step 1 Use the load chart to find the gross capacity for the boom length, boom angle, load radius, and quadrant of operation.

Step 2 Use the load chart to determine the parts of line needed for the lift and to determine the traveling block needed.

Step 3 Use the load chart to determine any load deductions. This may include anything below the boom tip, such as the traveling block, headache ball, lifting rope, and all load rigging.

Step 4 Subtract all noted deductions from the gross capacity to determine the net capacity.

2.5.2 Capacity with Attachments

To calculate the capacity with attachments, use the following steps:

Step 1 Use the load chart to find the gross capacity for the boom length, boom angle, load radius, and quadrant of operation.

Step 2 Use the load chart to determine the parts of line needed for the lift and to determine the traveling block needed.

Step 3 Use the load chart to determine the effective weight of the extension or jib and attachments. This applies to any extensions or jibs in a stored position on the main boom. It also applies to installed extensions or jibs when the main boom (and not the extension or jib) is being used for the lift. When installed, the effective weight of the extension or jib may be as much as twice its actual weight. Deduct the effective weight from the boom capacity.

Step 4 Use the load chart to determine any load deductions. This may include anything below the boom tip, such as the traveling block, headache ball, lifting rope, and all load rigging.

Load Charts

Each crane can have many different load charts to consult, depending primarily on the crane's configuration. For example, the Link-Belt 218 HSL lattice-boom crawler crane (shown here) is a fairly common model rated at 110 tons (99.8 metric tons). The technical specifications for this crane include 18 different load charts.

Figure Credit: Link-Belt Construction Equipment Company

Step 5 Subtract all noted deductions from the gross capacity to determine the net capacity.

2.5.3 Capacity Using Extensions or Jibs

To calculate the capacity of a crane using extensions or jibs, perform the following:

Step 1 Use the jib chart, boom chart, and chart notes to find the jib offset, boom angle, or jib-to-ground angle.

Step 2 Use the jib chart and/or the boom chart to determine the gross capacity.

Step 3 Use the jib chart to determine which headache ball or block to use. Extension or jib lifts are usually made with a single part of line.

Step 4 Use the jib chart to determine the load deductions. This may include anything below the boom tip, such as the headache ball or block, lifting rope, and all load rigging.

Step 5 Subtract all noted deductions from the gross capacity to determine the net capacity.

2.5.4 Critical Lifts

There are a number of factors in a lift that can place it into the category of a critical lift. Lifting hazardous materials, working in close proximity to power lines, or swinging a load in a congested work area are all possible reasons to classify a lift as critical. In addition, any lift that exceeds 75 percent of a crane's capacity has been historically considered a critical lift by many organizations. A critical lift is defined the same way in 29 *CFR 1926.751, Steel Erection.* Per *ASME Standard B30.5,* "the criteria to categorize a lift as critical on this basis are established by site supervision, project management, a qualified person, or company policies.

Lift planning and oversight shall be tailored to each hoisting operation and shall be sufficient to manage varying conditions and their associated hazards." *ASME Standard B30.5* also has an appendix entitled *Nonmandatory Appendix A, Critical Lifts* that provides guidance, including a number of scenarios that typically indicate the need for a critical-lift classification.

Critical lifts require a more detailed lift plan than a standard lift plan, and the plan must be documented in writing. A critical lift plan should include but not be limited to the following:

• The total weight to be lifted
• Crane placement location
• Identification of the crane(s) to be used, including its required configuration, and the percentage of its capacity the load represents in that configuration
• Sling and rigging component selection
• A dimensioned diagram of the lifting area
• A diagram of the rigging configuration

Once a plan has been assembled, all responsible job-site personnel, including the crane operator, must review, approve, and sign the plan before it can be implemented. A pre-lift meeting is also called to ensure all parties are aware of the concerns before the lift begins.

Additional Resources

ASME Standard B30.5, Mobile and Locomotive Cranes. Current edition. New York, NY: American Society of Mechanical Engineers.

29 *CFR* 1926.1400, **www.ecfr.gov**

Mobile Crane Safety Manual. 2014. Milwaukee, WI: Association of Equipment Manufacturers.

Cranes: Design, Practice, and Maintenance, Ing J. Verschoof. Second Edition. 2002. Hoboken, NJ: John Wiley and Sons, Inc.

IPT's Crane and Rigging Handbook, Ronald G. Garby. Current Edition. Spruce Grove, Alberta, Canada: IPT Publishing and Training Ltd.

North American Crane Bureau, Inc. website offers resources for products and training, **www.cranesafe.com**

2.0.0   Section Review

1. When working near an excavation, any weight-bearing component, such as an outrigger, must be positioned at least _____.

   a. 1.5 times the depth of the excavation away
   b. 2 times the depth of the excavation away
   c. 2.5 times the depth of the excavation away
   d. 3 times the depth of the excavation away

2. When using crane mats to support a crawler crane, how far should the mat extend beyond the ends and edges of the tracks?

   a. 18 inches (46 centimeters)
   b. 2 feet (0.6 meters)
   c. 4 feet (1.2 meters)
   d. 5 feet (1.5 meters)

3. The rated capacity of a crawler crane is typically _____.

   a. 65 percent of its tipping load
   b. 75 percent of its tipping load
   c. 85 percent of its tipping load
   d. 95 percent of its tipping load

4. Dragging a load along the ground from one side of the crane to the other will cause _____.

   a. load lag
   b. swing out
   c. effective loading
   d. side loading

5. Manufacturers use bold lines, shading, or asterisks to identify load-chart capacity figures that are based on _____.

   a. the quadrant of operation
   b. lifts with and without a jib
   c. the structural strength of the boom extension
   d. the structural strength of the ring gear drive

SUMMARY

Mobile cranes are common on modern construction sites and many other environments. Some cranes are used to lift and carry materials and equipment around the site. Others are used to lift and place materials during construction. Mobile crane operators must be familiar with the components of the crane in use and the factors affecting its lifting capacity. Operators must also be aware of overhead, ground level, and below-ground conditions that may present a danger to workers in the area and nearby property.

Range diagrams and load charts provide essential information for each and every crane. No crane should be operated without the appropriate load chart on hand. These critical data resources will be explored further in future training modules, since reading and interpreting the information on load charts is a vital operator skill.

Review Questions

1. Rough-terrain cranes generally travel at or below a speed of roughly _____.
   a. 10 mph (16 kph)
   b. 20 mph (32 kph)
   c. 30 mph (48 kph)
   d. 45 mph (72 kph)

2. Wheeled truck cranes without a load can typically climb a job site ramp with a _____.
   a. 40-percent grade
   b. 50-percent grade
   c. 60-percent grade
   d. 70-percent grade

3. In addition to the jib itself, the jib forestay and jib backstay are also attached to the _____.
   a. hoist line
   b. load moment
   c. hoist drum
   d. jib mast

4. The lowest portion of a lattice boom that is connected directly to the upperworks is called the _____.
   a. lead
   b. boom base
   c. jib
   d. boom butt

5. To move loose material and unload it easily, without the need for ropes or cables to operate the device, a crane operator can use a(n) _____.
   a. clamshell bucket
   b. self-dumping bin
   c. grapple
   d. rehandling clamshell bucket

6. The two primary types of pile-driving equipment handled by a crane are _____.
   a. pile hammers and pile drills
   b. pile drills and vibratory pile drivers
   c. pile hammers and sheet hammers
   d. vibratory pile drivers and pile hammers

7. An example of a Category II operational aid is a(n) _____.
   a. boom hoist limiting device
   b. luffing jib limiting device
   c. anti-two-blocking device
   d. jib angle indicator (for luffing jibs)

8. If a crane hoist sheave has four-part reeving, the hoist line is terminated at the _____.
   a. jib
   b. actuator
   c. boom
   d. hoist sheave

9. Cranes are required to be no more than one percent out of level, which equals _____.
   a. 3.6 degrees
   b. 2.31 degrees
   c. 1.14 degrees
   d. 0.57 degrees

10. When a crane's level is missing or unreliable, the crane can be leveled using the headache ball as a _____.
    a. plumb bob
    b. level
    c. counterweight
    d. protractor

11. For a rough-terrain crane, the maximum lift capacity is accessed when the boom is in which quadrant?
    a. Over the front
    b. Over the rear
    c. Over the side
    d. 360 degrees

12. The rated capacity of a wheeled truck crane when on outriggers is typically _____.
    a. 65 percent of its tipping load
    b. 75 percent of its tipping load
    c. 85 percent of its tipping load
    d. 95 percent of its tipping load

13. When a lattice boom is lowered to a smaller angle (closer to the horizontal plane), _____.
    a. it begins to sag in the middle
    b. the leverage applied by a load is reduced
    c. its capacity becomes nearly infinite
    d. a whip line cannot be used at all

14. When using a specific load chart for a crane, _____.
    a. you can assume the quadrant of operation is 360 degrees
    b. the effective weight of a jib will be the same as its actual weight
    c. interpolation is not allowed
    d. the net capacity is read directly from the chart

15. A critical lift plan must be reviewed and signed by _____.
    a. the site supervisor
    b. the crane manufacturer
    c. the lift director
    d. all responsible job site personnel

Trade Terms Quiz

Fill in the blank with the correct term that you learned from your study of this module.

1. A(n) _____ is a portable platform, typically made of large wooden timbers bolted together to support and spread the weight of a crane over a larger ground area.

2. A(n) _____ is a piece of standing rigging that is routed from the jib mast back to the main boom to help support the jib.

3. To pick up items such as a downed tree, a stack of tree limbs, or a large rock, the best attachments for the job would be _____.

4. The crawler crane assembly that consists of the carbody, ring gear drive, crawler frames, and tracks is called the _____.

5. Steel structures called _____ provide support for a pile hammer and help to align the hammer with the pile to be driven.

6. To prevent the load block or hook assembly from coming into contact with the boom tip, a(n) _____ is used.

7. A system of two or more pulleys with a rope or cable threaded between them to use as a lifting aid is called a(n) _____.

8. Loads that are not constant, but instead are consistently changing due to various factors in a lift are called _____.

9. The point at which the entire weight of an object is considered to be concentrated is called its _____.

10. The _____ is the lowest portion of a telescopic boom that houses the other telescopic sections, but does not itself extend.

11. Component used to guide or place tension on a belt or crawler-crane track are called _____.

12. To help oppose the weight of the load and improve stability, _____ are added to a crane.

13. Improper reeving of the boom tip sheave, where the descending rope is placed to one side of the sheave rather than in the center, leads to _____.

14. If crane documentation refers to a percentage of time that the crane is designed to operate over a 24-hour period, it is referring to the crane's _____.

15. Crane assemblies that include crawler tracks, track idlers, and track power sources of a crawler crane are called _____.

16. Soil and rock is used as _____ to level an area or fill voids, such as the perimeter of building foundations or trenches.

17. To determine an unknown value that lies between two known values, _____ is needed.

18. A(n) _____ is used to wind or unwind the rope for hoisting or lowering the load.

19. A device designed to allow flow in one direction but prevents fluid flow in the opposite direction is a(n) _____.

20. The _____ of an accessory such as a boom extension or jib reflects the effect of its weight on the lift, usually based on its position, rather than its actual weight.

21. _____ refers to a hard, compacted layer of subsoil, usually with a major clay component.

22. If a lifting operation will present a significantly increased level of risk beyond normal lifting activities, it will likely be identified as a(n) _____.

23. A(n) _____ extends or retracts using pressure from the hydraulic system.

24. A(n) _____ refers to the base of a wheeled crane that provides crane movement.

25. The feet of outriggers on wheeled truck cranes are called _____.

26. Motors powered by fluid pressure, called _____, are often used to power crawler-crane tracks.

27. Extendable or fixed members known as _____ are attached to a crane base to stabilize and support a crane.

28. The dynamic effects on a stationary or mobile body as imparted by the forcible contact of another moving body or the sudden stop of a fall are called _____.

29. When you apply a lever, you are using the mechanical advantage in power known as _____.

30. Extension attached to booms point to provide added length for reaching and lifting loads are called _____.

31. A heavy, round weight called a(n) _____ is often attached to a load line to provide sufficient weight to allow the load line to unspool from the drum when there is no live load.

32. When reeving, the space between the sheave and the frame of a block through which the rope is passed is called the _____.

33. The _____ is a piece of standing rigging that is routed from the far tip of the jib back to the jib mast, holding the tip of the jib up.

34. According to 29 *CFR* 1926.1401, the _____ of a crane can also be referred to as the superstructure.

35. A(n) _____ is a secondary hoisting rope usually of lower capacity than that provided by the main hoisting system.

36. A structure mounted on the main boom that provides a point of connection for the jib forestay and jib backstay on lattice-boom cranes is called the _____.

37. A boom constructed of steel angles or tubing to create a relatively lightweight but strong, rigid structure is referred to as a(n) _____.

38. Ropes or strands of a specified length with fixed end connections, used to support a lattice boom or boom components, are called _____.

39. The _____ is usually reported to the operator as a percentage of the crane's capacity at the present set of conditions.

40. The part of a crawler-crane base mounting that carries the rotating upperworks is called the _____.

41. The total amount a crane can safely lift under a given set of conditions defines its _____.

42. When you adjust the boom angle by varying the length of the suspension ropes, you are _____ the boom.

43. A lattice boom with an opening in the boom structure near the far end, allowing the hoist lines to drop through the boom rather than over the end of the boom, is called a(n) _____.

44. The _____ of a crane is its gross capacity minus all noted capacity deductions.

45. A process known as _____ is used to multiply pulling or lifting capability by using wire rope routed through multiple pulleys or sheaves a number of times.

46. A metal that contains no iron, such as aluminum, is considered _____.

47. The distance from the center of the boom's mounting point (usually the ring gear drive) to the center of gravity of the load defines a crane's _____.

48. A trailer with a low frame for transporting very tall or heavy loads, such as large cranes, is called a(n) _____.

49. Single- or multiple-line _____ is used for whip, boom, and jib lines.

50. If the capacity of a crane is exceeded, the crane will turn over at a point called the _____.

51. When a line is reeved on the hoist sheaves, the resulting number of lines that are supporting the load block are referred to as _____.

52. The combined operating height and operating radius of a boom determine a crane's _____.

53. The direction of the boom relative to the base mounting or carrier body determines its _____.

54. A jib mounted on the end of a boom that can be positioned at different angles relative to the main boom is called a(n) _____.

55. The upperworks of a crane are mounted to the _____, which provides the pivot point for the entire assembly. It is sometimes referred to as the swing circle.

56. Wheels that have a groove for a belt, rope, or cable to run in re called _____.

57. The distance between the front and rear axles of a vehicle is called the _____.

58. Cribbing is another word for _____.

Trade terms

Anti-two-blocking device	Duty cycle	Jib backstay	Parts of line
Backfill	Dynamic loads	Jib forestay	Pendants
Base mounting	Effective weight	Jib mast	Quadrant of operation
Base section	Floats	Lattice boom	Reach
Block and tackle	Grapples	Leads	Reeving
Blocking	Gross capacity	Leverage	Ring gear drive
Boom torque	Hardpan	Load moment	Sheaves
Carbody	Headache ball	Lowboy	Swallow
Carrier	Hoist drum	Luffing	Telescopic boom
Center of gravity	Hoist reeving	Luffing jib	Tipping fulcrum
Check valve	Hydraulic motors	Net capacity	Upperworks
Counterweights	Idlers	Non-ferrous	Wheelbase
Crane mat	Impact loads	Open-throat boom	Whip line
Crawler frames	Interpolation	Operating radius	
Critical lift	Jibs	Outriggers	

Richard Laird

Mobile Crane and Heavy Equipment Lead Instructor

Associated Builders and Contractors Pelican Chapter

Building upon a strong work ethic and an expansive base of experience in the petrochemical industry, Richard maintains his excellence by continually improving his own skills as well as training and certifying the next generation of Mobile Crane and Rigging/Signal persons, with a vigilant focus on worker safety.

Please give a brief synopsis of your construction career and your current position.

After working 34 years in the petroleum chemical industry, I retired from ExxonMobil as the refinery lift specialist in the rigging and mobile crane operation. My work also included procuring cranes and negotiating work orders with the major crane rental companies in the area. I now provide training and certification to companies, and work with Associated Builders and Contractors (ABC) in Baton Rouge as the Lead Instructor for mobile crane and heavy equipment operators. I am also involved with NCCER as a Mobile Crane Practical Examiner (CPE), and a Rigging/Signal Person Practical Examiner.

How did you get started in the construction industry?

Like every young man wanting to have a steady job and money in the bank, I successfully launched my career by answering an ad in the paper, and began working with Humble Oil and Refining Company in Baton Rouge, now known as ExxonMobil Refinery.

Who or what inspired you to enter the construction industry?

I enjoyed watching cranes and riggers work together to move equipment. The more I saw this, the more I became inspired and knew this was what I wanted to do every day.

How has training in construction impacted your life and career? What types of training have you completed?

As I progressed through my own training and mastered skills, I was offered the opportunity to do tutoring work for ABC. As a result, I became an instructor for mobile crane operators and, eventually, a Practical Examiner.

Why do you think credentials are important in the construction industry?

There is a great need in the industry today for construction people to be credentialed. Not only does this bring a more well-rounded and verifiably skilled person to the craft, but it also encourages a safety-focused culture and promotes worker awareness of the need to protect themselves and others on the job.

What do you enjoy most about your career?

The most rewarding part of my career is being able to encourage and watch both young men and women making career changes that generate job satisfaction and improved financial positions in their lives.

Would you recommend construction as a career to others? Why?

The construction environment presents an opportunity for people to find well-paying jobs that bring great personal satisfaction, and being able to provide a service that is of value and a benefit to others.

What advice would you give to someone who is new to the construction industry?

I would advise anyone beginning to work in the construction trades to be aware that there are many opportunities available for continued training and competency that can fast-track their career goals.

How do you define craftsmanship?

Craftsmanship is what happens when a tradesman advances the quality of his or her work to the next level and then beyond, achieving the goals of being well-trained and proficient in a given craft. A craft professional consistently produces excellence in their work by not short-cutting tasks and always maintaining a safe work environment for themselves and others.

Trade Terms Introduced in This Module

Anti-two-blocking device: Two-blocking refers to a condition in which the lower load block or hook assembly comes in contact with the boom tip, boom tip sheave assembly or any other component above it as it is being raised. If this occurs, continuing to apply lifting power to the cable can result is serious equipment damage and/or failure of the hoist line. An anti-two-blocking device, therefore, prevents this condition from occurring.

Backfill: Soil and rock used to level an area or fill voids, such as the perimeter of building foundations or trenches. Areas with fresh backfill may not be stable enough to support a crane.

Base mounting: A crawler crane assembly consisting primarily of the carbody, ring gear drive, crawler frames, and tracks.

Base section: The lowest portion of a telescopic boom that houses the other telescopic sections but does not extend.

Block and tackle: A system of two or more pulleys, which form a block, with a rope or cable threaded between them, reducing the force needed to lift or pull heavy loads.

Blocking: Wood or a similar material used under outrigger floats to support and distribute loads to the ground. Also referred to as *cribbing*.

Boom torque: A twisting force applied to the crane boom, typically resulting from imbalanced reeving of the boom tip sheave assembly ropes.

Carbody: The part of a crawler-crane base mounting that carries the rotating upperworks.

Carrier: The base of a wheeled crane that provides crane movement and supports the upperworks.

Center of gravity (CG): The point at which the entire weight of an object is considered to be concentrated, such that supporting the object at this specific point would result in its remaining balanced in position.

Check valve: A valve designed to allow flow in one direction but closes as necessary to prevent flow reversal.

Counterweights: Weights added to the crane, usually on the end opposite the boom, to help counter the weight of the load and improve stability.

Crane mat: A portable platform, typically made of large wooden timbers bolted together, used to support and spread the weight of a crane over a larger ground area.

Crawler frames: Crane assemblies comprised of the crawler tracks, track idlers, and track power sources of a crawler crane. Also called *tread members* or *track assemblies*.

Critical lift: Defined in *ASME Standard B30.5* as a hoisting or lifting operation that has been determined to present an increased level of risk beyond normal lifting activities. For example, increased risk may relate to personnel injury, damage to property, interruption of plant production, delays in schedule, release of hazards to the environment, or other significant factors.

Duty cycle: An expression of equipment use over time. In the case of mobile cranes, an 8-, 16-, or 24-hour rating expressed as a percentage.

Dynamic loads: A load on a structure (in this case, a crane) that is not constant, but consistently changing as the result of one or more changes in various factors. Also referred to as shock loading, significant dynamic loads can be applied to a crane through abrupt motions and lifting a load from its support too quickly.

Effective weight: The weight of an accessory such as a boom extension or jib that reflects the effect of its weight on the lift, usually based on its position, rather than its actual weight. For example, a jib folded and stored on the main boom will have different effective weight than when it is installed on the main boom tip.

Floats: The portion of outriggers that touches the ground; the feet of the outriggers.

Grapples: Devices used to pick up bulk items, containers, rocks, trees and tree limbs, etc. Grapples typically have several jaws that operate like fingers to pick up material, using mechanical or hydraulic power.

Gross capacity: The total amount a crane can safely lift under a given set of conditions. The gross capacity includes but is not limited to the load block, ropes, and rigging as well the primary load.

Hardpan: A hard, compacted layer of subsoil, usually with a major clay component.

Headache ball: A heavy round weight often attached to a load line to provide sufficient weight to allow the load line to unspool from the drum when there is no live load. Larger versions of headache balls are used to swing into structures to demolish them.

Hoist drum: A drum is a cylindrical component around which a rope is wound. The hoist drum is used to wind or unwind the rope for hoisting or lowering the load; the part of a crane that spools and unspools the lifting line.

Hoist reeving: The reeving pattern applied to the hoist sheaves. Single- or multiple-line hoist reeving is used for whip, boom, and jib lines.

Hydraulic motors: Motors powered by hydraulic pressure provided by an external pump. Hydraulic motors are often used to power the tracks of crawler cranes, instead of complex drive systems connected directly to the diesel engine.

Idlers: Pulleys, wheels, or rollers that do not transmit power, but guide or place tension on a belt or crawler-crane track.

Impact loads: The dynamic effects on a stationary or mobile body as imparted by the forcible contact of another moving body or the sudden stop of a fall.

Interpolation: The process of estimating or calculating unknown values between two known values.

Jibs: Extensions attached to the boom point to provide added boom length for reaching and lifting loads. Jibs may be in line with the boom, offset to another angle, or adjustable to a variety of angles. A jib is sometimes referred to as a *fly*.

Jib backstay: A piece of standing rigging that is routed from the jib mast back to the main boom to help support the jib.

Jib forestay: A piece of standing rigging that is routed from the far tip of the jib back to the jib mast, holding the tip of the jib up.

Jib mast: A structure mounted on the main boom that provides a fixed distance for the point of connection of the jib forestay and jib backstay. Also referred to as a *jib strut*.

Lattice boom: A boom constructed of steel angles or tubing to create a relatively lightweight but strong, rigid structure.

Leads: Steel structures that provide support for a pile hammer and help to align and position the hammer with the pile to be driven. The hammer can travel up or down in the leads as necessary.

Leverage: The mechanical advantage in power gained by using a lever.

Load moment: The force applied to the crane by the load; the leverage of the load, opposing the leverage of the crane. The load moment is calculated by multiplying the gross load weight by the horizontal distance from the tipping fulcrum to the center of gravity of the suspended load. The load moment is usually reported to the operator as a percentage of the crane's capacity at the present set of conditions. As those conditions change, such as the boom angle, the load moment changes as well.

Lowboy: A trailer with a low frame for transporting very tall or heavy loads. A typical lowboy has two drops in deck height: one right after the gooseneck connecting it to the tractor, and one right before the wheels. This allows the trailer deck to be extremely low compared with common trailers.

Luffing: Changing a boom angle by varying the length of the suspension ropes.

Luffing jib: A jib mounted on the end of a boom that can be positioned at different angles relative to the main boom.

Net capacity: The weight of the item(s) that can be lifted by the crane; the gross capacity of a crane minus all noted capacity deductions.

Non-ferrous: Having no iron. Ferrous metals, such as steel, contain iron and are magnetic as a result.

Open-throat boom: A lattice boom with an opening in the boom structure near the far end, allowing the hoist lines to drop through the boom rather than over the end of the boom.

Operating radius: The distance from the center of the boom's mounting point (usually the ring gear drive) to the center of gravity of the load.

Outriggers: Extendable or fixed members attached to a crane base that rest on ground supports at the outer end to stabilize and support the crane.

Parts of line: When a line is reeved more than once, the resulting number of lines that are supporting the load block.

Pendants: Ropes or strands of a specified length with fixed end connections, used to support a lattice boom or boom components. According to 29 *CFR* 1926.1401, a pendant may also consist of a solid bar.

Quadrant of operation: The direction of the boom relative to the base mounting or carrier body.

Reach: The combined operating height and radius of a boom, or the combination of boom and jib.

Reeving: A method often used to multiply the pulling or lifting capability by using wire rope routed through multiple pulleys or sheaves a number of times.

Ring gear drive: Sometimes referred to as the swing circle. An assembly that provides the point of attachment and pivot point for the upperworks of a crane. The ring gear is typically driven by hydraulic pressure, allowing the upperworks to rotate on a set of bearings that reduce friction and transfer the weight of the upperworks (and any load) to the carbody.

Sheaves: Wheels that have a groove for a belt, rope, or cable to run in. The terms *sheave* and *pulley* are often used interchangeably.

Swallow: The space between the sheave and the frame of a block, through which the rope is passed.

Telescopic boom: A crane boom that extends and retracts in sections that slide in and out, powered by hydraulic pressure.

Tipping fulcrum: The point of crane contact with the ground where it would pivot if it were to tip over; the fulcrum of the leverage applied by the load. Depending on the attitude and type of crane, the tipping fulcrum may be the edge of one crawler assembly, one or more outriggers, or similar locations.

Upperworks: A term that refers to the assembly of components above the ring gear drive; the rotating collection of components on top of the base mounting or carrier; may also be referred to as the *house,* or as the *superstructure* as defined in 29 *CFR* 1926.1401.

Wheelbase: The distance between the front and rear axles of a vehicle.

Whip line: A secondary hoisting rope usually of lower capacity than that provided by the main hoisting system. When a whip line exists, it is typically out at the tip of a jib, while the main hoist line is closer to the crane and operated from the tip of the main boom.

Additional Resources

This module presents thorough resources for task training. The following reference material is recommended for further study.

ASME Standard B30.5, Mobile and Locomotive Cranes. Current edition. New York, NY: American Society of Mechanical Engineers.

29 CFR 1926.1400, **www.ecfr.gov**

Mobile Crane Safety Manual (AEM MC-1407). 2014. Milwaukee, WI: Association of Equipment Manufacturers.

Cranes: Design, Practice, and Maintenance, Ing J. Verschoof. Second Edition. 2002. Hoboken, NJ: John Wiley and Sons, Inc.

IPT's Crane and Rigging Handbook, Ronald G. Garby. Current Edition. Spruce Grove, Alberta, Canada: IPT Publishing and Training Ltd.

North American Crane Bureau, Inc. website offers resources for products and training, **www.cranesafe.com**

Figure Credits

Link-Belt Construction Equipment Company, Module opener, Figures 1, 5, 7–9, 10A, 14–16, 28, SA04

Topaz Publications, Inc., Figures 2, 4

The Manitowoc Company, Inc., Figures 6, 12, 13, 47, 48

Manitex, Inc., Figure 10B

© Hellen Sergeyeva/Shutterstock.com, Figure 11

Liebherr USA, Co., SA01

Smiley Lifting Solutions - www.spydercrane.com, SA02

© iStockphoto.com/dane-mo, Figure 19

© iStockphoto.com/Paul Vasarhelyi, Figure 20

Bigfoot Crane Company Inc., Figure 21

© iStockphoto.com/Bradford Martin, Figure 22

© AngelPet/Shutterstock.com, Figure 23

Kenco Construction Products, Inc., Figure 24

Lifting Technologies, Figure 25

Carolina Bridge Co., Figures 26, 27

SANY America, Inc., Figure 29

DICA Outrigger Pads, SA03

Section Review Answer Key

Answer	Section Reference	Objective
Section One		
1. a	1.1.2	1a
2. c	1.2.2	1b
3. d	1.3.1	1c
4. a	1.4.0	1d
5. c	1.5.0	1e
Section Two		
1. a	2.1.1	2a
2. b	2.2.1	2b
3. b	2.3.2; Table 4	2c
4. d	2.4.1	2d
5. c	2.5.0	2e

NCCER CURRICULA — USER UPDATE

NCCER makes every effort to keep its textbooks up-to-date and free of technical errors. We appreciate your help in this process. If you find an error, a typographical mistake, or an inaccuracy in NCCER's curricula, please fill out this form (or a photocopy), or complete the online form at **www.nccer.org/olf**. Be sure to include the exact module ID number, page number, a detailed description, and your recommended correction. Your input will be brought to the attention of the Authoring Team. Thank you for your assistance.

Instructors – If you have an idea for improving this textbook, or have found that additional materials were necessary to teach this module effectively, please let us know so that we may present your suggestions to the Authoring Team.

NCCER Product Development and Revision

13614 Progress Blvd., Alachua, FL 32615

Email: curriculum@nccer.org
Online: www.nccer.org/olf

❏ Trainee Guide   ❏ Lesson Plans   ❏ Exam   ❏ PowerPoints   Other _____

Craft / Level: _____   Copyright Date: _____

Module ID Number / Title: _____

Section Number(s): _____

Description: _____

_____

_____

_____

Recommended Correction: _____

_____

_____

_____

Your Name: _____

Address: _____

_____

Email: _____   Phone: _____

53101-18
Crane Communications

OVERVIEW

This module focuses on the methods and modes of communication required in crane operations. General information about the communication process is also presented to help workers better understand the mechanics of communication in all environments. Signal persons are relied upon to properly communicate both verbally and nonverbally with crane operators, and crane operators must learn how to interpret verbal messages and hand signals provided by a signal person.

Module Twelve

Trainees with successful module completions may be eligible for credentialing through the NCCER Registry. To learn more, go to **www.nccer.org** or contact us at 1.888.622.3720. Our website has information on the latest product releases and training, as well as online versions of our *Cornerstone* magazine and Pearson's product catalog.

Your feedback is welcome. You may email your comments to **curriculum@nccer.org**, send general comments and inquiries to **info@nccer.org**, or fill in the User Update form at the back of this module.

This information is general in nature and intended for training purposes only. Actual performance of activities described in this manual requires compliance with all applicable operating, service, maintenance, and safety procedures under the direction of qualified personnel. References in this manual to patented or proprietary devices do not constitute a recommendation of their use.

53101-18
CRANE COMMUNICATIONS

Objectives

When you have completed this module, you will be able to do the following:

1. Describe the communication process and identify barriers to effective communication.
   a. Describe the basic communication process.
   b. Identify common barriers to effective communication.
2. Identify and interpret the OSHA regulations related to crane communications and explain how to communicate with crane operators verbally and nonverbally.
   a. Identify and interpret construction-related OSHA regulations associated with crane communications and signaling.
   b. Describe the equipment used for verbal communications and how to communicate with and direct a crane operator verbally.
   c. Explain how to communicate with and direct a crane operator nonverbally.

Performance Tasks

Under the supervision of your instructor, you should be able to do the following:

1. Demonstrate proper crane-communication techniques using a handheld radio or another acceptable verbal-signaling device.
2. Demonstrate each standard hand signal depicted in 29 *CFR* 1926.1400, Subpart CC, Appendix A.
3. Direct an operator to move and place a load using the appropriate hand signals.
4. Direct an operator to move and place a load using voice communication.

Trade Terms

Abstraction
Blind lift
Bridge
Consensus standard
Dedicated spotter
Diver tender

Line of sight
Nonverbal communication
Open mike
Paraphrasing
Trucks

Industry Recognized Credentials

If you are training through an NCCER-accredited sponsor, you may be eligible for credentials from NCCER's Registry. The ID number for this module is 53101-18. Note that this module may have been used in other NCCER curricula and may apply to other level completions. Contact NCCER's Registry at 888.622.3720 or go to **www.nccer.org** for more information.

Contents

Figures

1.0.0 THE COMMUNICATION PROCESS

Objective

Describe the communication process and identify barriers to effective communication.
  a. Describe the basic communication process.
  b. Identify common barriers to effective communication.

Trade Terms

Abstraction: Any form of verbal, graphical, or written communication representing a generalized and nonspecific idea or quality of a thing, action, or event.

Nonverbal communication: All communication that does not use words. This includes appearance, personal environment, use of time, and body language.

Paraphrasing: Expressing the perceived meaning of something read or heard in one's own words, generally to ensure clarity. Paraphrasing is an important component of active listening.

The ability of workers to communicate effectively is essential to the safe operation of cranes. The clarity of the information being exchanged is very important, and can be difficult to maintain due to distractions and noise that accompany crane operation and other construction activities. The techniques presented in this module can assist anyone that provides or responds to crane operating signals in strengthening their ability to communicate effectively.

Before presenting the detailed requirements of crane signaling and communication, it is beneficial to understand the general communication process. This knowledge helps workers understand how and why information may not be accurately transferred and take the necessary steps to avoid problems. The sections that follow provide general information about communication that are valuable both on and off the job.

1.1.0 Exchanging a Message

The communication process (*Figure 1*) consists of a message being sent and, ideally, received and interpreted precisely as intended. A message may be sent verbally or through nonverbal communication.

The challenge for those involved in operating cranes is not only to communicate the right information to co-workers, but to do so effectively.

The communication process consists of the following three general components:

- Sending the message
- Receiving the message
- Feedback

1.1.1 Sending the Message

There are four elements involved in sending a message. First, the sender prepares the message that is intended for communication. Next, the sender considers possible internal barriers that may affect the message. This includes the sender's own experiences, the terms used, and even the sender's feelings toward the receiver. External barriers, such as noise, must also be considered. Third, the sender encodes the message; that is, the sender puts the message into spoken words (verbal), written words (nonverbal), or gestures (nonverbal) that he or she wants to use. Finally, the sender sends the message through the chosen method.

1.1.2 Receiving the Message

There are also four elements involved in receiving a message. The receiver will first hear and/or see the message that was sent. Second, the influence of any active external or internal barriers takes effect. Possible internal barriers, for example, may include the receiver's experience level, the receiver's understanding of the terms used by the sender, the receiver's attitude toward the job, or even the way the receiver feels about the sender. Third, the receiver often decodes the

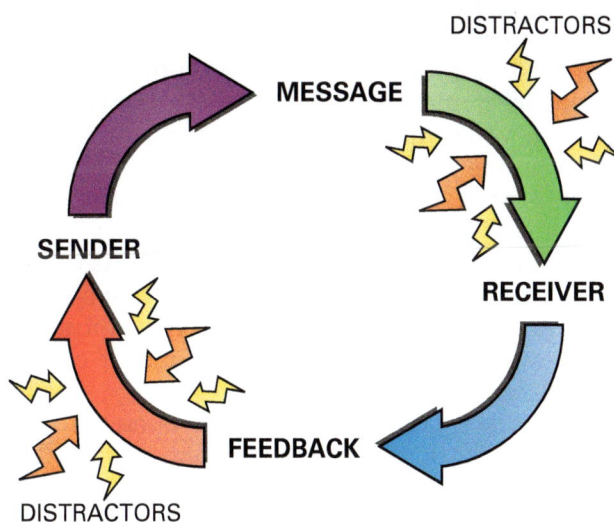

Figure 1 The communication process.

message through a rapid set of mental images. For instance, when a sender speaks the term *two-block*, a receiver does not generally imagine the letters that form the word. Instead, a mental image represented by the words appears. Many different mental images might arise among those who hear this word—perhaps a crane boom, a wire rope, or a load block might be briefly pictured. Fourth, the message is interpreted from the receiver's perspective. At this point in the process, there is no way to determine that the message was received as intended. To determine this, the receiver must provide feedback to the sender.

1.1.3 Feedback

Feedback, which may take several forms in general communication, provides essential information about the success in communicating a message. To get feedback, ask or look for a response. A simple yes or no may be all that is needed for simpler messages. Other times, a question may be asked in response to the message, indicating a lack of understanding or a need to verify the information. Nonverbal behaviors can also provide important clues about the clear reception of a message. Facial expressions and body language often indicate when the receiver clearly understands the message or is unsure of its meaning. Even a nonverbal response such as a nod of the head or a thumbs-up signal can be considered feedback.

Paraphrasing is a communication tool that often helps ensure clear communication between a sender and a receiver. It can be used when the message is not clear, or when verification is desired to avoid a mistake. Paraphrasing means to express the perceived meaning of something read or heard back to the sender using different words. For example, another worker may say, "I don't understand that guy. Some days he is on fire, but he's asleep at the wheel on other days." To ensure understanding, the other party may ask, "So you're not sure if he is consistently focused on his job?" This provides an opportunity for the receiver to find out if the message was interpreted correctly, and provides feedback to the first worker about how the message was interpreted. Paraphrasing is a valuable tool that can be used to reduce or eliminate miscommunication in many cases, including situations outside of the workplace.

Feedback to a message in crane operations is usually in the form of crane movement. Ideally, the crane operator repeats a signal back to the sender to ensure accuracy. The incorrect response to a signal may be uneventful, but it can also be

disastrous. For this reason, crane signaling goes further to formally close the communication loop with a three-step process:

- The signal person signals the crane operator.
- The operator repeats the signal back to the signal person.
- The signal person confirms that the crane operator has interpreted the signal correctly.

The terms and signals used to communicate with crane operators must be learned and practiced until they are deeply embedded in memory and are as natural to use as any other part of speech.

1.2.0 Communication Barriers

The purpose of effective communication in any environment is to ensure that the receiver accurately understands the material or information provided by the sender. The existence of certain barriers or filters increases the potential for poor communication. A barrier is something that can halt communication altogether. An example would be a jackhammer starting up nearby when you are speaking to someone. A filter is an internal, personal screening mechanism that receives a message and alters its meaning in some way. A filter can be a person's knowledge or prior experience, or even emotions. For example, if a signal person has sent the same signal incorrectly to a crane operator seven times in a row, the operator most likely will develop a filter that renders future uses of the same signal untrustworthy. By being aware of barriers and filters, lift teams can avoid them and ensure clear understanding of the information they are attempting to communicate. The following sections describe some of the barriers and filters that can damage effective communications, including those related to crane operations.

1.2.1 Lack of Common Experience

One potential barrier to effective communication is the lack of common experience. Operators are likely to find that their co-workers have many different backgrounds. Some individuals may have worked exclusively in one type of construction, such as bridge/road, structural steel, or residential, and are unable to immediately relate to another area of construction. To prevent misunderstandings, the sender and receiver first need to determine their experience level. Then, the sender should develop comparisons between the different types of work. For example, an operator who is accustomed to working on a building construction site might not understand some of the terminology used on bridge projects. Indeed, the word *bridge* itself has a completely different meaning in the world of overhead cranes than it does in bridge construction.

1.2.2 Language

Today's construction workplace is a reflection of the global marketplace. Some employees may not be fluent in the principal language of the jobsite or organization. A language barrier can certainly complicate the process of communication in any environment. The challenge of clear communication between individuals that do not share a native language is very common today, and a challenge that will likely continue to grow. Even when a non-native speaker has advanced knowledge of the vocabulary in a language, their pronunciation of some words may be difficult for a native speaker to understand.

One option is for all workers to become fluent in all the languages spoken on the job. Of course, that is an impractical solution. However, it is beneficial to understand some of the basic concepts and words of the language(s) you hear on the job regularly. A few words spoken in the language of the listener may be all it takes to clarify the instruction or statement.

Signal persons and crane operators must be fluent in a standardized language that is commonly understood by both. Using the standardized crane hand signals as outlined in 29 *CFR* 1926, Subpart CC, Appendix A is a reliable method of bridging language barriers in crane operations.

1.2.3 Overuse of Abstractions

An abstraction is a generalized, nonspecific concept or idea. To avoid confusion, speak in common, accepted, standardized terms, and be specific. Many abstractions are too general in nature, and therefore do not provide enough detail to the receiver. Be aware of the abilities of your co-workers and use appropriate terms and examples that suit their communication needs. Remember that signal persons are also responsible for the safe operation of the crane, so it is crucial that all communications between the operator and signal person are clearly understood. For example, if a signal person tells an operator to move a load "a little to the left"—an example of an abstraction—one operator may move it a few inches while another may move it several feet. The correct verbal instruction would include how far the load needs to be moved to the left. This removes the abstraction.

1.2.4 Fear

Fear may be one of the greatest barriers to effective communication. The fear of showing ignorance, the fear of disapproval, the fear of losing status, and the fear of judgment are all common barriers. Co-workers may have hidden anxieties or fears about their own abilities. They may lack confidence or be afraid of appearing ignorant, thus avoiding communication when they should speak out.

Around the World

Hand Signals for Everyone

More than ever before, the world is a global economy supported by a global workforce. Workers from China operate in the United States, workers from the United States operate in Brazil, and workers from Brazil might operate in Italy. This is true for crane operators as well. If someone has a taste for travel, an opportunity to serve as a crane operator can be found in almost any country.

Although it is certainly helpful, an English-speaking crane operator is not expected to master Portuguese, French, or any other language to function on the job. The most important language to know as a crane operator in a foreign country, regardless of your native language, is the language of hand signals.

The International Organization for Standardization (ISO) develops many different standards that help to ensure that products and services are safe and reliable, regardless of their origin. ISO is engaged in the world of crane operations through *ISO Standard 16715, Cranes—Hand Signals Used with Cranes*. The goal is to create a standard of hand signals for jobsites with an international workforce. If you plan to work outside of the Unites States, this is an important standard to be familiar with. In the United States, 29*CFR* 1926,*Subpart CC, Appendix A* is the US standard for crane hand signals.

Provide a threat-free environment by being encouraging and nonjudgmental. It is important that communication flow smoothly. During meetings of all kinds, including pre-lift meetings, encourage everyone to ask questions without fear of judgment and ridicule.

Fear can affect communication in crane operations as well. When confidence is lacking, a worker may avoid displaying the Stop or Emergency Stop signal on the job when a hazard clearly exists or is developing. A signal person may also hesitate to display a common signal because the worker is not confident it is the correct signal. This is yet another reason why crane signals must be learned and practiced until they become second nature. Signal persons must be prepared to stop a lift at any time if things are not progressing as expected.

1.2.5 Environmental Factors

Environmental factors such as noise and weather often interfere with the communication process on the job. Pile-driving operations (*Figure 2*) are a perfect example because the noise involved can be a barrier to hearing what is being said. In inclement weather, it may be difficult to see clearly or listen attentively. In good weather, environmental factors such as the sun in your eyes can pose a problem. However, subtle factors may also affect the communication process. The background color of a wall or structure, the position of the signal person, and the location of the load in relation to the signal person can all interfere with effective communication.

Both signal persons and crane operators should constantly strive to identify and eliminate interference from the communication process. As stated earlier, clarity in communication and adhering to standard signal sets is a crucial part of crane operations, since the feedback is usually in the form of crane movement. As a result, miscommunication can lead to equipment damage and/or personal injury.

Figure 2 Pile-driving operations.

NCCER – *Heavy Highway Construction Level One* 53101-18

Additional Resources

Interplay: The Process of Interpersonal Communication, Ronald Adler, Lawrence Rosenfeld, and Russell Proctor. 13th Edition. New York, NY: Oxford University Press.

NCCER Module 00107-15, *Basic Communication Skills*.

1.0.0 Section Review

1. Which of the following is an example of an internal barrier that may influence the receipt of a message?
   a. Noise on a jobsite
   b. The receiver's attitude toward the job
   c. Glare on the windshield of the crane
   d. The weather

2. Which of the following is a true statement?
   a. A barrier may affect how a message is interpreted, but a filter keeps it from being heard.
   b. The experience level of an individual does not represent any sort of communication barrier.
   c. A non-native speaker of a language may have problems with accurate pronunciation, even with an extensive knowledge of the vocabulary.
   d. Learning all the languages spoken by non-native speakers on a jobsite is a practical solution to a language barrier.

SECTION TWO

2.0.0 CRANE COMMUNICATIONS

Objective

Identify and interpret the OSHA regulations related to crane communications and explain how to communicate with crane operators verbally and nonverbally.

a. Identify and interpret construction-related OSHA regulations associated with crane communications and signaling.
b. Describe the equipment used for verbal communications and how to communicate with and direct a crane operator verbally.
c. Explain how to communicate with and direct a crane operator nonverbally.

Performance Tasks

1. Demonstrate proper crane-communication techniques using a handheld radio or another acceptable verbal-signaling device.
2. Demonstrate each standard hand signal depicted in 29 *CFR* 1926.1400, Subpart CC, Appendix A.
3. Direct an operator to move and place a load using the appropriate hand signals.
4. Direct an operator to move and place a load using voice communication.

Trade Terms

Blind lift: Any lift involving a load that is out of the direct view of the operator. Blind lifts are generally always categorized as critical lifts.

Bridge: In relation to overhead cranes, the part of an overhead crane consisting of one or more girders or beams and the supporting trucks. The bridge is the overhead, weight-bearing structure along which the trolley(s) and load block assembly travels.

Consensus standard: A set of proprietary guidelines published and agreed to by a consensus (representative majority) of members of a given industry. While not legally binding, they are often cited in governmental regulations, such as OSHA standards.

Dedicated spotter: An individual qualified as a signal person who is charged with monitoring the separation between power lines and the equipment, load line, and load, so that the minimum approach distance is not compromised per OSHA standards.

Diver tender: One or more individuals assigned to attend to a diver's needs, including providing assistance in equipment preparation and managing the diver's cables and hoses.

Line of sight: The straight-line path between an observer's eyes and the thing being observed.

Open mike: In electronic communications, the condition where a radio's Transmit button or switch is held continuously without releasing it, even during pauses in speaking.

Trucks: In relation to overhead cranes, a mechanical assembly consisting of a frame, wheels, bearings, and axles that support the bridge of an overhead crane and provide the ability for it to move along a set of parallel tracks.

The methods and modes of communication vary widely in mobile crane operations. The method of communication refers to whether the communication is verbal or nonverbal. The mode is defined by the means of performing the communication. Modes may include bullhorns, radios, hand signals, flags, etc.

2.1.0 OSHA Standards and Requirements

29 *CFR* 1926, Subpart CC, *Cranes and Derricks in Construction* is the primary safety standard related to crane operations in the construction environment. The directives and guidelines it contains are enforceable as federal laws. They are not simply suggestions or examples of best practices. The specific sections that are dedicated to crane communications and the role of signal persons are 29 *CFR* 1926.1419, *Signals—General Requirements* through 1926.1422, *Signals—Hand-Signal Chart*. In addition, 29 *CFR* 1926.1428, *Signal Person Qualifications* provides the information indicated by its title.

There is also one very important communication-related statement found in 29 *CFR* 1926.1417(y), as follows: "The operator must obey a stop (or emergency stop) signal, irrespective of who gives it." This is a very important provision, meant to ensure that any potentially unsafe condition can be recognized and the operation stopped by anyone that recognizes it. All members of a lift team have the authority to display the Stop or Emergency Stop signal, and the crane operator is required to obey it. Even a bystander can display the signal if a hazardous condition becomes apparent, and the crane operator must obey. This is rare, however, as those directly involved with a properly manned lift operation are more likely to recognize a hazard before a casual observer.

The remainder of this section presents detailed information related to crane communications from the relevant OSHA standards. In the case of crane communications, *ASME Standard B30.5, Mobile and Locomotive Cranes*, has not been incorporated by reference into the OSHA standards. As a result, the primary focus of this section is on the enforceable OSHA standards. However, the hand signal set and most of the signaling-related provisions of *ASME Standard B30.5* parallel the OSHA standard.

2.1.1 29 CFR 1926.1419, Signals—General Requirements

Per 29 *CFR* 1926.1419, there are situations in which a signal person must be provided. These situations are outlined as follows:

- When any part of the path of a load being moved, including its origin and destination, is outside of the view of the crane operator
- When the equipment is traveling but the driver's view in the direction of travel is blocked
- When either the crane operator or the personnel handling the load determine that it is necessary

A dedicated spotter is required under certain conditions when a crane is working in the vicinity of power lines. Although they are referred to as *spotters* in 29 *CFR* 1926.1411, *Power Line Safety*, and are assigned to focus their attention on the crane's proximity to the power lines, they must stay in continuous contact with the crane operator. To a certain extent, then, the spotter is serving as an auxiliary signal person, and must meet the signal person qualifications of 29 *CFR* 1926.1428 as stated in 29 *CFR* 1926.1401, *Definitions*. Verbal communication is best for these situations, as the information that a spotter may need to relay to the crane operator may not have any kind of standard hand signal assigned to it.

The standard allows for signals between the signal person(s) and crane operator to be by hand, by voice, or by an audible means other than voice. OSHA simply directs that the chosen method of communication be appropriate for the site conditions. New signals, other than hand, voice, and audible signals, may be used in some conditions if they meet specific OSHA requirements. Note that signal persons are required to provide the signals from the crane operator's perspective of direction. Signal persons must signal in the direction of required crane movement.

Hand signals are designed to be universal and do not require any special equipment. When hand signals are used, they must be those found in 29 *CFR* 1926, Subpart CC, Appendix A. This is referred to as the *Standard Method*. There are two exceptions to the rule of using the OSHA Standard Method:

- When the use of the Standard Method is determined to be infeasible
- When an attachment or required crane movement is not covered by the Standard Method

When an attachment or required crane movement is not covered by the Standard Method, a nonstandard hand signal can be identified and used. OSHA requires that the signal person, crane operator, and lift director discuss any nonstandard hand signals before the operation begins and agree to the signals and their use. Note that there is a difference between the terms *nonstandard signals* and *new signals*. New signals refer to a new method or mode of communicating, while a nonstandard signal refers only to a hand signal that is not part of the Standard Method.

Regardless of the method or mode used, the crane operator must bring the operation to a safe stop if the line of communication is interrupted. For example, if another construction vehicle suddenly blocks the operator's view of the hand signals, the operation must be safely halted until the line of sight is restored. The crane operator can also stop the operation temporarily when any unsafe condition is encountered and discuss the issue with the signal person. The operation resumes only after both parties agree that the issue has been satisfactorily resolved.

29 *CFR* 1926.1419(j) is related to the responsibility of others to display the Stop or Emergency Stop signal. It also refers the reader back to 29 *CFR* 1926.1417(y), which outlines the crane operator's responsibility to obey the signal, regardless of its source. This standard states that "anyone who becomes aware of a safety problem must alert the operator or signal person by giving the Stop or Emergency Stop signal."

A signal person can be in contact with and provide signals to more than one crane or derrick at a time. A system to identify which crane is being signaled must be developed and used. This is generally satisfied by a specific signal chosen and agreed upon by both parties to identify each crane before the functional signal is communicated.

2.1.2 29 CFR 1926.1420, Signals—Radio, Telephone, or Other Signals

OSHA allows a variety of devices to be used to communicate electronically in crane operations. In some cases, hand signals are not practical or appropriate. Any electronic means of communication must be tested at the site to ensure the equipment is fully functional and reliable before the operation begins.

As a general rule, electronic communication must take place on a dedicated channel between the signal person and the operator, with the following exceptions allowed:

• When multiple cranes are using the same signal person, or when multiple signal persons are in use, they may share the channel.
• When a crane is being operated on or near railroad tracks and the crane's movement needs to be coordinated with trains or other equipment moving on the tracks, the channel can be shared.

The final provision of this section is that the crane operator's method of receiving signals electronically must be hands-free. It is acceptable for the operator to use a push-to-talk system, where a button must be pressed to speak, as long as reception is hands-free.

In some cases, the signal person keeps the radio microphone keyed to continuously communicate with the crane operator. An open mike is often used when the signal person is speaking progressive information, such as distance to touchdown, during the lift. However, an open mike can prevent receipt of a Stop signal from others on the channel, depending on the equipment involved. This practice may differ from company to company and project to project. Open-mike procedures should be discussed and clarified in the pre-lift meeting. Some radios have an Emergency button on them, which allows Stop signals to be transmitted even when another mike is keyed.

Everyone involved in a lift must exercise some radio discipline. Perhaps you have been using a different radio channel to speak to another craft or your supervisor. Before entering or reentering the crane operations channel, take a few moments to monitor the conversations occurring before speaking. This practice will prevent talking over or interrupting an important communication in progress.

2.1.3 29 CFR 1926.1421, Signals—Voice Signals; Additional Requirements

The OSHA standard in 29 *CFR* 1926.1421 requires crane operators, signal persons, and lift directors to discuss and agree on the verbal signals that will be used. The team will only need to meet and discuss the matter again if a different worker is added or substituted, there seems to be confusion about a signal, or a signal needs to be changed or added. This standard also outlines the structure of voice signals. Each voice signal must contain these components that are spoken in the following order:

• Function and direction, such as "Hoist Up" or "Hoist Down"
• Distance and/or speed, such as "10 feet and slowly"
• Function, followed by the Stop command, such as "Swing Stop" or "Load Stop"

Providing distance and/or speed information is required. It is typically modified and repeated as the function progresses. A typical voice command from a signal person sounds something like this: "Swing right, 30 feet… 20 feet… 10 feet… 5 feet… swing stop." Another example is "Load down slow… slow… load stop." Note that a verbal instruction such as "Swing left… swing left… swing left… swing stop" does not meet the requirements since no speed or distance information is provided. It is also important to remember to supply the distance as the amount remaining before the load reaches the desired point, rather than as the distance it has already moved. Crane operators typically prefer that the signal person paint a real-time picture of progress and movement. Give the operator as much information as you think the operator needs—if you are providing too much, the operator will tell you.

This section of the OSHA standard ends by requiring that the crane operator, signal person(s), and lift director be able to effectively communicate in the language being used.

2.1.4 29 CFR 1926.1422, Signals—Hand-Signal Chart

Per 29 *CFR* 1926.1422, hand-signal charts must be posted on the equipment in use, typically the crane itself, or in a conspicuous location in the area of the hoisting operation. Most mobile cranes have a hand-signal chart posted on or near the cab of the crane in an easily visible location to satisfy this requirement. Where overhead or tower cranes are in use, the hand-signal chart should be posted in a more accessible location than the equipment itself generally provides.

2.1.5 29 CFR 1926.1428, Signal Person Qualifications

Organizations that employ signal persons have a responsibility to ensure that they are properly qualified as outlined in 29 *CFR* 1926.1428. Employers can choose one of two options to satisfy this OSHA standard:

- The signal person already possesses or obtains documentation from a third-party qualified evaluator (such as NCCER) showing that the individual has demonstrated the necessary skills and meets the qualification requirements.
- The employer provides a qualified evaluator from within the organization that ensures the signal person can demonstrate the necessary skills and then provides the supporting documentation.

This standard also requires employers to maintain the required documentation of signal-persons' skills at the site where they are employed. The documents must indicate which type of signals, such as hand signals or radio communications, that each individual is qualified to perform.

Employers also have a responsibility to ensure that a signal person's performance continues to be sound and accurate. If the performance of a previously-qualified signal person indicates that the individual is not properly qualified, the employer must not allow the worker to continue serving as a signal person until retraining has been completed and the skills are reevaluated through one of the two options outlined above.

The standard provides the following qualification requirements for signal persons:

- Knowledge and understanding of the type of signals being used. For hand signals, the individual must know the Standard Method signals provided in *Appendix A* of 29 *CFR* 1926,*Subpart CC*.
- Competence in the application of the signals used.
- A basic understanding of equipment operation and its limitations. This includes the crane dynamics related to swinging and stopping the movement of loads, and the boom deflection that results from hoisting a load.
- Knowledge and understanding of the requirements found in 29 *CFR* 1926.1419 through 1926.1422, as well as 1926.1428. The requirements of each of these standards have been covered in this section.
- Documentation that these requirements have been met through the successful completion of either an oral or a written test, and a practical

test. These tests are administered by the employer's qualified evaluator or a third-party organization such as NCCER.

2.1.6 Miscellaneous Requirements

There are several additional sections of the OSHA standard that mention the role or actions of signal persons. One such section is 29 *CFR* 1926.1431,*Hoisting Personnel*. An occupant of a hoisted personnel platform can serve as the signal person. If the signal person, when needed, is operating from a location other than the personnel platform, the occupants of the platform must remain in direct contact with the signal person at all times. This same standard also requires a pre-lift meeting to be held when personnel are to be hoisted, regardless of the environment. If a signal person will be used during the lift, he or she must attend the pre-lift meeting. If any workers assigned to the operation are later replaced, including the signal person, another meeting is required.

Another requirement is related to cranes that are supporting one or more divers in the water, found in 29 *CFR* 1926.1437(j). Note, however, that this standard clearly applies to floating cranes and derricks, or to land cranes positioned on a floating vessel. When a crane is devoted to diver entry and exit, it cannot be used for any other purpose until the diver(s) is/are safely out of the water. Divers require a diver tender who monitors them at all times to ensure safety. Their primary responsibility is to manage the bundle of cables and hoses, known as the diver's umbilical. The diver tender generally serves as the crane signal person when the crane is directly supporting the diver, and must be a qualified signal person to do so. The standard permits hand signals between the diver tender and the crane operator as long as a clear line of sight is maintained. Otherwise, signals must be transmitted electronically. If the diver is not directly connected to the crane and is instead swimming freely, a dive supervisor is likely in direct contact with both the diver and crane operator.

When cranes are working near power lines, a dedicated spotter is often required to ensure that no part of the crane or the load enters the safe zone established around the lines. The dedicated spotter is an individual separate from that of the primary signal person. Per OSHA's definition of a dedicated spotter, the individual must meet the same requirements as a signal person. These requirements are found in 29 *CFR* 1926.1428. In 29 *CFR* 1926.1410(d)(2)(i), dedicated spotters are

Know Your Craft

In the late 1990s, a young 24-year-old man was killed on his first day on the job, working at Shoreham Docks in the United Kingdom. He was assigned to assist with unloading bags of aggregate from the hold of a ship. The load block of a crane was descending through the opening in the hold from above when it struck the young man in the head, killing him instantly.

There were a number of errors that lead up to this tragedy. Although he had never done such work before, he received no training before taking his place on the crew. He was not even issued a hard hat by his employer, which could very well have saved his life that day. However, there was already an accident waiting to happen: the crane operator could not see into the hold, and the signal person calling the shots did not speak any English and was not familiar with standard hand signals. The language barrier prevented the signal person and crane operator from even considering a conversation related to signals.

Although many lifts may seem very simple and the idea of a serious accident seems remote, there is always the potential for a tragedy to occur. Being an effective signal person goes beyond just knowing what each hand signal means. It also means having the ability to evaluate and re-evaluate what is happening moment by moment, looking forward into the lift process without losing focus on the here and now. Learning to do that is a process. But the process does begin with knowing how to properly signal a crane without having to dig deep into your memory to find the correct signal. Hand signals should be practiced extensively. Practicing the signals in the mirror can help you determine if the hand and arm positions are appropriate. Practice by looking at the signals and identifying them to yourself, and practice making the signals while someone else calls them out rapidly. Practice in class and practice at home. Whatever you do, don't serve in the role of a signal person until you know the signals. You want to be thinking about what to direct the crane to do next, and not how to make the signal.

---

further required to be in continuous contact with the crane operator, rather than communicating through the signal person. The spotter's responsibility is to monitor the separation of the crane, load line, and load from the power lines. No communication is generally necessary unless one of the three components comes too close to, or enters, the safe zone around the power lines. There are other significant requirements related to the use of a dedicated spotter, but they are not related to signaling or communication.

There are occasions when more than one crane is required to manage and position a single load. When more than one crane is involved, clear communication becomes even more critical. According to 29 *CFR* 1926.1432, one designated person who meets the OSHA criteria as both a qualified person and a competent person shall serve as the lift director. Alternatively, a competent person can be assisted by one or more qualified people. In any case, when multiple cranes are used, a pre-lift planning meeting must be held beforehand with all workers participating in the lift. One signal person is generally assigned to control both cranes, but complex situations may lead the lift director to use more than one. A nonstandard signal must be developed and agreed upon to identify which crane is being signaled, since this type of signal is not a part of OSHA's Standard Method.

Whenever multiple signal persons are used, regardless of the reason, it is important for all members of the lift team to know and understand

the plan and how it will be implemented. Lifts involving multiple signal persons increase the complexity of the process. A thorough pre-lift meeting must be held to discuss the responsibilities. A blind lift is a good example of when it may be necessary to use multiple signal persons. If one signal person cannot observe both the complete load path and the crane, additional signal person or persons may be required to relay information. Note that relay signaling is not typically preferred and provides an additional opportunity for errors. Lifting operations should therefore move more slowly. When one signal person hands off responsibility to a second, it should be clear that the responsibility for the load has been transferred. The method of hand-off should be clearly discussed and agreed upon during the pre-lift meeting.

> **NOTE**
>
> Relay signaling automatically creates some lag time in signaling that can create problems and must be taken into consideration.

2.2.0 Verbal Crane Communications and Equipment

Verbal modes of communication vary depending on the requirements of the situation. Some of the most common devices used are portable radios,

---

often referred to as *walkie-talkies* (*Figure 3*). Compact, low-power, inexpensive pairs of units can enable a crane operator and signal person to communicate verbally in some environments. They do meet OSHA standards since the operator can hear the signals hands-free, but they usually require a hand to press a button when speaking.

There are some disadvantages to using low-power and inexpensive equipment in an industrial setting. One disadvantage is interference. With basic, low-quality units, the frequency used to carry the signal may have many other users. A crowded channel can cause signal disruptions and lead to accidents. Another disadvantage is the effect of background noise. When attempting to transmit in a noisy area, the person may transmit unintended noise, resulting in a garbled, unintelligible signal for the receiver. On the receiving end, the individual may not be able to hear the transmission clearly due to a high level of background noise.

> **NOTE**
>
> Wind and other external factors can create noise, distortion, and feedback in low-fidelity radio communications microphones. You can reduce these effects by cupping your hand around the microphone.

To overcome the shortcomings associated with low-power, handheld units, more expensive units with the ability to use specific, dedicated frequencies and transmit at a higher power level may be needed. An electronic communication standard referred to as DECT (Digital Enhanced Cordless Telecommunications), developed in the 1990s, provides improved range as well as encrypted and secure communications. Only DECT-compatible devices can operate on the specific frequencies allocated to them by the US Federal Communications Commission (FCC), and the frequencies are far enough away from those of other devices to eliminate cross-frequency interference issues. DECT-compatible headsets (*Figure 4*) offer high-quality communications for a large number of users on a single dedicated system. Pairing, or connecting, a group of headsets to a local communications hub, also shown in *Figure 4*, establishes a private audio network. Each headset can be configured for full-duplex operation, meaning that the user can both listen to others and transmit at will, completely hands-free. They can also be configured for listening only (broadcast mode), but returned quickly to full-duplex operation when necessary. Note that the over-ear headsets also provide hearing protection.

Cell phones may also qualify as an electronic device that can be used for crane communications. With a compatible headset of good quality to make them hands-free, preferably with functional noise-cancelling technology, they can provide good service. However, it is important to note that signal strength on some networks may be weak or even non-existent in some work areas. As a result, it is difficult to rely on them exclusively. DECT-compatible wireless communication networks can be established just about anywhere, but even this equipment has limitations related to line of sight, elevation, and distance between hub stations and headsets.

Hardwired communications systems are also still available. These units overcome some of the

WALKIE-TALKIES

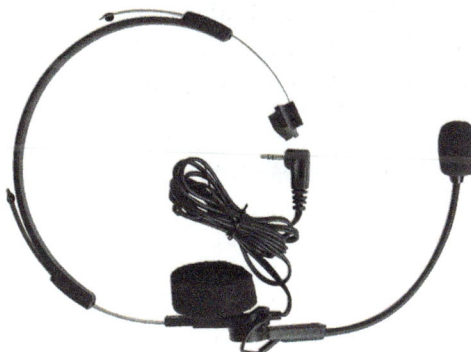

ACCESSORY HEADSET WITH BOOM MICROPHONE

Figure 3 Walkie-talkies with an accessory headset.

(A) HEADSET AND MICROPHONE

(B) HARDHAT-MOUNTED HEADSET

(C) WIRELESS COMMUNICATIONS HUB

Figure 4 Examples of DECT-compatible wireless communications gear.

disadvantages of radio use. When using this type of system, interference from another unit is unlikely because this system does not use a radio frequency to transmit information. Like a telephone system, occasional interference may be encountered if the wiring is not properly shielded from very strong radio transmissions or other electromagnetic interference. Hardwire systems, however, are not very portable or practical, especially when the crane must be moved often. Very long wires are often required and they are easily entangled and/or damaged on an active jobsite. As a general rule, hardwired systems are too clumsy to deploy and operate on most of today's jobsites, and the wires themselves may represent a separate hazard.

Regardless of the type of equipment used, it is important that signal persons remember and use the proper format for all verbal signals transmitted to a crane operator: function and direction, distance and/or speed, and function followed by a Stop command. These skills must be demonstrated during a practical examination. Signal persons should be familiar with the equipment in use, rather than try to figure it out for the first time as a lift progresses.

Remember that all electronic communication devices must be tested before a lift begins. Workers charged with the responsibility of maintaining the communications equipment should ensure that the batteries are kept freshly charged, and that spare batteries are always available on site. Shutting down a complex lift operation temporarily due to a lack of batteries is a very expensive, time-consuming, and often embarrassing way to learn this lesson.

2.3.0 Nonverbal Communications

Although OSHA's Standard Method of hand signaling is the most common type of nonverbal communication used in crane operations, several other nonverbal modes have also been used. One possible mode is the use of a distinct audible signal, such as a siren, buzzer, and/or whistle, in which the number of repetitions and duration of the sounds convey the message. An example of the use of audible signals in crane operations that can often be justified is related to work inside large tanks. Workers must often be inside the tank, receiving components lowered through an opening by a crane. Hand signals are not possible, and the tank itself might interfere with wireless communications. In this case, workers inside the tank might be forced to use a hammer to make a series of rapping sounds on the tank walls to communicate with the outside. However, a hard-wired means of verbal communication is certainly the more practical and safer alternative.

Another nonverbal mode, although rare, is the use of signal flags. This mode might require the use of different colored flags, or a specific positioning of the flags, to communicate the desired message. When there is considerable distance between the signal person and crane operator, this may seem to be a practical solution. However, verbal communication using reliable radio equipment likely provides a better solution, especially since a line of sight evidently does exist if flags are an option. When a line of sight exists, verbal electronic communications function at their greatest range (up to 1,600 feet or 500 meters with some DECT-compatible devices). The disadvantage of a flag-communication mode is that 29 *CFR* 1926.1419(d) requires that any new signals must be shown to be equally effective as the prescribed voice, audible, or Standard Method hand signals, or comply with an equally-effective national consensus standard. Since there is no apparent national consensus standard for flags, their use can be difficult to justify in most cases.

The most common mode of nonverbal communication is the use of the Standard Method of hand signals provided by OSHA. Although *ASME Standard B30.5* also provides a standard set of hand signals, it is the OSHA standard that is enforceable by law. The signals of the two sets are nearly identical, although the drawings used to depict the signals are drawn with slight differences. The required OSHA hand signals are shown in a sequence of drawings shown in *Figure 5A* through *Figure 5S*.

Figure 5A Hoist.

HOIST

With arm vertical, forefinger pointing up, move hand in small horizontal circle.

OPERATOR ACTION: Slowly pull the hoist control lever back, controlling the ascent speed of the load (block or ball) with the control lever position. Keep the engine rpm constant until the desired lift is complete, then slowly return the control to the center position.

If the load block or ball is near the boom point, exercise caution to avoid two-blocking.

EXPECTED MACHINE MOVEMENT: The load attached to the block or ball rises vertically, accelerating and decelerating smoothly.

Figure 5B Lower.

LOWER

With arm extended downward, forefinger pointing down, move hand in small horizontal circle.

OPERATOR ACTION: Slowly push the hoist control lever for the desired hoist forward, controlling the descent speed of the load (block or ball) with the control lever position. Keep the engine rpm constant until load lowering is complete, then slowly return the control to the center position.

Do not allow the block or ball to contact the ground or any surface that can cause slack in the load line(s).

EXPECTED MACHINE MOVEMENT: The load block or ball smoothly lowers vertically.

RAISE BOOM

Arm extended, fingers closed, thumb pointing upward.

OPERATOR ACTION: Slowly pull the boom control lever back, controlling the speed of the boom raising movement with the control lever position. Keep the engine rpm constant until the desired position is reached, then return the control to the center position.

Exercise caution to avoid any obstructions to boom movement in the vertical plane, such as trees or power lines.

If the load or load block is close to the ground in front of the machine, the operator may be required to hoist the load to avoid contacting the crane.

EXPECTED MACHINE MOVEMENT: The boom rises, increasing the hook height and reducing the overall machine height clearance. The operating radius is slowly decreased, thus possibly increasing machine capacity and stability.

Figure 5C Raise boom.

RAISE BOOM AND LOWER THE LOAD

Arm extended, fingers closed, thumb pointing up, flex fingers in and out as long as load movement is desired.

OPERATOR ACTION: This requires a two-hand operation. Use your left hand to push the hoist control lever forward, while using your right hand to pull the boom control lever back. Keep the engine rpm constant until the desired position is reached, then slowly return both controls to their center position.

Move both controls independently to maintain the load (block or ball) at an even distance from the ground, with movement horizontal to the level ground plane.

Keep the engine rpm at a high level to ensure sufficient oil flow to sustain smooth load movement. Exercise caution to avoid any obstructions to boom movement in the vertical plane, such as trees or power lines.

EXPECTED MACHINE MOVEMENT: The boom rises, reducing overall machine height clearance, as the load moves horizontally toward the crane. The operating radius is slowly decreased, thus possibly increasing machine capacity and stability.

Figure 5D Raise boom and lower the load.

Figure 5E Lower boom.

LOWER BOOM

Arm extended, fingers closed, thumb pointing downward.

OPERATOR ACTION: Slowly push the boom control lever forward, controlling the speed of the boom lowering movement with the control lever position. Keep the engine rpm constant until the desired position is reached, then return the control to the center position.

Exercise caution to avoid any obstructions to boom movement in the vertical plane, such as trees or power lines.

EXPECTED MACHINE MOVEMENT: The boom will lower, decreasing the hook height and reducing the overall machine horizontal clearance. The operating radius is slowly increased, thus possibly decreasing machine capacity and stability.

LOWER BOOM AND RAISE THE LOAD

Arm extended, fingers closed, thumb pointing down, flex fingers in and out as long as load movement is desired.

OPERATOR ACTION: This requires a two-hand operation. Use your left hand to pull the hoist control lever back, while using your right hand to push the boom control lever forward. Keep the engine rpm constant until the desired position is reached, then slowly return both controls to their center positions.

Move both controls independently to maintain the load (block or ball) at an even distance from the ground, with movement horizontal to the level ground plane.

The engine rpm must be kept at a high level to ensure sufficient oil flow to sustain smooth load movement.

EXPECTED MACHINE MOVEMENT: The boom lowers, increasing overall machine height clearance, as the load moves horizontally away from the crane. The operating radius is slowly increased, thus possibly reducing both machine capacity and stability.

Figure 5F Lower boom and raise the load.

Figure 5G Extend telescoping boom.

EXTEND TELESCOPING BOOM

Both fists in front of body at waist level, with thumbs pointing outward.

OPERATOR ACTION: Using your left hand, push the telescope control lever forward, controlling the boom extension speed with the control position and keeping the engine rpm constant until the desired boom length is reached. Slowly return the controls to the center position. Boom extension may also be accomplished by using the left foot to slowly rock the left foot pedal forward until the desired boom length is reached. Then, slowly return the foot pedal to the center position.

Take care not to bring the hook block or ball too close to the boom head when extending the boom to avoid two-blocking.

EXPECTED MACHINE MOVEMENT: Boom sections telescope out. The load radius is increased, possibly decreasing machine capacity and stability. The load (block or ball) rises vertically.

Figure 5H Retract telescoping boom.

RETRACT TELESCOPING BOOM

Both fists in front of body at waist level, with thumbs pointing toward each other.

OPERATOR ACTION: Using your left hand, pull the telescope control lever back, controlling the boom retraction speed with the control position and keeping the engine rpm constant until the desired boom length is reached. Slowly return the controls to the center position. Boom retraction may also be accomplished by using the left foot to slowly rock the left foot pedal rearward until the desired boom length is reached. Then, slowly return the foot pedal to the center position.

EXPECTED MACHINE MOVEMENT: Boom sections retract. The load radius is decreased, possibly increasing machine capacity and stability. The load (block or ball) lowers vertically.

SWING

Arm extended, point with index finger in direction of boom swing. (Swing left is shown as viewed by the operator.) Use appropriate arm for desired direction.

OPERATOR ACTION: Push the far left swing control lever forward to swing toward the boom, swinging left for right side operator position and right for left side operator position. For a centrally located operator position, control lever movement is the same as for the left side operator position. Pull rearward to reverse the action.

Keep the engine rpm constant. For inexperienced operators, keep the rpm at a lower level than is used for other craning operations.

Acceleration and deceleration should be slow and steady, with swing speed being adjusted near the end of the swing so that the final desired swinging position is not overrun, thus causing the load to swing like a pendulum. Note that when the signal person uses his or her left arm to swing, the swing is to the operator's right.

Exercise caution to avoid obstructions in the path of the swing. The signal person should warn the operator of obstructions, especially when the swing is to the boom side and the operator's vision is limited.

EXPECTED MACHINE MOVEMENT: The boom moves about the center of rotation with the load (block or ball) swinging in an arc, either toward the right or left, while remaining approximately equidistant to a level plane.

Figure 5I Swing.

MOVE SLOWLY
(ANY SIGNALED MOTION)

Use one hand to give any motion signal and place the other hand motionless over the hand giving the motion signal.

OPERATOR ACTION: Perform the action indicated by the hand signal in a slow manner. Pictured is hoist slowly, directing the operator to pull the hoist control lever for the desired hoist toward the operator until the load (block or ball) ascends slowly. Keep the engine rpm constant until the desired lift is complete, then slowly return the hoist control lever to the center position.

Other move slowly signals (not pictured) include hoist down slowly, raise boom slowly, etc.

EXPECTED MACHINE MOVEMENT: Machine movement will vary depending on the signal being given.

Figure 5J Move slowly.

Figure 5K Use main hoist.

Figure 5L Use auxiliary hoist (whipline).

Figure 5M Stop.

USE MAIN HOIST

Tap open hand, palm down, on head and then use regular hand signals to show the desired action.

OPERATOR ACTION: Grasp the hoist up/down control lever for the main hoist and await further signaling from the signal person.

EXPECTED MACHINE MOVEMENT: None. This signal is used only to inform the operator that the signal person has chosen the main hoist for the action to be performed as opposed to the auxiliary hoist.

USE AUXILIARY HOIST (WHIPLINE)

Tap elbow with open palm of one hand, then use regular hand signal to show desired action.

OPERATOR ACTION: Grasp the hoist up/down control lever for the auxiliary hoist and await further signaling from the signal person.

EXPECTED MACHINE MOVEMENT: None. This signal is used only to inform the operator that the signal person has chosen the auxiliary hoist for the action to be performed, as opposed to the main hoist.

STOP

Arm extended, palm down, move arm back and forth horizontally.

OPERATOR ACTION: Return all activated controls to the center or neutral position in a smooth motion at a slow to moderate speed to prevent pendulum action of the load (block or ball).

EXPECTED MACHINE MOVEMENT: None. All movement of the machine ceases.

NOTE: The Stop and Emergency Stop signals are the only signals that may be given by anyone, and the crane operator is required to obey.

Figure 5N Emergency stop.

EMERGENCY STOP

Both arms extended, palms down, move arms back and forth horizontally.

OPERATOR ACTION: Immediately return all activated controls to the center or neutral position as quickly as is safe.

EXPECTED MACHINE MOVEMENT: None. All movement of the machine ceases.

NOTE: The Stop and Emergency Stop signals are the only signals that may be given by anyone, and the crane operator is required to obey.

Figure 5O Dog everything.

DOG EVERYTHING

Clasp hands in front of body at waist height.

OPERATOR ACTION: Ensure that no controls are activated, then engage all positive locking devices including hoist pawls, swing brakes, and house locks.

EXPECTED MACHINE MOVEMENT: None.

Figure 5P Travel / tower travel.

TRAVEL / TOWER TRAVEL

Arms are straight and extended horizontally, moved back and forth in a pushing motion away from the body. All fingers point straight up.

OPERATOR ACTION: Move the crane in the direction indicated.

TRAVEL – BOTH TRACKS (CRAWLER CRANE)

Position both fists in front of body and rotate them around each other, indicating the direction of travel (forward or backward).

OPERATOR ACTION: Decrease the engine speed to idle. Move the slide pinion to the travel position. Hold down the deadman control button. Move the main drive control lever forward and back slightly to fully engage the slide pinion, depending on the grade of the travel surface (uphill, downhill, or level). Position the travel locks per the manufacturer's directions. Engage the swing lock. Position the boom angle as shown in the manufacturer's travel tables. Move the steering clutch control to the "straight" position, and move the main drive control lever in the desired direction to travel forward or backward.

EXPECTED MACHINE MOVEMENT: Machine travels in the direction chosen.

Figure 5Q Travel — both tracks (crawler crane).

TRAVEL – ONE TRACK (CRAWLER CRANE)

Lock track on side of raised fist. Travel opposite track in direction indicated by circular motion of other fist, rotated vertically in front of body.

OPERATOR ACTION: Decrease the engine speed to idle. Move the slide pinion to the travel position. Hold down the deadman control button. Move the main drive control lever forward and back slightly to fully engage the slide pinion. Depending on the grade of the travel surface (uphill, downhill, or level), position the travel locks per the manufacturer's directions. Engage the swing lock. Position the boom angle as shown in the manufacturer's travel tables. Move the steering half lock control to either engage for a gradual turn or disengage for a sharp turn. Move the steering clutch control to the desired position to turn right or left. Move the main drive control lever in the desired direction to turn right or left.

EXPECTED MACHINE MOVEMENT: Machine turns in the direction chosen.

Figure 5R Travel — one track (crawler crane).

TROLLEY TRAVEL
(TOWER OR OVERHEAD CRANE)

Elbow bent with the palm towards the body and fingers closed; thumb pointing in the direction of motion and moved horizontally back and forth.

OPERATOR ACTION: Use the required controls to move the trolley of a tower or overhead crane in the direction indicated.

EXPECTED MACHINE MOVEMENT: The trolley assembly moves in the direction indicated.

Figure 5S Trolley travel (tower or overhead crane).

Remember that the hand-signal chart must also be posted conspicuously at the jobsite. These hand signals are recognized by the industry as the standard hand signals to be used on all jobsites. This helps ensure that there is a common core of knowledge and a universal meaning to the signals when lifting operations are being conducted. As discussed previously, this helps to eliminate a significant barrier to effective communication. These same signals apply to tower cranes, including luffing-boom tower cranes (*Figure 6*), as well as mobile cranes.

Additions or modifications may be made for crane functions not covered by the OSHA-illustrated hand signals, such as the deployment of outriggers. The operator and signal person must agree upon any signals and their meaning that are not illustrated in the Standard Method before the lift begins. These signals cannot conflict or be easily confused with any Standard Method signal.

There are two hand signals pictured in *ASME Standard B30.5* that are not included in the OSHA Standard Method. Since they are not pictured in the OSHA Standard Method set of signals, they are technically considered nonstandard. However, they are also a part of a national consensus standard and are commonly used. These two signals (*Figure 7*) are one-handed alternates for the standard telescopic boom retraction and extension signals, typically used when the signal person is also manning a tag line or has one hand otherwise occupied.

ASME Standard B30.2, Overhead and Gantry Cranes, also offers some hand signals (*Figure 8*) that are devoted to the operation of overhead cranes and similar equipment. Again, they are not part of the OSHA Standard Method set of signals, but are endorsed by a national consensus standard. Therefore, they are the best choice when such signals are needed. The signal for Bridge Travel is the same as that for Travel/Trolley Travel used with other cranes. The bridge of an overhead crane is defined as the part of

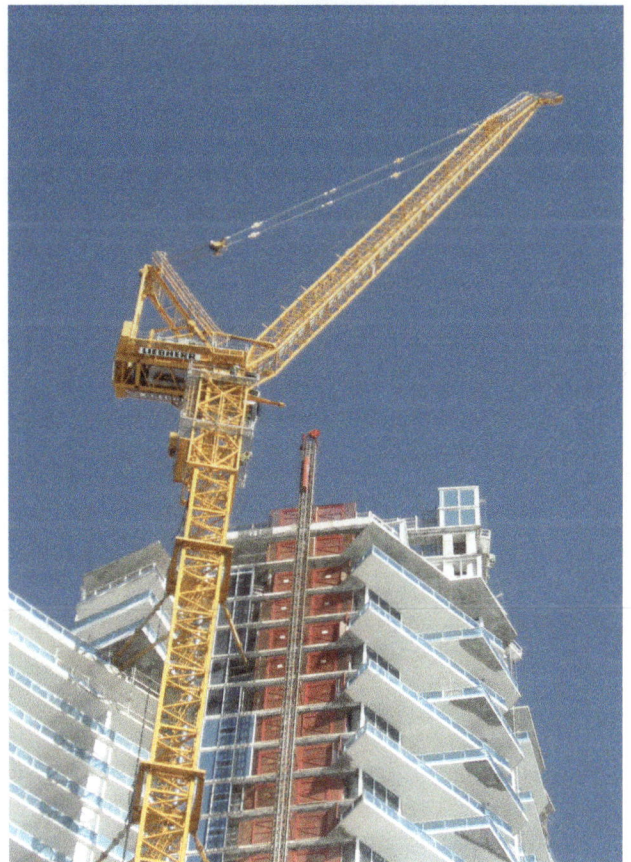

Figure 6 Luffing-boom tower crane.

One fist in front of chest with thumb tapping chest.

(A) EXTEND BOOM

One fist in front of chest, thumb pointing outward and heel of fist tapping chest.

(B) RETRACT BOOM

Figure 7 Alternate one-hand signals for telescopic boom retraction and extension.

a crane consisting of one or more girders and the supporting trucks that carry the trolley(s). This basically describes the large, weight-bearing structure above the hook. The trolley assembly moves along the bridge, from side-to-side.

Remember that whenever nonstandard signals are used, OSHA requires the signal person and crane operator to agree on their use and meaning before the operation begins. If it is necessary to switch from hand signals to verbal signals during an operation, all crane motions must be stopped before doing so.

A signal person's position at a lift site is very important, especially when hand signals are used. He or she must be in full view of the lift operation, as well as in the clear view of the crane operator. Signal persons should always wear high-visibility clothing or a vest that contrasts with the surroundings. High-visibility gloves may also be in order, so the crane operator can clearly see the position and/or motion of the hands.

As a general rule, the best possible position for the signal person to be positioned is with the line of sight perpendicular to the boom. In this position the signal person is visible to the operator, as well as able to observe the load and any boom deflection that may occur. Choose the side of the boom that presents the fewest visual distractions or obstructions.

Another common nonverbal signal set is actually one used by the crane operator when moving a mobile crane. These signals are outlined in *ASME Standard B30.5*, Section 5-3.3.7. It is common for the operator to use the following audible travel signals, using the crane's horn:

- *Stop* – One audible signal
- *Forward* – Two audible signals
- *Reverse* – Three audible signals

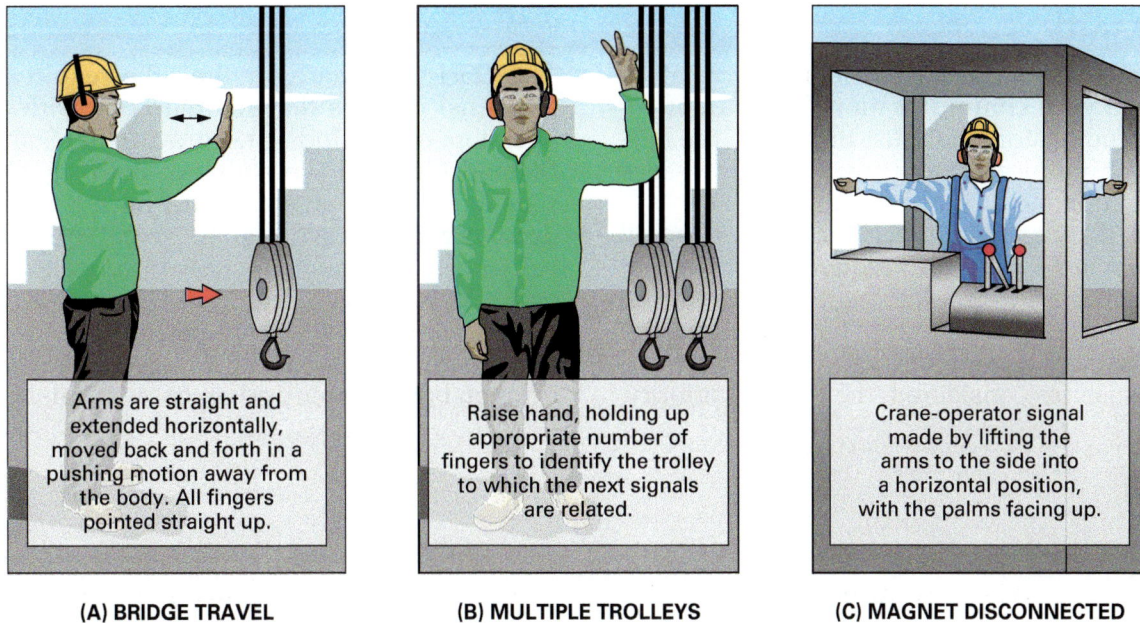

(A) BRIDGE TRAVEL	(B) MULTIPLE TROLLEYS	(C) MAGNET DISCONNECTED
Arms are straight and extended horizontally, moved back and forth in a pushing motion away from the body. All fingers pointed straight up.	Raise hand, holding up appropriate number of fingers to identify the trolley to which the next signals are related.	Crane-operator signal made by lifting the arms to the side into a horizontal position, with the palms facing up.

Figure 8 Additional signals for overhead cranes.

Additional Resources

ASME Standard B30.2, Overhead and Gantry Cranes. Current edition. New York, NY: American Society of Mechanical Engineers.

ASME Standard B30.5, Mobile and Locomotive Cranes. Current edition. New York, NY: American Society of Mechanical Engineers.

29 *CFR* 1926, Subpart CC, **www.ecfr.gov**

2.0.0 Section Review

1. In regards to a signal person signaling more than one crane at a time, _____.

    a. OSHA standards prohibit it under any circumstances
    b. crane operators must exit their crane cab when they are not the active crane
    c. the Standard Method of hand signaling cannot be used
    d. signals to identify which crane is being signaled must be determined and agreed upon

2. Which of the following is a true statement about verbal communications with a crane?

    a. Cell phones legally qualify as a verbal communication device, but reliable signal strength at every jobsite is difficult to predict.
    b. Hardwired communication equipment is no longer allowed by OSHA standards.
    c. Low-powered, inexpensive walkie-talkie equipment is the best choice for every environment.
    d. DECT-compatible devices are designed to seek out and use any frequency that shows the least amount of signal traffic.

3. OSHA's chart of Standard Method hand signals must be posted _____.

    a. in the lift plan
    b. in a conspicuous location on the jobsite
    c. on each signal person's clothing
    d. on the load being lifted

SUMMARY

There are many different aspects and variables associated with communication. To be an effective communicator, one must minimize the barriers to communication and ensure that messages are transmitted and received clearly.

OSHA standards found in the *Code of Federal Regulations (CFR)* apply to many industries and are enforceable by law. Although there are other national consensus standards that apply to crane operations and communication, the OSHA standards must be considered the higher priority.

The OSHA standards outline the requirements for signal persons that communicate with crane operators of all kinds, and provide the Standard Method for hand signals used. Both signal persons and crane operators must have a clear understanding of each signal, verbal or nonverbal, and its meaning in order to function safely and effectively as a team. Like crane operators, OSHA requires that signal persons possess specific qualifications and knowledge that is tested through both written and practical means.

1. Internal and external barriers to communication affect _____.

   a. only the sender
   b. only the receiver
   c. both the sender and the receiver
   d. only the party to which the message is related

2. Expressing the perceived meaning of something one has read or heard from a sender back to the sender in your own words is referred to as _____.

   a. paraphrasing
   b. passive listening
   c. message filtration
   d. negotiation

3. Verbal communication barriers include language differences, overuse of abstractions, and _____.

   a. body language
   b. feedback
   c. the use of clear language
   d. a lack of confidence

4. The Standard Method of hand signals is found in which standard?

   a. 29CFR 1926, Subpart CC, Appendix A
   b. 29CFR1926.1428
   c. *ASME Standard B30.5*
   d. *ASME Standard B30.2*

5. The main advantage of a hardwired crane-communication system is _____.

   a. low cost
   b. portability
   c. less radio interference
   d. ease of use

6. A signal person's position at the lift site should be _____.

   a. behind the crane where the operator can see the signal person in the cab's rearview mirrors
   b. in full view of the lifting operation and crane operator
   c. following the load from directly underneath
   d. in full view of the crane operator and within the swing radius of the crane

7. What action is signaled by tapping the palm of your hand on the top of your head, and then following up with signals for other functions?

   a. Lower
   b. Use Main Hoist
   c. Use Auxiliary Hoist (whipline)
   d. Raise Boom and Lower Load

8. When the signal person gives the Dog Everything signal, the crane operator _____.

   a. pulls the telescope control lever toward himself/herself
   b. decreases the engine speed to a fast idle
   c. moves the slide pinion to the travel position and holds down the deadman control button
   d. engages all positive locking devices including hoist pawls, swing brakes, and house locks after ensuring that no controls are activated

9. When preparing to drive a mobile crane in reverse, how many audible travel signal(s) must be given using the crane's horn?

   a. One
   b. Two
   c. Three
   d. Four

10. Alternate one-hand signals to signal telescopic boom retraction and extension are _____.

    a. part of OSHA's Standard Method
    b. found in *ASME Standard B30.5*
    c. found in both the OSHA and ASME crane-related standards
    d. used only if directly approved by an OSHA representative

Trade Terms Quiz

Fill in the blank with the correct term that you learned from your study of this module.

1. The trolley of a typical overhead crane moves back and forth along the _____.

2. When a signal person continuously holds a radio's Transmit button or switch down, even during pauses in speech, it is referred to as an _____.

3. Repeating the perceived meaning of a received message using different words describes _____.

4. A _____ describes the lifting of a load that is out of sight of the crane operator.

5. Giving someone a thumbs-up signal is a good example of _____.

6. Giving the verbal order, "Swing right a hair" is a type of communication _____.

7. A _____ is responsible for managing the umbilical of a submerged diver.

8. Effective nonverbal communications between the signal person and the crane operator depend on a clear _____.

9. The moving, upright portions of an overhead crane that support the bridge are referred to as _____.

10. When working near power lines with a crane, a _____ is often needed to ensure that the minimum approach distance is not compromised by the crane, the load line, or the load.

11. *ASME Standard B30.5, Mobile and Locomotive Cranes* is an example of a(n) _____.

Trade Terms

Abstraction	Dedicated spotter	Open mike
Blind lift	Diver tender	Paraphrasing
Bridge	Line of sight	Trucks
Consensus standard	Nonverbal communication	

Robert Capelli
Senior Health, Safety, and Environmental Manager, Orion Marine Construction

With a wealth of experience, insight, and extensive training from the U.S. Marine Corps, Robert continues to inspire excellence in his role as Senior Health, Safety and Environmental Manager with a large marine construction company.

Please give a brief synopsis of your construction career and your current position.

My first job after the Marine Corps was as a safety specialist in a roadway maintenance facility operated by the Florida Department of Transportation, which was a great segue into construction. The knowledge and experience I gained there was invaluable. From that foundation, I then worked for a company that had road construction, utility, demolition, and mining divisions. The diversity of work and operations in that one company alone gave me experience that would have taken four times as long to obtain elsewhere and has served me well in dealing with industry business owners and regulators over the years. I now work for a marine construction company as their safety manager.

How did you get started in the construction industry?

After I left active duty in 1991, I planned on going to school to become a firefighter/paramedic, but I ran into problems during school enrollment and was forced to look for work. As it turned out, the safety position I was offered with the Florida Department of Transportation used portions of the risk assessment training program I experienced in the Marines and the fire service.

Who or what inspired you to enter the construction industry?

The need to provide for my family was the motivation and inspiration to accept a position in the construction industry. I had a completely different career path in mind and was not even considering construction as an option, but because of the income and benefits I would lose if I didn't

take the job I was offered, the choice was easy to make. That was my inspiration, and it was a good choice.

How has training in construction impacted your life and career? What types of training have you completed?

My training certificate folder is several inches thick, and I have been very fortunate that my careers and volunteer work have overlapped and provided training not available to most safety professionals. The knowledge gained also revealed opportunities to me that were not available to others. After active duty, I continued serving in the Marine Corps Reserve and was the officer in charge of an environmental response team that performed hazardous material storage and remediation activities. Additionally, I was the fire chief of a local volunteer fire department, providing fire and rescue services.

In 2015, I earned a Master of Public Health degree, with a concentration in health, safety and environment. Many courses I have taken applied to more than one field. Hazardous material and environmental classes transcended all three fields. Urban search and rescue technician, confined space, fall protection, rigging, rope rescue, emergency medical technician, pump operator, and fire inspector classes all have applications in both the fire service and construction industries.

One may have twenty years of experience in construction, but is that one year of experience twenty times over, or twenty years of progressive experience with training and growth in the industry? The answer to this question is fairly evident when I perform the rigging practical examinations, where we always screen the candidates to ensure they are not overstating their qualifications. Many of them have struggled even on the basic rigger practical exam because they have not taken any additional training classes and have stayed within their industry segment. This leads to workers that cannot solve problems any more difficult than being confronted with a two-legged bridle sling.

Why do you think credentials are important in the construction industry?

Credentials are important in any profession as a way to validate experience and knowledge in a given field. In many cases, employers seek out credentialed people because they have verifiable knowledge and skills the employer needs.

The construction industry is very transient, with construction workers chasing work from one project to another. A person's reputation and who they know is often the key to getting hired for that next project. That works well if hiring is performed at the local level, but when hiring is centralized or the project is very large, recruiting managers have to sort through hundreds of applications to fill slots. Having an industry recognized credential, such as the NCCER Certified Plus credential, makes an individual stand out to a hiring manager.

The value of the right credentials presented to acknowledge your skills is huge. Obtaining industry-recognized credentials, such as those provided by NCCER, will help you get ahead.

What do you enjoy most about your career?

I thrive on the variety of work, with every project having its own challenges, and the need to solve problems and develop unique solutions is what keeps the job interesting. I have worked with some really great craft professionals who taught me the business of construction as well as the mechanics of the field, and every one of them had great problem-solving skills.

Would you recommend construction as a career to others? Why?

Absolutely, yes! This industry offers everyone who applies themselves to the effort an opportunity to provide for their family and continue growing as a professional. Academia has perhaps done a generation of Americans a disservice by saying you can only get ahead in life with a college degree. The theory that a four-year college degree is the one true path to success has been debunked over the years, and there are many unskilled workers with college degrees that are not gaining significant value from that education.

Starting out in construction with a broad orientation to the trades and applying yourself over the years will get you ahead. At some point, vocational school training or a bachelor's degree may be necessary to help you progress to the next level, but it is unnecessary at the start of your career. The opportunities to travel, learn new crafts, and move into management positions are unlimited.

What advice would you give to someone who is new to the construction industry?

Apply yourself, learn, seek opportunities, and don't become stagnant in your job. Approach every task by asking yourself, "What could go wrong here?". If you see something wrong or unsafe, fix it or tell a supervisor who can fix it. There is nothing so important that it has to be done right now at the expense of injuring someone, or worse. We always seem to find time to correct things, so why not do it right the first time? In addition, your reputation will follow you forever, so protect it.

How do you define craftsmanship?

Craftsmanship encompasses the physical result of applying your focus, training, and skills to the work effort in the best way possible.

Trade Terms Introduced in This Module

Abstraction: Any form of verbal, graphical, or written communication representing a generalized and nonspecific idea or quality of a thing, action, or event.

Blind lift: Any lift involving a load that is out of the direct view of the operator. Blind lifts are generally always categorized as critical lifts.

Bridge: In relation to overhead cranes, the part of an overhead crane consisting of one or more girders or beams and the supporting trucks. The bridge is the overhead, weight-bearing structure along which the trolley(s) and load block assembly travels.

Consensus standard: A set of proprietary guidelines published and agreed to by a consensus (representative majority) of members of a given industry. While not legally binding, they are often cited in governmental regulations, such as OSHA standards.

Dedicated spotter: An individual qualified as a signal person who is charged with monitoring the separation between power lines and the equipment, load line, and load, so that the minimum approach distance is not compromised per OSHA standards.

Diver tender: One or more individuals assigned to attend to a diver's needs, including providing assistance in equipment preparation and managing the diver's cables and hoses.

Line of sight: The straight-line path between an observer's eyes and the thing being observed.

Nonverbal communication: All communication that does not use words. This includes appearance, personal environment, use of time, and body language.

Open mike: In electronic communications, the condition where a radio's Transmit button or switch is held continuously without releasing it, even during pauses in speaking.

Paraphrasing: Expressing the perceived meaning of something read or heard in one's own words, generally to ensure clarity. Paraphrasing is an important component of active listening.

Trucks: In relation to overhead cranes, a mechanical assembly consisting of a frame, wheels, bearings, and axles that support the bridge of an overhead crane and provide the ability for it to move along a set of parallel tracks.

Additional Resources

This module presents thorough resources for task training. The following reference material is recommended for further study.

ASME Standard B30.2, Overhead and Gantry Cranes. Current edition. New York, NY: American Society of Mechanical Engineers.

ASME Standard B30.5, Mobile and Locomotive Cranes. Current edition. New York, NY: American Society of Mechanical Engineers.

Interplay: The Process of Interpersonal Communication, Ronald Adler, Lawrence Rosenfeld, and Russell Proctor. 13th Edition. New York, NY: Oxford University Press.

NCCER Module 00107-15, *Basic Communication Skills*.

29 CFR 1926, Subpart CC, **www.ecfr.gov**

Figure Credits

© iStock.com/VasilySmirnov, Module opener

Carolina Bridge Co., Figure 2

Motorola Solutions, Inc., Figure 3

Sonetics Corporation, Figure 4

Roy Laney, SME, Figure 6

Section Review Answer Key

Answer	Section Reference	Objective
Section One		
1. b	1.1.2	1a
2. c	1.2.2	1b
Section Two		
1. d	2.1.1	2a
2. a	2.2.0	2b
3. b	2.3.0	2c

NCCER CURRICULA — USER UPDATE

NCCER makes every effort to keep its textbooks up-to-date and free of technical errors. We appreciate your help in this process. If you find an error, a typographical mistake, or an inaccuracy in NCCER's curricula, please fill out this form (or a photocopy), or complete the online form at **www.nccer.org/olf**. Be sure to include the exact module ID number, page number, a detailed description, and your recommended correction. Your input will be brought to the attention of the Authoring Team. Thank you for your assistance.

Instructors – If you have an idea for improving this textbook, or have found that additional materials were necessary to teach this module effectively, please let us know so that we may present your suggestions to the Authoring Team.

NCCER Product Development and Revision

13614 Progress Blvd., Alachua, FL 32615

Email: curriculum@nccer.org
Online: www.nccer.org/olf

❏ Trainee Guide ❏ Lesson Plans ❏ Exam ❏ PowerPoints Other _____

Craft / Level: _____ Copyright Date: _____

Module ID Number / Title: _____

Section Number(s): _____

Description: _____

_____

_____

_____

_____

Recommended Correction: _____

_____

_____

_____

_____

Your Name: _____

Address: _____

_____

Email: _____ Phone: _____

Glossary

Abstraction: Any form of verbal, graphical, or written communication representing a generalized and non-specific idea or quality of a thing, action, or event.

American Association of State Highway and Transportation Officials (AASHTO): An organization representing the interest of all state government highway and transportation agencies throughout the United States. This organization establishes design standards, materials-testing requirements, and other technical specifications concerning highway planning, design, construction, and maintenance.

American Society of Testing Materials (ASTM): A national organization that establishes standards for testing and evaluation of manufactured and raw materials.

Anti-two-blocking device: Two-blocking refers to a condition in which the lower load block or hook assembly comes in contact with the boom tip, boom tip sheave assembly or any other component above it as it is being raised. If this occurs, continuing to apply lifting power to the cable can result is serious equipment damage and/or failure of the hoist line. An anti-two-blocking device, therefore, prevents this condition from occurring.

Apprenticeship: A system of providing training and mentoring for entry-level trade workers.

Apron: A movable section on the forward wall of the bowl on a scraper.

Aquifer: An underground layer of water-bearing permeable rock or unconsolidated materials through which water can easily move.

Articulated: Two parts connected by a joint so as to move independently.

Auger: A screw conveyor that is used to move bulk material such as asphalt.

Average: The middle point between two numbers or the mean of two or more numbers. It is calculated by adding all numbers together, and then dividing the sum by the quantity of numbers added. For example, the average (or mean) of 3, 7, 11 is 7 (3 + 7 + 11 = 21; 21 ÷ 3 = 7).

Avoidance zone: An area both above and below one or more power lines that is defined by the outer perimeter of the prohibited zone. As the name implies, any part of the crane should avoid this area whenever possible, and may not enter the area except under special circumstances.

Backfill: Soil and rock used to level an area or fill voids, such as the perimeter of building foundations or trenches. Areas with fresh backfill may not be stable enough to support a crane.

Balance point: The location on the ground that marks the change from a cut to a fill. On large excavation projects there may be several balance points.

Banked: Any soil mass that is to be excavated from its natural position.

Base mounting: A crawler crane assembly consisting primarily of the carbody, ring gear drive, crawler frames, and tracks.

Base section: The lowest portion of a telescopic boom that houses the other telescopic sections but does not extend.

Basket hitch: A common hitch made by passing a sling around a load or through a connection and attaching both sling eyes to the hoist line.

Bedding material: Select material that is used on the floor of a trench to support the weight of pipe. Bedding material serves as a base for the pipe.

Bedrock: The solid layer of rock under Earth's surface. Its solid-rock state distinguishes it from boulders.

Berm: A raised bank of earth.

Bird caging: A deformation of wire rope that causes the strands or lays to separate and balloon outward like the vertical bars of a bird cage.

Blind hole: A hole that does not penetrate the material completely, leaving a hole with a bottom.

Blind lift: Any lift involving a load that is out of the direct view of the operator. All blind lifts are generally categorized as critical lifts.

Block and tackle: A system of two or more pulleys, which form a block, with a rope or cable threaded between them, reducing the force needed to lift or pull heavy loads.

Blocking: Wood or a similar material used under outrigger floats to support and distribute loads to the ground. Also referred to as *cribbing*.

Boom torque: A twisting force applied to the crane boom, typically resulting from imbalanced reeving of the boom tip sheave assembly ropes.

Boot: A special name for laths that are placed by a grade setter to help control the grading operation. The boot can also be the mark on the lath, usually 3, 4, or 5 feet above the finish grade elevation, which can be easily sighted. This allows the grade setter to check the grade alone instead of having to use another person to hold a level rod on the top of the grade stake.

Bowl: The area on a scraper where soil is stored when it is scraped from the surface.

Bridge: In relation to overhead cranes, the part of an overhead crane consisting of one or more girders or beams and the supporting trucks. The bridge is the overhead, weight-bearing structure along which the trolley(s) and load block assembly travels.

Bridle hitch: A type of hitch that consists of two or more slings that support the load common lifting point.

Caisson: A watertight chamber commonly used in building or repairing bridge pilings.

Capillary action: The tendency of water to move into free space or between soil particles, regardless of gravity.

Carbody: The part of a crawler-crane base mounting that carries the rotating upperworks.

Carrier: The base of a wheeled crane that provides crane movement and supports the upperworks.

Center of gravity: The point at which the entire weight of an object is considered to be concentrated, such that supporting the object at this specific point would result in its remaining balanced in position; the point where all of an object's weight is evenly distributed.

Change order: A formal instruction describing and authorizing a project change.

Check valve: A valve designed to allow flow in one direction but closes as necessary to prevent flow reversal.

Choker hitch: A hitch made by passing a sling around the load, and then passing one eye of the sling through the other. The one eye is then connected to the hoist line, creating a choke-hold on the load.

Cofferdam: a structure built to keep water away from a construction area.

Cohesive: The ability to bond together in a permanent or semipermanent state. To stick together.

Combustible: A substance that readily ignites and burns.

Competent person: As defined by OSHA, an individual who is capable of identifying existing and predictable hazards in the surroundings or working conditions which are unsanitary, hazardous, or dangerous to employees, and who has the authorization to take prompt corrective measures to eliminate such hazards; in the context of trenches, a person with knowledge of soil types and soil stability who will inspect the trench whenever required.

Consensus standard: A set of proprietary guidelines published and agreed to by a consensus (representative majority) of members of a given industry. While not legally binding, they are often cited in governmental regulations, such as OSHA standards.

Consolidation: To become firm by compacting the particles so they will be closer together.

Conspicuous location: A particularly noticeable spot, as would be appropriate for posting an important sign or tag to ensure it is seen.

Constant: A value in an equation that is always the same; for example pi is always 3.14.

Contour lines: Imaginary lines on a site/plot plan that connect points of the same elevation. Contour lines never cross each other.

Counterweights: Weights added to the crane, usually on the end opposite the boom, to help counter the weight of the load and improve stability.

Crane mat: A portable platform, typically made of large wooden timbers bolted together, used to support and spread the weight of a crane over a larger ground area.

Crawler frames: Crane assemblies comprised of the crawler tracks, track idlers, and track power sources of a crawler crane. Also called a *tread member* or *track assembly*.

Cribbing: Alternately stacked timbers used to support heavy loads.

Critical lift: As defined in *ASME Standard B30.5*, a hoisting or lifting operation that has been determined to present an increased level of risk beyond normal lifting activities. For example, increased risk may relate to personnel injury, damage to property, interruption of plant production, delays in schedule, release of hazards to the environment, or other significant factors.

Cross braces: The horizontal members of a shoring system installed perpendicular to the sides of the excavation, the ends of which bear against either uprights or walers.

Culvert: A drain or channel that crosses under a road.

Dedicated spotter: An individual qualified as a signal person who is charged with monitoring the separation between power lines and the equipment, load line, and load, so that the minimum approach distance is not compromised per OSHA standards.

Density: Ratio of the weight of material to its volume.

Dewater: To remove water from a site.

Dipper stick: A pivoting section that connects the bucket to the boom on a hydraulic excavator.

Diver tender: One or more individuals assigned to attend to a diver's needs, including providing assistance in equipment preparation and managing the diver's cables and hoses.

Duty cycle: An expression of equipment use over time. In the case of mobile cranes, an 8-, 16-, or 24-hour rating expressed as a percentage.

Dynamic loads: A load on a structure (in this case, a crane) that is not constant, but consistently changing as the result of one or more changes in various factors. Also referred to as *shock loading*, significant dynamic loads can be applied to a crane through abrupt motions and lifting a load from its support too quickly.

Easement: A legal right-of-way provision on another person's property (for example, the right of a neighbor to build a driveway or a public utility to install water and gas lines on the property). A property owner cannot build on an area where an easement has been identified.

Effective weight: The weight of an accessory such as a boom extension or jib that reflects the effect of its weight on the lift, usually based on its position, rather than its actual weight. For example, a jib folded and stored on the main boom will have different effective weight than when it is installed on the main boom tip.

Elasticity: The property of a soil that allows it to return to its original shape after a force is removed.

Elevation view: A drawing giving a view from the front or side of a structure.

Equalizer beam: A beam used to distribute the load weight on multi-crane lifts. The beam attaches to the load below, with two or more cranes attached to lifting eyes on the top.

Equalizer plates: A type of rigging plate that has three or more holes, used to level loads when sling lengths are unequal.

Erosion: The removal of soil from an area by water or wind.

Expansive soil: A soil that expands and shrinks with moisture. Clay is an expansive soil.

Fines: Very small particles of soil. Usually particles that pass the No. 200 sieve.

Flaggers: Workers who are specially trained to direct traffic through and around a work zone.

Flammable: Capable of burning.

Floats: The portion of outriggers that touches the ground; the feet of the outriggers.

Friable: Crumbles easily.

Gantry: A framed overhead structure supported by legs on each end, used to cross over obstructions. Gantries can be portable or permanent, providing support for hoisting equipment or raising and supporting lighting, cameras, and similar equipment.

Grapple: A device used to pick up bulk items, containers, rocks, trees and tree limbs, etc. Grapples typically have several jaws that operate like fingers to pick up material, using mechanical or hydraulic power.

Gross capacity: The total amount a crane can safely lift under a given set of conditions. The gross capacity includes but is not limited to the load block, ropes, and rigging as well the primary load.

Groundwater: Water beneath the surface of the ground.

Hardpan: A hard, compacted layer of subsoil, usually with a major clay component.

Haul road: A compacted dirt road used to move material and equipment on and off the site.

Haul truck: A name that is sometimes used to describe a rigid-frame dump truck or a mining truck.

Hauling line: The portion of a rope or chain on hoisting equipment that the operator uses to raise or lower the load. Also known as a *hauling part*.

Headache ball: A heavy round weight often attached to a load line to provide sufficient weight to allow the load line to unspool from the drum when there is no live load. Larger versions of headache balls are used to swing into structures to demolish them.

Heat index: A value that combines temperature and humidity to establish an equivalent value that represents the effect of humidity on perceived temperature.

High-voltage proximity warning device: An early-warning device that senses the electric fields created by high-voltage power lines and alerts the crane operator and/or the lift team to the hazard.

Hoist drum: A drum is a cylindrical component around which a rope is wound. The hoist drum is used to wind or unwind the rope for hoisting or lowering the load; the part of a crane that spools and unspools the lifting line.

Hoist reeving: The reeving pattern applied to the hoist sheaves. Single- or multiple-line hoist reeving is used for whip, boom, and jib lines.

Horizon: Layers of soil that develop over time.

Humus: Dark swamp soil or decaying organic matter. Also called *peat*.

Hydraulic breaker: A hydraulic attachment for an excavator that is used for breaking boulders and other solid objects.

Hydraulic motor: A motor powered by hydraulic pressure provided by an external pump. Hydraulic motors are often used to power the tracks of crawler cranes, instead of complex drive systems connected directly to the diesel engine.

Hypotenuse: The long dimension of a right triangle and always the side opposite the right angle.

Idlers: Pulleys, wheels, or rollers that do not transmit power, but guides or places tension on a belt or crawler-crane track.

Impact load: The dynamic effect on a stationary or mobile body as imparted by the forcible contact of another moving body or the sudden stop of a fall.

In situ: In the natural or original place on site.

Independent wire rope core (IWRC): Wire rope with a core consisting of wire rope, as opposed to a fiber or single-stranded core; considered to be the most durable for rigging applications.

Inorganic: Derived from other than living organisms, such as rock.

Insulating link: An electrical insulating device used on the crane hook to protect workers in contact with the load from the danger of electrocution in the event the crane contacts a powerline. The link can also provide some level of protection for the crane if the load alone contacts a power line.

Interpolation: The process of estimating or calculating unknown values between two known values.

Invert: The lowest portion of the interior of a pipe, also called the flow line.

Jib: An extension attached to the boom point to provide added boom length for reaching and lifting loads. The jib may be in line with the boom, offset to another angle, or adjustable to a variety of angles. A jib is sometimes referred to as a *fly*.

Jib backstay: A piece of standing rigging that is routed from the jib mast back to the main boom to help support the jib.

Jib forestay: A piece of standing rigging that is routed from the far tip of the jib back to the jib mast, holding the tip of the jib up.

Jib mast: Also referred to as a jib strut. A structure mounted on the main boom that provides a fixed distance for the point of connection of the jib forestay and jib backstay.

Knuckle boom: A term sometimes used for a boom and stick combination that resembles a knuckle at the pivot point of the boom and stick.

Lattice boom: A boom constructed of steel angles or tubing to create a relatively light-weight but strong, rigid structure; a type of boom used on cranes and excavators that has a crisscross pattern of braces that enable the machine to lift heavy loads.

Leads: Steel structures that provide support for a pile hammer and help to align and position the hammer with the pile to be driven. The hammer can travel up or down in the leads as necessary.

Leverage: The mechanical advantage in power gained by using a lever.

Line of sight: The straight-line path between an observer's eyes and the thing being observed.

Liquid limit: The amount of moisture that causes a soil to become a fluid.

Load moment: The force applied to the crane by the load; the leverage of the load, opposing the leverage of the crane. The load moment is calculated by multiplying the gross load weight by the horizontal distance from the tipping fulcrum to the center of gravity of the suspended load. The load moment is usually reported to the operator as a percentage of the crane's capacity at the present set of conditions. As those conditions change, such as the boom angle, the load moment changes as well.

Loadbearing: A base designed to support the weight of an object of structure.

Loading: Applying a force to soil. A building can be a permanent load at a site, and a truck can be a passing load on a roadway.

Locked out: Of machinery or equipment, de-energized and removed from service during periods of maintenance or repair, rendering the equipment safe.

Lowboy: A trailer with a low frame for transporting very tall or heavy loads. A typical lowboy has two drops in deck height: one right after the gooseneck connecting it to the tractor, and one right before the wheels. This allows the trailer deck to be extremely low compared with common trailers.

Luffing: Changing a boom angle by varying the length of the suspension ropes.

Luffing jib: A jib mounted on the end of a boom that can be positioned at different angles relative to the main boom.

Machine guarding: The process of protecting an operator from the moving parts of a machine.

Minimum breaking strength (MBS): The amount of stress required to bring a rigging component to its breaking point. The MBS is a factor in determining a components' rated load capacity.

Minimum clearance distance: The OSHA-required distance that cranes, load lines, and loads must maintain from energized power lines. This OSHA term is synonymous with the ASME term *prohibited zone*.

Monuments: Physical structures that mark the locations of survey points.

Net capacity: The weight of the item(s) that can be lifted by the crane; the gross capacity of a crane minus all noted capacity deductions.

Non-ferrous: Having no iron. Ferrous metals, such as steel, contain iron and are magnetic as a result.

Nonverbal communication: Within communications between individuals, refers to unspoken communication through behaviors such as hand gestures, facial expressions, eye movement, physical touch, animation, and even tone of voice. Nonverbal communication can also include subtler messages such as an individual's dress, posture, or maintained distance from other people.

Open mike: In electronic communications, the condition where a radio's Transmit button or switch is held continuously without releasing it, even during pauses in speaking.

Open-throat boom: A lattice boom with an opening in the boom structure near the far end, allowing the hoist lines to drop through the boom rather than over the end of the boom.

Operating radius: The distance from the center of the boom's mounting point (usually the ring gear drive) to the center of gravity of the load.

Optimum moisture: The percent of moisture at which the greatest density of a particular soil can be obtained through compaction.

Organic: Derived from living organisms, such as plants and animals.

Outriggers: Extendable or fixed members attached to a crane base that rest on ground supports at the outer end to stabilize and support the crane; stabilizer legs that can be extended to widen the stance of a piece of equipment to keep it from tipping or rolling.

Parallel: Two lines that are always the same distance apart even if they go on into infinity (forever is called infinity in mathematics).

Parallelogram: A two-dimensional shape that has two sets of parallel lines.

Paraphrasing: Expressing the perceived meaning of something read or heard using different words, generally to ensure clarity. Paraphrasing is an important component of active listening.

Parts of line: The resulting number of lines that are supporting the load block, when a line is reeved more than once.

Peat: Dark swamp soil or decaying organic matter. Also called humus.

Pendants: Ropes or strands of a specified length with fixed end connections, used to support a lattice boom or boom components. According to *OSHA Standard 1926.1401*, a pendant may also consist of a solid bar.

Personal fall arrest system (PFAS): A system designed to prevent fall injuries by securing a worker to a solid structure.

Personal flotation device (PFD): A sleeveless jacket composed of bouyant or inflatable material and used to prevent drowning; commonly called a *life vest*.

Pinch points: The area in which two moving equipment parts come together.

Plan view: A drawing that represents a view looking down on an object.

Plastic limit: The amount of water that causes a soil to become plastic (easily shaped without crumbling).

Plasticity: The range of water content in which a soil remains plastic or is easily shaped without crumbling.

Pneumatic: Inflated with compressed air.

Power takeoff (PTO): A system found on construction tractors that uses a shaft to transfer power from the tractor to an attachment.

Prohibited zone: An area of specific dimensions, based on the voltage of a power line(s) that no part of the crane is allowed to enter during normal operations. Special considerations and preparations are required if the crane's task must place any part of it within the prohibited zone. The prohibited zone is a term used by ASME that is synonymous with the term *minimum clearance distance* used by OSHA.

Property lines: The recorded legal boundaries of a piece of property.

Quadrant of operation: The direction of the boom relative to the base mounting or carrier body.

Quadrilateral: A four-sided, closed shape with four angles whose sum is 360 degrees.

Rated load: The maximum working load permitted by a component manufacturer under a specific set of conditions. Alternate names for rated load include *working load limit* (WLL), *rated capacity*, and *safe working load* (SWL).

Reach: The combined operating height and radius of a boom, or the combination of boom and jib.

Rebar: A contraction of reinforcing bars. Used to reinforce concrete structures.

Recloser: A device that functions much like a circuit breaker, or in conjunction with a circuit breaker, in power distribution and transmission systems that automatically recloses the circuit after a fault has been detected and the circuit has been opened. Reclosers allow the power system to be re-energized quickly after a transient (temporary) condition, such as a tree limb falling across power lines and then falling to the ground, has occurred. If the fault reoccurs upon closure, the circuit will typically remain open until the situation has been addressed by power line workers or operators.

Reeving: A method often used to multiply the pulling or lifting capability by using wire rope routed through multiple pulleys or sheaves a number of times.

Request for information (RFI): A form used to question discrepancies on the drawings or to ask for clarification.

Rigging links: Links or plates with two holes used as termination hardware to appropriate lifting points.

Right-of-way: A type of easement which designates land as reserved for transportation or other restricted use.

Ring gear drive: Sometimes referred to as the *swing circle*. An assembly that provides the point of attachment and pivot point for the upperworks of a crane. The ring gear is typically driven by hydraulic pressure, allowing the upperworks to rotate on a set of bearings that reduce friction and transfer the weight of the upperworks (and any load) to the carbody.

Ripper: A towed attachment with teeth used on dozers, motor graders, and other machines to loosen heavily compacted soil and soft rock.

Saddle: The portion of a hook directly below the center of the lifting eye.

Scarifying: Using an attachment with teeth on a motor grader to loosen soil in front of the moldboard.

Screed: A blade-like component on a paver that levels and smoothes asphalt or concrete as it is applied.

Sedimentation: Soil particles that are removed from their original location by water, wind, or mechanical means.

Setback: The distance from a property line in which no structures are permitted.

Settlement: To become firm by compacting the particles so they will be closer together.

Sheave: A wheel with a groove for a belt, rope, or cable to run in. The terms *sheave* and *pulley* are often used interchangeably.

Shielding: A structure that is able to withstand the forces imposed on it by a cave-in and thereby protect employees within the structure.

Shock loading: A sudden, dramatically increased load imposed on a crane and rigging, usually as the result of momentum from the load that occurs due to swinging side-to-side, dropping the load and then stopping it suddenly, and similar actions that create momentum.

Shooting-boom excavator: A term sometimes used to describe a telescoping-boom excavator.

Shoring: Structures used to brace the sides of a trench.

Shrinkage: Decrease in volume when soil is compacted.

Sling angle: The angle formed by the legs of a sling with respect to the horizontal plane when tension is placed on the rigging.

Slipform paver: A type of concrete paver that evenly spreads bulk concrete that is dumped in front of it.

Spoil: Soil removed from a trench.

Spreader beam: A beam or bar used to distribute the load of a lift across more than one point to increase stability. They are often used when the object being lifted is too long or large to be lifted from a single point, or when the use of slings around the load may crush the sides.

Spur track: A relatively short branch leading from a primary railroad track to a destination for loading or unloading. A spur is typically connected to the main at its origin only (a dead end).

Squared: Multiplied by itself.

Squirt boom: A component on some telescoping-boom forklifts that allows the fork carriage to be moved in the horizontal plane while the boom remains stationary.

Standard lift: A lift that can be accomplished through standard procedures, allowing load-handling and lift team personnel to execute it using common methods, materials, and equipment.

Standards: As defined by OSHA, standards are statements that require conditions, or the adoption or use of one or more practices, means, methods, operations, or processes, that are reasonably necessary or appropriate to provide safe or healthful employment and places of employment. Standards developed by some organizations are voluntary in nature, while OSHA standards and those they incorporate by reference are enforceable by law.

Stations: Designated points along a line or a network of points used to survey and lay out construction work. The distance between two stations is normally 100 feet or 100 meters, depending on the measurement system used.

Stormwater: Water from rain or melting snow.

String line: A tough cord or small diameter wire stretched between posts or pins to designate the line and elevation of a grade. String lines take the place of hubs and stakes for some operations.

Subsidence: Pressure created by the weight of the soil pushing on the walls of the excavation. It stresses the excavation walls and can cause them to bulge.

Substructure: The part of a bridge below the top of the bearings. The substructure of a bridge supports the superstructure.

Sump: A small excavation dug below grade for the purpose of draining or retaining subsurface water. The water is then usually pumped out of the sump by mechanical means.

Superstructure: The part of a bridge above the top of the bearings. The superstructure of a bridge supports the traffic load.

Swale: A shallow trench used to direct the flow of water.

Swallow: The space between the sheave and the frame of a block, through which the rope is passed.

Swell: Increase in volume when soil is excavated.

Swell factor: The ratio of the banked weight of a soil to the loose weight of a soil.

Tagline: A rope attached to a lifted load for the purpose of controlling load spinning and swinging, or used to stabilize and control suspended attachments.

Tamping roller: A name sometimes used to describe a sheepsfoot roller.

Telescopic boom: A crane boom that extends and retracts in sections that slide in and out, powered by hydraulic pressure.

Temporary traffic control (TTC) zone: Traffic controls enacted when the normal use of a road is disrupted by construction work.

Temporary traffic control (TTC): Device and plan used to safely divert traffic when the normal use of the road is disrupted due to construction.

Tipping fulcrum: The point of crane contact with the ground where it would pivot if it were to tip over; the fulcrum of the leverage applied by the load. Depending on the attitude and type of crane, the tipping fulcrum may be the edge of one crawler assembly, one or more outriggers, or similar locations.

Topographic survey: The process of surveying a geographic area to collect data indicating the shape of the terrain and the location of natural and man-made objects.

Trench box: A structure which is placed into a trench to brace the trench walls and prevent cave-ins.

Trucks: In relation to overhead cranes, a mechanical assembly consisting of a frame, wheels, bearings, and axles that support the bridge of an overhead crane and provide the ability for it to move along a set of parallel tracks.

Undercarriage: The lower frame of an excavator that supports the turntable and has the tracks or wheels attached.

Uniform Construction Index: The construction specification format adopted by the Construction Specification Institute (CSI). Known as the CSI format.

Upperworks: A term that refers to the assembly of components above the ring gear drive; the rotating collection of components on top of the base mounting or carrier; may also be referred to as the house, or as the superstructure as defined in *OSHA Standard 1926.1401*.

Uprights: The vertical members of a trench shoring system placed in contact with the earth and usually positioned so that individual members do not contact each other. Uprights placed so that individual members are closely spaced, in contact with, or interconnected to each other, are often called sheeting.

Variable: A value in an equation that depends on the factors being considered; for example, the lengths of the sides of a triangle may vary from one triangle to another.

Vertical hitch: A simple hitch that uses one end of a sling to connect to a point on the load and the opposite end to connect to the hoist line. Also know as a *straight-line hitch*.

Voids: Open space between soil or aggregate particles. A reference to voids usually means that there are air pockets or open spaces between particles.

Walers: Horizontal members of a shoring system or coffer dam placed parallel to the excavation face whose sides bear against the vertical members of the shoring system or the earth. Also, supports for piles in a coffer dam.

Water table: The depth below the ground's surface at which the soil is saturated with water.

Well-graded: Soil that contains enough small particles to fill the voids between larger ones.

Wheelbase: The distance between the front and rear axles of a vehicle.

Whip line: A secondary hoisting rope usually of lower capacity than that provided by the main hoisting system. When a whip line exists, it is typically out at the tip of a jib, while the main hoist line is closer to the crane and operated from the tip of the main boom.

Index

Carriers
  defined, (21102):1, 3, 50
  rough-terrain cranes, (21102):6
  wheeled truck cranes, (21102):5
Carry-deck cranes, (21102):6
Caterpillar Accugrade GPS grade control system, (22210):36
Caterpillar Accugrade laser grade control system,
  (22210):35, 36
Caught-between hazards, (36110):22–24
Caution signs, (36110):4, 6, (75104):3–4
Caution tags, (75104):5
Center line, (22209):4, 17
Center line stakes, (22210):29–30
Center of gravity (CG)
  calculating, (38102):27–28
  defined, (21102):23, 50, (38102):1, 54, (75104):10, 14, 17
  factors affecting, (75104):14
  lift capacity and, (21102):24
  suspending a load near, (38102):7
CG. *See* Center of gravity (CG)
Chain
  applications, (38102):17
  characteristics, (38102):19
  common configurations, (38102):19
  inspection, (38102):20
  storage, (38102):20
Chain fall, (38102):41
Chain hoists
  characteristics, (38102):41–42
  electric, (38102):43, 44
  hook-style, (38102):43
  inspection and use of, (38102):43–44
  load capacity, (38102):43
  manual, (38102):43
  selection of, (38102):43–44
  spur-geared, (38102):42
  trolly-style, (38102):42–43
Change orders, (22209):1, 13, 36
Check valve, (21102):1, 15, 50
Cherry picker, (21102):6
Choker hardware, synthetic web slings, (38102):16
Choker hitch, (38102):11, 12, 25–26, 54
  double-wrap, (38102):26
Circles
  area calculations, (22207):17–18
  properties, (22207):17
Civil drawings
  abbreviations, (22209):A7, 18–21
  as-built drawings, (22209):13–14
  computer-aided design (CAD), (22209):3
  lines on
    break line, (22209):18
    center line, (22209):17
    contour lines, (22209):18
    cutting plane (section line), (22209):18
    dimension and extension lines, (22209):17
    hidden line, (22209):18
    leader line, (22209):17
    object lines, (22209):17
    phantom line, (22209):18
  project location map, (22209):2
  reading and interpreting
    highway plans, (22209):21–22
    method for maximum understanding, (22209):16–17
    site plans, (22209):19–21
  revision blocks, (22209):2, 3
  scale, (22209):2, 5
  specifications

format, (22209):25–31
  organization of, (22209):23–25
  purposes, (22209):23
  symbols, (22209):A6, 18, 20
  title blocks, (22209):2, 3
  title sheets, (22209):2
Clamps, (21102):12, 16
Clamshell buckets, (21102):10–11, 14
Clay loam, (22308):7
Clay soil
  classification, (22308):3, 5–6
  compacting, (22308):11
  defined, (22308):5
  loose and bank weight, (22308):15
  properties, (22308):7–8, 9
  trenching hazards, (22308):26
  weight-bearing capability, (22308):20
  weight per cubic yard, (22207):23
Clove hitch, (38102):34–36
Cofferdams, (36110):1, 3, 39
Cohesive, (22308):1, 8, 30
Cold planers, (36111):19
Cold-related hazards, (21106):18
Cold-related illness
  frostbite/frostnip, (36110):20
  hypothermia, (36110):20
  preventing, (36110):20
  treating, (36110):20
Colloid clays, (22308):5–6, 8
Combustible, (36110):1, 13, 39
Come-alongs, (38102):44–45
Communication. *See also* Hand signals
  ASME standards, (53101):7
  devices
    cell phones, (53101):11
    DECT-compatible devices, (53101):11, 13
    hardwired systems, (53101):11–12
    portable radios, (53101):10–11
    walkie-talkies, (53101):11
  electronic, (53101):8
  nonverbal
    audible signals, (53101):12, (75104):5
    defined, (53101):1, 29
    flags, (21106):14, (36110):29–30, (53101):13
    horn signals, (21106):21, (53101):22
  OSHA standards, (53101):6–7
  paraphrasing, (53101):1, 2, 29
  verbal, (53101):10–12
  voice signals, (21106):7–8, (53101):8
Communication barriers
  abstractions, (53101):1, 3, 29
  environmental factors, (53101):4
  fear, (53101):3–4
  lack of common experience, (53101):2–3
  language, (53101):2–3
Communication filters, (53101):2
Communication process
  message exchange steps
    feedback, (53101):1–2
    receiving the message, (53101):1–2
    sending the message, (53101):1
  receiver's role, (53101):1–2
  sender's role, (53101):1–2
Compaction
  road surfaces, (36101):2, 10, 11
  soil, (22308):2, 10, 11, 15–16, 20
Compaction equipment
  compactors, (36111):22